Scheduling
with *SureTrak*

Second Edition

David A. Marchman
Tulio A. Sulbaran

THOMSON

DELMAR LEARNING

Australia Canada Mexico Singapore Spain United Kingdom United States

Scheduling with *SureTrak,* Second Edition

David A. Marchman and Tulio A. Sulbaran

Vice President, Technology and Trades SBU:
Alar Elken

Editorial Director:
Sandy Clark

Acquisitions Editor:
Alison Weintraub

Developmental Editor:
Jennifer A. Thompson

Marketing Director:
Dave Garza

Channel Manager:
William Lawrensen

Marketing Coordinator:
Mark Pierro

Production Director:
Mary Ellen Black

Production Editor:
Barbara L. Diaz

Technology Project Manager:
Kevin Smith

Technology Project Specialist:
Linda Verde

Library of Congress Cataloging-in-Publication Data

Marchman, David A.
   Construction scheduling with SureTrak / David A. Marchman, Tulio A. Sulbaran.— 2nd ed.
       p.   cm.
   Includes index.
   ISBN 1-4018-6721-9
   1. Building—Superintendence.
   2. Production scheduling. 3. SureTrak.
   I. Sulbaran, Tulio A. II. Title.
     TH438.4.M38 2005
       690'.068—dc22   2005006984

NOTICE TO THE READER

Contents

SECTION 2   Scheduling   51

CHAPTER 3   Schedule Calculations   53

CHAPTER 4   Bar Chart Creation   75

Preface

COMPUTERS IN CONSTRUCTION SCHEDULING

The Advantages of Using Computers in Construction Scheduling

One of the primary advantages of using computers for construction scheduling is that the mathematical computations are instantaneous and error-free. The speed and accuracy of the mathematical scheduling computations and analysis of the information produced make computerized scheduling a valuable tool for construction project controls. The low cost and superb on-screen and hard copy graphics make it an effective communications and project controls tool for owners, contractors, subcontractors, suppliers, and vendors.

The Advantages of *SureTrak*

Primavera Systems, Inc.'s economical project management/scheduling software package is *SureTrak,* which is the little brother to Primavera Engineering & Construction (formerly P3 E/C for construction). This book explains how to use *SureTrak Project Manager 3.0.* The advantages of *SureTrak* are:

- Between *SureTrak* and its related packages, *Primavera Systems, Inc.* controls the largest share of the market for construction scheduling software in the United States.
- *Primavera Systems, Inc.* is among the top fifty computer software vendors in the United States.
- *Primavera Systems, Inc.* offers excellent training and customer support in the use of its software.

SureTrak Windows **System Requirements**

- Pentium PC
- 40 MB of free hard disk space
- 16 MB RAM, 32 MB recommended
- Windows NT, 97, 98, 2000
- VGA or higher monitor
- CD-ROM drive

Help Topics

The online help in *SureTrak* is extensive. The Help Topics can be accessed three ways for easy use:

- **Contents**—Displays *SureTrak* Help Contents by category (books).
- **Index**—Displays the **Search** dialog box that enables users to find Help information using keywords.
- **Find**—Enables user to search for specific words and phrases in Help Topics instead of searching by category.

Another helpful feature of *SureTrak* is that by pressing the F1 (first function) key, the Help Topics screen for that field is pulled up from any field within *SureTrak*. The tutorial is a very helpful walkthrough of the different functions of the product.

ABOUT THIS BOOK

This book is a graphic, step-by-step introduction to good construction project control techniques. This book shows construction scheduling as it has not been presented before. It includes the traditional theory on planning, scheduling, and controlling construction projects. Topics included are schedule development, activity definition, relationships, calculations, resources, costs, and monitoring, documenting, and controlling change. The difference between this and other scheduling texts is that this book takes the student through all these topics with an example construction project schedule using *SureTrak*. *SureTrak,* along with *Primavera Engineering & Construction*, is generally recognized as the premier construction scheduling software on the market today. The student is not only exposed to onscreen images, but is also shown how to manipulate hard copy prints for exceptional communication and demonstration results.

The step-by-step tutorials are the strength of this book. This book is designed for the classroom, but will work very well for people in the industry who need to learn *SureTrak* on their own.

WHO SHOULD USE THIS BOOK

This book was written for students of construction, construction technology, and for working professionals who want to gain a better understanding of scheduling and the use of *SureTrak*.

USING THIS BOOK

This book can be used either as a stand-alone text or in conjunction with another more traditional planning/scheduling book. This book is designed primarily as a problems-oriented lab manual to improve student use of *SureTrak*. Students should be able to use this book as a step-by-step guide while they are sitting at their desks running *SureTrak*. The chapters are organized with major headings in capitalized boldfaced print. Main topics are given at the beginning of the paragraph and are also in bold print. Boldfaced print is also used for the software commands making it very simple for the student to follow along using *SureTrak* while reading the book.

The construction professional can use the book to choose particular components of scheduling with *SureTrak* to read about. You do not have to start at the beginning of the book and go to the end. The detail provided in the Contents makes it easy to find specific information.

The strength of this book lies in its detailed examples of a sample schedule that uses scheduling information in numerous ways. This book can be used in conjunction with *SureTrak*'s Help Topics and tutorial for an efficient introduction to the subject of planning and scheduling and the use of *SureTrak*. This book presents scheduling primarily from the general contractor's point of view, but most of the information can be applied to the subcontractor, fabricator, owner, and construction manager.

USING THE EXERCISES

A sample problem and exercises at the end of each chapter help make this text more useful in a classroom setting. A student who completes the exercises using *SureTrak* will have a sound basis for producing construction schedules.

ABOUT THE AUTHORS

David Marchman teaches estimating, scheduling, project management, and other construction courses in the School of Construction at The University of Southern Mississippi. His teaching career is supported by extensive experience in the construction industry (residential, commercial, and industrial), including seven years with Brown & Root. He continues to consult with numerous companies on estimating, scheduling, and cost accounting. Professor Marchman has spent summers with Bill Harbert International Construction, Hensel Phelps Construction, and

Brice Building Company. He has taught numerous construction-related seminars and workshops for such companies as W. G. Yates Construction, Ivey Mechanical Corporation, and the Associated Builders and Contractors of Mississippi.

Dr. Tulio Sulbaran teaches estimating, scheduling, project management, and other construction courses in the School of Construction at the University of Southern Mississippi. His teaching career is supported by a doctorate degree from Georgia Institute of Technology and several years of international work experience in the A/E/C (architecture, engineering, and construction) industry in planning and managing engineering projects both in the field and from the office. Dr. Sulbaran continues to consult with numerous companies on scheduling and estimating, such as URS Corp, W. G. Yates Construction, Superior Asphalt, and Mississippi Power among others. He has been very active in organization such as Associated School of Construction (ASC), American Council for Construction Education (ACCE), American Society for Engineering Education (ASEE), and Associated Builders and Contractors (ABC) to mention a few. Additionally, he has taught numerous construction-related seminars and workshop as well as presented in both nationally and internationally recognized conferences.

ACKNOWLEDGMENTS

David Marchman wants to thank his wife Janet, daughter Dee Ann, and son Dane, who were a great inspiration and support in the writing process.

Dr. Tulio Sulbaran wants to express great gratitude to his undergraduate students, construction companies, and construction organizations for providing priceless input and opportunities for personal and professional growth. He wants to recognize the important role that his graduate students Claire Freeney, Mathew Carpenter, Blake Howell, Chad Marcum, Susan Rayborn, and Carlos Sterling played by helping him in other activities, which provided the much needed time to focus on this book. Further, he wants to give very special thanks to Prof. Marchman for the invitation to participate in the preparation of this book and for sharing numerous hours of enjoyable and productive work. Finally, all of this would not have been possible without the nurturing of his mother Alida Gonzalez, the support and understanding from his wife Virginia Sulbaran, and the joy from his son Tulio Nicolas and daughter Virginia Valentina.

The authors and Thomson Delmar Learning gratefully acknowledge the comments and suggestions of the members of the review panel who contributed to this text. Their efforts have proven valuable to its success.

Section 1

Planning

1 | Introduction to Scheduling

Objectives

Upon completion of this chapter, you should be able to:

- Describe the benefits of good scheduling practices
- Identify scheduling phases
- Contrast scheduling design-bid-build versus fast-track construction
- Describe scheduling levels
- Define, identify, and describe activities and their relationships
- Identify types of scheduling presentation formats

HISTORY OF SCHEDULING

Human Construction Capabilities

Construction scheduling is as old as human attempts to alter and control their environment. Prehistoric humans scheduled their lives around the cycles of nature that impacted their means of survival. Changes in seasons or long-term weather patterns, changes in herd movements, and changes in the availability of materials required a constant re-evaluation of the need to move and rebuild shelters and facilities.

As human's mastery of the environment progressed, so did the need for more detailed and accurate schedules. The Seven Wonders of the Ancient World are a tribute to humanity's ability to utilize tools, materials, and labor to alter and improve the environment we live in. Of those marvelous examples of construction, only the pyramids of Egypt still remain. Their construction required the efforts of thousands of skilled and manual workers and many years of planning and implementation. The techniques utilized by the Egyptians to construct these monuments are still not completely understood; however, it is widely assumed that the transportation of the huge stones necessary for the construction of the pyramids was heavily dependent on the cycles of the Nile River flood events. The commitment to build such enduring structures and the contributions of money, time, and other resources reflect the greatness of these societies.

An Example of Efficient Scheduling

Scheduling has always been an essential part of construction. Resources and time to complete projects are always limited. End users of every structure anxiously await its completion. Scheduling techniques, including cost management, have improved over the years to meet the demand for advanced structures. These scheduling techniques focus on defining the project in terms of activities. The duration of the activities and their relationship determine project duration. The activities may require the use of resources such as labor, material, and equipment.

A scheduling feat that preceded the advent of both the computer and modern scheduling techniques was construction of the Empire State Building in New York City, the tallest building in the world at the time. Because the site was located in downtown Manhattan, there was already tremendous congestion. Without space on the site to store construction materials, the builders had to bring in the structural steel nightly to be used the next day. Compounding the site-related difficulties were labor problems. In addition, the owner wanted to use the building as soon as possible. This clearly was a project demanding good scheduling techniques. Remarkably, thanks to efficient scheduling, the builders managed to erect one floor every three days, on average.

SCHEDULING IMPACTS ON PROJECT MANAGEMENT

Scheduling impacts all project management aspects. The purpose of project management is to achieve the project goals and objectives through the planned expenditure of resources that meet the project's quality, cost, time, and safety considerations. The following is a brief description of the project management considerations and how they are impacted by scheduling practices.

Quality

Good management assures that a project attain the level of quality as defined in the contract documents. Not only do quality and pride in workmanship go hand in hand and improve worker pride and project outcome, but they also lead to profitability through repeat business, referrals, and negotiated works.

Quality does not mean perfection. Quality means meeting the minimum requirements of the project as listed in the specifications. The contractor needs to remember that the owner is paying for, and expects, a certain level of quality. Therefore, setting unrealistic goals in the schedule will negatively impact the quality of the project during construction.

Cost

The ability to estimate and then to complete a project within budget goes to the very heart of good construction project management. The owner looks to project management for the efficient expenditure of resources to get the most results for the dollars invested. Since even small problems on the project can quickly turn into situations costing thousands of dollars, it is very important for project management to make the best use of the schedule to minimize the occurrence or impact of these construction problems. Furthermore, the schedule allows project management to assess the problem impact over all the interrelated activities that make up the project.

Time

Time is money! Time is of the essence! Time is usually just as critical to the contract and the owner's needs as are quality and cost constraints. The ability to determine a schedule and then to complete the project within the timeframe also goes to the very essence of good project management. A properly formulated and executed schedule can help both the owner and the construction manager to recognize and deal with disruptions far in advance of a pending catastrophic delay. Many times an owner chooses a constructor based upon the constructor's ability to marshal forces to complete a project in a timely manner, rather than basing the decision solely on the lowest cost. Getting the project in service and generating a return on investment as soon as possible is critically important to the owner.

Safety

Another integral component of good project management is safety. Accidents can be extremely costly, not only in their direct costs, but also because of their influence on the Workers' Compensation Experience Modification Rate (EMR). The rating of the construction company's safety record affects the amount the company must pay for insurance coverage for employees. Accidents also affect worker productivity, morale, and other direct and indirect costs. The scheduling of activities directly affects the risk that workers will be placed in during the construction. Thus, safety aspects are taken into consideration during the sequencing of activities that make up the schedule implemented by project management.

BENEFITS OF GOOD SCHEDULING PRACTICES

The benefits of good scheduling practices in construction can be organized in the following six categories.

Build the Project on Paper

Construction projects are becoming more and more complex and costly, requiring greater attention to the schedule. Construction schedules allow the planning and controlling of resources and time required to complete projects. Thus, the schedule is the vehicle that the project team uses to build the project in their minds and on paper before the actual construction begins and resources have been expected. The building of the project on paper is the essence of planning. During this planning stage, the scheduling team can explore different alternatives to identify the best sequencing of activities that leads to the most efficient use of resources and time.

Use Resources Efficiently

Any type of construction requires the expenditure of resources. Thus, the efficient expenditure of resources directly impact project profitability. The resources in construction are the labor, equipment, and materials that are expended during the project duration. Typically, the contractor will benefit from a schedule that minimizes the project's duration. An example of the benefits of reducing project duration is the reduction in indirect cost associated with temporary facilities, utilities, field supervision, and construction equipment. Efficient use of these resources at the job site will lower cost and improve the profitability of the project. Another example is a crane that might cost $1,000 per day to have at the job site. If the crane is only needed for 17 days rather than the 20 days, the project saves $3,000. The same work is accomplished, but a more efficient use of resources is obtained through the implementation of a better schedule. The accumulation of such savings can significantly reduce total project costs. Of course, if a schedule is shortened by means of overtime, overall costs may not be reduced at all!

Control the Project

Good project management means "project control" rather than a project "out of control." All projects have two primary control documents: the estimate and the schedule. The *estimate* defines the scope of the project in terms of cost associated with material quantities, labor work hours, and equipment hours needed. The *schedule* defines the scope of the project in terms of time associated with the completion of interrelated activities which require the expenditure of resources.

Since time is money, the resources and the timeframe over which they are expended are directly related. It is very important for the construction team to use the schedule to control project cost and duration. Effective project control combines and updates both the estimate and the schedule to achieve a clear understanding of the status of the project at any given time. Furthermore, effective construction project control today may require, due to different types of contracts, the close integration of planning, design, estimating, and scheduling to achieve efficient project delivery.

Increase Return on Investment

The old adage "time is money" is particularly true in construction and impacts both the owner/investor and constructor. Upon completion of a construction project, the owner/investor uses the facility for its intended purpose and therefore begins to receive a return on the investment. Additionally, upon completion of the project, the constructor resources are made available and could be used in another project to increase profitability of the construction company. Therefore, proper scheduling practices benefit all parties by increasing return on their particular investment.

Anticipate and Solve Problems

Scheduling enables the project team to anticipate problems before they occur at the construction job site. Estimating and, to an even greater extent, scheduling provide the project organization with an efficient mechanism for anticipating and solving problems during the planning process before resources are expended—rather than on the job site during the construction phase. The process of planning is intended to make the construction process run smoothly, minimize surprises, and prevent the expenditure of costly resources with poor results.

Act as a Communication Tool

Another benefit of good scheduling practice is to act as a communication tool to avoid unwelcome and costly surprises. Most construction projects require the services of many people with diverse expertise and various contractual relationships. The project manager needs to provide and receive information at every stage of the construction project. The same is true for superintendents, foremen, subcontractors, and the owner. For a project to run smoothly, all parties need to be "on the same page at the same time." This requires continual communication so that all participants know what the plan is and what their responsibilities, requirements, and expectations are. Keeping everyone informed is one of the critical benefits of the construction schedule.

An effective schedule needs to communicate the status of the project at any given time and must be as clear as possible. A schedule with too much detail is overly complicated and is usually ignored. A schedule with too little detail is meaningless as a management tool and is also ignored. Finding the right balance of detail and essential information is a responsibility of the scheduler.

SCHEDULING PHASES

Planning

Planning and scheduling efforts are done in conceptual project phases by the owner, the architect/engineer, or the construction manager to determine the expected project duration inserted in the contract. But, in a design-bid-build, planning for the contractor takes place during the stage prior to construction where the following must take place: decision making about the execution of the project; information gathering about the work to be accomplished; and identifying and defining activities. Good scheduling requires creativity and flexibility in identifying and defining activities and their interrelationships.

Decision Making.   Prior to starting construction, the management team must plan the execution of the project. Planning is a form of decision making since it involves choosing among alternative courses of action. The effective project plan is always preceded by a carefully formulated preplan. The preplan gives project participants an opportunity to consider how the project fits into the company's overall operational strategy.

Information Gathering.   Planning entails defining the work to be accomplished. Communication is required to gather information from many persons and places. For example, questions might include: When and from what source is the right crane available? Are there enough skilled masons available to complete the brickwork on time? What will be the impact on the project schedule of long lead-time items, such as an elevator for the project that will require special fabrication at a shop where there is a backlog?

Identifying/Defining Activities.   Proper scheduling requires a thorough knowledge of construction methods as well as the ability to visualize individual work elements and establish their mutual interdependencies. The team preparing the schedule must create a rough diagram that identifies and defines activities and their relationships to other activities. Essentially, the project is built on paper with activities and interrelationships as the building blocks. The entire project is first constructed in the minds of the scheduling team, before being put on paper. The relationships among activities, building methods, problem solving, and communications that define the plan are generated. Activity relationships are defined using information derived from the contractor's management team, the accounting department, the equipment department, the plans and specifications, a visit to the site, the nature of the work, the owner, the banker, the subcontractors, the governmental agencies involved and their requirements, and the suppliers.

Creativity.   Good planning requires being creative and not being bound by preconceived notions. Just because a company has always tackled a certain construction sequence or segment of work a particular way, it need not assume that this is the only or best way. By breaking away from established paradigms, models, or false constraints, planners can incorporate improved methods into the schedule. One way to incorporate new ideas and methods is brainstorming with key participants to solve a specific problem. For example, perhaps the question is "What is the best way to pour the elevated concrete column on a specific project?" Participants in the brainstorming session each state the first method or solution to the problem that comes to mind, no matter how far-fetched or nontraditional the approach. This and other methods to expand and change the way team members look at the project are very useful in the planning process.

Flexibility.   Good planning also requires flexibility. It is the natural tendency of first-time schedulers to build a long chain of activities, one after another, with no branches, and with only one event happening at a time. The most efficient way to shorten project duration is to have as many activities going on concurrently as possible without their hindering each other's progress. Instead of single crews working consecutively at the job site, there could be five crews working concurrently but not interfering with each other's progress. In Figure 1–1, the activity Rough Framing Walls comes after Pour Slab. They are scheduled consecutively because the first must be complete before the other can begin. The activities Rough Plbg, Ext Fin Carp, Rough Elect, Rough HVAC, and Inst Wall Insul are scheduled concurrently because they can be going on at the same time.

Interrelationships.   During the planning stage, durations are not applied to the schedule. The key development at this stage is naming (identifying/defining) activities and establishing relationships to the other activities. In establishing activity relationships, resource requirements must be considered. For example, an activity might be either cast-in-place concrete or precast concrete. Thus, some resource decision has

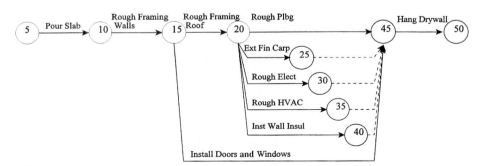

Figure 1–1   Activity on Arrow Diagram—Example of Consecutive and Concurrent Activities

already been made, but the activities are usually only defined along with their associated interrelationships and constraints. This is the phase of the scheduling process with the most potential for the creation and development of new approaches, systems, or methods for putting thework together. When the project has been defined on paper in the form of the rough schedule, the planning stage gives way to the scheduling phase.

Scheduling

Durations of Activities.   The second phase of the overall scheduling process involves filling in more precise estimates of the time and other resources in the rough schedule produced in the planning phase. The other resources might include labor, equipment needs, or division of responsibility (for example, crew versus subcontractor). By calculating the estimated duration of each activity, the scheduler can calculate the project duration. If the first pass yields a project duration that is too long, the original logic constraints may have to be adjusted (for example, by increasing crew size to complete an activity sooner, by modifying sequences of activities, or by making activities concurrent). The schedule is continually fine-tuned until it becomes satisfactory to all parties; it is then accepted as the project schedule.

Evolution of the Schedule.   Development of the schedule from the rough stage to the project schedule is an evolutionary process that requires communication and the approval of all parties involved—the contractor's management team, subcontractors, major suppliers, and, of course, the owner. The contractor must receive input from these parties, and the contractor wants each of the parties to accept or "buy into" the schedule. This process makes it "our schedule" instead of "your schedule." Typically the contract documents make the contractor the keeper of the schedule for the construction project. It is the contractor's responsibility to resolve scheduling disputes between the different subcontractors and the contractor's forces. Thus the process includes soliciting input from the different parties, drafting a rough schedule, circulating it for feedback, making modifications, resolving disputes, and reaching agreements. This process produces the project schedule that all parties can accept.

A properly prepared construction schedule always proceeds from a small number of general activities to a more detailed number of specific activities. For example, in the early stages of schedule preparation, a single activity may be listed as "concrete slabs." As the schedule is developed, this activity may be further broken down into "erection of forms," "placement of reinforcing steel," "concrete placement," "finishing," "curing," and "stripping." A good scheduler will seek to achieve a balance between a schedule that is so general that it does not yield adequate information and one that is so specific that it hides the truly important planning factors.

Monitoring the Schedule

Actual Progress.   The third phase in the scheduling process is the monitoring phase. The *data date* is the date on which a schedule is updated with current information. Typically, at fixed intervals throughout the project (usually at the end of each month) progress is determined as of that date. This is the cutoff date for comparing the actual project progress to the planned progress. Determining actual activity progress can be done in several different ways. The following describes the expenditure of resources. Monitoring durations/resources involves determining the physical progress in the field and inputting the progress of each activity. First, it is necessary to establish a database of the actual expenditure of resources for each activity. This information is compared to planned expenditure to determine the percent complete for each activity. The progress of each individual activity is established using the information from the database. Next, the schedule is recalculated with the updated information to determine if each activity is ahead of or behind schedule when compared with the plan (baseline) and if the overall project is ahead of or behind schedule. This process is called *updating the schedule*.

Controlling the Schedule

Schedule Changes.   The fourth and last phase in the scheduling process is the controlling phase. Controlling usually involves documenting and communicating changes to the plan and schedule. As projects develop, the sequence of activities originally planned may change. The reason may be schedule updates, changes in scope, material delays, lower or higher productivity, or some other factors. An example of a change may be the decision to use a different erection method or system, which essentially alters the schedule. This new, revised schedule then becomes the current project schedule. If the project is behind schedule, the contractor may have to "crash" or accelerate the schedule to make up for lost time by adding shifts, having laborers work overtime, or adding craftspeople to a crew. Keeping the schedule relevant and useful requires redrawing the plan to incorporate such changes in the relationships. The new, revised schedule will be better, since it is based on more current information.

Schedule Progress.   Usually schedule progress is the basis of monthly meetings with the owner/architect to determine the contractor's compliance with the contract and invoice for payment. The owner is presented with an updated schedule showing the data date, progress during the last thirty days, and the forecast for the next sixty days. The schedule is a critical document for determining whether the project is progressing according to the original plan. By using the current (updated) schedule, along with the target (original budgeted) schedule, it is easy to spot which activities are in trouble and whether the project itself is on schedule or in trouble. This concept is known as *management by exception*. Management and usually the owner want to know which parts of the schedule are in trouble so they can determine which activities to spend their time on and which activities are the most likely candidates for making up lost time. It does not do any good to put extra resources on an activity that is not critical to the completion of the project schedule. Controlling is the process of constantly modifying the schedule to make sure it is current with the latest plan for how the project will be constructed, and in case of deviation from the original schedule, suggesting and implementing a corrective action.

Documentation.   When changes are made, the construction team must document or record the changes for historical information and project backup. The historical information is used as a database for reference for future projects. The project backup is necessary for use in possible legal claims or settlement of project disputes. The importance of adequate documentation cannot be overemphasized. Many cases of construction litigation have been resolved in favor of the party having the most legitimate and thorough documentation.

SCHEDULING DESIGN-BID-BUILD

In design-bid-build construction, processes are consecutive rather than concurrent. There are breaks between the design, bid, and construction functions. The owner first formulates an idea for a potential project. After determining the feasibility of the project, the owner contracts with the architect and engineer to produce the project contract documents that define the project scope. Next, bids for construction services are accepted from contractors. The owner/architect selects the successful contractor's bid and awards a contract, and construction commences. Thus, the contractor knows the scope of the work, and the owner knows the price and duration of the work before the contract is awarded. For the constructor, scheduling lump-sum, linear projects deals primarily with scheduling field construction activities. The preconstruction activities of architecture/engineering design and the postconstruction activities of maintenance/operation are not part of the contractor's responsibility and therefore are not scheduled.

Field Construction

In controlling the field construction, the contractor is concerned with controlling time, resources, labor, materials, subcontractors, vendors, equipment, and money. It is critical to control such influences to the schedule as:

- The relationship of the activities to each other (which activities precede which, and which can occur at the same time)
- The size of crews
- The availability of labor
- Construction methods
- The cost of resources
- The types of construction equipment
- Work schedule (number of hours per day, shifts, weekends, holidays)
- Shop drawing review and approval
- Material deliveries
- Inspections
- Payment schedules

All these factors must be efficiently organized, sequenced, and controlled to optimize and maximize efficiencies.

Timing

On a lump-sum project, detailed scheduling typically begins once the contractor has signed the contract with the owner to construct the project. Although some scheduling takes place during the estimating/bidding phase, this is usually general in nature, for example, determining how long a superintendent will be needed on the project site.

Since producing the detailed schedule can cost a substantial amount, it is not begun until the contract is awarded, and it is completed as soon as possible. There are three primary reasons for preparing the schedule immediately. First, the contractor wants the schedule to control field operations. Second, the planning and scheduling process is a tool to help point out and solve problems before they arise in the field. Third, the owner/contractor contract may require the presentation of a project schedule to the owner/architect for use in monitoring the project.

The scheduling process used in setting up project controls for lump-sum, linear construction involves four phases: planning, scheduling, monitoring, and controlling.

SCHEDULING FAST-TRACK CONSTRUCTION

The previous section of this chapter dealt with sequential or design-bid-build construction. The scheduling for the usual negotiated/fast-track construction requires a different mind-set. The negotiated/fast-track construction is discussed in this section.

Integration of Activities

Fast-track, or phased, construction involves the integration of detailed design and construction. On larger projects, significant time can be saved by overlapping or concurrently designing and building. The primary disadvantage is that, since the design is not complete, the full scope of the work is not known before field work commences. These projects are, therefore, negotiated contracts, with the owner accepting a greater portion of the risk for project cost increases.

Team Approach to Construction

Another advantage of fast-track construction is the team approach to the design/construction process. Teamwork among the owner, designer, and constructor contrasts with the confrontational relationship that lump-sum construction typically leads to. The constructor is typically paid a fee for services and is part of a team. Under the lump-sum arrangement, the constructor is paid according to bid and receives money saved. This leads to the construction firm looking out for its own best interest and not that of the owner. Under the team concept, if the contractor's fee is fixed, money saved returns to the owner, or can be shared between owner and contractor, and the constructor is looking out for the best interest of the owner.

Team Approach to Design

Another advantage of fast-track construction is constructor involvment in the design process enhancements. Significant cost savings can be derived through "constructibility"—designing for ease and economy of construction. Another way to save is by value engineering. *Value engineering* is a systematic approach to evaluating a number of cost alternatives to choose the best one for the particular project. An added advantage of value engineering is that, in many construction contracts, money saved on the project is shared with the contractor.

A good rule of thumb is that 80% to 90% of a project's cost should be fixed by the time conceptual design (the sketching phase) is completed. The project cost is fixed to a large extent before the first detailed drawing is ever completed. This is because when certain project parameters are defined, the scope and therefore the cost of the project become fixed. Once the function, location, size, vertical or horizontal orientation, type of construction, and type of environmental controls are fixed, the project cost is essentially determined. Since the constructor's knowledge of cost and how the project goes together is likely to be greater than that of other members of the team, the constructor's input during the conceptual design phase is invaluable in controlling project cost and schedule.

Comparing the Scheduling of Design-Bid-Build versus Fast-Track Construction

The four phases of scheduling (planning, scheduling, monitoring and controlling) are the same for design-bid-build and fast-track construction. The primary difference is that the scheduling effort in the design-bid-build project focuses on the field construction effort, whereas the fast-track schedule integrates the conceptual design, detail design, project management, procurement, expediting, documentation, field construction, and possibly maintenance/operations. The entire project is looked at as a whole, with break-out schedules for each of the areas such as design, construction, and project management.

Scheduling Levels

The schedule should be prepared with different levels of details depending on the project phase. The earlier in the project the schedule is produced, the less detail will be required. Figure 1–2 organizes the construction project into three generic phases (prior to construction—bid, prior to construction—awarded, and during construction). During each of these phases, the level of details required changes due to different uses of the schedule.

Prior to Construction (Bid)

During this phase the construction company prepares a bid package and a schedule for preplanning purposes. This preplanning schedule is used to determine the overall duration of the project. This preplanning schedule also allows (1) verification if the timeframe required by the owner can be met, and (2) calculation of indirect cost such as temporary

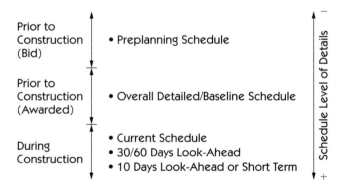

Figure 1–2   Level of Details in Construction Scheduling

facilities, supervision, and so on. This schedule can be put together very quickly because it does not contain a great level of detail.

Prior to Construction (Awarded)

This phase is when the constructor has been awarded the job but has not begun actual construction. This is the period where the constructor prepares an overall detailed/baseline schedule to manage field operations and to meet clients' requirements. This overall detailed/baseline schedule requires the construction team to build the project on paper defining activities, establishing interrelationships, and determining resource requirements.

During Construction

This phase is when actual construction occurs. During the construction phase, most constructors consider the overall detailed/baseline schedule established prior to construction (awarded) as a frame of reference for performance evaluation. This baseline schedule must be updated and it is called current schedule. The current schedule incorporates changes in the work and actual physical progress as the project is put in place. Good scheduling requires more detailed preparation as a particular activity gets closer to actual installation. These more detailed schedules are the 30/60 days and 10 days look-ahead. The 30/60 days schedule allows the construction team to concentrate on activities that will take place in the near future. It is also good practice to prepare a written ten-day (two-week) look-ahead schedule of all upcoming activities. Depending on the size and complexity of the project, the 10 days look-ahead schedule might require the input of a combination of the project manager, project engineer, superintendent, and foreman. This short-term schedule is usually a field function and is used for communication with the crews. The 10 days look-ahead is essentially a problem-solving exercise to save time and money by averting problems that might arise in the field.

ACTIVITY DEFINITION

One of the first steps in putting any schedule together is identifying the activities or tasks that must be completed to attain the project goals and objectives of the project team. Since this book is based on Primavera Systems, Inc.'s project management/scheduling software package, *SureTrak,* activity definitions and other relationships used in this book reflect the use of this software. This book is based on *SureTrak Project Manager,* Version 3.0.

Activity Defined

Construction projects are made up of a number of individual *activities* that must be accomplished in order to complete the project. The activity type in *SureTrak* is the *independent activity,* which requires time and usually resources to complete. Independent activities have specific characteristics. Following is a brief description of each characteristic.

Time Consumed. The task activity breaks the schedule down into more easily estimated smaller components. A task activity may consume weeks, days, or hours. Duration is a function of the scope of work for the activity and resources assigned to accomplish the work.

Resources Consumed. Usually resources must be expended for a task activity. The assigned resources should be scheduled according to the activity's base calendar. Labor is expended to install material and equipment resources to complete the finished structure. The scope of work, as defined by the estimate of labor, materials, and equipment, determines the duration of the activity.

Definable Start and Finish. Task activities consume time and resources and are tied to related activities by relationships. These relationships determine which activities must be complete before the activity in question can begin. The scheduler determines duration of the activity from the estimated quantities of materials to be placed and the size of the crew to place the materials. By knowing the relationships between the activities and their durations, the scheduler can determine the planned start and finish dates of the activity. Therefore, each activity is definable in terms of its planned start time, duration, and planned finish.

Assignability. Determining responsibility for each activity is critical to any scheduling effort, since any construction project involves bringing together many craftspeople, subcontractors, suppliers, and others to attain project completion. Task activities should be defined so that the responsibility for activity completion is clear and assignable to a single party. If the activity is to be completed by the contractor's own forces, it should be defined by the crew to identify the proper superintendent or foreman. If the activity is to be completed by a subcontractor or vendor, responsibility should be assigned to that party, so that as the project progresses, communication can be expedited using the schedule as the baseline indicator of the performance. The plan is something to measure against to determine performance.

Measurability. The duration and resources assigned to the activity must be measurable to determine whether the budget for the activity duration was met. How many days were actually spent? What was the actual physical progress in terms of days? How many resources were actually expended? What was the actual physical progress in terms of resources? The duration and resources are measured as of a particular date, and an evaluation is made about how the project is going in terms of the original plan. Is the project ahead of or behind the original schedule with respect to time? Is the project ahead of or behind the original schedule in terms of the expenditure of resources? Answering these questions is essential for monitoring and controlling the project. Since the activities are assigned, the party responsible for controlling the duration and the expenditure of resources is identified by activity, and communications to resolve project control problems are enhanced.

ACTIVITY IDENTIFICATION

The activity identification (activity ID) is the way the activity logic is identified to the computer. The activity ID is used by *SureTrak* (and most other scheduling software programs) to give the activity a short name or identifier that can be used for sort functions. In Figure 1–3, note that the Clear Site activity has an activity ID of 1000. An example of using this field as a sort field would be if an owner wanted to break a project into three phases (A, B, and C). All activities relating to phase A would start with the identifier A, such as A1000. All activities relating to phase B would start with the identifier B. This is a handy tool for quick activity identification.

SureTrak maintains a database of information about each activity, and the key or primary field used to sort the information is the activity ID field.

The activity ID itself can be a convenient means to sort activities. Take care when developing the naming format to make the activity ID a sortable field. Be consistent in naming all project activities to make the sort possible.

ACTIVITY DESCRIPTION

The activity description is a *SureTrak* field that is longer than the activity ID field and is used to describe the activity (see Figure 1–3). This important communications tool must be clear, concise, and have the same meaning to all parties using the schedule. This includes the contractor's forces, subcontractors, owner, and the architect/engineer.

The activity description must communicate the scope and location of the portion of the work that the activity encompasses. Because so much information is communicated, descriptions must be consistent in format, and abbreviations are commonly used. Abbreviations and procedures for naming activities should be consistent throughout the project, as this will make the schedule much easier to use. Whenever possible, use standard industry abbreviations, such as "Ftg" for footing and "Conc" for concrete.

ACTIVITY RELATIONSHIPS

The relationships among activities determine which other activities must come before, after, or can be going on at the same time as the activity being defined. Again, activity relationships are defined using information derived from the contractor's management team, the accounting department, the equipment department, the plans and specifications, a visit to the site, the nature of the work, the owner, the banker, the subcontractors, the governmental agencies involved and their requirements, and the suppliers. These relationships give the schedule its "logic" to enable the calculations to work. The activity relationships determine the interaction of the parties to the work and to each other.

TYPE OF SCHEDULE PRESENTATION

The construction schedules can be presented in different ways according to the needs of the users. The most common types of schedule presentation are Gantt (bar) chart, PERT diagram, CPM diagram, timescale logic diagram, and tabular report. Each has its own advantages and disadvantages, which will be discussed in this section.

Gantt (Bar) Charts

The Gantt chart is a convenient and easy-to-read method of viewing the schedule in bar chart form. Henry L. Gantt popularized this graphical representation in the early 1900s. Simply put, the horizontal axis represents a time scale of the project and the vertical scale lists the activities necessary to put the project together. These activity descriptions can be as broad or narrow as the scheduler needs to adequately describe the project. Figure 1–3 is an example of a *SureTrak* generated Gantt chart.

Advantages of the Gantt (Bar) Chart.   There are three primary advantages to using the Gantt chart. First, it is usually easy to read. It is apparent when the activities should take place. Anyone involved in the construction process—owner, architect, banker, bonding agent, contractor, subcontractor, or supplier—can interpret this simple document. A second advantage is that, because of its simplicity, it is a great communications tool. The third advantage is that it is easy to update the chart to show progress.

Disadvantages of the Gantt (Bar) Chart.   The primary disadvantage of the Gantt chart is that it does not show the interrelationships among the various activities. What happens if one of the activities is completed late? How is the rest of the project delayed? The impact of the delay can

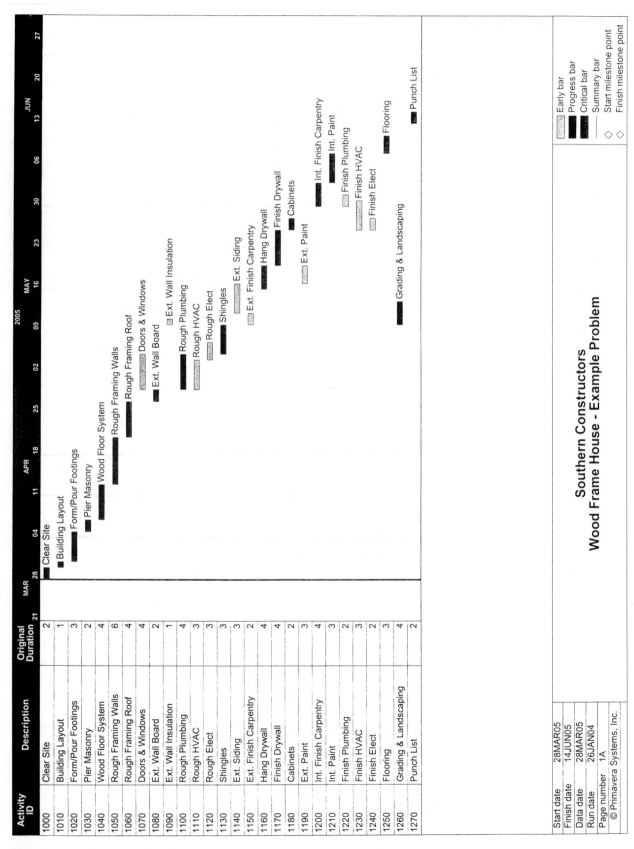

Figure 1–3 Gantt (Bar) Chart—Wood-Framed House

be evaluated by analysis of the Gantt chart, but since the relationship between the activities is not shown, the conclusions are open to debate. The logic of the interrelated activities may be very formalized, but it is not clearly and completely conveyed to the user of the Gantt chart. A great majority of the construction claims relating to schedules that are lost by contractors are lost because the contractor cannot prove the impact of schedule delays. A construction claim requires proof through documentation. For example, if the drywall subcontractor is late hanging the drywall, typically all interior work is delayed. This would have an impact on finishing the drywall, placing cabinets, completing the interior finish carpentry, and the rest of the schedule, but because the Gantt chart does not show the direct impact on these activities, there is room for argument and the claim of the contractor is more difficult to prove. A better tool for proving impact cost or ripple damages is a PERT diagram, which is discussed in the next section.

Gantt (Bar) Chart Format.   The Gantt chart format (along with the PERT and timescale logic diagrams) is one of *SureTrak*'s primary hard-copy graphical print formats. Once the scheduler has entered the activity information into *SureTrak*, any of the three print formats can be requested. The Gantt chart is a convenient vehicle for confirmation and dissemination of the information used for more complex formats. Figure 1–3 is an example of the *SureTrak* hard-copy print of a Gantt chart for a typical residential construction.

PERT Diagrams

The graphics for logic diagrams within *SureTrak* are organized for activity-on-node diagrams. The nodes (rectangles in Figures 1–4a and 1–4b) are the activities, and they are connected by arrows that show relationships between and among activities. The nodes may also contain information about the activities. The activity-on-node diagrams originated as the Program Evaluation and Review Technique (PERT) diagrams. PERT diagrams were developed for projects where activity durations and scope of work could not be determined with great accuracy, such as for new types of projects that had never been built before. The full extent of the work—or the relationships between the activities—was not understood. This method was originally used by the Special Projects Office of the Navy Bureau of Ordnance in the late 1950s and early 1960s in the development of the Polaris missile.

The node or box is the activity, and the arrows connecting the boxes show the relationships between the activities. Compare Figure 1–4a and Figure 1–4b to Figure 1–1 to see the difference between the activity-on-node and activity-on-arrow diagramming. The activity-on-node diagram is *SureTrak*'s PERT diagram. Figure 1–4a and Figure 1–4b are examples of a hard-copy print of *SureTrak*'s PERT diagram.

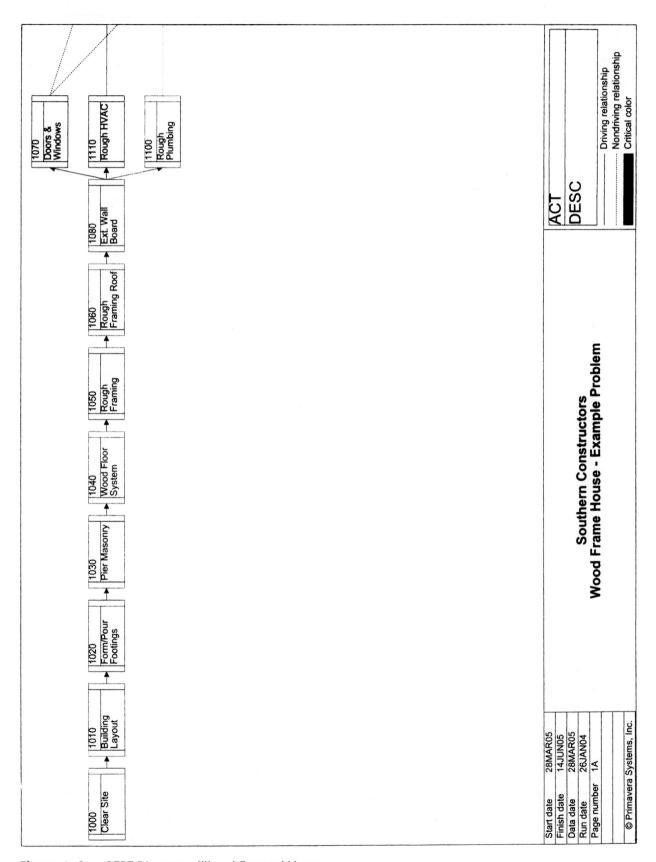

Figure 1–4a PERT Diagram—Wood-Framed House

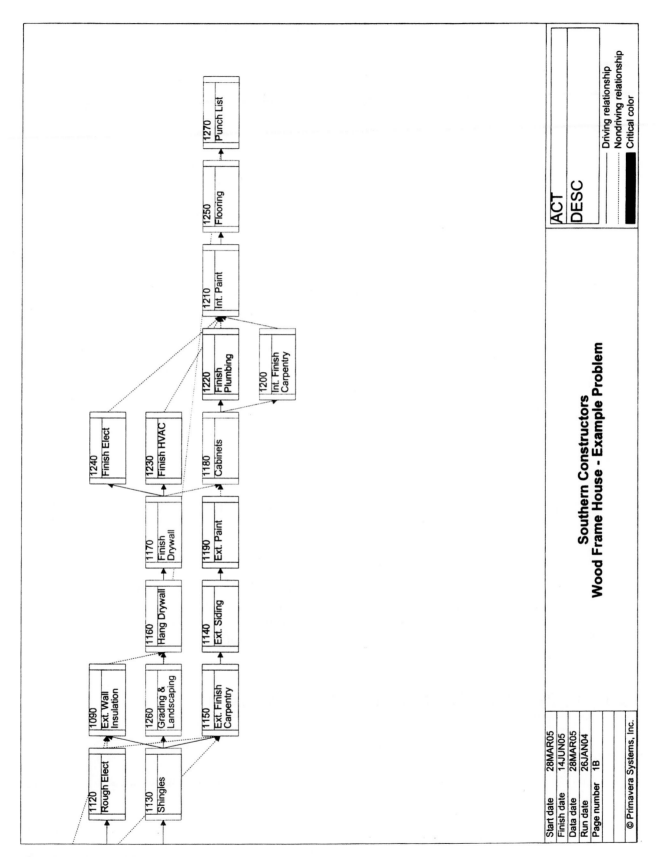

Figure 1–4b   PERT Diagram—Wood-Framed House

Critical Path Method (CPM) Diagrams

The activity-on-arrow diagramming method for scheduling was developed by the E. I. du Pont de Nemours Company in the late 1950s and was based on the Project Planning and Scheduling method (PPS). This form of scheduling was eventually called the Critical Path Method (CPM) (Figure 1–1).

SureTrak does not support arrow notation. The activity-on-node diagram is neater and more efficient from a graphical point of view. It is much easier to design graphics by putting information into a box, rather than associating it with a line of unknown length with which you are also trying to show logical relationships. Because of this, the activity-on-arrow diagrams are falling into disuse.

Timescaled Logic Diagrams

SureTrak's timescaled logic diagram (Figure 1–5) combines the advantages of the Gantt (bar) chart and the pure logic diagramming methods (PERT). Like the bar chart, it shows the activities' relationships to time, either in work days or calendar days. It also shows the relationships among activities.

The advantages of the timescaled logic diagrams are that they are easy to understand, like the bar chart, and they define the logic. For updating and documentation purposes, this is a tremendous advantage. The disadvantage, however, is that with increasing the size of the project, and therefore the number of activities, it becomes very hard to read and follow the logic. Figure 1–5 is an example of a hard-copy print of *SureTrak*'s timescaled logic diagram.

Tabular Reports

Sometimes a table or tabular report is the easiest way to communicate or update information. It is also easier to catalog in the project's historical database. Figure 1–6 is an example of a tabular report that lists logic information associated with each activity.

EXAMPLE PROBLEM: GETTING READY FOR WORK

Table 1–1 is a list of activities for getting ready for work. Figures 1–7 to 1–9 are completed samples of a pure logic diagram (PERT), a bar chart, and a timescaled logic diagram using the list of activities from Table 1–1.

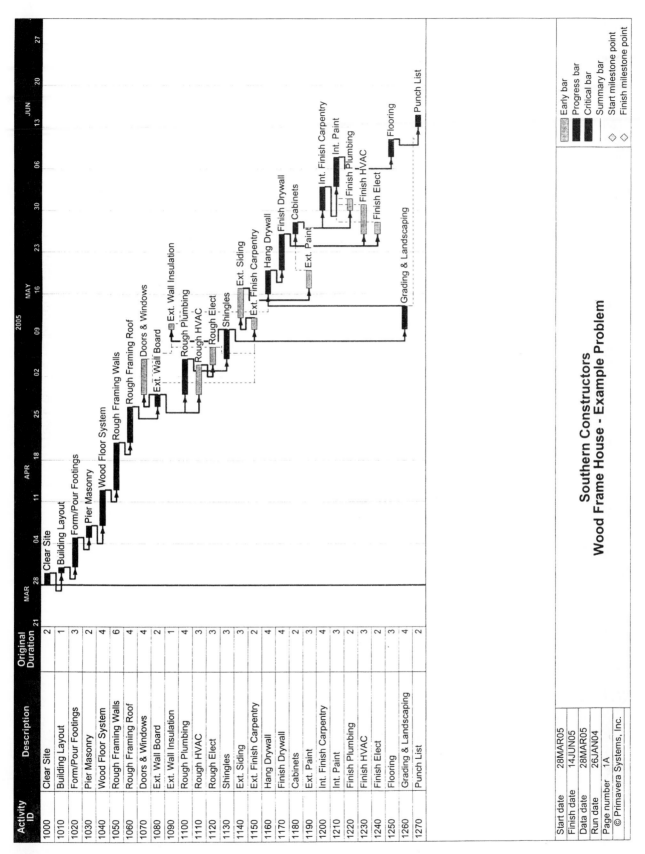

Figure 1–5   Timescaled Logic Diagram—Wood-Framed House

Activity ID	Description	Original Duration	Early Start	Early Finish	Total Float	Predecessors	Successors
1000	Clear Site	2	28MAR05	29MAR05	0		1010
1010	Building Layout	1	30MAR05	30MAR05	0	1000	1020
1020	Form/Pour Footings	3	31MAR05	04APR05	0	1010	1030
1030	Pier Masonry	2	05APR05	06APR05	0	1020	1040
1040	Wood Floor System	4	07APR05	12APR05	0	1030	1050
1050	Rough Framing Walls	6	13APR05	20APR05	0	1040	1060
1060	Rough Framing Roof	4	21APR05	26APR05	0	1050	1080
1070	Doors & Windows	4	29APR05	04MAY05	6d	1080	1090, 1150
1080	Ext. Wall Board	2	27APR05	28APR05	0	1060	1070, 1100, 1110
1090	Ext. Wall Insulation	1	10MAY05	10MAY05	3d	1070, 1120, 1130	1160
1100	Rough Plumbing	4	29APR05	04MAY05	0	1080	1120
1110	Rough HVAC	3	29APR05	03MAY05	4d	1080	1120
1120	Rough Elect	3	04MAY05	06MAY05	4d	1110	1090, 1150
1130	Shingles	3	05MAY05	09MAY05	0	1100	1090, 1150, 1260
1140	Ext. Siding	3	12MAY05	16MAY05	4d	1150	1190
1150	Ext. Finish Carpentry	2	10MAY05	11MAY05	4d	1070, 1120, 1130	1140
1160	Hang Drywall	4	16MAY05	19MAY05	0	1090, 1260	1170
1170	Finish Drywall	4	20MAY05	25MAY05	0	1160	1180, 1230, 1240
1180	Cabinets	2	26MAY05	27MAY05	0	1170, 1190	1200, 1220
1190	Ext. Paint	3	17MAY05	19MAY05	4d	1140	1180
1200	Int. Finish Carpentry	4	30MAY05	02JUN05	0	1180	1210
1210	Int. Paint	3	03JUN05	07JUN05	0	1200, 1220, 1230, 1240	1250
1220	Finish Plumbing	2	30MAY05	31MAY05	2d	1180	1210
1230	Finish HVAC	3	26MAY05	30MAY05	3d	1170	1210
1240	Finish Elect	2	26MAY05	27MAY05	4d	1170	1210
1250	Flooring	3	08JUN05	10JUN05	0	1210	1270
1260	Grading & Landscaping	4	10MAY05	13MAY05	0	1130	1160, 1270
1270	Punch List	2	13JUN05	14JUN05	0	1250, 1260	

Southern Constructors
Wood Frame House - Example Problem

Start date	28MAR05
Finish date	14JUN05
Data date	28MAR05
Run date	26JAN04
Page number	1A

© Primavera Systems, Inc.

Figure 1–6 Tabular Report—Wood Framed House

Activity Name	Duration (Minutes)	Activity Name	Duration (Minutes)
1. Turn Off Alarm	1	12. Place Underwear	1
2. Get Out of Bed	1	13. Place Shoes	1
3. Remove Pajamas	1	14. Place Shirt	1
4. Brush Teeth	2	15. Place Pants	1
5. Take Shower	5	16. Place Tie	1
6. Wash Hair	2	17. Take Vitamins	1
7. Make Coffee	2	18. Fix Cereal	1
8. Perk Coffee	5	19. Eat Cereal	3
9. Drink Coffee	10	20. Make Bed	2
10. Dry Body/Hair	3	21. Leave for Work	1
11. Comb Hair	1		

Table 1–1 Activity List—Getting Ready for Work

Activities	1	2	3	4	5	6	7	8	9	10	11	12	13	14	15	16	17	18	19	20	21	22	23	24	25	26
Turn Off Alarm	×																									
Get Out of Bed		×																								
Remove Pajamas				×																						
Brush Teeth														×												
Take Shower					×	×	×	×	×																	
Wash Hair								×	×																	
Make Coffee			×	×																						
Perk Coffee					×	×	×	×	×																	
Drink Coffee											×	×	×	×	×	×	×	×	×	×						
Dry Body/Hair											×	×	×													
Comb Hair														×												
Place Underwear																×										
Place Shoes																		×								
Place Shirt																	×									
Place Pants																×										
Place Tie																			×							
Take Vitamins														×												
Fix Cereal																				×						
Eat Cereal																						×	×	×		
Make Bed																									×	×
Leave for Work																										×

Figure 1–7 Example Problem—Bar Chart Format

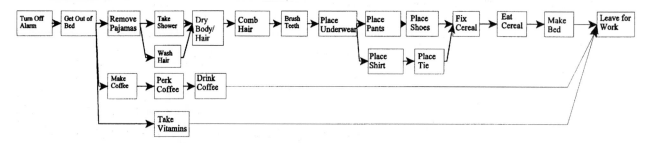

Figure 1–8   Example Problem—Pure Logic Diagram Format

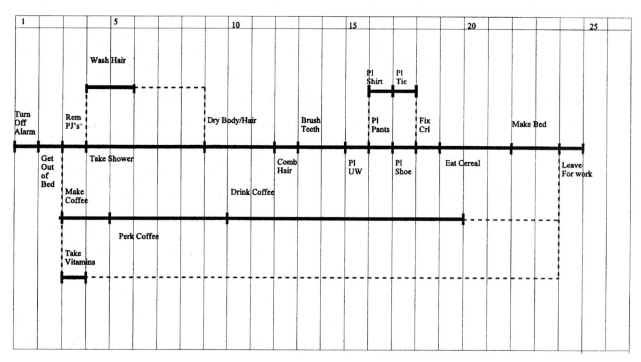

Figure 1–9   Example Problem—Timescaled Logic Diagram Format

SUMMARY

This chapter provided a brief overview of the history of scheduling and highlighted the human need to more efficiently control the time and resources necessary to complete construction projects. These construction projects rely on good scheduling practices that project management can use to better control quality, cost, time, and safety. Good scheduling practices benefits could be organized in six categories: build the project on paper, use resources efficiently, control the project, increase return on investment, anticipate and solve problems, and act as a communication tool. To obtain these benefits, the schedule generation involves four phases: planning, scheduling, monitoring, and controlling. These phases apply to both scheduling design-bid-build and fast-track projects. The primary difference between scheduling design-bid-build and fast-track projects is that in the former, the main focus is on the construction, while in the latter, the focus is on the project as a whole, including design, construction, and project management. The schedule is prepared with different levels of details depending upon the need of the user and the phase of the project. The most common scheduling levels are preplanning schedules, overall detailed/baseline schedules, current schedules, and 30/60 or 10 days look-ahead schedules.

The chapter then provided an introduction to the main component of a schedule, which is the activities. The activities are tasks that must be completed to achieve project goals and objectives. The activities have the following characteristics: consume time, consume resources, defined start and finish, assignable, and measurable. The computer recognizes the activities by their ID rather than their description. Activity description is a brief explanation used to communicate with schedule users. Additionally, these activities are interrelated using logic to determine which other activities must come before or after, or can be going on at the same time as the activity being defined.

Finally, the chapter presented the most common types of schedule presentation: Gantt (bar) chart, PERT diagram, CPM diagram, timescale logic diagram, and tabular report.

EXERCISES

1. **Benefits of Good Scheduling Practices**
   Describe in 3–5 lines each of the benefits of good scheduling practices.
2. **Scheduling Phases**
   Name the four scheduling phases.
3. **Scheduling Design-Bid-Build versus Fast-Track Projects**
   Compare in 3–5 lines the difference between scheduling design-bid-build and fast-track projects.
4. **Scheduling Activities**
   Summarize the identification, description, and relationship of scheduling activities.
5. **Scheduling Presentation**
   List the types of scheduling presentations.
6. **Building a Shed**
   Produce a Gantt (bar) chart, a PERT diagram, and a timescale logic diagram using the tabular list of activities from Table 1–2, and assume the precedence relationships between activities. Refer to the example problem for the format of each of these schedule presentation types.

	Activity Name	Duration (Days)		Activity Name	Duration (Days)
1.	Clear Site	1	11.	Place Interior Paneling	1
2.	Remove Topsoil	1	12.	Ext Trim Carpentry	2
3.	Form Slab	2	13.	Install Overhead Door	1
4.	Place Rebar/Embeds	1	14.	Rough Electrical	1
5.	Pour Slab	1	15.	Finish Electrical	1
6.	Prefab Wood Walls	2	16.	Place Shingles	2
7.	Erect Wood Walls	1	17.	Install Finish Carpentry	1
8.	Install Siding	2	18.	Place Topsoil/Grade	1
9.	Place Trusses	1	19.	Landscape	1
10.	Place Roof Sheathing	2			

Table 1–2 Activity List—Building a Shed

7. Purchasing a New Automobile

Produce a Gantt (bar) chart, a PERT diagram, and a timescale logic diagram using the tabular list of activities from Table 1–3, and assume the precedence relationships between activities. Use as many concurrent activities as possible. Refer to the example problem for the format of each of these schedule presentation types.

	Activity Name	Duration (Days)
1.	Decision - Type Car	5
2.	10 Models - Make List	1
3.	10 Models - Obtain Consumer Ratings	1
4.	10 Models - Obtain Pricing Publication	1
5.	10 Models - Talk to Vehicle Owners	1
6.	3 Models - Decision	2
7.	3 Models - Test Drive	1
8.	3 Models - Obtain Information	1
9.	3 Models - Compare Lease/Purchase Options	1
10.	1 Model - Decision	2
11.	1 Model - Negotiate Purchase Contract	1
12.	3 Institutions - Compare Financing	1
13.	Decide on Institution	1
14.	Obtain Financing	1
15.	Money for Down Payment	1
16.	Obtain Insurance	1
17.	Obtain Tag	1
18.	Drive Away with Purchased Vehicle	1

Table 1–3 Activity List—Purchasing a New Automobile

2 Work Breakdown Structure and Rough Logic Diagram

Objectives

Upon completion of this chapter, you should be able to:

- List the reasons for preparing a work breakdown structure
- Identify the information required to prepare a rough logic diagram and work breakdown structure
- Describe the steps to prepare a rough logic diagram
- Contrast the rough logic diagram of a design-bid-build versus fast-track construction
- Describe the types of activity relationships
- Prepare a rough logic diagram
- Estimate activity duration
- Describe factors affecting productivity

INTRODUCTION

A rough logic diagram is used to build a construction project on paper before the actual construction begins. During the preparation of the rough logic diagram, the scheduler defines the relationship (logic) between the activities. Thus, it is important for the scheduler to identify all the activities and their level of detail before preparing the rough logic diagram. This can be accomplished by preparing a work breakdown structure. Experienced schedulers seldom prepare a work breakdown structure prior to creating a rough logic diagram. However, it is recommended (especially for novices) to prepare a work breakdown structure to increase the level of completeness, reduce confusion, and ultimately save time by identifying all necessary activities.

Preparing the work breakdown structure consists of systematically dividing a project into its components. These components become more specific and smaller at each level of the work breakdown structure. Figure 2–1 shows a partial work breakdown structure for a Small Construction Project. The first level, Project, corresponds to the construction project that is being scheduled, in this case, the Small Construction Project. The second level, Sub-Project, corresponds to the

Level Description

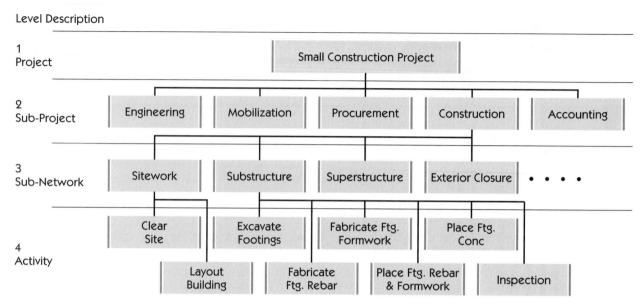

Figure 2–1 Work Breakdown Structure

company units that will be responsible for the areas of the project, in this case, engineering, mobilization, construction, and so on. The third level, Sub-Network, corresponds to the major packages of each area of the project; in this case, the Construction area has been broken down into: sitework, substructure, and so on. The fourth level, Activity, corresponds to the elements that will be performed to complete the major packages of each area and to ultimately complete the project; in this case, sitework has been broken down into two activities: clear site and layout building.

Information Gathering for the Rough Logic Diagram and Work Breakdown Structure

During the prebid or estimating phase of the project life cycle, the estimate becomes the focal point in gathering information relating to the project. Information must be gathered from many sources and incorporated into the estimate. The following is typical of the information that must be collected:

- Owner time constraints and other input
- Scope definition
- Building methods and procedures to be used
- Productivity rates, crew balances, and crew sizes
- Labor availability and wage rates
- Construction equipment to be used
- Construction equipment availability and use rates
- Material availability and prices

- Subcontractor availability and prices
- Fabricator availability and prices
- Project organizational structure
- Rough preliminary schedule
- Temporary facilities requirements
- Permit and test requirements
- Tax requirements
- Insurance requirements

In-Depth Planning.   If a contractor's bid wins the contract for a project, the estimate is used to generate the schedule. Now that the project is a real live job rather than just a proposal, more in-depth planning can take place. A logic diagram rather than a "quick and dirty" bar chart is used to fine-tune the information already gathered in the estimating stage.

Parties.   The schedule becomes the focal point of project information that is received from the project manager, superintendent, foremen, estimator, subcontractors, fabricators, suppliers, vendors, owner, and architect/engineer.

ROUGH LOGIC DIAGRAM PREPARATION

As discussed in Chapter 1, the four scheduling phases are planning, scheduling, monitoring, and controlling. The first step of the planning phase is preparing a rough logic diagram. During the rough logic diagram preparation, the approach to placing the project takes shape. Therefore, the scheduler should strike a balance between the intricacy of the project and the number of activities scheduled. Too few activities in the rough logic diagram results in a schedule that is insufficient in detail to be worthwhile. Likewise, too many activities results in confusion and needless detail. In either case, the schedule is likely to be ignored by those it intended to benefit. Once accepted, developed, and refined, the rough logic diagram becomes the construction schedule to which more durations and resources are applied.

Construction decisions concern:

- Construction methods
- Flow of materials
- Prefab versus on-site assembly of materials
- Types of construction equipment
- Crew size and balance
- Productivity expectations
- Subcontractor definition

Management decisions include:

- Work breakdown structure (WBS)
- Division of responsibility
- Division of authority
- Software to be used
- Level and distribution of reports
- Interface with other functions (payroll, accounting, and so on)

There are five steps in developing a rough logic diagram. The steps for preparing a construction-only schedule are as follows:

1. **Contractor's Initial Meeting.** The contractor's project team members have an initial meeting to discuss the entire project from beginning to end. The members include the estimator, project manager, possibly the superintendent, and the person creating the schedule. The team builds the project on paper, starting from the estimate. Either a tape recorder is used or the scheduler takes notes. Some people prefer to simply list the activities. In any case, the estimate is the benchmark from which to start.

2. **Rough Diagram Creation.** The scheduler uses the information from the initial meeting to draw the rough diagram (usually PERT) on paper, showing interrelationships between the activities. Some schedulers prefer to go straight to the software at this stage rather than to manually draw the rough diagram on paper. Those who have software with an integrator function can also use it to estimate cost and resources needed for each activity at this stage.

3. **Review.** The entire project team must "buy into" the rough diagram so that it is "their" schedule. It is critical for each member to agree as a team to the concept, methods, and procedures to be used in the construction of the project.

4. **Major Subcontractors and Suppliers Input.** Input from major subcontractors and suppliers of long lead-time material and/or equipment items is critical. With the contract documents, the general contractor or construction manager has overall responsibility for coordinating and scheduling the work. The contractor must coordinate all parties' work and resolve any disputes. Input is usually required of the subcontractor in the contractor-subcontractor contract. Through discussions with the subcontractor/suppliers, the contractor modifies the plan to reconcile differences. The subcontractors/suppliers then accept the contractor's plan as the project plan. Contractors need to make sure they have obtained information from any outside party that can impact the schedule.

5. **Project Schedule Acceptance.** After the contractor has reviewed the revised rough schedule with the subcontractors/suppliers, it becomes the official project schedule.

PREPARATION OF A ROUGH LOGIC DIAGRAM OF DESIGN-BID-BUILD VERSUS FAST-TRACK CONSTRUCTION

During the rough logic diagram preparation, the scheduler should consider the difference between a design-bid-build and a fast-track construction. As indicated in Chapter 1, the primary difference is that the scheduling effort in the design-bid-build project focuses on the field construction effort, whereas the fast-track schedule looks at the project as a whole.

In a design-bid-build construction, the scheduler preparing the rough logic diagram is limited because many decisions impacting the schedule have already been made before the scheduler becomes involved in the project. In the fast-track project, on the other hand, the scheduler preparing the rough logic diagram is involved in decisions on the project as a whole. Further discussion regarding design-bid-build and fast-track construction is provided in this section.

Design-Bid-Build Construction

Design-bid-build construction carries out the construction life cycle phases in consecutive rather than concurrent, order—each phase is completed before the next begins. Building the project with this constraint takes longer. Because each step of the process is defined, finished, and usually paid for before the next begins, and because this process is usually based on lump-sum contracts, the owner transfers the risk of cost overruns to other parties. The owner knows the cost of each phase before committing to pay for it. The owner also knows that because the project can be canceled at any time, she is only at risk for the work that has been released.

No Scheduler Input. In a lump-sum contract (design-bid-build construction), the contractor's contract typically includes the construction, commissioning, and closeout phases of the construction life cycle but no input during the conceptual or detailed design phases. This is unfortunate because who knows more about minimizing construction cost by efficient design, constructibility, materials selection, and using prefabricated materials than someone who is involved in the actual building process every day? Many times, the designers operate in a vacuum during the design phase, relying only on their own past experience to select materials and construction methods. Significant savings can usually be realized by involving contractors, subcontractors, and material suppliers at this stage of the project. Using construction management professional services may be a good solution, as it incorporates the construction experience in the design process before a contractor is available.

Steps in the Design-Bid-Build Construction.   The following is a sequential listing of the steps typically followed with lump-sum construction in the United States:

1. Owner determines a need.
2. Owner contacts architect/engineer for conceptual design.
3. Conceptual design is completed.
4. Owner approves conceptual design or sends it back for modification.
5. Owner contracts architect/engineer for detailed design contract documents.
6. Detailed design is completed.
7. Owner approves detailed design or sends it back for modification.
8. Owner puts contract documents out for bid.
9. Bids are received from contractors and negotiated.
10. Owner and contractor sign contract.
11. Construction proceeds.
12. Project is commissioned.
13. Project is closed out.

Negotiated, Fast-Track Construction

With a fast-track type project, the constructor (construction manager) is typically involved in all construction life-cycle phases except possibly maintenance. The constructor has a negotiated contract with the owner and acts as the owner's agent in a fiduciary relationship. The constructor is looking out for the owner's best interest. The constructor has input in defining the owner's needs during both the conceptual and detailed design phases. The contractor is primarily concerned with minimizing construction cost by efficient design, constructibility, materials selection, and the use of prefabricated materials while the critical decisions about these factors are being made. The constructor will produce the conceptual estimate and cost studies of different design scenarios. Having a member of the design team who is a cost-conscious, knowledgeable constructor can be a tremendous advantage to the owner in producing a successful project.

Steps in Fast-Track Construction.   The following set of steps is typical of fast-track construction in the United States using the agency construction management type contract:

1. Owner has a need.
2. Owner contacts construction manager for services, which include control of design, estimating, and scheduling. The construction manager may be the architect or engineer as well as the contractor.
3. Construction manager works with architect/engineer to produce conceptual design.
4. Construction manager estimates costs to fine-tune conceptual design until it meets the owner's return-on-investment requirements.
5. Owner/construction manager approves conceptual design.

6. Owner/construction manager contracts with architect/engineer for detailed design contract documents, then breaks down the design into packages or phases that can be completed and put out for construction bid.

7. Owner/construction manager approves detailed design one package at a time.

8. Owner/construction manager puts contract documents out for bid, one package at a time.

9. Owner and successful package contractor sign contract for single package.

10. Owner/construction manager brings all the packages through the steps of detailed design, bid, award, and construction.

11. Construction progresses simultaneously with further detailed design. The construction manager monitors the schedule, controls multiple contractors, and approves pay requests. The construction manager acts as the general contractor at the job site.

12. Project commissioning is usually handled one package or system at a time.

13. Project closeout is handled by the construction manager in much the same way as the general contractor would handle it.

With the fast-track approach to construction services, the owner assumes more of the risk because construction is proceeding before the design documents are completed. The construction manager is the owner's agent, and therefore typically does not sign a lump-sum contract with the owner, unless it is a Construction Management (CM) at-risk contract. The real advantage to the owner is the time savings of producing the detailed design and construction concurrently rather than consecutively. This approach can cut in half the overall time needed to bring a project "on line," thus saving time and money.

TYPES OF ACTIVITY RELATIONSHIPS

Logic refers to the entire network of relationships between and among activities. This network controls the scheduling of activities. When you prepare the schedule, you establish relationships between activities. These relationships describe which activities depend on others. The relationships between activities are defined as predecessor, successor, or concurrent relationships. *A predecessor activity must be completed before a given activity can be started. A successor activity cannot start until a given activity is completed. Concurrent activities* are logically independent of one another and can be performed at the same time. Relationships between a particular activity and its predecessor and successor activities can vary. The options, as shown in Figure 2-2, are:

- Finish to Start (FS)
- Start to Start (SS)
- Finish to Finish (FF)
- Start to Finish (SF)

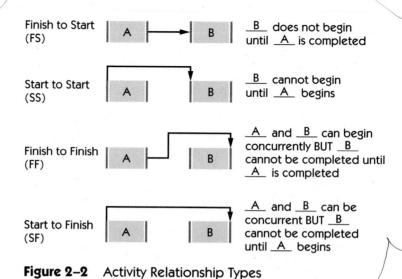

Figure 2–2   Activity Relationship Types

The Finish to Start (FS) relationship means the predecessor activity must finish before the successor activity can start. The Start to Start (SS) relationship means the successor activity cannot start until the predecessor activity has started. The Finish to Finish (FF) relationship means the successor activity cannot finish until the predecessor activity has finished. The Start to Finish (SF) relationship means the predecessor activity must start before the successor activity can finish. Figure 2–3 is an example of a rough logic diagram for a small portion of a construction project. In this example, the type of relationship between the activities is noted on the relationship arrow between the activities.

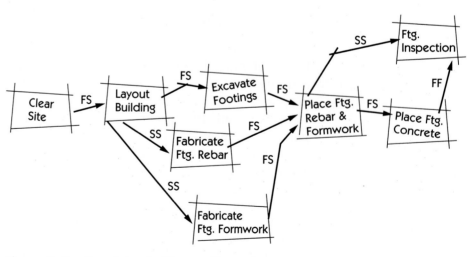

Figure 2–3   Rough Logic Diagram Example

Use of Schedules

A critical part of any scheduling effort is deciding how the schedule will be used to control a project. The following issues should be defined before the rough diagram is prepared:

- Type of project
- Purpose of the schedule
- Software requirements
- Parties involved: owner; construction manager; architect/engineer; contractors, major subcontractors, minor subcontractors, and sub-subcontractors; suppliers of long lead-time items
- Authority/responsibility of involved parties (Who is "keeper of the schedule"?); responsibility for resolving disputes and interferences
- Needs of the involved parties from a scheduling point of view
- Type, frequency, and depth of reports to be provided
- Schedule update requirements (What is the time frame for providing updated information in what format, and who is to provide the information?)
- Project change requirements
- Resources to be controlled: labor, materials, subcontractors, construction equipment
- Cash flow requirements
- Payment (schedule of values) requirements

Time Units

The Day

The time unit used for most construction schedules is the day.

The determination of the minimum planning unit depends on the total expended duration of the project, as well as the required level of detail. In the typical construction schedule with days as the unit of measure, when activities are being defined, no block of work is assigned a duration of less than a day to complete. If the activity takes less than a day, it is either rounded up to a day or combined with another task/activity.

The Hour

Sometimes the hour is a more convenient unit than the day. This is the usual unit of choice for a "turnaround" schedule as used in a paper mill, chemical plant, refinery, or other facility in operation. The facility must be brought down or "off-line" for maintenance or to add or modify plant systems. Typically, the owner loses thousands of dollars per day while the plant is out of operation. Making the modifications as quickly as possible and getting the plant going again is essential to the owner. Each hour is critical.

CONCURRENT RATHER THAN CONSECUTIVE LOGIC

Besides trying to improve efficiencies, schedulers also need to improve the logic of activity interrelationships and sequences when putting the rough diagram together. Instead of having just one thing at a time happen at the job site, as many things, crews, or work functions as possible need to be happening without interfering with each other and without risking exposure or damage by project components being in place too early. What is the shortest and most efficient way to accomplish the work?

ESTIMATING ACTIVITY DURATION

The duration of an activity is a function of the quantity of work to be done by the activity and the rate of production at which the work can be accomplished. The formula is:

$$\text{Activity Duration} = \frac{\text{Quantity of Work}}{\text{Productivity Rate}}$$

Take, for example, a masonry activity with a quantity of work of 10,000 regular masonry blocks to be placed by a crew of three masons, two laborers, and a mixer. The crew can place 800 blocks per day. Thus, duration = 10,000 blocks/800 blocks per day = 12.5 or 13 days.

RESOURCE AVAILABILITY

Critical to estimating activity duration is the availability of resources, including labor, materials, equipment, subcontractors, and suppliers/fabricators.

Driving Resources

Most activities have certain driving resources that control activity duration. For example, in the preceding masonry activity, the three masons are the driving resource. Their craft determines activity duration. When the ratio of laborers and mixers reaches maximum efficiency for three masons, no matter how many more laborers and mixers are added, the performance of the three masons will not increase.

Reduction of Duration

You can reduce the duration of 13 days for the masonry activity in three ways. First, increase the productivity rate of the three masons to more than 800 blocks per day by improving a method or system for placing the block. Second, add more masons, with enough laborers and mixers to ensure full production. Third, have the masons work overtime (more hours per day).

Extended Scheduled Overtime

Extended scheduled overtime can ultimately have a negative impact because of lost productivity. It is also important to consider the higher cost of overtime pay. Studies have shown that consistently relying on overtime by the same crew actually results in less productivity than is normally achieved in an eight-hour day.

As a rule, the more resources that are available to put the activity in place, the less time it takes to place the activity. If any of the resources necessary to place the activity are missing or are not handled properly, the activity duration calculation is impacted.

QUANTITY OF WORK

The contract documents define the scope of the project. Using the plans and specifications, the estimators survey quantities of materials necessary to complete the project. The estimate may not provide a bill of materials in enough detail to enable purchasing of all materials for the project, but it will provide enough detail to bid the project. The estimate prepared at this stage relies solely on the contract documents. If insufficient detail or ambiguity exists in the contract documents, the estimate will usually be overly conservative to account for uncertainties.

To produce a *quantity survey*, the estimators organize by type of work. Building contractors usually use Construction Specifications Institute (CSI) cost code structure. The estimators assign work definition by spec division, by phase, and then by the item or unit of work within the phase.

Contractors usually produce a quantity survey only for work to be completed by the contractor's own forces. For subcontracted work, contractors usually depend on the market for the best price.

A code of accounts is a coding/numbering system to categorize work for controlling/tracking purposes. The contractor organizes the code of accounts and quantity survey that each type of crew will accomplish into identifiable areas or phases, usually organized according to CSI format. The major CSI phase headings are:

01000	General Requirements	09000	Finishes
02000	Sitework	10000	Specialties
03000	Concrete	11000	Equipment
04000	Masonry	12000	Furnishings
05000	Metals	13000	Special Construction
06000	Wood and Plastics	14000	Conveying Systems
07000	Thermal and Moisture Protection	15000	Mechanical
08000	Doors and Windows	16000	Electrical

The cost code structure is organized by craft designation so that work boundaries are understood by all parties involved. For example, all masonry-related work is categorized in the 04000s.

PRODUCTIVITY RATE

Productivity Rate by Activity

The quantity survey defines the amount of materials for a particular type of work, such as the square footage of a slab to be finished. The next step is to assign a productivity rate to determine the duration and resources necessary for the activity. The rate is the quantity of work accomplished per work hour (for example, 30 square feet of concrete finished per work hour, or 30 SF/WH), or quantity of work accomplished per crew hour (for example, 120 square feet/crew hour based on a four-person crew, or 120 SF/CH). The rate per man-hour (work hour) has to be adjusted for the size of the crew to determine duration.

Sources of Productivity Information

Sometimes if no formal estimate is prepared, the scheduler may have to come up with the productivity rate. The following sources may be useful:

- Company records for producing the same type of work on previous projects (historical data)
- Published information in reference text
- Observation and measurement of work performance as it is being put in place on another project
- Qualified expert opinion

FACTORS AFFECTING PRODUCTIVITY

The amount of time it takes to accomplish a unit of work can vary appreciably for the same type of work from project to project and is the reason construction labor costs are unpredictable. This variability in performance results from differences in communications, supervision/ proper preplanning, layout of the work, crew balance, skill/craftsmanship, mental attitude of workers, purchasing practices, working conditions, environmental factors, continuously scheduled overtime, safety practices, work rules, and availability of work.

Communications

Good writing communications cannot be stressed enough when discussing productivity. It is common practice for the superintendent to communicate orally with the foreman about the work to be accomplished, referring to the plans, specs, and shop drawings. The foreman in turn communicates orally to craftspersons the direction, methods, and layout for accomplishing the work. This chain of oral communication leads to a great deal of rework, exemplified by the comment "I built what I thought you wanted."

Some supervisory personnel seem to almost take pride in not communicating information to their subordinates. This attitude of "I'm the only one who knows all the answers" promotes distrust and low morale among employees. The effective supervisor, however, has learned that open and effective communications with everyone involved in the construction process makes the job much easier and cuts down on needless finger pointing. A good subconscious thought for this manager to constantly ask is "If the owner of this project came onto the job site and asked any one individual what they were working on (and why), would that person know the answer?"

Many contractors now put communications in writing, using 10 day look-ahead and job-assignment sheets for in-depth planning and scheduling. An emphasis on sketches and drawings further reduces dependency on oral communications. The improved documentation improves productivity.

Supervision and Preplanning

Good supervision and proper preplanning also improve productivity. Commitment to cost and schedule control by top management leads to improvements in training, safety, scheduling, estimating, and purchasing and emphasizes quality on all projects.

Efficient Layout of the Work

Efficient layout is critically important in the proper execution of the work. Efficient layout includes:

- Storing materials so they can be located quickly and easily when necessary
- Minimal handling of materials (on many work items, more time is spent handling the materials than actually putting the materials in place)
- Using the right equipment for the job
- Providing efficient access to tools, utilities, drinking water, and sanitary facilities
- Where possible, working at waist level to reduce unnecessary motion and fatigue

Proper Crew Balance

A construction crew has a proper balance of workers for various aspects of each activity. In a masonry crew, for example, the ratio of the workers mixing the mortar to the laborers transporting the mortar and stacking the block for placement and to masons actually placing the block must be balanced properly to efficiently accomplish the work. The ratio of highly paid skilled workers to the lower paid helpers is important for peak efficiency.

Skill/Craftsmanship

Training construction craftspersons to improve their skills is necessary because of ever-changing technology. Improvements in construction equipment are changing the way work is accomplished and increasing productivity. Because a construction company's primary asset is its employees, nurturing and training them to increase performance is a wise investment.

Mental Attitude of Workers

People are not machines. The mental attitude of workers seriously affects productivity. Management should strive to promote worker feelings of pride in the company, faith in a secure future, confidence in good management, the company's respect for employees, the company's regard for employees' opinions, company growth, and the potential for individual growth. When employees get satisfaction from working with a company and their individual needs are being met, they are more productive at the job site.

Many construction companies have improved the attitude of their workers by adopting a Total Quality Management (TQM) philosophy. This organizational and operational approach, originally implemented in the manufacturing sector, has proven to be beneficial for the construction industry as well. The Construction Industry Institute in a study of seventeen major contractors who had developed a TQM operating philosophy found they had four major things in common:

- Greater productivity
- Increased customer satisfaction
- More profits
- Higher employee morale

As well as the obvious benefits of increased morale, many construction companies have documented savings in the form of better safety records and a significant reduction in "reworks."

Purchasing Practices

A construction company's purchasing practices can have a tremendous impact on job site productivity. Using prefabricated assemblies or pre-assembled units or taking labor off-site to a manufacturing-type controlled environment can have great impact on worker hours spent at the job site. A lower average wage rate, lower-level craft skills required, increased availability of specialized tools and jigs, the ability to work under a roof in a controlled environment, and availability of local labor can all be great advantages. A scheduling advantage is that the fabrication can be concurrent with work at the job site, reducing the overall time for the project to be completed. Prefabricating materials in a controlled environment rather than at the job site can also have a huge impact on cost.

Working Conditions

In construction, the worker is typically exposed to the elements. Extremes can slow productivity. Conditions that influence job site productivity include some that cannot be controlled—heat, cold, rain, humidity, dust, wind, and odor, and some that can be controlled—noise, climbing up or down, bending low or reaching high, and the number of people on site.

Some companies have taken innovative approaches to address the day-to-day working conditions of their employees. Texas Instruments Corporation (TI), for example, initiated radical changes to the way construction was carried out during the building of many of its newer facilities. Texas Instruments implemented a TQM policy to be followed by all the general contractors and subcontractors working on TI projects. One of the requirements was the construction of a clean, air-conditioned lunchroom for the workers at every one of their job sites. The only stipulation was that no food or drink (other than water) was allowed on the actual construction site. This one change, along with other elements of the TQM philosophy, resulted in higher employee morale, increased profits reported by all the contractors involved, and a reduction in "reworks" from 10% to 2%.

Continuously Scheduled Overtime

Many studies have shown that continuously scheduled overtime has a disastrous impact on productivity. The construction industry Business Round Table study entitled "Cost Effectiveness Study C-3" (November, 1980) showed that workers putting in sixty-hour weeks for nine straight weeks accomplished the same amount of work in the ninth week as in a normal forty-hour workweek without the overtime. This means that paying for the extra twenty hours accomplished nothing. Compounding this false economy is the enormous cost of paying overtime for work in excess of forty hours per week. Spot overtime can be effective, but continuously scheduled overtime is decidedly ineffective economically.

Safety Practices

Safety practices have direct and indirect influences on productivity. The direct influence is that when someone is hurt, work at the job site usually stops or is at least impacted by the disturbance. It is the topic of conversation. Everyone is concerned. The company has at least temporarily lost the services of an employee. The loss also affects the crew balance and hence the productivity of a work unit.

The indirect impact relates to the effect accidents have on workers' morale and feelings of personal safety and security. Accidents affect the way employees think of the quality of the company's management and ability to manage the job site. Accidents also affect the employee's pride in the company and desire to stay with the company.

Work Rules

Sometimes, labor constraints can have a negative impact on productivity. Union constraints that can limit management's ability to organize for maximum productivity include jurisdictional disputes, production guidelines, limits on time studies, limits on piecework, and limits on prefabrication.

Availability of Work

If times are good and plenty of construction work is available, employees know that other jobs are readily available. There is not as much pressure to produce at the existing job. Conversely, if things are tight, there is much more pressure to keep the existing job, or possibly to draw out the current work as long as possible. Work availability can be a double-edged sword, depending on how it is perceived by the workers.

ACCURACY OF ESTIMATING ACTIVITY DURATION

Schedulers sometimes make subjective judgments about the duration of activities rather than taking the time to refer to the estimate and perform an actual calculation based on productivity rates and quantity of work. Because time is money, the accuracy of a contractor's schedule depends directly on the reliability of activity duration estimates.

SUMMARY

This chapter provided an overview of work breakdown structure as a way to identify project activities before preparing a rough logic diagram. Some of the benefits of preparing a work breakdown structure before preparing a rough logic diagram include increased level of completeness, reduction of confusion, and, ultimately, saving time by identifying all necessary activities.

Several pieces of information required to prepare a rough logic diagram were discussed. Emphasis was placed on the importance of the balance between the intricacy of a project and the number of activities included in the schedule. Five steps were presented to prepare a rough logic diagram including contractor's initial meeting, rough diagram creation, review, major subcontractor and supplier input, and project schedule acceptance.

Differences between preparing a rough logic diagram for a design-bid-build and fast-track projects were presented. During the preparation of a rough logic diagram in a design-bid-build project, the scheduler is limited by decisions made prior to the bidding process. In a fast-track project, however, the scheduler is involved in the decisions on the project as a whole, certainly impacting the rough logic diagram. Another critical aspect during the preparation of the rough logic diagram is identifying the type of project, purpose of the project, and intended use of the schedule.

The chapter then explained the relationship network and the four types of activity relationships in a rough logic diagram: Finish to Start (FS), Start to Start (SS), Finish to Finish (FF), and Start to Finish (SF). Special consideration was given to the time units such as days and hours, activity duration, resource availability, and productivity.

EXAMPLE PROBLEM: ROUGH MANUAL LOGIC DIAGRAM

Table 2–1 lists twenty-eight activities (activity ID and description) for a house put together as an example for student use (see the wood-framed house drawings in the Appendix).

Act ID	Act Description
A1000	Clear Site
A1010	Building Layout
A1020	Form/Pour Footings
A1030	Pier Masonry
A1040	Wood Floor System
A1050	Rough Framing Walls
A1060	Rough Framing Roof
A1070	Doors and Windows
A1080	Ext Wall Board
A1090	Ext Wall Insulation
A1100	Rough Plumbing
A1110	Rough HVAC
A1120	Rough Elect
A1130	Shingles
A1140	Ext Siding
A1150	Ext Finish Carpentry
A1160	Hang Drywall
A1170	Finish Drywall
A1180	Cabinets
A1190	Ext Paint
A1200	Int Finish Carpentry
A1210	Int Paint
A1220	Finish Plumbing
A1230	Finish HVAC
A1240	Finish Elect
A1250	Flooring
A1260	Grading & Landscaping
A1270	Punch List

Table 2–1   Activity List—Wood-Framed House

EXERCISES

1. **Work Breakdown Structure**
   List three reasons for preparing a work breakdown structure.
2. **Rough Logic Diagram**
   Describe in 3–5 lines each of the steps in preparing a rough logic diagram.

3. **Rough Logic Diagram of Design-Bid-Build versus Fast-Track Projects**
   Compare in 3–5 lines the difference between a rough logic diagram for design-bid-build and one for fast-track projects.
4. **Types of Activity Relationships**
   Draw four pairs of construction activities demonstrating the four different types of activity relationships (FS, SS, FF, SF).
5. **Types of Activity Relationships**
   Prepare a rough logic diagram based on the logic statements provided in Figure 2–4 through Figure 2–9.

A.

1.	A	Must Precede	B, C, D
2.	B	Must Precede	E
3.	C, D	Must Precede	F
4.	E	Must Precede	G, H
5.	F	Must Precede	H
6.	G, H	Must Precede	I

Figure 2–4   Exercise 5A—Logic Statements

B.

1.	A	Must Precede	B, C
2.	B	Must Precede	D, E
3.	C	Must Precede	F
4.	D	Must Precede	G
5.	E, F	Must Precede	H
6.	G	Must Precede	I, J
7.	H	Must Precede	J
8.	I, J	Must Precede	K

Figure 2–5   Exercise 5B—Logic Statements

C.

1.	A	Must Precede	B, C, D
2.	B, C	Must Precede	E
3.	C, D	Must Precede	F
4.	D	Must Precede	G
5.	E, F	Must Precede	H
6.	G, F	Must Precede	I
7.	H, I	Must Precede	J
8.	I	Must Precede	K
9.	J, K	Must Precede	L

Figure 2–6   Exercise 5C—Logic Statements

D.

1.	A	Must Precede	C, D
2.	B	Must Precede	D, E
3.	C	Must Precede	F
4.	D	Must Precede	G
5.	E	Must Precede	H, I
6.	F, G	Must Precede	J
7.	G, H	Must Precede	K
8.	I	Must Precede	L
9.	J, K	Must Precede	M
10.	K	Must Precede	N
11.	K, L	Must Precede	O
12.	M, N	Must Precede	P
13.	O	Must Precede	Q
14.	P, Q	Must Precede	R

Figure 2–7   Exercise 5D—Logic Statements

E.

1.	A	Must Precede	B, C
2.	B	Must Precede	D, E
3.	C	Must Precede	F
4.	D	Must Precede	G, H
5.	E, F	Must Precede	I
6.	F	Must Precede	J, K
7.	G, H, I	Must Precede	L
8.	I	Must Precede	M
9.	J, K	Must Precede	N
10.	L, M	Must Precede	O
11.	M, N	Must Precede	P
12.	O, P	Must Precede	Q
13.	P	Must Precede	R

Figure 2–8   Exercise 5E—Logic Statements

F.

1.	A	Must Precede	C, D
2.	B	Must Precede	E
3.	C	Must Precede	F, G
4.	D	Must Precede	G, H
5.	E	Must Precede	H, I, J
6.	F	Must Precede	K
7.	G	Must Precede	K, L, M
8.	H	Must Precede	L, M, N
9.	I	Must Precede	M, N
10.	J	Must Precede	O
11.	K	Must Precede	P, Q
12.	L	Must Precede	R
13.	M	Must Precede	Q, R
14.	N	Must Precede	R, S, T
15.	O	Must Precede	S, T
16.	P	Must Precede	U
17.	Q	Must Precede	V
18.	R	Must Precede	V, W
19.	S, T	Must Precede	W, X
20.	U	Must Precede	Y
21.	V	Must Precede	Y, Z
22.	W	Must Precede	Z
23.	X	Must Precede	A'
24.	Y, Z	Must Precede	B'
25.	A'	Must Precede	C'

Figure 2–9   Exercise 5F—Logic Statements

G. Small Commercial Concrete Block Building
Prepare a rough logic diagram for the small commercial concrete block building for which drawings are included in the Appendix. Follow these steps:
  A. Prepare a work breakdown structure (minimum of thirty activities).
  B. Establish activity relationships.
  C. Create the rough logic diagram.

H. Large Commercial Building
Prepare a rough logic diagram for the large commercial building for which a drawing is included in the Appendix. Follow these steps:
  A. Prepare a work breakdown structure (minimum of eighty activities).
  B. Establish activity relationships.
  C. Create the rough logic diagram.

Section 2

Scheduling

3 Schedule Calculations

Objectives

Upon completion of this chapter, you should be able to:

- Define important scheduling terms
- Calculate forward pass
- Calculate backward pass
- Calculate total float
- Calculate free float
- Complete a data table
- Correlate ordinal and calendar days

DEFINITIONS OF IMPORTANT TERMS

The mathematical calculations of a precedence (activity-on-node) diagram follow a few simple rules. To make these calculations, you need to understand several terms and their definitions.

Activity. A task activity has five specific characteristics. It consumes time, consumes resources, has a definable start and finish, is assignable, and is measurable. Construction projects are made up of a number of individual activities (events) that must be accomplished to complete the project.

Node. The node is the box (or other shape) containing activity information. Nodes contain varying amounts of information per the requirements of the diagram. The nodes are connected by arrows showing relationships among the activities. The nodes used in the diagrams in this chapter contain activity description, duration, early start, late finish, and float (Figure 3–1).

Precedence (Activity-on-Node) Diagrams. The nodes are the activities, and they are connected by arrows that show relationships among activities. The nodes contain information about the activities. This information may contain any or all of the following: identification, description, original duration, remaining duration, early start, late start, early

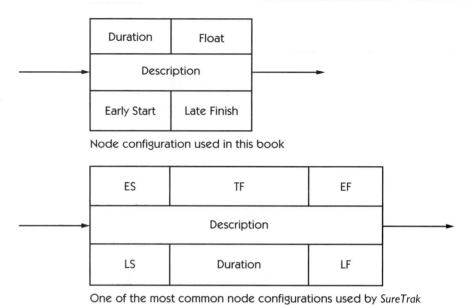

Node configuration used in this book

One of the most common node configurations used by *SureTrak*

Figure 3–1   Node Configurations

finish, late finish, and total float. Because *SureTrak* is written to support activity-on-node diagramming, this is the diagramming method used in this book.

Critical Path.   The critical path is the continuous chain of activities with the longest duration; it determines the project duration. Note that the critical path can change throughout the duration of a project. Activities that were originally secondary at the start of the project can become critical if they are delayed or exceed their total float. A noncritical activity can exceed its planned duration if the activity is within its total float without becoming critical.

Forward Pass.   Calculations for the forward pass start at the beginning of the diagram and proceed forward to the end. They determine early start, early finish, and project duration. For a finish-start relationship, all predecessor activities must be complete before an activity can start.

Early Start.   This is the earliest possible time that an activity can start according to the relationships assigned.

Early Finish.   This is the earliest time that an activity can finish and not prolong the project.

Backward Pass.   Backward pass calculations start at the end of the diagram and work backward to the beginning, following logic constraints. They determine the late finish and late start of each activity.

Late Finish.   This is the latest time that an activity can finish and not prolong the project.

Late Start.   This is the latest time that an activity can start and not prolong the project.

Float.   The amount of time difference between the calculated duration of the activity chain and the critical path. Float permits an activity to start later than its early start and not prolong the project. Float is sometimes called slacktime and may be classified as total or free.

Total Float.   Total float is the measure of leeway in starting and completing an activity. It is the number of time units (hours, days, weeks, years) that an activity (or chain of activities) can be delayed without affecting the project end date. It is a shared property among activities on a certain path or chain.

Free Float.   Free float is also referred to as activity float because, unlike total float, free float is the property of an activity and not the network path of which an activity is part. Free float is the amount of time the start or finish of an activity can be delayed without delaying the start of a successor activity (depending on the relationship).

FORWARD PASS

Calculating Early Start

To calculate an activity's early start (forward pass), add the activity's duration to the early start of the preceding activity. All predecessor activities to the activity must be complete. Activity R (Figure 3–2) is the first activity and therefore has no predecessors and an early start of 1 (Figure 3–3). The early start of G is 5 (Figure 3–3), or 1 (early start of R, the preceding activity) plus 4 (duration of R, the preceding activity).

Larger Value Takes Precedence in the Early Start

Where two chains converge on a single activity, such as Activity J (Figure 3–3), the larger early start value controls the path because that path has the longer duration and *all* prior activities must be finished. The two choices are 8 days from Activity G and 11 days from Activity C. The 11 controls because it is the larger value. See Table 3–1 for the early start calculations for Figure 3–3. The early start formula is:

$$\text{Early Start} = (\text{Highest Value of Predecessor Early Start}) + (\text{Predecessor Duration})$$

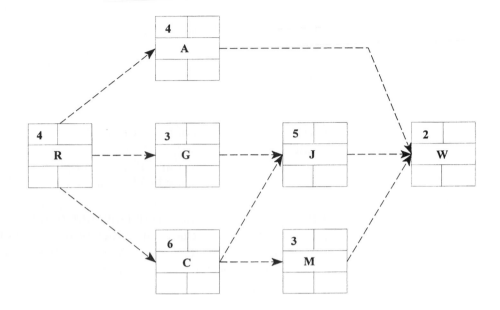

Figure 3–2 Precedence Diagram—Durations

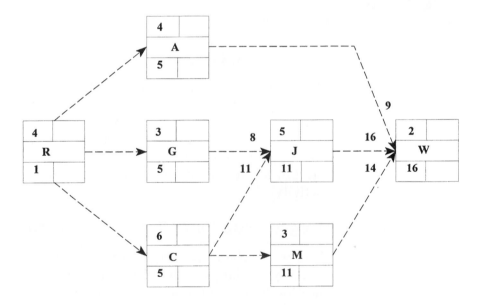

Figure 3–3 Forward Pass—Early Starts

Activity	Preceding Activity	Early Start of Preceding Activity	Duration of Preceding Activity	Early Start of Activity (*Controls)
R				1
A	R	1	4	5
G	R	1	4	5
C	R	1	4	5
J	G	3	5	8
	C	6	5	11*
M	C	6	5	11
W	A	5	4	9
	J	11	5	16*
	M	11	3	14

Table 3–1   Early Start Calculations

BACKWARD PASS

Calculating Late Finish

The backward pass is used to determine activity late finishes. As the name implies, calculations begin at the diagram end and pass in a backward direction to the beginning. The formula for the late finish of an activity is the late finish of the following activity minus the duration of the following activity. The late finish of activity W in Figure 3–4 is the end of day 17. Because W is the last activity, it has to be on the critical path. W has an early start of 16 and a duration of 2. This means that Activity W will be performed on Days 16 and 17. The late finish of W is on Day 17 (16 + 2 − 1). The late finish of J is 15, or 17 (late finish of following activity) minus 2 (duration of following activity).

Smaller Value Takes Precedence in Late Finish

When two paths converge on a single activity, the smaller value is used. The backward pass to C from the path through J has a late finish of 10 (15 − 5). The path to C through M has a late finish of 12 (15 − 3). Because 10 is the smaller value, it is used as the late finish of C. See Table 3–2 for the late finish calculations for Figure 3–4. The early late finish formula is:

$$\text{Late Finish} = (\text{Lowest Value of Successor Late Finish}) - (\text{Successor Duration})$$

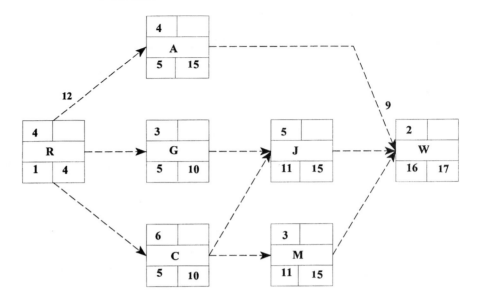

Figure 3–4 Reverse Pass—Late Finishes

Activity	Following Activity	Late Finish of Following Activity	Duration of Following Activity	Late Finish (*Controls)
W				17
M	W	17	2	15
J	W	17	2	15
C	M	15	3	12
	J	15	5	10*
G	J	15	5	10
A	W	17	2	15
R	C	10	6	4*
	G	10	3	7
	A	15	4	11

Table 3–2 Late Finish Calculations

FLOAT

Construction schedulers use the float to determine which activities are critical and which activities have "slack" or "fluff."

Flattened-requirements float associated with noncritical activities can be used to "flatten" or level the requirements for resources, including personnel, materials, construction equipment, and cash. For example, if two activities have a scheduled early start of the same day, and both require a carpentry crew, completing the critical activity first reduces the total number of carpenters needed. Such sequencing can similarly reduce material, equipment, and cash flow requirements. Noncritical activities do not have the same priority as critical activities and do not determine the critical path (project duration). It is important to note that if the float is used up, the critical path changes and previously noncritical activities may become critical.

Formulas to Calculate Float

The formulas for calculating total float require late start and early finish calculations. The formulas are:

$$\text{Late Start} = \text{Late Finish} - \text{Duration} + 1$$
(Determined from backward pass)

$$\text{Early Finish} = \text{Early Start} + \text{Duration} - 1$$
(Determined from forward pass)

There are three formulas for calculating activity total float:

$$\text{Total Float} = \text{Late Finish} - (\text{Early Start} + \text{Duration}) + 1$$

$$\text{Total Float} = \text{Late Start} - \text{Early Start}$$

$$\text{Total Float} = \text{Late Finish} - \text{Early Finish}$$

Given the information contained in the nodes in Figure 3–5, the first formula is the most suitable. All three of these formulas must yield the same value, or a mistake has been made. See Table 3–3 for total float calculations for Figure 3–5. By examining Table 3–4 for the example problem, you can see that the three formulas return the same value. The three formulas for calculating total float hold true unless the activity is in physical progress. The float calculation for activities that are partially completed will be covered in the chapters on activity updating.

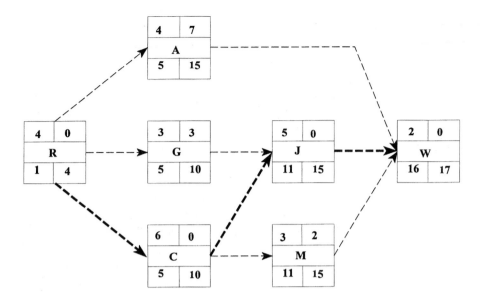

Figure 3–5 Float

Activity	Late Finish	Early Start	Duration	Total Float
R	4	1	4	0
A	15	5	4	7
G	10	5	3	3
C	10	5	6	0
J	15	11	5	0
M	15	11	3	2
W	17	16	2	0

Table 3–3 Total Float Calculations

Activity	Duration	Early Start	Late Start	Early Finish	Late Finish	Total Float
R	4	1	1	4	4	0
A	4	5	12	8	15	7
G	3	5	8	7	10	3
C	6	5	5	10	10	0
J	5	11	11	15	15	0
M	3	11	13	13	15	2
W	2	16	16	17	17	0

Table 3–4 Data Table

DATA TABLE

Table 3–4 is a classic scheduling data table (tabular sort) sorted by early starts. The table includes late starts and early finishes. Again the formulas for late starts and early finishes are:

Late Start = Late Finish − Duration
Early Finish = Early Start + Duration

CALENDARS

So far in this chapter, the scheduling units have been ordinal (numeric). To be more useful, ordinal days can be converted to calendar days (Figure 3–6). Calendar days account for nonworkdays, including Saturdays, Sundays, holidays, and a possible allowance for rain or other bad weather. These calendar decisions must be made to convert from ordinal days to calendar days (Figure 3–7).

After allowing for nonworkdays, the ordinal days can now be correlated with calendar days. The project starts on the tenth day of the first month (Figure 3–8).

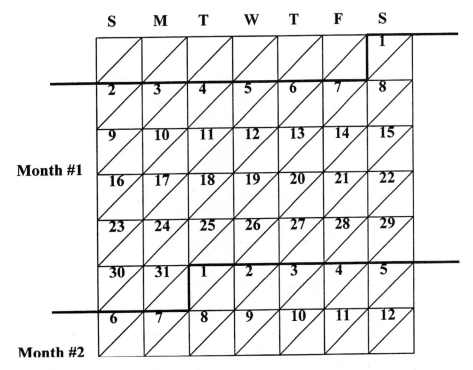

Figure 3–6 Calendar Days

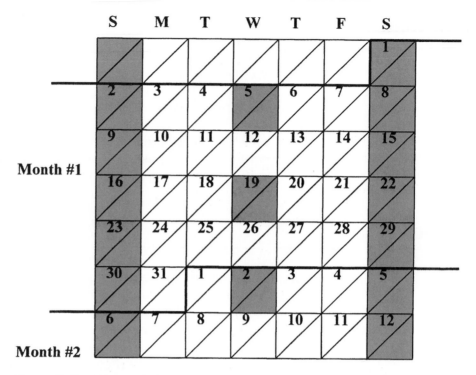

Figure 3–7 Nonworkdays

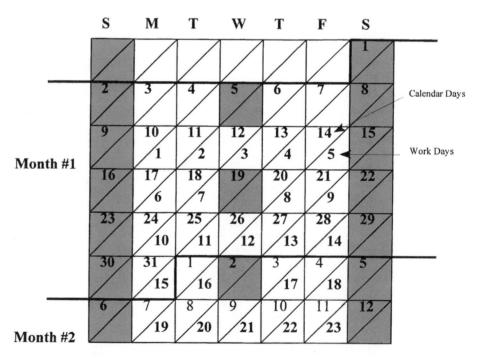

Figure 3–8 Work/Calendar Days

SureTrak allows the user to assign an activity to any one of a number of possible calendars. Having different calendar days available for progress changes the ordinal to calendar day correlation. The concept of changing (having multiple) calendars will be covered in Chapter 4.

EXAMPLE PROBLEM #1: CALCULATIONS

Figure 3–9 illustrates the forward pass, backward pass, and float calculation. Table 3–5 shows the same information in tabular format.

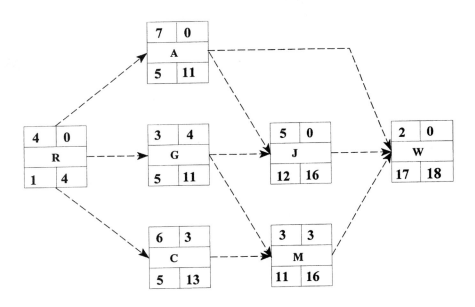

Figure 3–9 Example Problem—Precedence Diagram

Activity	Duration	Early Start	Late Start	Early Finish	Late Finish	Total Float
R	4	1	1	4	4	0
A	4	5	5	11	11	0
G	3	5	9	7	11	4
C	6	5	8	10	13	3
J	5	12	12	16	16	0
M	3	11	14	13	16	3
W	2	17	17	18	18	0

Table 3–5 Example Problem—Data Table

EXAMPLE PROBLEM #2 : CALCULATIONS

Figure 3–10 illustrates the forward pass, backward pass, and float calculation. Table 3–6 shows the same information in tabular format.

Note: The "Ftg. Inspection" activity's dates are based on the fact that the Finish-to-Finish relationship it has from "Place Ftg. Concrete" is driving and takes precedence over the Start-to-Start relationship it has from "Place Ftg. Rebar & Formwork." When you enter these activities in *SureTrak,* you will see how *SureTrak* favors the Finish-to-Finish relationship in this situation.

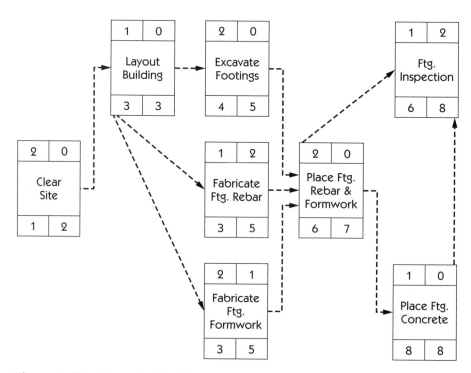

Figure 3–10   Example Problem

Activity	Duration	Early Start	Late Start	Early Finish	Late Finish	Total Float
Clear Site	2	1	1	2	2	0
Lay out Building	1	3	3	3	3	0
Excavate Footings	2	4	4	5	5	0
Fabricate Ftg. Rebar	1	3	5	3	5	2
Fabricate Ftg. Formwork	2	3	4	4	5	1
Place Ftg. Rebar & Formwork	2	6	6	7	7	0
Place Ftg. Concrete	1	8	8	8	8	0
Ftg. Inspection	1	6	8	8	8	2

Table 3–6   Example Problem—Data Table

SUMMARY

This chapter provided important term definitions such as activity, node, precedence (activity-on-node) diagrams, critical path, forward pass, early start, early finish, backward pass, late finish, late start, float, total float, and free flow. You learned about forward, backward, and float calculations and were introduced to data table and calendars, as well as *SureTrak* calculations.

EXERCISES

1. **Scheduling Important Terms**
   Define in 4–6 lines the following terms: activity, forward pass, backward pass, early start, late start, early finish, late finish, float, and critical path.

2. **Ordinal versus Calendar Days**
   Describe in 3–5 lines the difference between ordinal and calendar days.

3. **Hand versus *SureTrak* Calculations**
   Explain the difference between hand calculations and *SureTrak* calculations for early start, late finish, late start, and early finish.

Complete the precedence diagrams and data tables for Exercises 4 to 11.

4.

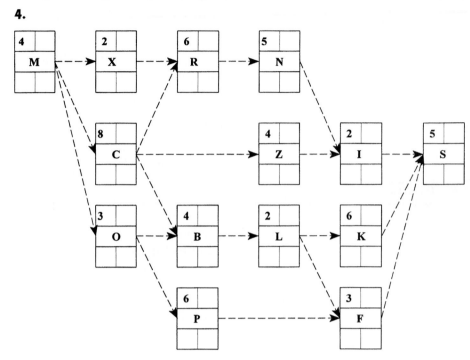

Figure 3–11 Exercise 4—Precedence Diagram

Activity	Duration	Early Start	Late Start	Early Finish	Late Finish	Total Float
M						
X						
C						
O						
R						
B						
P						
N						
Z						
L						
I						
K						
F						
S						

Table 3–7 Exercise 4—Data Table

5.

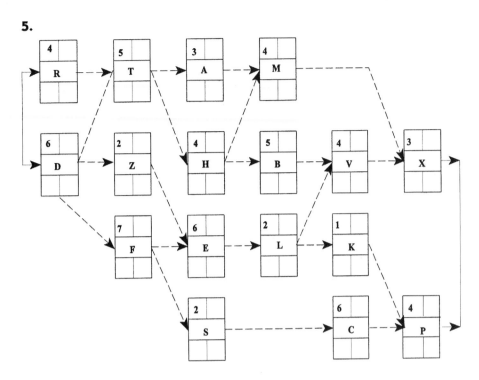

Figure 3–12   Exercise 5—Precedence Diagram

Activity	Duration	Early Start	Late Start	Early Finish	Late Finish	Total Float
R						
D						
T						
Z						
F						
A						
H						
E						
S						
M						
B						
L						
V						
K						
C						
X						
P						

Table 3–8   Exercise 5—Data Table

6.

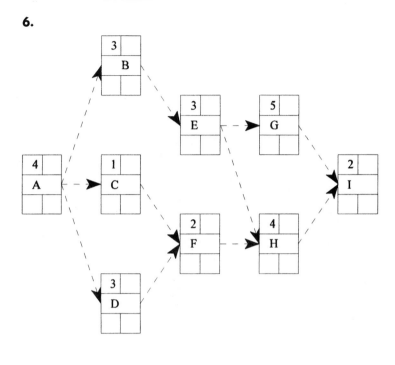

Figure 3–13   Exercise 6—Precedence Diagram

Activity	Duration	Early Start	Late Start	Early Finish	Late Finish	Total Float
A						
B						
C						
D						
E						
F						
G						
H						
I						

Table 3–9   Exercise 6—Data Table

7.

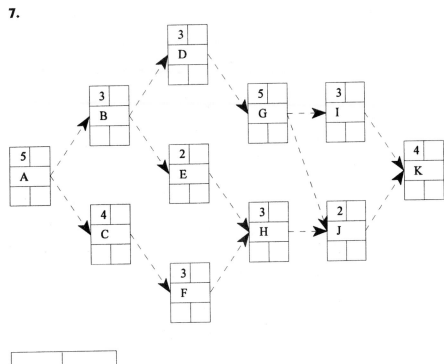

Figure 3-14   Exercise 7—Precedence Diagram

Activity	Duration	Early Start	Late Start	Early Finish	Late Finish	Total Float
A						
B						
C						
D						
E						
F						
G						
H						
I						
J						
K						

Table 3-10   Exercise 7—Data Table

8.

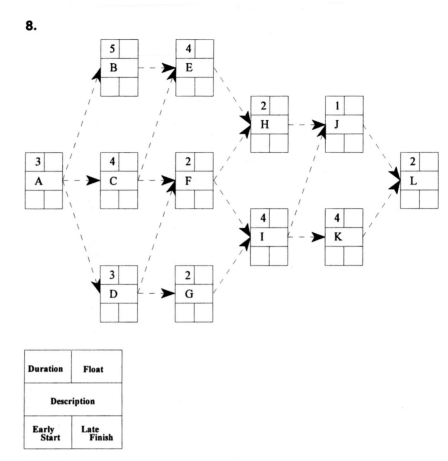

Figure 3–15   Exercise 8—Precedence Diagram

Activity	Duration	Early Start	Late Start	Early Finish	Late Finish	Total Float
A						
B						
C						
D						
E						
F						
G						
H						
I						
J						
K						
L						

Table 3–11   Exercise 8—Data Table

9.

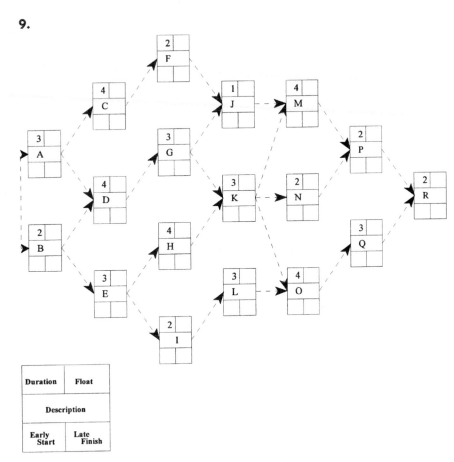

Figure 3–16   Exercise 9—Precedence Diagram

Activity	Duration	Early Start	Late Start	Early Finish	Late Finish	Total Float
A						
B						
C						
D						
E						
F						
G						
H						
I						
J						
K						
L						
M						
N						
O						
P						
Q						
R						

Table 3–12   Exercise 9—Data Table

10.

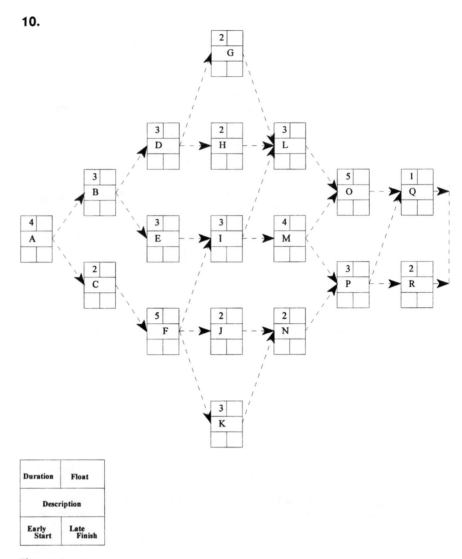

Duration	Float
Description	
Early Start	Late Finish

Figure 3–17   Exercise 10—Precedence Diagram

Activity	Duration	Early Start	Late Start	Early Finish	Late Finish	Total Float
A						
B						
C						
D						
E						
F						
G						
H						
I						
J						
K						
L						
M						
N						
O						
P						
Q						
R						

Table 3–13   Exercise 10—Data Table

11.

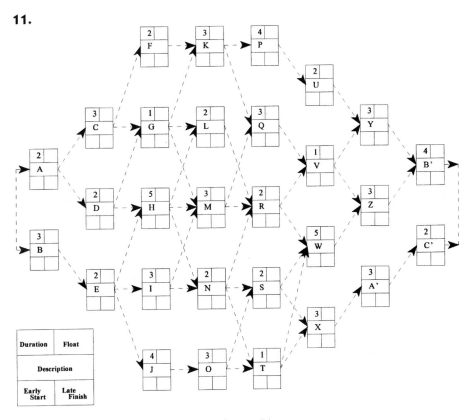

Figure 3–18   Exercise 11—Precedence Diagram

Activity	Duration	Early Start	Late Start	Early Finish	Late Finish	Total Float
A						
B						
C						
D						
E						
F						
G						
H						
I						
J						
K						
L						
M						
N						
O						
P						
Q						
R						
S						
T						
U						
V						
W						
X						
Y						
Z						
A'						
B'						
C'						

Table 3–14   Exercise 11—Data Table

Bar Chart Creation

Objectives

Upon completion of this chapter, you should be able to:

- Start *SureTrak*
- Create a new project
- Add activities to a project
- Modify project timescale
- Define relationships between activities
- Calculate the schedule
- Edit activities
- Close and open a project
- Delete a project
- Modify project overview information
- Create a new project using the wizard
- Modify project options
- List the reasons for preparing a work breakdown structure
- Identify the information required to prepare a rough logic diagram and work breakdown structure
- Describe the steps to prepare a rough logic diagram
- Contrast the rough logic diagram of a design-bid-build and a fast-track construction
- Describe the types of activity relationships
- Prepare a rough logic diagram
- Estimate activity duration
- Describe factors that affect productivity

STARTING *SURETRAK*

To start *SureTrak*, double-click on the application icon (Figure 4–1) located on the computer desktop. A welcome screen (Figure 4–2) appears when *SureTrak* is started (unless a user name and password must be entered).

To go to the project manager windows, click on **Close** and the *SureTrak* project manager window will be available to work in (Figure 4–3).

Figure 4–1   *SureTrak* Desktop Icon

Figure 4–2   *SureTrak* Welcome Screen

Menu

Toolbar

New

Open

Exit *SureTrak*

Tutorial

Help

Figure 4–3   *SureTrak* Project Manager Screen

Project Manager: Menu Bar Options

Pull-down menu options on the *SureTrak* opening screen are:

File Click on this menu option to create a new project or to open an existing project.

Tools Contains **Wizards, Basic Scripts, Custom Tools, Project Utilities, Customize,** and **Options.**

Help Provides the Help index.

Toolbar Options

Toolbar options on the *SureTrak* entrance screen are:

New Click on this option to add a new project to the system. (An alternative to the **File** menu option.)

Open Click on this option to open an existing project. (An alternative to the **File** menu option.)

Exit *SureTrak* Can be used to exit *SureTrak* as an alternative to the *SureTrak* **Project Manager** pull-down menu.

The next four buttons to the right of the Exit *SureTrak* button in Figure 4–3 are not identified. They relate to options identified in the **Tools** main pull-down menu and are discussed later in the book.

Tutorial Provides access to *SureTrak*'s internal tutorial.

Help Provides the Help index. The online help in *SureTrak* is extensive. The "?" button appears on both the opening screen and the bar chart screen. Clicking on the Help button produces the **Help Contents** menu.

Help Contents

The menu files are:

Contents Displays *SureTrak* help Contents by category (books).

Search Displays the **Help Topics** dialog box that enables you to find Help information using keywords. Use either **Index** or **Find. Find** enables you to search for specific words and phrases in **Help Topics** instead of searching by category.

Another convenience of *SureTrak* is that you can pull up the Help Topics screen from almost any field within *SureTrak* by pressing the F1 (first function) key.

CREATE A NEW PROJECT SCHEDULE

From the toolbar, click on **File** and then click on **New** and the dialog box shown in Figure 4–4 appears. This screen provides the option of using the wizard to prepare the schedule. Click on **No** to proceed to the **New Project** dialog box (Figure 4–5). The **New Project** dialog box information fields are:

Current Folder The current directory (Figure 4–5) happens to be *c:\..\suretrak\projects,* but it could be any other as well. Having a separate folder to store project files (as opposed to program files) is convenient. The current directory can be changed by simply clicking on the **Browse** button.

Project name The name is the primary sort field in project file listings (when opening an existing schedule). The name acts as a short identification for the project. For a *SureTrak* project, enter up to eight alpha/numeric characters with no spaces. All project files will share this name. Because *SureTrak* creates approximately fifteen new files when a project is created, this makes their identification much simpler. The project name identified here will also appear at the top of the project screen. If you are creating a *Project Groups*, *Concentric (P3)*, or *Finest Hour* project type, use exactly four characters.

Template Click on this option to use an existing project as a starting point for the project you are creating. The template project (Figure 4–5) must be in the TEMPLATE directory in the *SureTrak* program directory.

Type The choices are **SureTrak**, **Project Groups,** and **Concentric (P3).** Choose **SureTrak** when creating a stand-alone project.

If you are creating a project group or a member project, select **Project Groups.** If you will be exchanging project data with *P3 (Primavera*

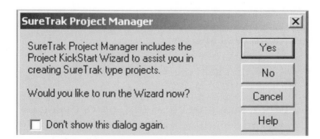

Figure 4–4 *SureTrak* Project Manager Window

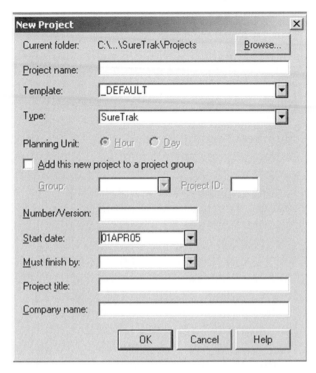

Figure 4–5   New Project Dialog Box

Project Planner) users, select **Concentric (P3).** Note that there is an inherent difference between the philosophy of the makeup of a *SureTrak* and a Concentric (P3). All of *SureTrak*'s quantity fields, resources, costs, revenues, and so on are calculated hourly. *P3*'s primary fields, on the other hand, can be calculated daily. This concept is further discussed in the **Planning unit** field.

Planning unit The **Planning unit** choices are Hour and Day. This option is not available (grayed out) for *SureTrak* schedules. It is only for Concentric (*P3*), Project Groups, and Finest Hour types (see **Types** field). The default can be changed by going to the **Tools** main pull-down menu and selecting **Options.** The **Duration** default can be changed under the Project tab. We discuss this later in the chapter.

On many large construction projects, the day is the ideal unit. On projects that are estimated by productivity units or man-hours, it is more convenient to use hours as the basis. An industrial turn-around project is highly schedule intensive, so an hourly planning unit may be appropriate. Note that *SureTrak*'s quantity, resource, costs, and other fields are always calculated using hourly units. So, *SureTrak* is built around using hourly rates, whereas *P3* works with units per time period (not just units per hour), and you can define units per hour, day, or week. In this book, we have chosen the unit of a day.

Add this new project to a project group Mark this checkbox to designate a project as part of a project group. This function breaks down a large project into more manageable units. If a large project includes five buildings, each of the buildings can be a member project under a project group. Information such as the Resource Dictionary can be put in at the project group level and will filter down to the member project level. Information input at the member project level can be merged into reports at the project group level.

<u>N</u>umber/Version   For updating and control purposes, you may find it necessary to copy a schedule. The **<u>N</u>umber/Version** field is a way to keep track of the relationships of the multiple schedules on a project. Changes can be made to the schedule and then compared to the original or an earlier version of the schedule. Note that the change in the **<u>N</u>umber/Version** does not change the file name.

Start Date   The **Project start** (Figure 4–6) is a required entry that identifies the date on which work can start on this project. The motion bar to the right of the pull-down calendar is used to select the month. Click below the slide button on the motion bar to advance the calendar one month at a time. Click above the slide button to back up the calendar a month at a time. Click on the appropriate date, then click on the **OK** button to select the project start date.

Must <u>f</u>inish by   This is an optional entry. If there is a finish date, that date can be input using the pull-down calendar the same way the

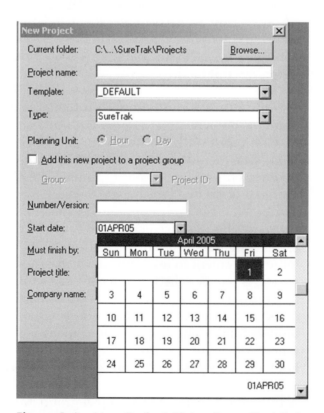

Figure 4–6   New Project Dialog Box—Start Date

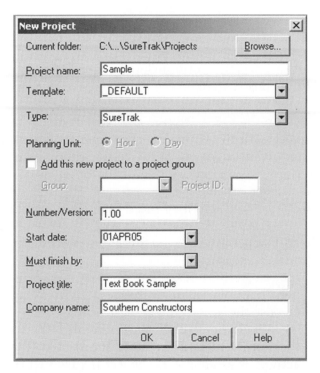

Figure 4–7   New Project Dialog Box—Completed

project start date was input. If there is no required finish date, this field can be left blank and *SureTrak* will calculate the finish date based on activities, durations, and logic restraints.

Project title   The **Project title** field (Figure 4–7) is a larger field than the **Project name** field. By default, the project title is printed on hard copy on both tabular and graphic reports.

Company name   By default the **company name** is printed on hard copy of both tabular and graphic formats. If you do not want to specify a company name, you can leave this field blank.

NEW PROJECT DIALOG BOX BUTTONS

The **New Project** dialog box toolbar options are:

OK   Click on the **OK** button and the new project is added to the projects directory and the necessary files are created (Figure 4–7).

Cancel   Click on the **Cancel** button and the **Add a new project** dialog box is canceled without creating a new project.

Help   Takes you to the **Help** menu relating to **New Project.**

ACTIVITIES

Figure 4–8 is the example logic diagram that is used throughout the rest of the book to explain the concepts in each chapter.

Click on the **OK** button in the **New Project** dialog box to create a new project with the blank bar chart screen (Figure 4–9) ready for input of schedule information. This Classic Schedule Layout screen is organized in days. If you choose one of the other options, the Classic Schedule Layout screen changes accordingly. The first step in creating the schedule is to create activities and define relationships and durations.

To add an activity, use the following procedure.

ID An Activity ID (identification) is a string of letters and/or numbers that uniquely identifies an activity. When entering activities, you can use the auto-sequencing feature, which numbers each new activity in increments of 10 starting with 1000. To add an activity, click in the **Act ID** (activity identification) field (Figure 4–10), and the new activity is established. The **Act ID, Description,** and **Orig Dur** (original duration) can either be input directly into these fields as shown in Figure 4–9, or the Activity Form can be used as shown in Figure 4–10. To access the Activity Form, click on the **View** main pull-down menu and select **Activity Form.**

SureTrak automatically establishes the first activity with an ID of 1000. Each activity has a unique name, so *SureTrak* can use the unique designator to identify relationships used in establishing project duration, float, and so on. Because this is an alpha/numeric field, the same

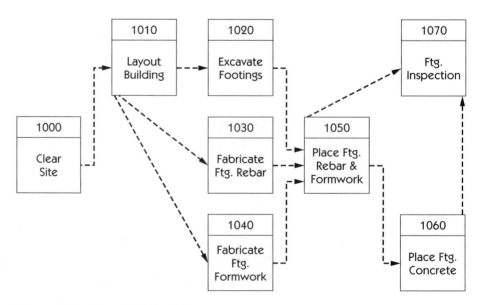

Figure 4–8 Sample Schedule

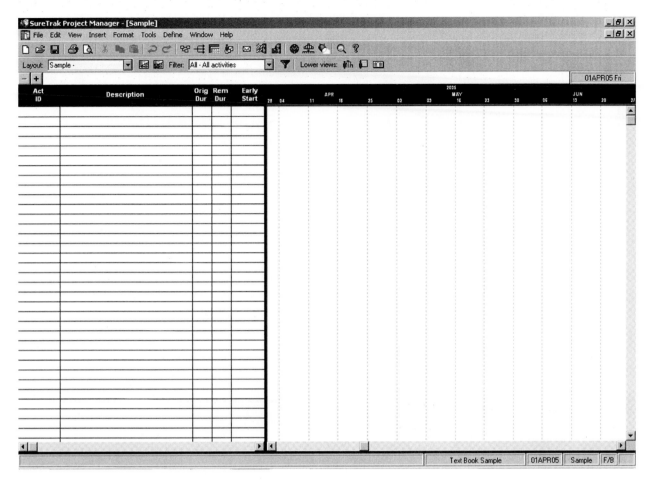

Figure 4–9   Onscreen Bar Chart (Gantt)

number of characters must be used so that each activity is evaluated on a similar basis.

When *SureTrak* assigns a new ID, it checks whether the ID already exists in the project. If it exists, *SureTrak* searches in increments of 10 until it finds an unused ID. If you delete an activity, that ID becomes available for a new activity. If you want *SureTrak* to increment IDs by a value other than 10, choose the **Tools** main pull-down menu, **Options,** and then the **Default** tab. Type the value you want in the **Increase activity ID by** field. The **ID** field can be used to sort activities into ID categories within a project.

Description   This alpha/numeric field is used to name or describe the activity. It is usually not a sort field within *SureTrak*. Use short, concise descriptions—usually two or three words in a noun + verb format. After entering the **Description**, click on the **Orig Dur** field on the bar chart or **Duration** on the Activity Form.

Orig Dur   This stands for Original Duration or the basis of the original schedule. If the scheduling unit is the day, then input the number

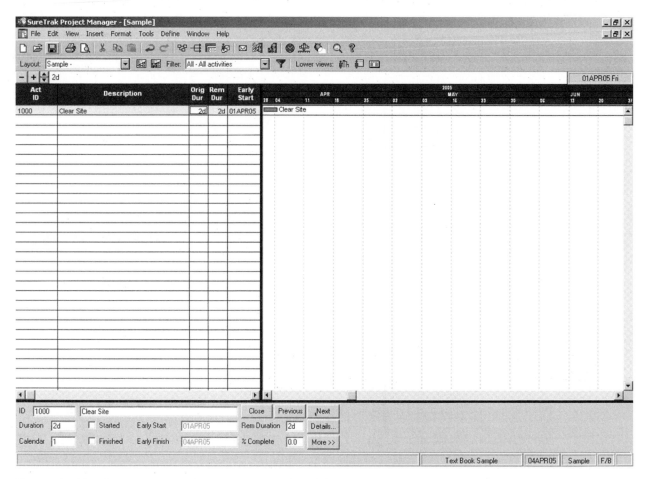

Figure 4–10 Onscreen Bar Chart (Gantt) and Activity Form

of days for the activity duration. If you need to edit the activity, double-click the activity on the bar chart, and the Activity Form screen will appear, filled with the activity information. Any information other than the activity ID can readily be changed. Figure 4–11 shows all the new activities added for the Sample Schedule. Note that *SureTrak* has organizing options (**Format, Organize**) to arrange the activities. So, if the activities are sorted by some criterion other than activity ID, the activity will not necessarily appear where it was initially placed. The original **Duration** field from the exposed Activity Form (see Figure 4–10) can also be used to input the original duration.

Calendar The **Calendar** field from the exposed Activity Form (see Figure 4–10) is used to define the calendar by which the activity will be scheduled. A base calendar defines when its activities can be worked on. Any activities to which you assign a particular base calendar can be worked on during the workdays in that base calendar, and they cannot be worked on during the nonworkdays in that base calendar. To create a base calendar, choose **Define,** and then **Calendars.**

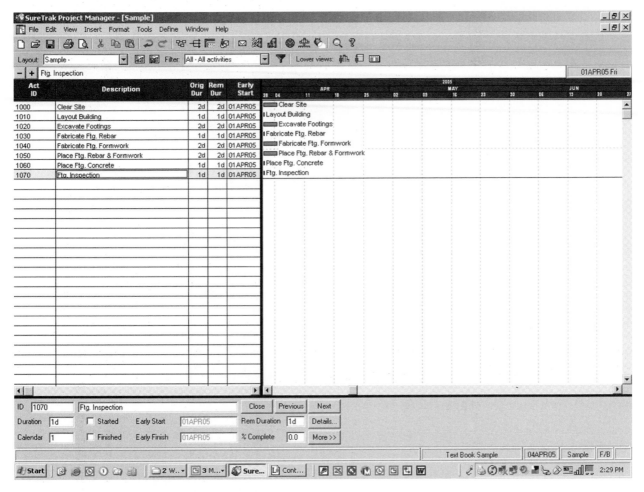

Figure 4–11 Onscreen Bar Chart (Gantt)—All Activity

MODIFY TIMESCALE

Because the sample project is of short duration, the timescale of the bar chart can be modified to make the information easier to read.

Timescale Definition Dialog Box

The **Timescale** dialog box (Figure 4–12) can be accessed in one of two ways. The first is to click on the **Format** pull-down menu, and then click on the **Timescale** option. The other method is to double-click anywhere on the shaded timescale bar itself.

Density The slide bar under **Density** controls the screen timescale format. Click on the arrows or click and hold the slide button and slide it to either side to modify the timescale of the onscreen bar chart. The scale in Figure 4–13 was modified to show only part of a month

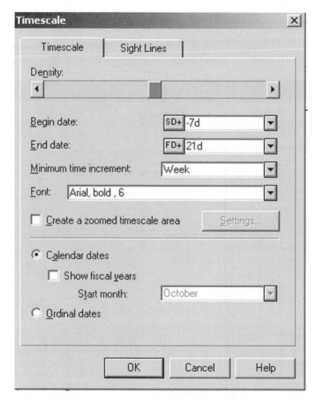

Figure 4–12 Timescale Dialog Box

instead of three. When the timescale is modified and the **OK** button is clicked on the **Timescale** dialog box, *SureTrak* modifies the appearance of the activities to relate to the new timescale.

The other modifications that can be made using the **Timescale** dialog box are:

Begin date The value of SD_+-7d is shown in Figure 4–12. This indicates that the onscreen bar chart starts at a minus seven days from the **Start date** given when the schedule was created. To increase or decrease the number, there are two options. First, click on the down arrow button to the left of this field (**SD+** in Figure 4–12). The **Begin date** can be tied to four options: **Calendar date, Start date, Data date,** and **Finish date.** The other option is to simply type in a new number in the place of the -7d or click on the down arrow to the right of the field and a pop-up calendar will appear, from which you can select a date.

End date The default as shown on Figure 4–12 is FD_+21d. This indicates that the onscreen bar chart has a minimum length of at least the **Finish date** of the project when scheduled plus 21 days. For the **End date,** the same four options are available as for the **Begin date.**

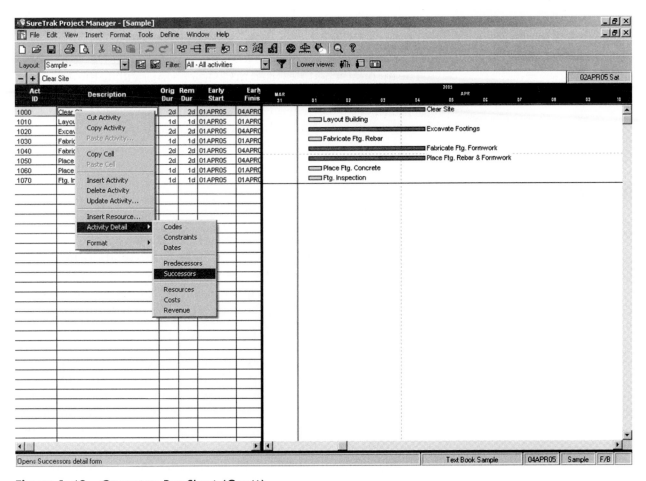

Figure 4–13   Onscreen Bar Chart (Gantt)

<u>M</u>inimum time increment The timescale option shown in Figure 4–12 is weeks. When Figure 4–13 was executed, days were selected. Notice the difference shown on the timescale bar. Figure 4–12 shows only the Mondays (beginning of the week) that are available. Figure 4–13 shows all days. The timescale can be changed to hour, day, month, quarter, or year as the onscreen bar chart scale. To change the time frame, use the down arrow button.

<u>F</u>ont Modifying this font option changes only the font used for the letters on the timescale bar. To change the type, size, and bold options of the font, click the down arrow button for the available choices.

Create a Zoomed Timescale Area This is a new option in *SureTrak* V.3. By clicking in the box (Figure 4–12), the zoomed timescale area is activated. This allows a particular segment of your schedule to have a different timescale with a greater level of detail. To specify the dates of the segment that will have a greater level of detail, you can input a beginning and ending date by clicking on the settings button.

$\underline{C}$alendar dates or $\underline{O}$rdinal dates Usually it is easier to think of your schedule in terms of calendar days. But sometimes, particularly for the beginning scheduler, it is simpler to see the ordinal (or numerical) workdays. The manual schedules used in Chapter 3 to calculate the examples all used ordinal dates. As an example, the first ordinal day of this project is September 6 (calendar day). By clicking on the **Ordinal dates** selection box, the timescale bar is changed from calendar to ordinal days.

RELATIONSHIPS

All the activities in our sample project start on the same date, the first day of the project (see Figure 4–11). This is because no relationships have been established between the activities yet. After establishing the activity ID and giving the activity description and a duration, the next step is to establish relationships among the activities. The procedure for establishing relationships is as follows:

Successor/Predecessor An activity relationship can be either a predecessor or a successor relationship. If Activity A must occur before B, A is a predecessor to B and B is a successor to A. Either of these relations can be identified to *SureTrak*. If one is identified, the program automatically assumes the other. Click on the activity to which relationships are to be established. In Figure 4–13, Activity 1000, Clear Site, is selected. Right-click and select **Activity Detail,** then **Successors** or **Predecessors** (see Figure 4–13). In Figure 4–14, **Successors** was chosen and the **Successors** dialog box appears. To add and define a relationship, click on the + button.

Successor To identify the activity to which the relationship will be established, click on the down arrow button (Figure 4–15). A pop-up screen will appear with activity IDs and descriptions. To establish the relationship, click on the selected successor activity.

Type When activity 1010 is selected as a successor to activity 1000, the default relationship (**Type**) is a FS - Finish to Start relationship (Figure 4–16). Activity 1000 must be finished before activity 1010 can start. Click in the **Type** field and then click on the down arrow on the toolbar to see a pull-down menu of other relationship types (Figure 4–17). There are four relationship types—the other three are SS - Start to Start, FF - Finish to Finish, and SF - Start to Finish. Clicking on one of the other options changes the default **FS** relationship. Multiple successors to an activity are shown in Figure 4–18. This can be done by simply repeating the preceding steps.

Lag Lag is a delay or offset time in a relationship or resource assignment. If you specify two days of lag in a Finish to Start (FS) relationship, the successor activity starts two days after the predecessor

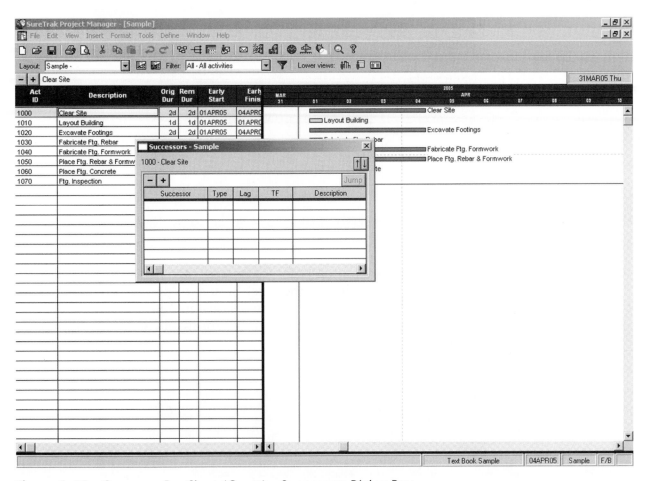

Figure 4–14   Onscreen Bar Chart (Gantt)—Successors Dialog Box

finishes. In Figure 4–19, activities 1010 and 1030 did not have a pure start to start the relationship. Activity 1030 will start 2 days after the start of activity 1010. A lag of 2 days is imposed on the relationship. Click in the **Lag** field of the activity that should have lag, and then an increase/decrease number field button appears on the toolbar. Click

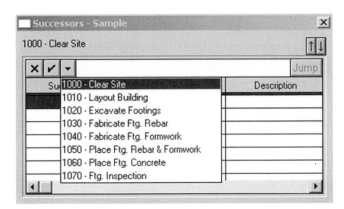

Figure 4–15   Successors Dialog Box—Choosing Successor Activity

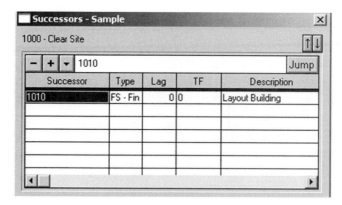

Figure 4–16   Successors Dialog Box—Successor Activity Chosen

on the increase button twice (or type **2**); then press the **Enter** key on the keyboard or click on the check button to enter the 2 days of lag for activity 1030. For example, lag time is given to allow elevated concrete to cure to a certain level before the formwork wrecking/reshoring activity can begin.

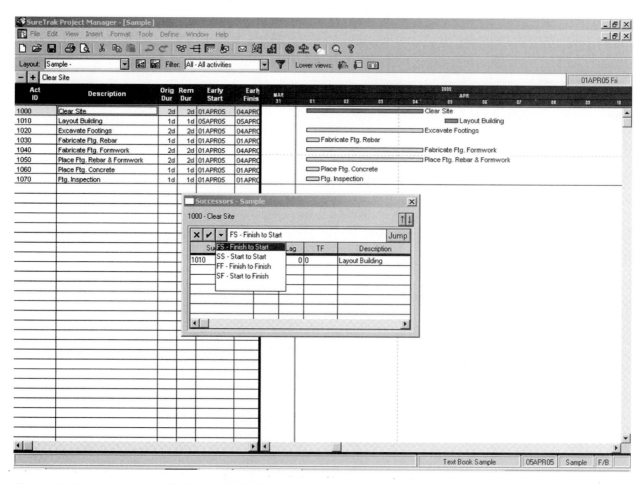

Figure 4–17   Successors Dialog Box—Relationship Type Options

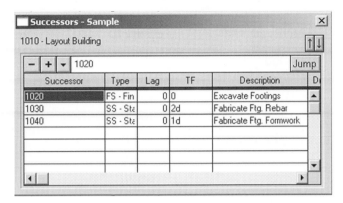

Figure 4–18   Successors Dialog Box—Adding Other Relationships

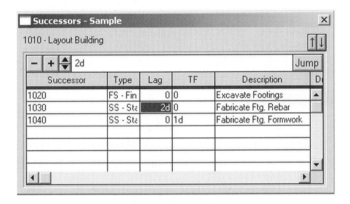

Figure 4–19   Successors Dialog Box—Lag

Relationships Another way to establish relationships is to draw them on the screen. Click on the **View** main pull-down menu, and then select **Relationships.** The successor and predecessor relationships assigned are shown onscreen. Figure 4–23, later in this chapter, is the sample schedule showing the relationships. Compare Figure 4–23 with relationships to Figure 4–22 without relationships.

To establish new relationships, place the cursor at the beginning or end of an activity; this causes the pitchfork symbol to appear. Hold the left mouse button down and drag this symbol from activity to activity to establish the same relationships as using the methods just described. *Note: the pitchfork icon can also be selected from the main toolbar.* This method for establishing relationships is discussed in detail in the chapter on PERT diagramming later in the book. To correct an error made in inputting activity relationships, see the "Edit Activities" section.

Table 4–1 is a list of the activities, the successor activity IDs, type relationships, and the lags input for the Sample Schedule. These are calculated in the next section.

Activity ID	Successor Activity ID	Relationship	Lag
1000	1010	FS	0
1010	1020	FS	0
	1030	SS	0
	1040	SS	0
1020	1050	FS	0
1030	1050	FS	0
1040	1050	FS	0
1050	1060	FS	0
	1070	SS	0
1060	1070	FF	0
1070	—	—	—

Table 4–1   Sample Schedule Activity Relationships

CALCULATE THE SCHEDULE

Notice that in Figure 4–20, *SureTrak* scheduled the activities (compare Figure 4–20 to Figure 4–11) as their relationships were established. This was done because *SureTrak's* automatic schedule calculation was on. When the automatic scheduling calculation is off, *SureTrak* provides three ways to begin the schedule process.

The Schedule Dialog Box

Method 1.   Click the **Tools** main pull-down menu (Figure 4–20) and then click on **Schedule.**

Method 2.   Press the **F9** function key on the keyboard. *Note: using the F9 option does not permit the user to choose options for the calculation.*

Method 3.   Click on the toolbar button that looks like a round red clock. This method works the same way as the **F9** function key.

The **Schedule** dialog box is displayed (Figure 4–21). The data date that was entered when the project was added can now be modified if necessary. The **Project data date** establishes the heavy black vertical line (Figure 4–22) that reflects the date that will be the basis of the schedule, which can be either the schedule start date or an interim update date. As the schedule is updated, the **Project data date** will change with each update.

The **Logic** option (Figure 4–21) is critically important when inputting progress into the schedule. Figure 4–23 shows the logic for the original activity relationships. When the schedule is updated, these relationships have to be overridden with actual progress. The **Out-of sequence** field is

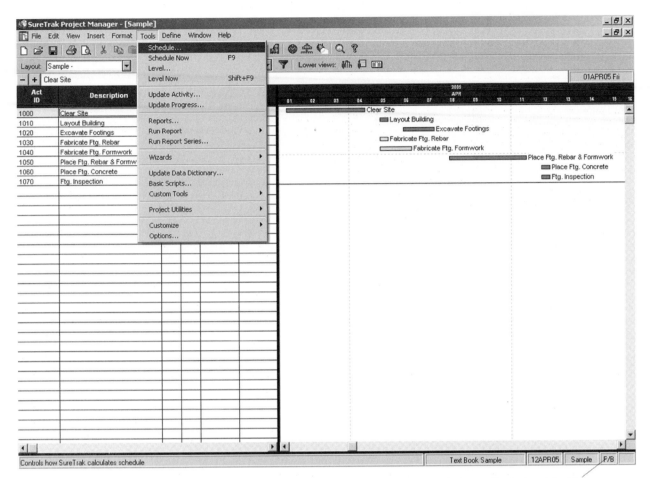

Figure 4–20   Tool Pull-Down Menu

used to determine whether the original activity relationship logic figures in the mathematical forward/backward pass calculations or the actual progress durations input during schedule update controls. The two options are **Retained logic** and **Progress override.** When you choose **Retained logic,** *SureTrak* does not schedule the remaining duration of a progressed activity until all its predecessors are complete. **Progress override** is used to update the schedule.

Figure 4–21   Schedule Dialog Box

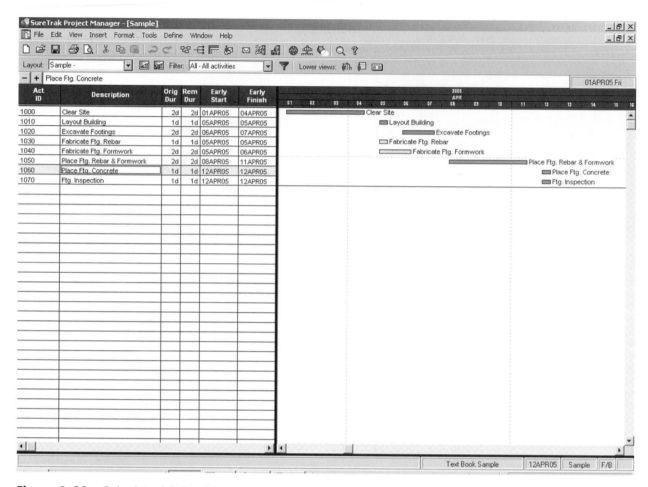

Figure 4–22 Calculated Schedule

At the bottom of the **Schedule** dialog box is the **Automatic Schedule Calculation** control. With this function on, *SureTrak* recalculates each time you add or delete an activity or relationship, or change an activity duration or relationship type. *Note: the second to last box on the status bar indicates whether automatic scheduling is on.* **F/B** indicates it is on and calculating the forward and backward passes. The **FWD** indicates it is on and calculating the forward pass only. If this box is blank, the automatic scheduling calculation is off. The automatic scheduling function can be turned off by clicking **Off.**

Next, click on the **OK** button and the schedule is calculated.

EDIT ACTIVITIES

Activity descriptions, original durations, relationships, or other input data can easily be changed or added to an existing activity. Click on the activity to be modified. Turn the **Activity form** on (if it is off) by

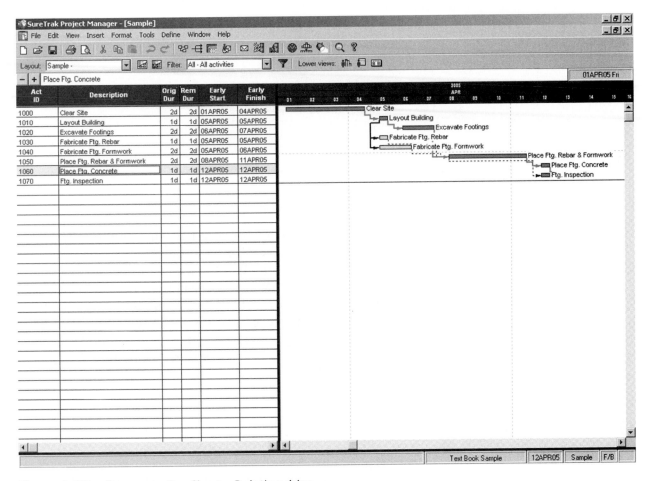

Figure 4–23   Onscreen Bar Chart—Relationships

clicking on **View** from the main pull-down menu, and then clicking on **Activity form.**

Any of the fields for the highlighted activity can be modified. While still in this edit mode with the **Activity form** turned on, other activities can be edited by clicking on them.

CLOSE A PROJECT

SureTrak provides three methods to close an existing project:

- **Method 1** Click on the **File** main pull-down menu. Click on **Close**. This closes the particular schedule, but does not shut down *SureTrak*.
- **Method 2** Click on the **SureTrak Project Manager** pull-down menu. Click on **Close**. This not only closes the particular schedule, it also shuts down *SureTrak*.
- **Method 3** Simultaneously press the **Alt** and the **F4** keys. This closes the particular schedule and shuts down *SureTrak*.

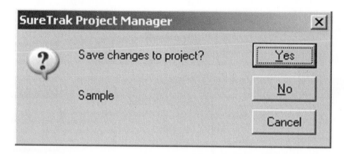

Figure 4–24 Save Changes to Project Dialog Box

No matter which of the three methods for closing the schedule you choose, the **Save changes to project?** dialog box will appear (Figure 4–24). Here you have a chance to save or abandon the changes you have made since the last save. If you try to close a project that has not been saved since the last change made to it, *SureTrak* prompts you to save the project. To save your changes and close the project, click on **Yes.** To close the project without saving changes, click on **No.** *Note: You can control how often SureTrak automatically saves your projects by selecting the **Tools** main pull-down menu, then **Options**, and then the **General** tab.*

OPEN A PROJECT

To open an existing project rather than creating a new project as described in this chapter, click on the **File** main pull-down menu and select **Open,** or use the Open button from the toolbar (refer to Figure 4–3). The **Open Project** dialog box of existing projects appears (Figure 4–25). You can change directories and drives when selecting a project. Click

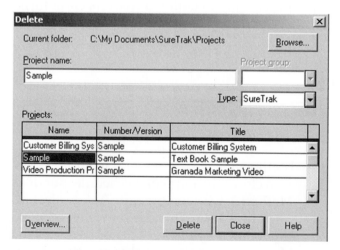

Figure 4–25 Open Project Dialog Box

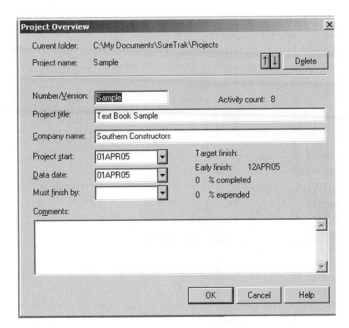

Figure 4–26   Project Overview Dialog Box

on the **Overview** button to view the **Project Overview** dialog box (Figure 4–26). Compare Figure 4–26, project information for the existing project, to Figure 4–7, the same project information input as the project was added. Sometimes it is necessary to change project information in the **Project Overview** dialog box after a project has been created. You can also go directly to the **Project Overview** dialog box while you are already in an open project by chasing **File** main pull-down menu and selecting **Project Overvie_w_.**

TOOLS

Some of the features of the **Tools** main pull-down menu have already been discussed in this chapter. Here we discuss some of the other features that simplify the use of *SureTrak*.

Project Utilities

First, look at the **Project Utilities** selection in the **Tools** main pull-down menu (Figure 4–27). Clicking on **Project Utilities** provides four options for file management:

- **Check-in/Check-out** This feature is used to check a member project out of or back into a project group and keep track of which member projects have been checked out. This is very handy when member projects are being updated.

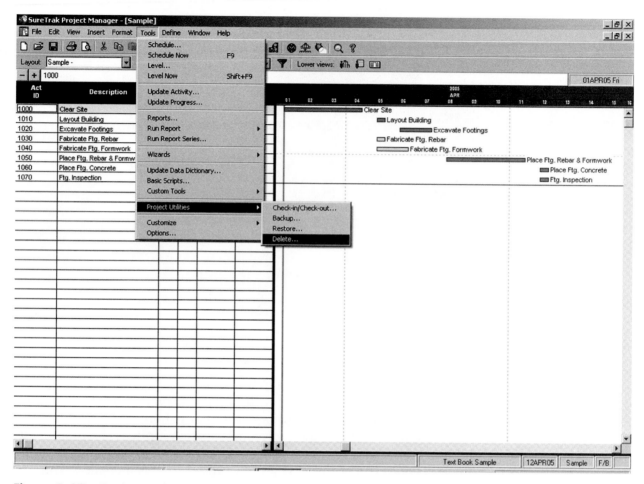

Figure 4–27   Tools—Project Utilities

- **Backup** Use this option to make a copy of a project. The primary difference between this option and the **Save As** option from the **File** main pull-down menu is the Compressed Format feature. This feature puts the *SureTrak* project files into a more concise format that takes up less disk space, making it more efficient for storing projects that you might not use for a while.
- **Restore** Use this option to restore a copy of a backed-up or archived project. This can be done from a normal or compressed format.
- **Delete** As an example of good file management, when you back up a project in a compressed format on a disk for permanent storage, delete the original project from the hard disk. By using the **Delete** feature rather than *Windows Explorer*, *SureTrak* deletes all files it creates with that name. Due to the number of files created for each project, this is a more efficient way to delete a project.

Custom Tools

Next, look at the **Custom Tools** selection from the **Tools** main pulldown menu (Figure 4–28). The **Custom Tools** menu lists some commonly used

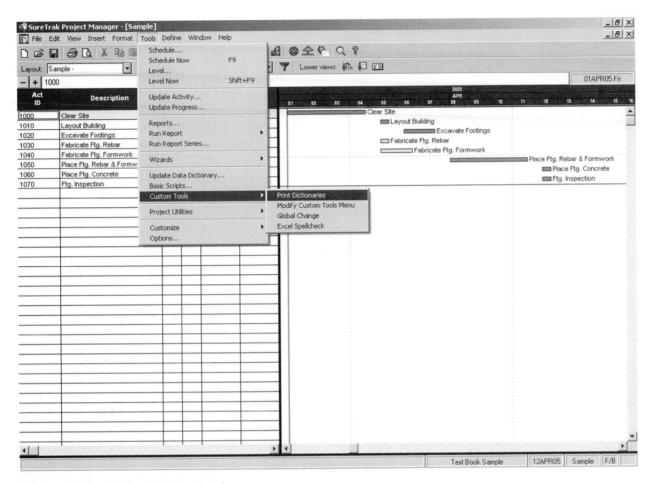

Figure 4–28   Tools—Custom Tools

scripts or executable files that were added to this menu using the **Modify Custom Tools Menu** selection. Click on **Custom Tools** to see four useful options for using *SureTrak*.

- **Print Dictionaries** This is a handy option for getting a hard copy printout of the options under the **Define** main pull-down menu; these include the resources, WBS codes, activity codes, and activity IDs dictionaries.
- **Modify Custom Tools Menu** You can customize this list so that it contains other Basic scripts or executable files such as *Notepad, Microsoft Word,* or *Microsoft Excel.* The icons for these applications can be added to the toolbar and launched from within *SureTrak.*
- **Global Change** This option is used to make global changes to activity remaining durations and percent complete, to delete activity assignments, to change resource assignments, or to modify resource/cost data.
- **Excel Spellcheck** You can use this tool within defined fields such as the activity descriptions to check for spelling errors.

Customize

Next, look at the **Customize** selection from the **Tools** main pull-down menu (Figure 4–29). The Customize menu lists some tools for modifying the onscreen *SureTrak* image, toolbar, and status bar.

- **Default Font** Click on this option to see the **Font** dialog box (Figure 4–30). Here you can choose a default font (typeface) for the contents of the project window. If a project is open, you are establishing the default font for the project. If no projects are open, you are establishing the default font for all future projects.
- **Status Bar** Refer to Figure 4–29 for the location of the status bar. The Status Bar option lets you select up to three items of information for the status bar. Here, information is displayed about the entire project, not individual activities. Most of the items in this list are from information input to the **Project Overview** dialog box when the project was created. Figure 4–29 shows the project title, finish date, and layout selections for the status bar.
- **Toolbar** Click on this option to produce the **Toolbar** dialog box (Figure 4–31). Use this dialog box to control which icons appear on the

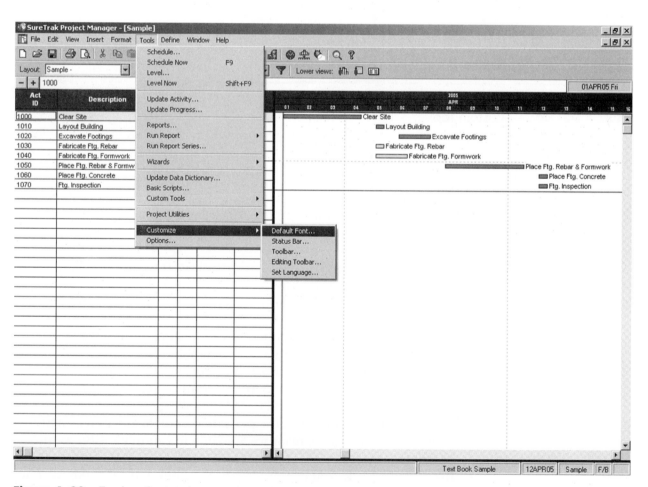

Figure 4–29   Tools—Customize

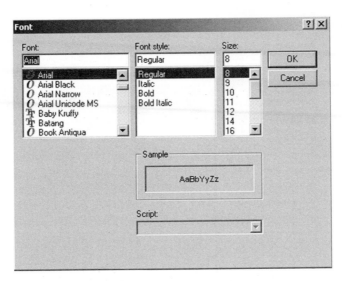

Figure 4–30   Font Dialog Box

main toolbar, and in what order. Icons are shortcuts. When you click on an icon, you execute a command or series of commands just as though you had chosen them from the menu.

- **Editing Toolbar** To expose the editing toolbar, select **Editing Toolbar** from the **View** main pull-down menu (Figure 4–32). The Editing Toolbar dialog box controls which icons appear on the editing toolbar. Generally, you should put icons you use every time you use *SureTrak* on the main toolbar, and use special function icons on the editing toolbar.

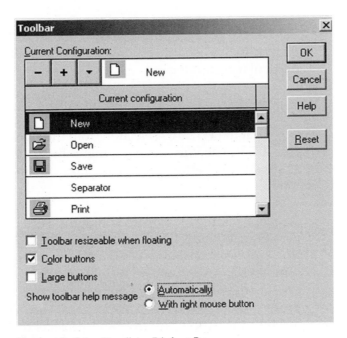

Figure 4–31   Toolbar Dialog Box

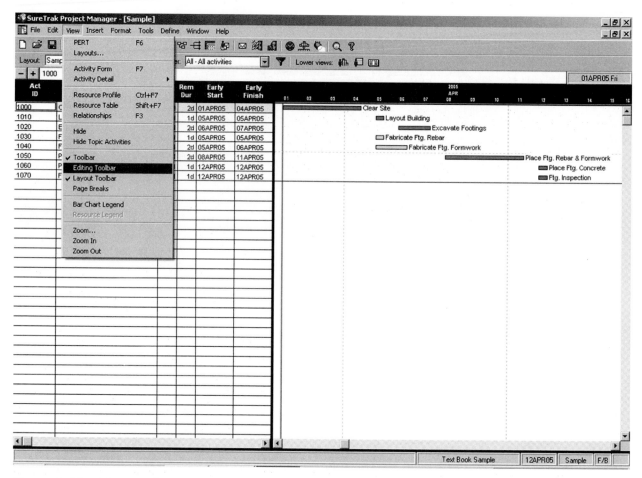

Figure 4–32   View—Editing Toolbar

- **Set Language** This option allows you to pick a language other than American English for certain text items for all *SureTrak* projects on the particular computer.

Wizards

Now look at the **Wizards** selection from the **Tools** main pull-down menu (Figure 4–33). The Wizards Tools menu lists some commonly used built-in templates for performing basic *SureTrak* functions in a very efficient manner.

- **Project KickStart Wizard** Making this selection produces the **Project KickStart Wizard** dialog box (Figure 4–34). Use this wizard to create a new project simply and easily. You answer questions about specific aspects of the new project, and the wizard assigns project phases and goals names activities, and defines and assigns resources.
- **Project Group Wizard** This wizard (see Figure 4–33) walks you through the process of grouping projects together so they can share resources, coding systems, and even relationships. To manage a group

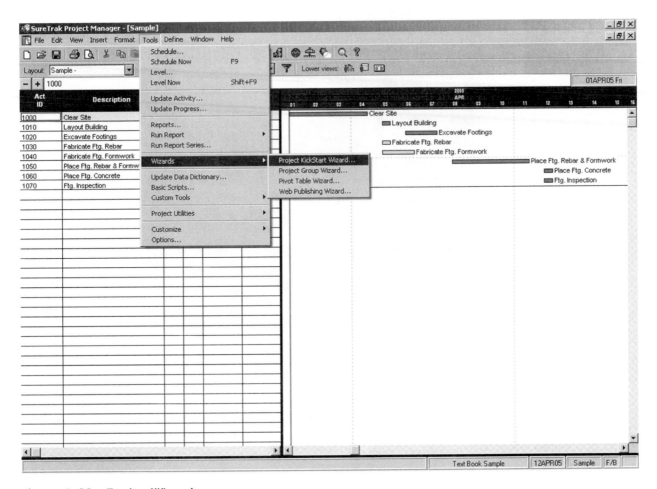

Figure 4–33   Tools—Wizards

of projects together, create the project group first, and then create the projects that will be members of this group. Make all these projects Project Group type, and then use the **Project Group Wizard** to collect the projects into the group.

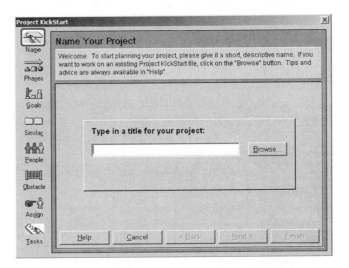

Figure 4–34   Project KickStart Wizard Dialog Box

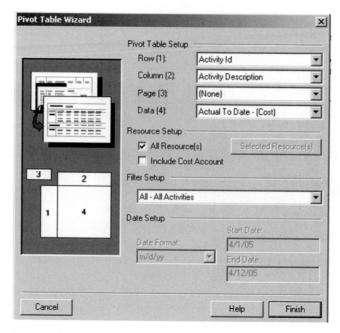

Figure 4–35   Pivot Table Wizard Dialog Box

- **Pivot Table Wizard** This wizard (see Figure 4–33) walks you through the process of creating an *Excel* pivot table. The table can be used to organize the information by choosing the items and fields you want to appear in the pivot table. This information can be summarized by time period (hour, day, week, month, or year). You can also create tables that organize and total *SureTrak* activity and resource data by activity code such as Responsibility. Pivot tables make it easier to analyze information. Clicking on the **Pivot Table Wizard** produces the **Pivot Table Wizard** dialog box (Figure 4–35). By defining the information to be included in each of the pivot table fields and then clicking **Finish,** *Excel* automatically launches with the created pivot table.
- **Web Publishing Wizard** This wizard enables you to publish project information on the World Wide Web or on an office intranet for others to access.

Options

Finally, look at the **Options** selection from the **Tools** main pull-down menu (see Figure 4–27). The **Options** dialog box (Figure 4–36) controls many of the basic *SureTrak* formatting features. Because the **Options** dialog box controls so many of the elements of *SureTrak*, it is referred to regularly throughout the book. Here, we discuss each of the tabs in the dialog box. The six tabs are:

1. Project
2. Resource
3. General
4. Defaults
5. View
6. File Locations

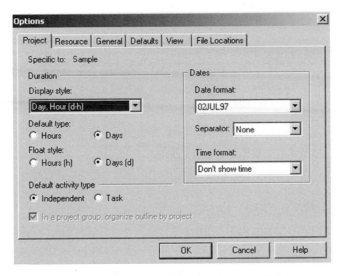

Figure 4–36   Options Dialog Box—Project Tab

Project.   Figure 4–36 shows the **Project** tab of the **Options** dialog box. To establish project options for one project, open the **Project** tab while the project is active. For Figure 4–35, the sample project is active, so the choices made impact this project only. To establish default project options for future projects, open the **Project** tab while no projects are open. Next, determine the default **Duration** unit of measure. For the examples in this book, **Day (d)** is chosen. When you type in 1 in the duration field, *SureTrak* knows this is one day, not one hour, week, or month. The choices are:

- Week, Day, Hour (w-d-h)
- Week, Day (w-d)
- Day, Hour (d-h)
- Day (d) (the option used in this book)
- Hour (h) (the basis for *SureTrak* calculations)

Again, note that all of *SureTrak*'s quantity fields (resources, costs, revenues, and so on) are calculated hourly. So, choosing an hourly (h) display style has inherent advantages.

Float style options are **Hours (h)** and **Days (d)**.
Default Activity Type options are **Independent** and **Task**:

- **Independent** *SureTrak* schedules an independent activity when each resource assigned to it can work. Each resource is scheduled to work on the activity according to its own calendar.
- **Task** An activity is scheduled according to the calendar assigned to it in the Activity Form rather than according to the resource calendars assigned to it.

The **Dates** section has the following options:

- **Date Format** Specify the order of the components of the date. For example, you could select that the date is shown day/month/year or month/day/year.
- **Separator** Specify the separator between the components of the date. For example, slash (/) dd/mm/yy or comma (,) dd,mm,yy.
- **Time format** Specify how time should be displayed: either with AM and PM (Show 12:00 hour) to distinguish between morning and afternoon or as the 24-hour clock (Show 24:00 hour).

Resource.   Figure 4–37 shows the **Resource** tab of the **Options** dialog box. **Autocost Rules** are operations that *SureTrak* performs automatically when the user updates resource information. See Chapter 12 for a detailed discussion of *SureTrak* autocost uses. Here the **Autocost Rules** are turned on. In the **Resource Data** section, you can specify the number of decimals to be shown in the cost calculation.

General.   Figure 4–38 shows the **General** tab of the **Options** dialog box.

The **Detail forms** section consists of:

- **Close detail forms when closing activity form** When this option is selected, *SureTrak* closes all opened detail forms when the Activity Form is closed.
- **Restore detail forms when opening the project** When this option is selected, *SureTrak* shows the detailed forms as they were left the last time that the project was saved and closed.
- **Allow multiple detail forms to be open at once** When this option is selected, *SureTrak* allows several detail forms to be opened at the same time.

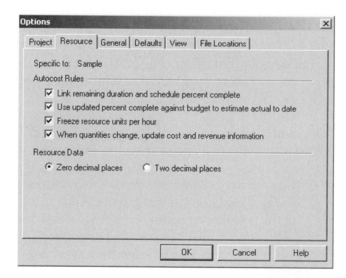

Figure 4–37   Options Dialog Box—Resource Tab

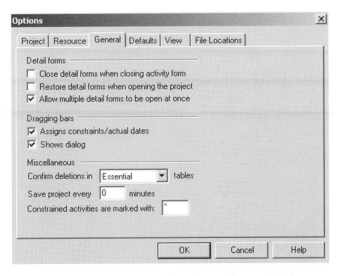

Figure 4–38 Options Dialog Box—General Tab

The **Dragging Bars** section consists of:

- **Assigns constraints/actual dates** This checkbox determines how *SureTrak* responds when you move (drag) activity bars on the screen. Mark the checkbox to have *SureTrak* assign a constraint or actual date when you move an activity bar, or clear the checkbox if you want *SureTrak* to move activity bars back to the logic calculated dates when you reschedule the project.
- **Shows dialog** Mark this checkbox if you want *SureTrak* to display a dialog box each time you move (drag) activity bars that tells you that it assigned a constraint or actual date to the moved activity.

The **Miscellaneous** section consists of:

- **Confirms deletions in ____ tables** Specify whether you want to be prompted to confirm your intention to delete something from *any* table (for example, an activity), only a table from which a deletion is *unrecoverable* (you cannot click on **Cancel** to abandon the change), or *never*.
- **Save project every ____ minutes** If you want *SureTrak* to save your project automatically, specify a time interval here. If *SureTrak* automatically saves it and you later exit the project without saving it, your project will be saved as of the time of the last automatic save.
- **Constrained activities are mark with ____** This option allows you to indicate the character that *SureTrak* will use to highlight the activities that have been constrained.

Defaults.   Figure 4–39 shows the **Forms** tab of the **Defaults** dialog box.

The **Activity defaults** section consists of:

- **Display activity form when adding activities in the bar view** Mark this box if you want to add activities using the Activity Form.

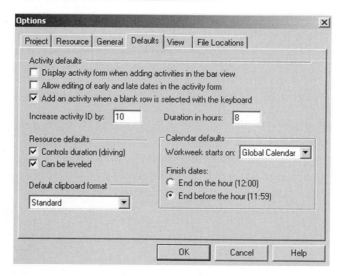

Figure 4–39   Options Dialog Box—Defaults Tab

- **Allow editing of early and late dates in the activity form** *SureTrak* calculates early and late dates for each activity based on the logic provided (activity predecessors and successors). If you mark this box, you can edit these dates temporarily on the Activity Form, although the next time you schedule the project, *SureTrak* will recalculate them based on logic. Adding a constraint is the way to keep the date assigned.
- **Add an activity when a blank row is selected with the keyboard** When this option is selected, you can add new activities by pressing the Down Arrow key followed by the Enter key.
- **Increase activity ID by** Specify a value for *SureTrak* to use when incrementing activity IDs. By default, *SureTrak* increments the activity ID by 10 for each new activity.
- **Duration in hours** Specify a default activity duration for *SureTrak* to use when you add activities.

The **Resource defaults** section consists of:

- **Controls duration (driving)** When this option is selected, *SureTrak* sets the resources as driving. In other words, the quantity of resources determines the duration of the activity.
- **Can be leveled** Mark to automatically set resources to level when you create resources.

The **Calendar defaults** section consists of:

- **Workweek starts on** Specify the day on which the workweek starts; for example, Monday or Sunday. Additionally, you can select the global calendar and the workweek will begin based on the settings specified in the global calendar. Calendar definition is discussed in a later chapter.
- **Finish dates** Specify the time in which finish dates end.

View.   Figure 4–40 shows the **View** tab of the **Options** dialog box.

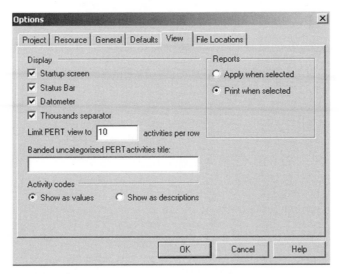

Figure 4–40   Options Dialog Box—View Tab

The **Display** section consists of:

- **Startup screen** Mark this checkbox if you want to see the *SureTrak Welcome Screen* each time you open *SureTrak*. Clear the checkbox if you do not want to see the screen.
- **Status Bar** Mark the checkbox to show the status bar at the bottom of the onscreen project window. Clear the box to hide it.
- **Datometer** Mark the checkbox to show the datometer. Clear the checkbox to hide it.
- **Thousands separator** Mark the checkbox to display numeric figures above 999 with a comma separator (for example, 1,000). Clear this checkbox to remove the commas.
- **Activity codes** When you display activity codes as activity columns, choose whether the code value or description appears by default.
- **Reports** When you select a report from the **Tools, Run Report** menu, you can specify whether *SureTrak* displays a **Print** dialog box so you can print the report, or immediately applies the report settings to the current layout.

File Locations. Figure 4–41 shows the **File Locations** tab of the **Options** dialog box. This tab shows where *SureTrak* files are stored.

EXAMPLE PROBLEM: ONSCREEN BAR CHART

Table 4–2 is a tabular list of twenty-eight activities (activity ID, description, duration, and successor relationships) for a house put together as an example for student use (see the wood frame house drawings in the Appendix). Figure 4–42 is the onscreen bar chart of the house based on the list provided in Table 4–2.

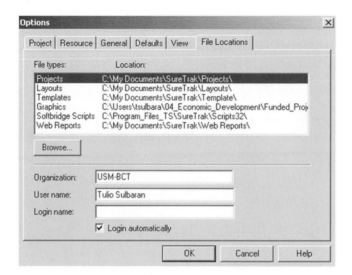

Figure 4–41   Options Dialog Box—File Locations

Activity ID	Activity Description	Duration (Days)	Successors
1000	Clear Site	2	1010
1010	Building Layout	1	1020
1020	Form/Pour Footings	3	1030
1030	Pier Masonry	2	1040
1040	Wood Floor System	4	1050
1050	Rough Framing Walls	6	1060
1060	Rough Framing Roof	4	1080
1070	Doors & Windows	4	1090, 1150
1080	Ext Wall Board	2	1070, 1100, 1110
1090	Ext Wall Insulation	1	1160
1100	Rough Plumbing	4	1130
1110	Rough HVAC	3	1120
1120	Rough Elect	3	1090
1130	Shingles	3	1090, 1150, 1260
1140	Ext Siding	3	1190
1150	Ext Finish Carpentry	2	1140
1160	Hang Drywall	4	1170
1170	Finish Drywall	4	1180, 1230, 1240
1180	Cabinets	2	1200, 1220
1190	Ext Paint	3	1210
1200	Int Finish Carpentry	4	1210
1210	Int Paint	3	1250
1220	Finish Plumbing	2	1210
1230	Finish HVAC	3	1210
1240	Finish Elect	2	1210
1250	Flooring	3	1270
1260	Grading & Landscaping	4	1270
1270	Punch List	2	

Table 4–2   Activity List with Durations and Relationships—Wood Frame House

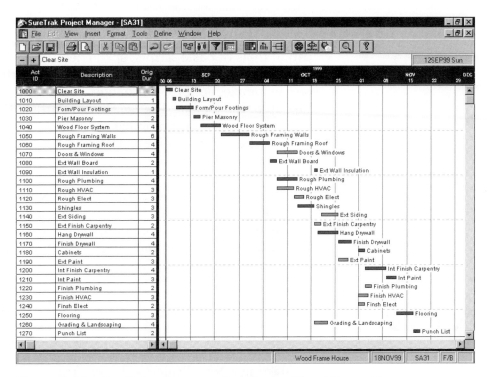

Figure 4–42   *SureTrak*—Onscreen Bar Chart—Wood Frame House

SUMMARY

This chapter provided an overview of creating a project schedule within *SureTrak*. The scheduler starts the *SureTrak* using the **SureTrak Project Manager** screen and the **New Project** dialog box to create a new project schedule.

The scheduler creates new activities using the Activity Form. Information identified for each activity includes activity ID, description, original duration, and calendar. Also within the **SureTrak Project Manager** screen, the timescale is modified using timescale density, begin date, end date, font, minimum time increment, and ordinal dates.

The scheduler creates activity relationships using the **Successors** dialog box. The successor ID, identification, relationship type, and lag fields are used to define the activity relationships.

The scheduler then calculates the schedule using the **Schedule** dialog box. The project data date is defined as the heavy blue vertical line and reflects the date that is the basis of the schedule. Schedule calculations use retained logic and forward and backward calculation.

The scheduler then can use the newly created schedule by editing activities, closing the schedule, and opening the schedule back up. The scheduler also can project utilities that were used to check in and out, back up, restore, and delete schedules. The scheduler also can use the custom tools of basic scripts, spell check, global changes, and printing dictionaries to make modifications to the schedule. The scheduler can use customize options to change the default font, and change the makeup of the toolbar. The *SureTrak* wizards are commonly used built-in templates for performing basic *SureTrak* functions in a very efficient manner. These wizards include Project KickStart, Project Group, Pivot Table, and Web Publishing.

The scheduler can use the **Options** dialog box to control many of the elements of *SureTrak*. The default project time unit, float style, default activity type, resource calculation rules, dates format, display of forms, screen view modifications, and file directory options are controlled using the **Options** dialog box.

EXERCISES

1. **Starting a New Project with *SureTrak***
   Start *SureTrak* and begin a new project with the following description:

Project name:	ComBldgSouth
Template:	_Default
Type:	SureTrak
Number/Version:	1
Start date:	Nov 06 2007
Project Title:	Small Commercial Building in the South
Company Name:	Southern Constructors

   Define in 4–6 lines the following terms: activity, forward pass, backward pass, early start, late start, early finish, late finish, float, and critical path.

2. **Adding Activities to a Project**
   Add the activities listed in Table 4–3 to the project created in Exercise 1.

Activity ID	Description	Duration
1000	A	3d
1010	B	4d
1020	C	6d
1030	D	2d
1040	E	5d
1050	F	2d

   Table 4–3   Activities for the Small Commercial Building in the South

3. **Defining Relationships between Activities**
   Define the relationships and include the lags listed in Table 4–4 to the activities added to the project.

Activity ID	Successor Activity ID	Relationship	Lag
1000	1010	FS	0
1010	1020	FS	3
	1030	SS	0
1020	1040	FS	2
1030	1050	SS	0
1040	1050	FF	0
1050	—	—	—

Table 4–4   Activity Relationships for the Small Commercial Building in the South

4. Creating an onscreen *SureTrak* bar chart for Exercise 4 in Chapter 3.
5. Creating an onscreen *SureTrak* bar chart for Exercise 5 in Chapter 3.
6. Creating an onscreen *SureTrak* bar chart for Exercise 6 in Chapter 3.
7. Creating an onscreen *SureTrak* bar chart for Exercise 7 in Chapter 3.
8. Creating an onscreen *SureTrak* bar chart for Exercise 8 in Chapter 3.
9. Creating an onscreen *SureTrak* bar chart for Exercise 9 in Chapter 3.
10. Creating an onscreen *SureTrak* bar chart for Exercise 10 in Chapter 3.
11. Creating an onscreen *SureTrak* bar chart for Exercise 11 in Chapter 3.
12. **Preparing an onscreen bar chart for the Small Commercial Concrete Block Building**
    Prepare an onscreen bar chart for the small commercial concrete block building located in the Appendix. This exercise should include the following steps:
    1. Prepare a list of activity descriptions (minimum of sixty activities).
    2. Establish activity durations.
    3. Establish activity relationships.
    4. Create the onscreen *SureTrak* bar chart.
13. **Preparing an onscreen bar chart for a Large Commercial Building**
    Prepare an onscreen bar chart for the large commercial building located in the Appendix. This exercise should include the following steps:
    1. Prepare a list of activity descriptions (minimum of 150 activities).
    2. Establish activity durations.
    3. Establish activity relationships.
    4. Create the onscreen *SureTrak* bar chart.

Bar Chart Format

Objectives

Upon completion of this chapter, you should be able to:

- Select layouts
- Change zoom magnification
- Insert new columns
- Modify characteristics of the activity bars
- Modify characteristics of the relationship lines
- Change sight lines
- Vary row height
- Adjust the screen colors
- Define values for activity codes
- Organize activities
- Summarize activities
- Filter activities
- Add clip art to the activity bar area
- Add text to the activity bar area
- List all objects included in the bar area

FORMATTING WITH SCHEDULING SOFTWARE

One of the major advantages of using the computer for scheduling construction project management applications is the ease of modifying onscreen and hard copy printouts to format the schedule to communicate information in the clearest way possible. This includes changing the appearance and content of columns, bars, row heights, sight lines, and so on; organizing the information using activity codes; and selecting particular activities included in the schedule using filtering. Clip art and extra text can also be added. All this flexibility assures that *SureTrak* can produce a schedule in the format and with the information to fit the multiple scheduling needs of a complex project.

LAYOUTS

Layouts is the term used by *SureTrak* for keeping multiple configured formats of a particular schedule. Click on the **View** main pull-down menu and select **Layouts** (Figure 5–1). The **Layouts** dialog box will appear (Figure 5–2). In the middle of the dialog box is a button named **Format Selected Layout** that contains formatting options for changing the onscreen bar chart and hard copy printouts of *SureTrak*. All these formatting options are covered in this chapter. The advantage to accessing them from the **Layouts** dialog box is to make and save multiple layouts of a schedule quickly.

In Figure 5–2, the project name for the sample schedule used in this book is Sample. *SureTrak* automatically saves that as the default name of the layout for the sample schedule. (Sample) is the **Name** of the **Layouts** in use (the highlighted layout). You could have a longer description for the layout by filling out the **Description** field.

When a schedule is closed, you are always asked if you want to **Save changes to project?** If you answer yes in this dialog box, all format

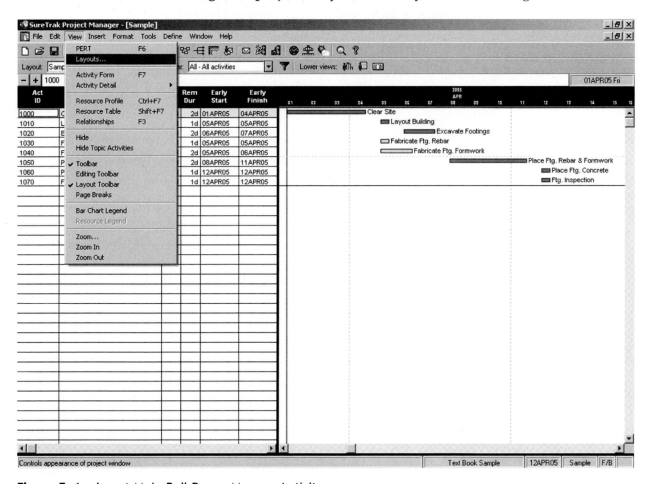

Figure 5–1   Insert Main Pull-Down Menu—Activity

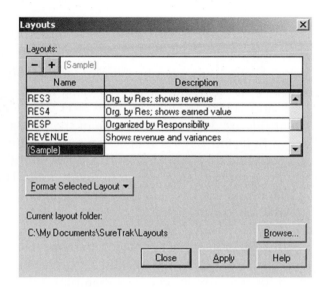

Figure 5–2   Layouts Dialog Box

changes to the current layout are kept. If you answer no, the format changes because the last time the schedule was saved is lost.

Four layout control options are available under **Layouts**:

- The first option is to use an existing layout. For example, if you want the **RESP** layout, click on that layout and highlight it. Then click on the **Apply** button. To change back to the original layout, click on (Sample), then select the **Apply** button and you are back to your originally formatted layout.
- The second option is to change an existing layout. Select its name and highlight it. Then click the **Format Selected Layout** button to change the corresponding part of the layout.
- The third option is to delete an existing layout. Select its name, and then click on the − button.
- The fourth option is to create a new layout. Select the layout you want to base the new layout on. Then click on the + button, and enter a **Name** and **Description** for the new layout. Use the **Format Selected Layout** button to change the new layout. The advantage of this option is that you can save multiple layouts of the same schedule. For example, you may want a separately named and saved layout for viewing and printing the bar chart, another for viewing resources, and another for viewing costs.

To just see how a layout looks, click on the **Apply** button. To get back to the previous layout, select the previous layout.

To save any changes to a layout, after closing the **Layouts** dialog box, select the **File** main pull-down menu, then **Save,** or close the schedule and save it. Layouts are not just job specific. When you create or modify a layout in *SureTrak*, all projects can use that layout.

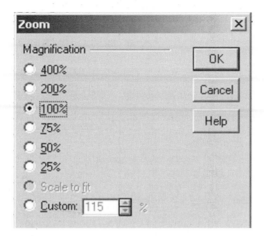

Figure 5–3   Zoom Dialog Box

ZOOM

Another feature provided by *SureTrak* for modifying the onscreen bar chart is the **Zoom** feature. This feature allows you to modify the magnification of the information provided on the screen. Click on the **View** main pull-down menu and select **Zoom.** The **Zoom** dialog box appears (Figure 5–3). The standard choices are **25, 50, 75, 100, 200,** and **400%** of the original-sized configuration. The other choices are **Scale to fit** for schedules larger than a screen, which will automatically size the schedule to fit on a screen, and the **Custom** size, where the sizing can be tinkered with until you meet the exact size requirements you want. To get back where you started, simply click on **100%.** Figure 5–4 is the result of the **Magnification** of **200%.** Compare Figure 5–4 with magnification to Figure 5–1 without magnification.

The other two zoom options under the **View** main pull-down menu are the **Zoom In** and **Zoom Out** selections. Click on **Zoom In** once to produce a screen sized at 150% of the original. Click on it again to produce a screen sized at 200% of the original. Click on **Zoom Out** to produce a screen sized at 75% of the original. Click on it again to produce a screen sized at 50% of the original.

COLUMNS

Look at the Columns and Column title location in Figure 5–4. To modify the onscreen columns (Figure 5–5) configuration, select **Columns** from the **Format** main pull-down menu. The **Columns** dialog box appears (Figure 5–6). Before modifying a column, look at the columns as they are presently configured for the Sample Schedule. Close the **Columns** dialog

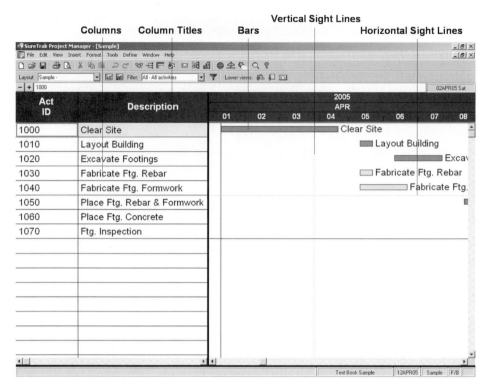

Figure 5–4 Sample Schedule—Zoomed

box to return to the onscreen bar chart. Click and drag the mouse on the vertical split bar to expose the columns behind the bar chart (Figure 5–7). All the columns in Figure 5–7 are in **Column specifications** in Figure 5–6. The **Column data** options are:

Activity ID	Constraint Start
Activity Codes (all defined activity codes will appear)	Constraint Start Type
	Cost at Completion
Activity Type	Cost to Complete
Actual Cost to Date	Cost Variance
Actual Duration	Description
Actual Finish	Driving Relationship
Actual Quantity to Date	Driving Resource
Actual Start	Early Finish
Blank	Early Start
Budgeted Cost	Earned Value Cost
Budgeted Quantity	Earned Value Quantity
Calendar	Expected Finish Constraint
Completion Variance Cost	Finish Variance
Completion Variance Quantity	Free Float
Constraint Finish	Late Finish
Constraint Finish Type	Late Start

(continued on page 120)

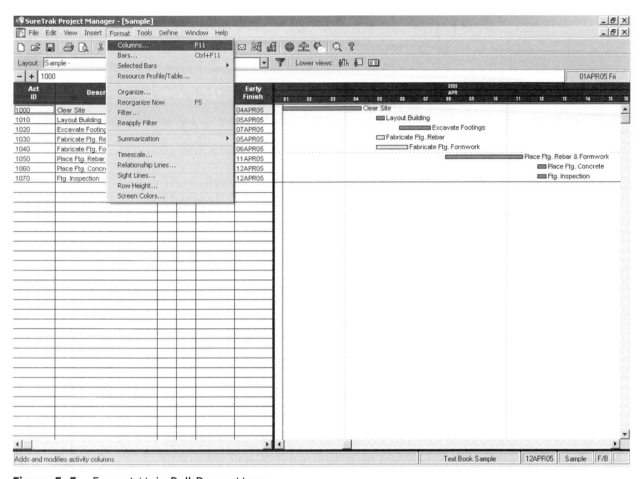

Figure 5–5   Format Main Pull-Down Menu

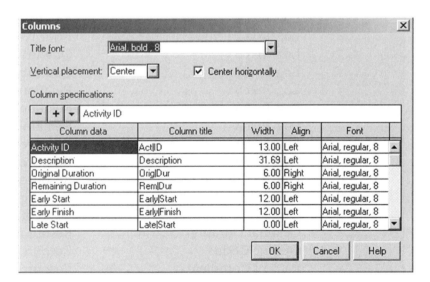

Figure 5–6   Columns Dialog Box

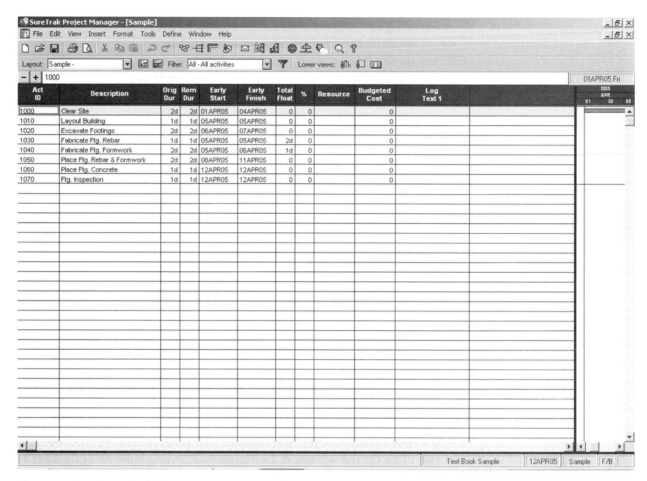

Figure 5–9 Sample Schedule—Column Added

BARS

Look at the bars location in Figure 5–4. To modify the onscreen bars configuration, select **Bars** from the **Format** main pull-down menu (refer to Figure 5–5). The **Bars** dialog box appears (Figure 5–10). By clicking on the down arrow, the data items can be displayed. The data item options are:

- Early Bar
- Late Bar
- Target Bar
- Unleveled Bar
- Free Float Bar
- Total Float Bar
- Early Start Point
- Early Finish Point
- Late Start Point
- Late Finish Point

- Target Start Point
- Target Finish Point
- Unleveled Start Point
- Unleveled Finish Point
- Free Float Point
- Total Float Point
- Finish Variance Bar
- Level Delay Bar
- Finish Variance Point
- Level Delay Point

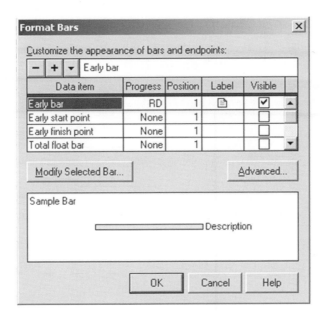

Figure 5–10   Format Bars Dialog Box

Select the **Data item** you want, and then click on the **Visible** field. The **Sample** view screen at the bottom of the dialog box allows you to preview any changes before accepting them and returning to the onscreen bar chart. Figure 5–11 is the result of making the **Early start point,** the **Late finish point,** and the **Total float bar** visible.

Note that in Figure 5–11, only the **Description** data item field is added to the early bar. Suppose you want to add the log text field that you added when the columns were reformatted earlier in the chapter. Click on the label cell and the **Label Text** dialog box appears (see Figure 5–12). Click on the cell below the description as shown in Figure 5–12. Click on the down arrow and to see the possible labels choices. The choices for possible information to include on the bar are the same as the **Column data** options. From these options, select the **Log Text 1** option. Click on the **OK** button to accept the change. In Figure 5–13, the columns are exposed, and the **Log Text 1** field is selected. Type in the text to appear on the schedule. Note the appearance of the onscreen bar chart in Figure 5–14 with log appearing. The entries in the column field can be made at any time. To turn this field off and have the bars appear as they did earlier in Figure 5–4, with the **Early bar** graphic element selected, click on the label text item to be removed (**Log Text 1**) and then click on the - button. Now the bar appears as it did before.

By clicking on **Modify Selected Bar,** you can modify the color, border, shape, pattern, and size of the bar (see Figure 5–15). You can also modify the style, pattern, and color by clicking on **Advanced** (see Figure 5–16).

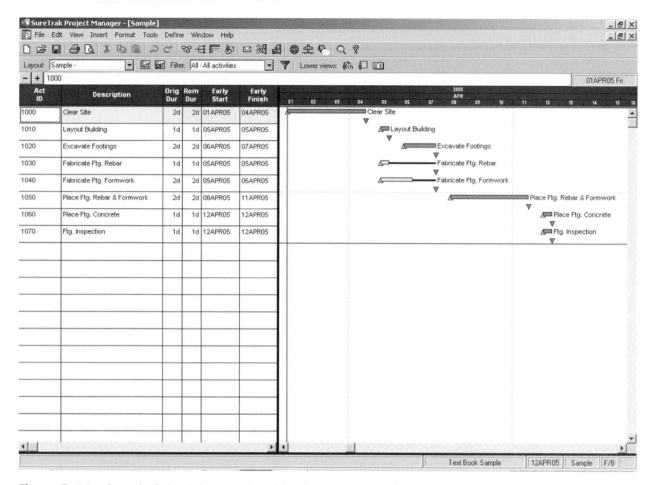

Figure 5–11   Sample Schedule—Bar Graphic Elements Added

Figure 5–12   Bars Dialog Box—Label Text

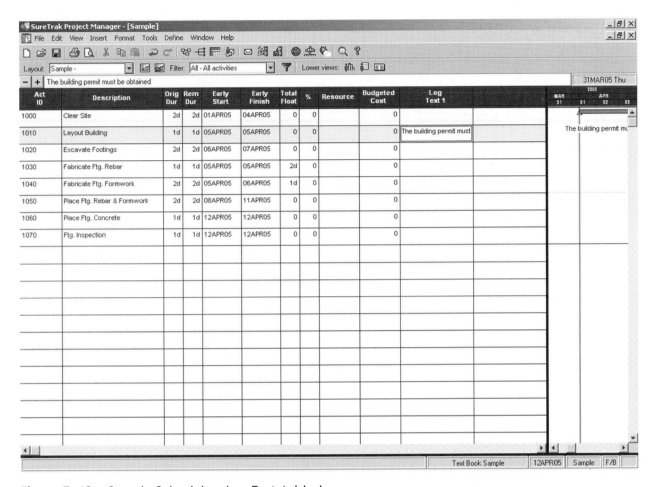

Figure 5–13   Sample Schedule—Log Text Added

RELATIONSHIP LINES

The **Relationships** feature in *SureTrak* allows you to modify the onscreen bar chart. As discussed earlier, the successor and predecessor relationships assigned can be shown onscreen, which is handy for evaluating the logic relationships for whatever reason. To actually see all the relationship lines on the screen rather than in a table or dialog box makes the logic relationships much easier to understand. To do this, click on **View** and select **Relationships.** Figure 5–17 is the sample schedule with relationships showing. Compare Figure 5–17, with the relationships showing, to Figure 5–4 without relationships.

To reformat the relationship lines, select **Relationship Lines** from the **Format** main pull-down menu (refer to Figure 5–5). The **Format Relationship Lines** dialog box appears (Figure 5–18). The types of relationship lines (**Critical, Driving,** or **Nondriving**) can be differentiated by **Line type, Color,** and whether **Visible** or not.

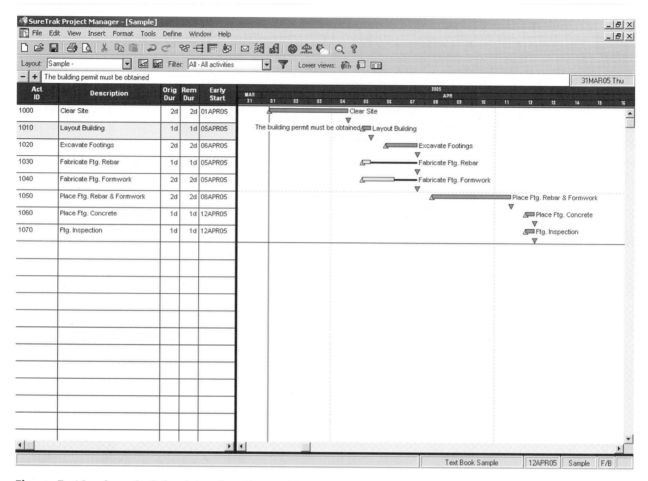

Figure 5–14   Sample Schedule—Log Text Added

Figure 5–15   Format Bars Dialog Box—Modify Bar Elements

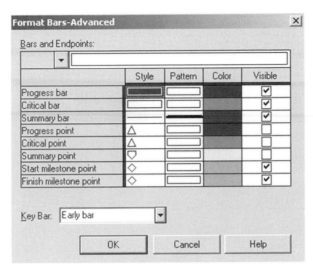

Figure 5–16   Format Bars Dialog Box—Format Bars-Advanced

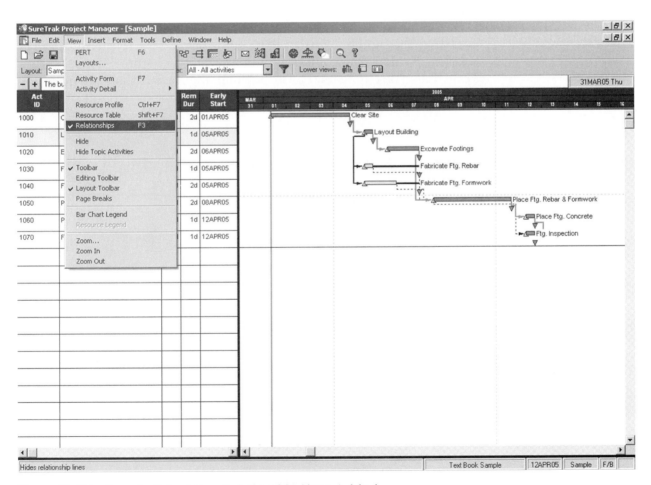

Figure 5–17   Sample Schedule—Relationship Lines Added

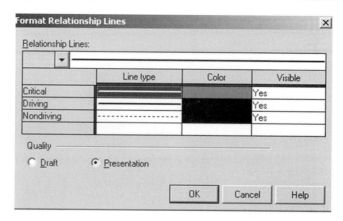

Figure 5–18   Format Relationship Lines Dialog Box

SIGHT LINES

Sight lines are *SureTrak* formatting options that make the onscreen bar chart and hard copy printouts much easier to read. The onscreen sight lines are labeled in Figure 5–4.

To change the sight line format to meet the user tastes, select **Sight Lines** from the **Format** main pull-down menu (refer to Figure 5–5). The **Time Scale** dialog box appears (Figure 5–19). In Figure 5–5, there are no major horizontal or vertical sight lines. There is a minor horizontal sight line

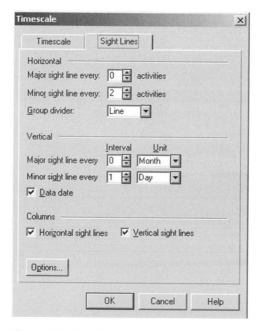

Figure 5–19   Sight Lines Dialog Box

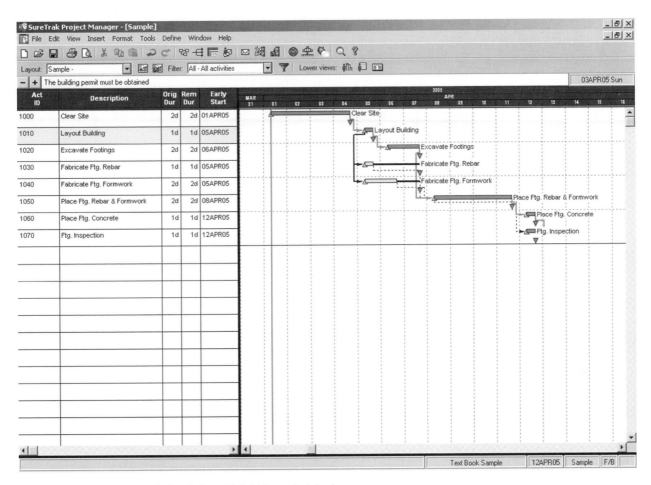

Figure 5–20   Sample Schedule—Sight Lines Added

after every fifth activity, or after Activity 1040. There is a minor vertical sight line after every week. These configurations were changed to produce the screen in Figure 5–20. Again, there are no major horizontal or vertical sight lines. There is a minor horizontal sight line after every second activity, and a minor vertical sight line after every day. This is a matter of taste. Some feel that the screen is too busy, others feel that it is much easier to interpret an activity's relationship to the timescale with more sight lines.

Row Height

Changing row height provides more spacing between activities or draws attention to particular activities. To demonstrate this formatting option of *SureTrak*, increase the row height of Activities 1010, 1030, and 1050. To

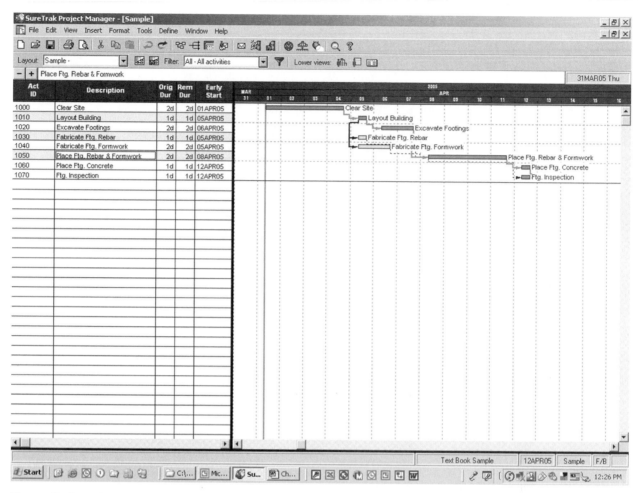

Figure 5–21 Sample Schedule—Selected Activities

select these three activities, hold down the Control (Ctrl) key while clicking on the three activities (Figure 5–21). These three activities are now highlighted. Next, select **Row Height** from the **Format** main pull-down menu (refer to Figure 5–5). The **Row Height** dialog box appears (Figure 5–22). Deselect the **Automatic size** checkbox, and increase the **Row**

Figure 5–22 Row Height Dialog Box

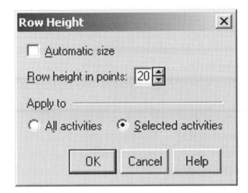

Figure 5–23   Row Height Dialog Box—Apply to Selected Activities

height in points to 20 (Figure 5–22). Select the **Apply to Selected activities** checkbox (see Figure 5–23). Figure 5–24 is the result of these changes to the onscreen bar chart. If you want the change to apply to all activities, click on the **Apply to All activities** checkbox.

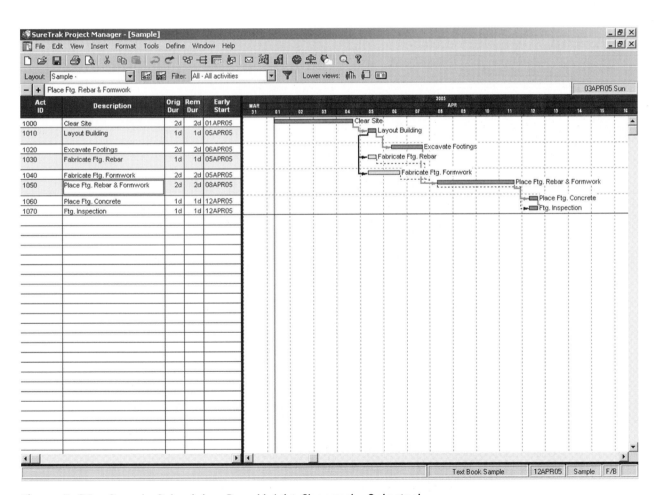

Figure 5–24   Sample Schedule—Row Height Changed—Selected

SCREEN COLORS

A *SureTrak* formatting option that gives the onscreen bar chart and hard copy printouts a personal touch is the ability to change screen and print colors. The ability to change the color of *SureTrak* components is a powerful tool in presentations and helps create a company image. To change the entire screen color scheme or that of individual components, select **Screen Colors** from the **Format** main pull-down menu (see Figure 5–5). The **Screen Colors** dialog box appears (Figure 5–25).

The top portion of the dialog box is used for changing the entire color scheme of the onscreen bar and hard copy prints. The **Sche̲me** options are:

> Ocean Breeze
> Irish Isle
> Rugby
> Frankfurter
> Pastel
> Monochrome
> Spice
> December
> Fluorescent
> Tweed
> Southwest
> Leather

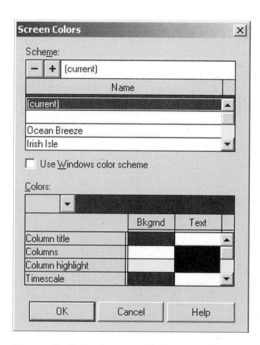

Figure 5–25   Screen Colors Dialog Box

Use the down arrow to scroll down the pick list. Select a **Scheme,** and the new scheme samples are shown in the **Colors** section. Even if an overall color scheme is selected, the **Bkgrnd** (Background) and **Text** colors can be changed for individual components of the onscreen bar chart and hard copy printouts. The individual components that you can change are:

Column title
Columns
Column highlight
Timescale
Bars
Resource table
Resource profile
Spotlight

Click on the individual field to be changed, and then click on the down arrow under **Colors** to bring up the color palette. Finally, click on the color you want. When the selection process is complete, click on the **OK** button to take the new changes back to the onscreen bar chart and hard copy printouts, or click on **Cancel** to return to the scheme before changes.

ACTIVITY CODES

Sometimes it is useful to organize the project by another sort rather than by activity ID. For instance, you may want to look at the project by crew—laborer crew versus carpentry crew. Click on the **Define** main pull-down menu and select **Activity Codes** to the **Activity Codes** dialog box (Figure 5–26).

Notice that under **Codes** there are already defaults in the system. Figure 5–26 shows Responsibility, Area, Phase, and Mail Codes. These fields can be modified or new fields can be added. If you want to sort the sample project by responsibility, click to highlight the **RESP** field. To define the values for the responsibility field, click on the + button under **Values,** enter an identifier under **Value,** and then add a full **Description** (Figure 5–27). Here the carpentry and laborer crews were input and the **Close** button was clicked.

If another *SureTrak* schedule has been completed and you are pleased with the way the **Activity Codes** are defined, you can use the **Transfer** function. Many companies use the same or a similar coding system for every project. It is much faster to transfer these codes to each new project than to reenter then. The **Transfer Activity Codes** dialog box can be used to copy activity codes from any other project to the current project. The Activity Codes Dictionaries and codes for the selected project completely replace the dictionaries and codes for the current project.

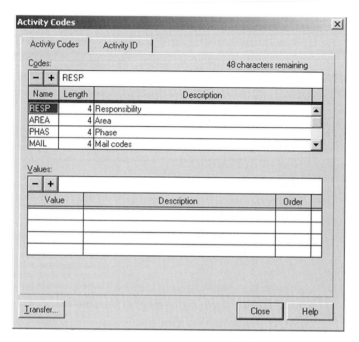

Figure 5–26   Activity Codes Dialog Box

Activity Form

The next step is to identify each activity by responsibility. Click on the first activity to be identified. In Figure 5–28, Activity 1000 has been highlighted. Click on the **View** main pull-down menu and select **Activity Form.** If you access the **Activity Form** and not all of the form is

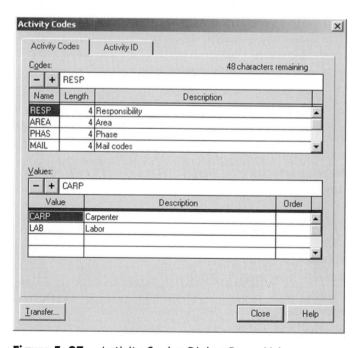

Figure 5–27   Activity Codes Dialog Box—Values

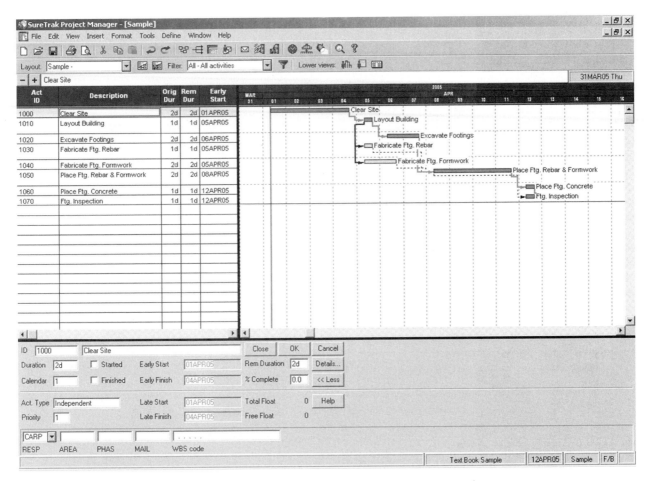

Figure 5–28   Sample Schedule—Activity Form

exposed, move your cursor to the dark view control line. The cursor image will change from an arrow to two short double lines. By pressing and holding down the left mouse button, you expose all of the **Activity Form.**

Click in the **RESP** field at the bottom of the form. A down arrow accesses a select box of the activity codes. **CARP** was selected for Activity 1000. Next click on the rest of the activities to identify their responsibility requirements (see Figure 5–29).

ORGANIZE

To view the onscreen bar chart by responsibility, click on the **Format** main pull-down menu. Select **Organize** and the **Organize** dialog box appears (Figure 5–30). With the **Activity data item** selected, click on the + button to define **Group by.** The menu as defined in Activity Codes will appear. Select **RESP,** as shown in Figure 5–31. Notice that the groups can

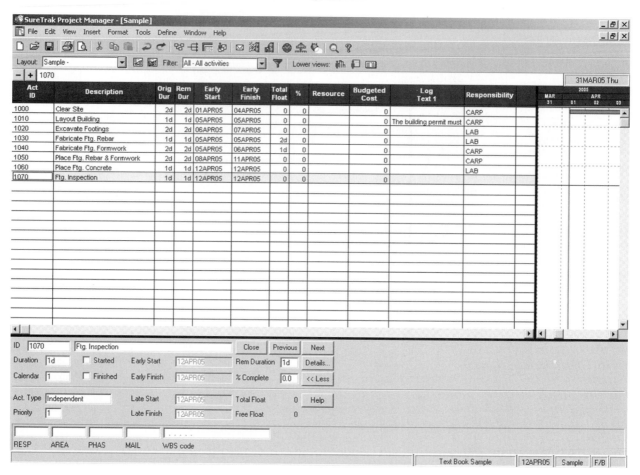

Figure 5–29 Sample Schedule—Activity Responsibility

Figure 5–30 Organize Dialog Box

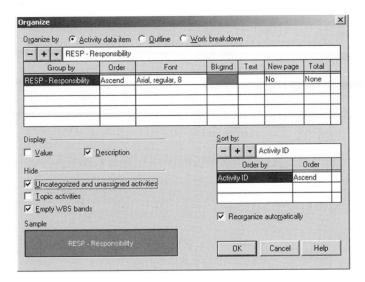

Figure 5–31   Organize Dialog Box—Group by

be customized by **Order, Font, Bkgrnd** (background color), **Text** (text color), and so on. Here, the **Reorganize auto̲matically** checkbox is checked at the bottom of the dialog box. Click on the **OK** button to accept the organizational change.

Compare Figure 5–32, sorted by responsibility, to Figure 5–29, sorted by activity ID. Activities 1000, 1010, 1040, and 1050 were defined as Carpentry Crew responsibilities. Activities 1020, 1030, and 1060 were defined as Laborer Crew responsibilities. Also notice that Activity 1070 is not shown because it was not assigned a responsibility unit. The communications advantages are obvious. The sort capabilities of *SureTrak* make it a great communications tool. Once the information is input, with minor changes, it can be resorted to provide different members of the project team with the scheduling data sorted in the best way for them to use it.

SUMMARIZE

Sometimes it is useful to summarize the project by sorted information. This way, instead of looking at hundreds of detail activities, project management personnel can look at a few summary activities that are the result of all the detail activities. For example, you may want to look at summary of the project by responsibility. Click on the **Format** main pull-down menu, select **Summarization** (see Figure 5–5), and then **Format Summary Bars.** The **Summary Bars** dialog box appears (Figure 5–33). Here the **O̲ne summarized bar** and the **Show n̲onwork time as a neck in the bar have been selected.**

Next, click on the **Format** main pull-down menu, select **Summarization** (refer to Figure 5–5), and then **Summarize All.** The **Summarize All**

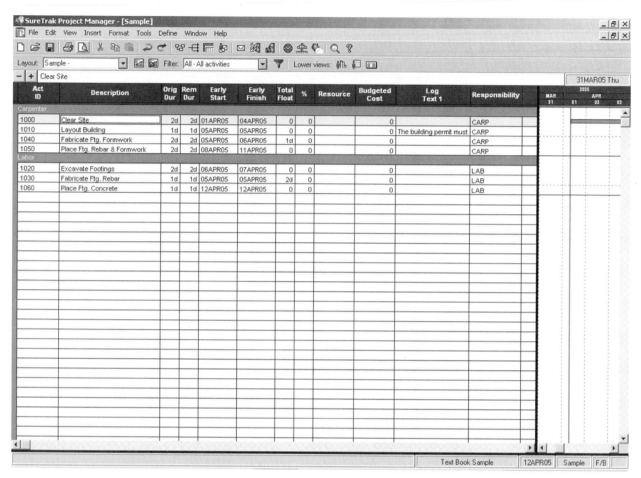

Figure 5–32 On-Screen Bar Chart—Organized by Responsibility

dialog box appears (Figure 5–34). Here the options chosen were to apply the summarization to **All bands,** rather than just to selected activities. Next, you can choose how to summarize the schedule activities. If you have grouped activities at several levels (for example by phase and then responsibility) specify whether *SureTrak* should summarize up to a particular level, leaving outer levels unsummarized, or whether it should

Figure 5–33 Summary Bars Dialog Box

Figure 5–34   Summarize All Dialog Box

summarize all levels of a grouping. **RESP** was chosen in Figure 5–34 in the **Summarize to** field. Figure 5–35 is the result of the summarization process for the Sample Schedule by responsibility. It shows a summary bar for all carpentry crew activities (summarized by responsibility) and necking for nonwork periods (weekends). The options of the **Summarize to** field are whatever project groupings are defined in **Activity Codes,** from the **Define** main pull-down menu. To unsummarize the onscreen bar chart and return it to its original format, select **None** in the **Summarize to** field, and click on the **OK** button.

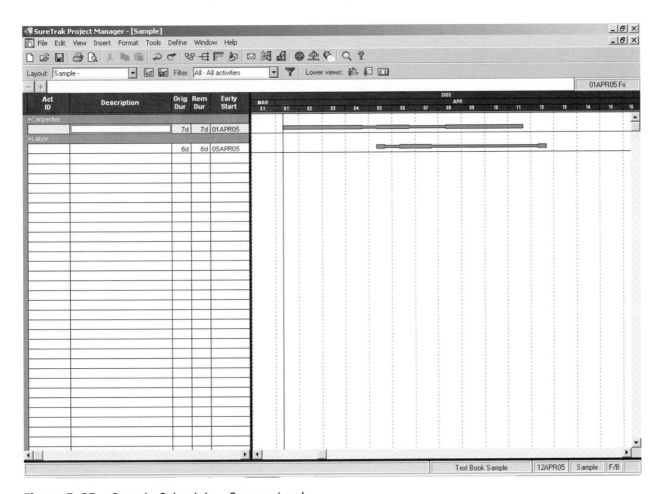

Figure 5–35   Sample Schedule—Summarized

FILTER

Sometimes it can be useful to filter the project activities and show onscreen and on hard copy printouts only a portion of the activities sorted by some identifier. This way, instead of looking at hundreds of detail activities, project management personnel can look at only the activities meeting the sort criteria. For example, you may want to look at only Carpentry responsibility of the Sample Schedule. Click on the **Format** main pull-down menu and select **Filter** (see Figure 5–5). The **Filter** dialog box appears (Figure 5–36). The built-in predefined filter options are:

ID	Description
All	All Activities
1WK	One week lookahead
3MON	Three month lookahead
4WK	Four week lookahead
CNST	Activities with constraints
CRIT	Critical activities
CST>	Cost is greater than?
DLAY	Delayed activities
DONE	Completed activities
INDP	Independent activities
MEET	Meeting activities
MILE	Milestone activities
NEAR	Near critical activities
NET+	Positive net income
NET−	Negative net income
NONE	Activities with no progress
OVER	Overbudget activities
PROG	Work in progress
RES	Activities for resources
STRT	Activities that start between? and?
TASK	Task activities
TODO	Activities underway or with no progress
TOPX	Topic activities
XTOP	Nontopic activities

The TOPX or topic activities (just one of the sample filters that have been predefined for *SureTrak* users) lets you define the filter criteria of the activities to be displayed. Select the **Topic activities (TOPX)** filter and click on the **Modify** button of the Filter dialog box. The **Filter Specification** dialog box appears (Figure 5–37). Here again, you can filter by multiple levels of criteria. Under **Level 1 must meet,** the + button was selected to establish a criterion to filter the activities. The filter options available are the same as those available in the column data options, and the possible information fields are added to the bar chart bar. Figure 5–37 shows the **Select if**

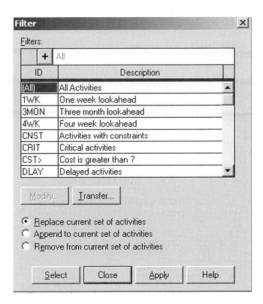

Figure 5–36 Filter Dialog Box

option of **RESP - Responsibility** was selected. Under **Is** the **Equal to** option is selected, and under **Low value, CARP** is selected. This means that from all the possible activities, you are filtering the activities that meet the criterion of having CARP identified as **RESP - Responsibility.** Click on **OK** in the **Filter Specification** dialog box, and **Apply** in the **Filter** dialog box. Figure 5–38 is the result. The advantage of giving your carpentry foreman a printout containing only his information is that the foreman does not have to waste time sifting through all the other information.

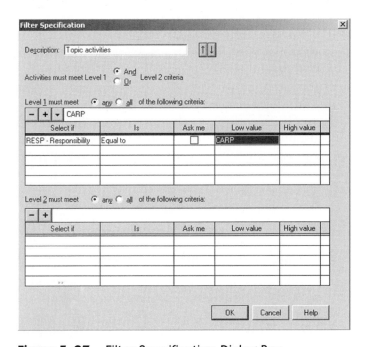

Figure 5–37 Filter Specification Dialog Box

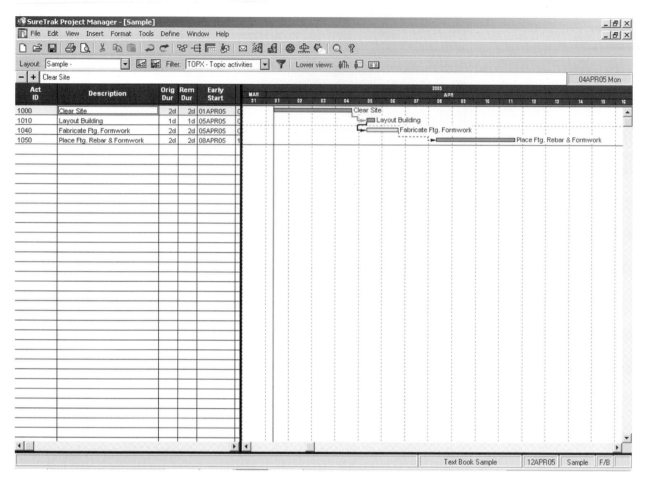

Figure 5–38   Sample Schedule—Filtered by Responsibility

Compare Figure 5–38, which shows the activities filtered by the CARP responsibility, to Figure 5–32, which shows the activities organized by responsibility, to Figure 5–4, where the activities are only sorted by activity ID. Again, being able to format the *SureTrak* onscreen bar chart and hard copy printouts to sort, organize, and filter activities is a powerful communications tool that makes the scheduling information more usable by all parties involved in the scheduling process. Note that you can edit any of these filters or even create your own.

CLIP ART

In Chapter 8, we will discuss putting images in the Logo field of the Title Block. The clip art (object/picture) discussed in this section is for inclusion in the body of the onscreen bar chart or the hard copy printout.

A powerful feature provided by *SureTrak* for modifying the onscreen bar chart image is the ability to include clip art on the screen or hard copy

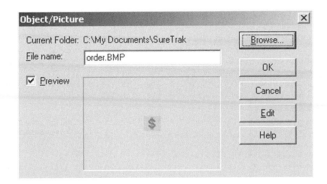

Figure 5–39   Object/Picture Dialog Box

printout. The image can be turned on or off depending on whether you want to see it or not. To add a clip art image, click on **Insert** in the main pull-down menu and select **Object/Picture.** The **Object/Picture** dialog box appears (Figure 5–39). Notice that when a file is selected (order.BMP in Figure 5–39), the preview screen at the bottom of the dialog box shows a preview of the image. When you click on the **Browse** button the options under **Files of Type** are:

- Bitmaps (*.bmp)
- Metafiles (*.wmf)
- Microsoft Word Files (*.doc)
- Microsoft Excel Files (*.xl*)
- Video for Windows Files (*.avi)
- Sound Files (*.wav)
- CorelDRAW Files (*.cdr)
- Visio Files (*.vsd)
- All Files (*.*)

Click on the **OK** button to accept the image and return to the onscreen bar chart (Figure 5–40). Note the small dark boxes surrounding the clip art image. Place the mouse on one of these boxes to turn it into a two-way arrow that can be used to increase/decrease the size of the image. Also, by clicking the mouse anywhere inside the image, and dragging/ dropping the mouse button, the image can be moved to any location.

TEXT

Another feature provided by *SureTrak* for modifying the onscreen bar chart image is the ability to add text that is not part of an activity on the screen or hard copy printout. The added text block can be turned on or off depending on whether you want to see it or not. To add a text block, click on the **Insert** main pull-down menu and select **Text/Hyperlink.** The **Text/Hyperlink** dialog box appears (Figure 5–41). Type the text you want

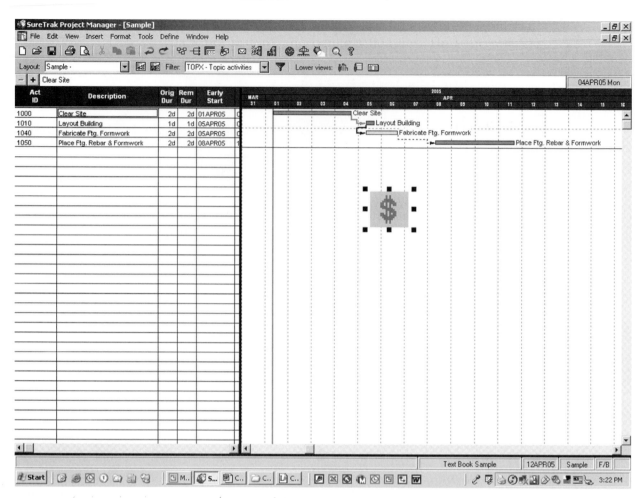

Figure 5–40 Sample Schedule—Clip Art Included

in the **Text** field. The **Font** and **Text color** down arrows can be used to change the appearance of the onscreen text image. Click on the **Show box** checkbox to show a border around the text image, and the **Box color** to change the background color of the text block. Click on the **OK** button to accept the text block and return to the onscreen bar chart (Figure 5–42).

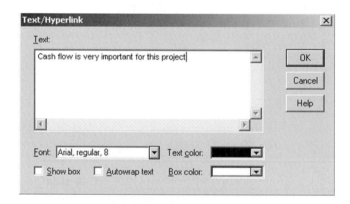

Figure 5–41 Text/Hyperlink Dialog Box

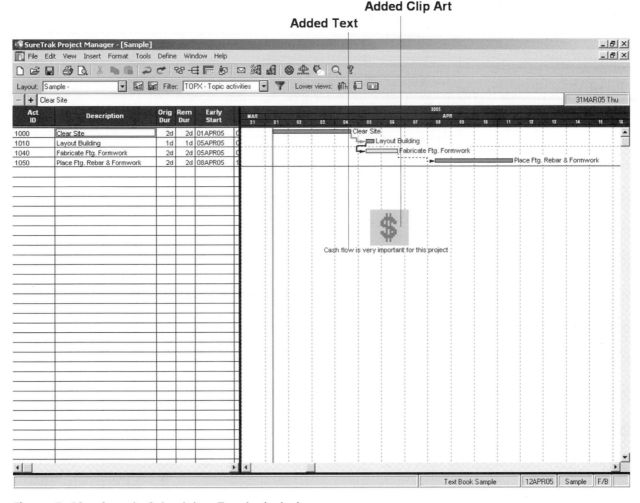

Figure 5–42   Sample Schedule—Text Included

The resizable block of text can be dragged anywhere on the onscreen bar chart or linked to any activity. Once you insert a text block, you can click on the box to select it, and either drag it to reposition it or drag on any corner or edge to resize it. You can also choose the **Insert** main pull-down menu and select **Attach Object** to attach the text block to any activity.

You can turn off (hide) either the clip art or text block additions to the onscreen bar chart and hard copy printouts without deleting the files. Click on the **Insert** main pull-down menu and select **List Objects.** The **List Objects** dialog box appears (Figure 5–43). Click on the **Hide all objects** checkbox at the bottom of the dialog box to temporarily turn off (hide) all objects. Click on the **Disable all objects** checkbox at the bottom of the dialog box to temporarily "park" the objects at their current locations so that they cannot be dragged elsewhere. Notice that the two objects just created (one clip art file and one text block) are listed under **Objects.** To modify either object, click on it. If you want to delete it, click

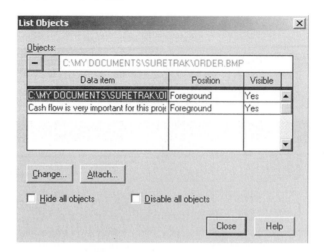

Figure 5–43 List Objects Dialog Box

on the - button; if you want to render it temporarily invisible, select **No** under **Visible**. Another option is to make the object **Foreground** (on top of) or **Background** (below) bars, relationship lines, labels, and other onscreen and hard copy printout bar chart items.

SUMMARY

This chapter provided an overview of formatting the onscreen bar chart within *SureTrak*.

The scheduler uses the Layouts SureTrak formatting feature to keep multiple configured formats of a particular schedule available for use. The Zoom feature allows the scheduler to modify the magnification of the onscreen bar chart to control the amount of the schedule exposed to view. The Columns formatting feature allows the scheduler to select and format the data that will appear in *SureTrak*'s column fields.

The Bars formatting feature allows the scheduler to select and format the data that will appear in *SureTrak*'s bars. The Relationship Lines formatting feature allows the scheduler the ability to make display and format relationship lines for displaying schedule logic. The Sight Lines formatting feature allows the scheduler to make the onscreen and hard copy *SureTrak* output easier to interpret by using horizontal and vertical lines.

The Row Height formatting feature allows the scheduler to vary the onscreen activity row height between selected or all activities. The Screen Colors formatting feature allows the scheduler to vary the onscreen and hard color scheme of the schedule for visual impact.

The Activity Codes formatting feature allows the scheduler great latitude in sorting and displaying schedule information. The Organize formatting feature allows the scheduler to select how *SureTrak* will sort and display the schedule information. The Summarize formatting feature allows the scheduler to summarize numerous activities into a single activity and then reverse the process when desired. The Filter formatting feature allows the scheduler to show only certain activities chosen by some selection criteria.

The Clip Art formatting feature allows the scheduler to insert art into the schedule to make a point. The Text feature can also be used to make the schedule communication better by providing customized printouts independent of the activities.

All of these *SureTrak* formatting features enable the scheduler to customize the schedule to maximize the information communicated and the visual impact of the schedule.

EXERCISES

1. **Selecting Layouts**
   Start *SureTrak* and open the project created in Chapter 4 (Exercise 3) named ComBldgSouth. Change the layout to DAY—Organized by Day.

2. **Changing Zoom Magnification**
   Modify the Zoom feature of the onscreen *SureTrak* bar chart created in Exercise 1 of this Chapter.

3. **Inserting New Columns**
   Modify the Columns feature of the onscreen *SureTrak* bar chart created in Exercise 2 of this Chapter by inserting a new column named Field Comments.

4. **Modifying Activity Bars**
   Modify the Bars feature of the onscreen *SureTrak* bar chart created in Exercise 3 of this Chapter by increasing the size to 10.

5. **Changing the Sight Lines**
   Change the Sight Lines feature of the onscreen *SureTrak* bar chart created in Exercise 4 of this Chapter by setting the horizontal minor sight lines to 2 and the vertical minor sight line to Day.

6. Create Activity Codes feature for the onscreen *SureTrak* bar chart created in Exercise 5 of this Chapter.

7. Use the Organize feature for the onscreen *SureTrak* bar chart created in Exercise 6 of this Chapter.

8. Use the Summarize feature for the onscreen *SureTrak* bar chart created in Exercise 7 of this Chapter.

9. Include Clipart and text for the onscreen *SureTrak* bar chart created in Exercise 8 of this Chapter.

10. Modify the **Small Commercial Concrete Block Building OnScreen Bar Chart** Using the onscreen bar chart for the small commercial concrete block building created in Exercise 12 of Chapter 4, use the following *SureTrak* features to modify its appearance:
- Zoom
- Columns
- Bars
- Sight lines
- Row height

11. Modify the Large Commercial Building OnScreen Bar Chart Using the onscreen bar chart for the large commercial building created in Exercise 13 of Chapter 4, use the following *SureTrak* features to modify its appearance:
- Screen colors
- Activity codes
- Organize
- Summarize
- Filter
- Clip art
- Text

Resources

Objectives

Upon completion of this chapter, you should be able to:

- Describe the necessity of controlling resources
- Create a new resource dictionary
- Assign resources to activities
- Display resource profiles
- Display resource tables
- Modify the appearance of resource profiles/tables
- Define driving resources
- Set resource limits
- Level resources

NECESSITY OF CONTROLLING RESOURCES

The contractor must be able to control resources—labor hours, bulk materials, construction equipment, and permanent equipment. The ability to get the greatest bang for the buck in putting the resources in place usually determines a contractor's success. Control of resources involves:

- Maximizing resources
- Paying attention to cost and time
- Controlling waste
- Paying attention to detail
- Preplanning
- Paying attention to efficiencies

Resource loading the schedule (defining the people, materials, and equipment you plan to use) is an effective way to control resources. Loading the labor resources identifies (weeks in advance) the exact activities to be worked on as of a particular day as well as the crafts and number of workers per craft that are required. Knowing the subcontractors' labor plans is also an effective way to measure subcontractor performance.

Controlling resources involves resource limits and leveling. A *limit* is the maximum amount of a resource available at one time. If a person serves a unique function on a project, the limit is one. If five workers of one type are available at any one time, then the limit is five. Obviously, resource limitations impact the scheduling of the project and the relationship of activities to each other.

Resource *leveling* is the redistribution of resources to eliminate resource conflict. If a resource is overallocated (more is assigned than is available for a given time period), *SureTrak* can be directed to reschedule the activities (modify the activity logic) so that the resources are not overcommitted (the resource does not exceed its limit).

Remember that time is money. A contractor who uses only a cost system to control resources is not getting the complete picture in the effective use of the resources. The entire scheduling process should be looked at from the point of view of efficient use of time, money, and resources.

DEFINE RESOURCES AND REQUIREMENTS

The scheduler has a great deal of flexibility in determining the resource requirements of a project. Resources are typically thought of as labor, material, or construction equipment. *In this chapter, we analyze only direct labor.* We begin by defining the labor requirements of each activity, and then we evaluate the requirements of the entire project and refine it if necessary.

Defining Resources

The resources must be defined before the requirements of each activity can be addressed. Click on the **Define** main pull-down menu, and then click on the **Resources** option (Figure 6–1). The **Define Resources** dialog box appears (Figure 6–2).

Transfer

There are two ways to define the resources in the dictionary for your particular project. The first—and easiest—is to transfer the defined resources from another project. By clicking on the **Transfer** button, the **Transfer Resources** dialog box appears. It includes a projects directory for scanning *SureTrak* files. Select a project in the **Project name** field, and then click on the **OK** button. The defined resources for the selected project become the basis of the resources for the new project. Obviously, after you have developed a dictionary that meets the needs of the type project you intend to build, there is no need to reinvent the wheel and redefine the resource dictionary. Using a previously defined dictionary speeds up the process.

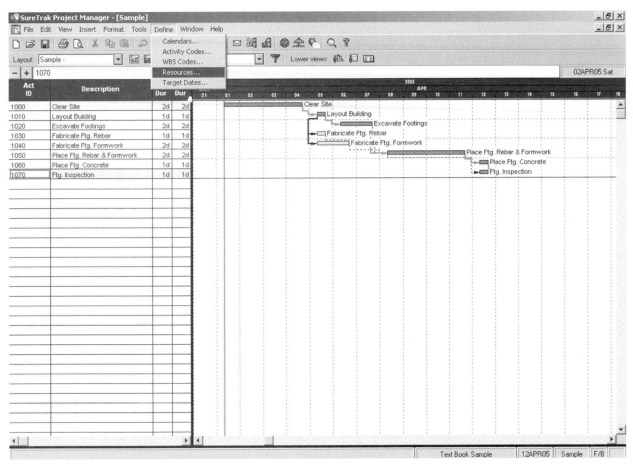

Figure 6–1   Sample Schedule—Onscreen Bar Chart

Creating a New Resource Dictionary

The second method of defining resources is to create a new resource dictionary. In this chapter, we will create a new resource dictionary instead of using **Transfer** to copy an existing dictionary from an earlier project.

Figure 6–2   Define Resources Dialog Box

Click on the + icon bar option of the **Define Resources** dialog box (Figure 6–2). The top portion of the dialog box is used to define the resources for the project.

Resource The **Resource** column shows the name (eight-character maximum) for the resource that will be assigned to the activity. Type the resource abbreviation (see CARP 1 in Figure 6–3). Press the **Enter** key to add the resource to the dictionary. Other information fields are available to define the resource.

Description This is used to enter the full description of the resource. Carpenter 1 was added in Figure 6–3.

Units Specify the unit of measure for this resource or the increment in which you assign and use it. Resources assigned in units of time (hours, days, weeks) should be measured in hours (HR). Choose your own units for materials and similar resources such as SQFT, TONS, LS, EA, and so on. Regardless of the units you input, remember that *SureTrak* calculates quantity, resource, cost, etc., fields hourly. *SureTrak* always does things in hours.

Cost How much one unit of this resource costs is entered here. This field can be left at 0.00 if costs are not generated for the resource.

Revenue How much income you expect from using or producing one unit of this resource is entered here. This field can be left at 0.00 if revenue is not generated for the resource.

Driving A *driving resource* determines the duration of an activity (the resource is "driving" the activity). In other words, for nondriving resources, the activity determines the resource duration (or work) and for driving resources, the resource determines the activity duration. A driving resource can be defined in the **Define Resources** dialog

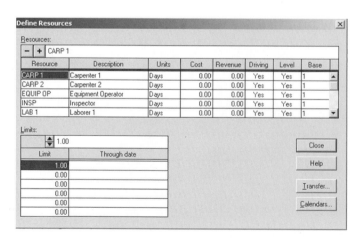

Figure 6–3 Define Resources Dialog Box—Resource Field

box or the activity **Resources** dialog box. Because a resource may not be the controlling one for all activities for which it is used, be careful when defining it in the Resource Dictionary. Additional care should be also taken whether or not the **Driving** column is set to yes or no. Changing this column's value only affects the activities this resource is assigned to after the change. In other words, changing the value of this **Driving column** will not affect the existing resource assignments.

Level Leveling is the redistribution of resources to eliminate resource conflicts. If a resource is overallocated (more is assigned than is available for a given time period), *SureTrak* reschedules the activities so that resources are not overcommitted. Indicate whether this resource should be leveled when you level resources for the project. The leveling process is initiated by selecting the **Tools** main pull-down menu, and then selecting **Level.** If the quantity of this resource is limited and that limitation can affect the duration of the activity, choose **Yes** under the **Driving** column, and define the **Limits** in the bottom half of the **Define Resources** dialog box.

Base Specify which calendar this resource is based on. A base calendar defines when its activities can be worked on. Any activities to which you assign a particular base calendar can be worked on during the workdays in that base calendar, and cannot be worked on during the nonworkdays in that base calendar. To create a base calendar, choose **Define,** and then **Calendars.**

The bottom portion of the **Define Resources** dialog box is used to define the limits of resources for the project.

Limits If a resource is to have a limit put on it (the quantity of this resource is limited and that limitation can affect the duration of the activity), select the **Resources** in the top half of the dialog box, and then set the quantity limit under **Limits.** Specify the maximum amount of the resource available at once.

Through Date If the available quantity of this resource changes during the course of the project, use this field to establish time periods.

ASSIGNING RESOURCES

When the crafts (resources) have been defined in the resource dictionary, you can specify the requirements of each activity. The **Resources** form can be accessed in one of three ways. One way is to click on the **View** main pull-down menu, click on **Activity Form,** then click on the **Details** button, and select **Resources.** The second method is to right click, select **Activity Detail,** and then click **Resources** (Figure 6–4). The third method

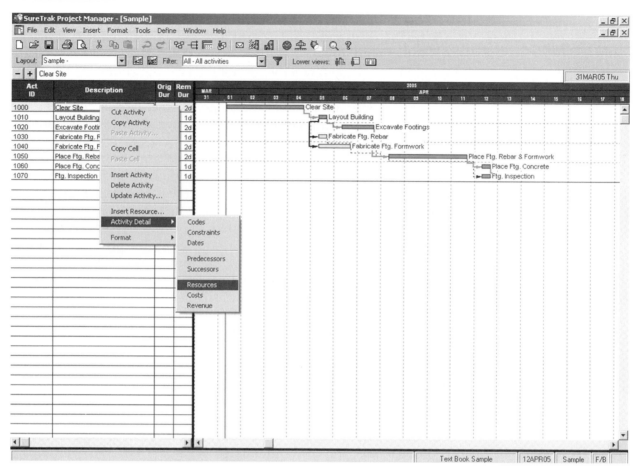

Figure 6–4   Activity Detail Option Box—Resources

is to click on the **View** main pull-down menu, select **Activity Detail,** and then click on **Resources.** See Table 6–1 for a listing of the resources and the units per day as input into the sample schedule.

Resources Dialog Box

Use the **Resources** dialog box (Figure 6–5) to specify activity resource requirements within *SureTrak*. To add a resource requirement to an activity, click on the **+** button and then the down arrow to open the resource listing as defined in the resource dictionary. Click on the desired resource. In Figure 6–5, CARP 1 (Carpenter - Level 1) is selected.

To complete the **Units per hour** field (Figure 6–6), use the up/down arrow to specify the number of units, and then press the **Enter** key. From Table 6–1, one carpenter was needed for activity 1000. Because *SureTrak* does everything in hours, you need to figure out the units per hour rate. One carpenter divided by 8 hours per day is 0.12 units per hour.

In this book, we look at production per day. The **Units** of **Days** were chosen in the **Define Resources** dialog box (refer to Figure 6–3). Keeping

Activity ID	Description	Resource	Units per Day
1000	Clear Site	CARP 1	1
1010	Building Layout	CARP 1	1
		CARP 2	2
		LAB 1	1
1020	Excavate Footings	CARP 1	1
		LAB 2	3
		EQUIP OP	1
1030	Fabricate Ftg Rebar	LAB 1	1
		LAB 2	1
1040	Fabricate Ftg Formwork	CARP 1	1
		CARP 2	1
		LAB 2	1
1050	Place Ftg Rebar & Formwork	CARP 1	1
		LAB 2	1
		EQUIP OP	1
1060	Place Ftg Conc	CARP 1	1
		LAB 1	1
		EQUIP OP	1
1070	Ftg Inspection	INSP	1

Table 6–1   Sample Schedule Resource List

it in a simple hour format is the way many *SureTrak* users use the system. If the **Units** in Figure 6–3 were stated in HR or hours, then the **Units per hour** in Figure 6–6 would be 1. This is probably easier to see, but it depends on your preference.

Notice that each time the units per hour is input and accepted, *SureTrak* automatically calculates the **Budgeted quantity** (.12 units per hour times

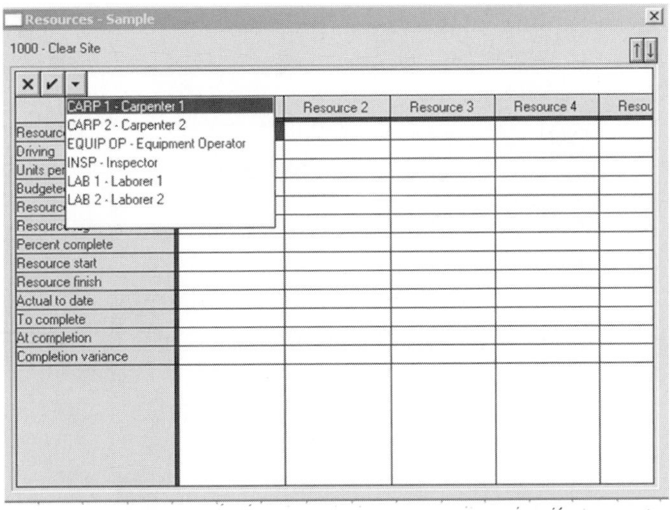

Figure 6–5   Resources Dialog Box

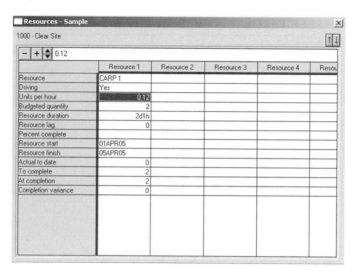

Figure 6–6   Resources Dialog Box—Resource Field

8 hours per day times a **Resource duration** of 2d or 2 days), the **To complete** (same 2 days), and the **At completion** (same 2 days) quantities. The budgeted quantity, the to complete, and at completion are always the same until progress is made on the activity; then, actual resource usage is input using these fields.

As you can see from Figure 6–7, you can specify more than one resource per activity. This figure shows the requirements for Activity 1010, Building Layout. CARP 1, CARP 2, and LAB 1 are specified.

Next Activity

Instead of closing the **Resources** dialog box after inputting the resource requirements for each activity, simply click on the next activity and a

Figure 6–7   Resources Dialog Box—Activity 1010

blank **Resources** dialog box appears for that activity. The **Resources** dialog box does not have to be closed until all resource requirements are defined. *Note: If the __Activity Form__ is active, the **Previous** and **Next** buttons can be used to scroll through the activities.*

RESOURCE PROFILES

The purpose of resource profiles and tables is to create graphical representations of the resource requirements of the project. Placing the resource profile or table under the bar chart makes visual interpretation and analysis simpler.

In the previous sections, you saw how to use the **Resources** dialog box to complete the resource dictionary and define the individual activity requirements. Now that all the information is in the system, you can use the information. To do this, look at either resource profiles or tables. To access resource profiles, click on the **View** main pull-down menu (Figure 6–8), and then click on **Resource Profile**. The Resource Profile appears at

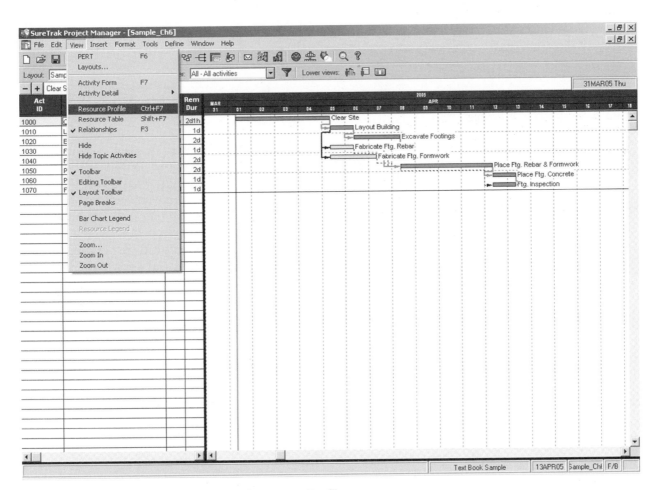

Figure 6–8   *View Pull-Down Menu—Resource Profile*

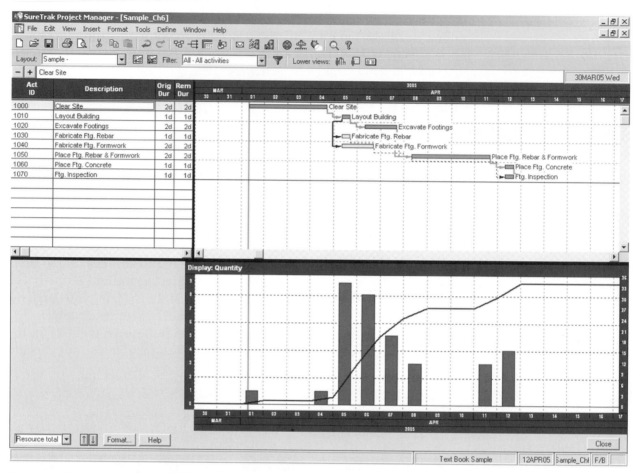

Figure 6–9 Resource Profile

the bottom of the screen (Figure 6–9). The resource profile can also be accessed by clicking on the Resource Profile button on the toolbar.

A profile is a side view or cross section. A *resource profile* is a graphical representation of resources across time. In the resource selection screen in Figure 6–9, **Resource total** is selected. All resources (all carpenters, laborers, equipment operators, and so on), as defined by the resource dictionary and on the **Resources** dialog box of the individual activities, are shown.

The profile shows:

April 01	1 craftsperson
April 02–03 (Weekend)	0 craftsperson
April 04	1 craftsperson
April 05	9 craftsperson
April 06	8 craftsperson
April 07	5 craftsperson
April 08	3 craftsperson
April 09–10 (Weekend)	0 craftsperson
April 11	3 craftsperson
April 12	4 craftsperson

Obviously, this resource profile is not flat. The wide range of needs from one to nine people in a given week may create a staffing problem. Floats can be used to shift activities around to "flatten" the peak resource requirements.

Modify Appearance

The appearance of the Resource Profile can be modified using the **Format Resource Profile/Table** dialog box. Click on the **Format** button and the **Format Resource Profile Display Options** dialog box appears (Figure 6–10). There are three tabs: **Display, Profile,** and **Table.**

The **Display** tab (Figure 6–10) has the following options.

> **Display** Click on the down arrow to select from multiple options. In this chapter on resource control, **Quantity** is selected (Figure 6–10). Quantity represents how much of a resource is currently assigned and how much has been used. Other possible choices for this field are **Costs, Revenue, Net, Earned Value, Budget, BCWS, Performance,** and **Cash Flow.**

> **Calculate** Determines how *SureTrak* calculates resource values. **T̲otal** is the sum of the resource values per time period. This is good for providing detailed resource values for each time period. **P̲eak** is the highest quantity used per hour during the selected timescale interval. This is good for checking on potential trouble spots.

> **Average** The total resource values per hour for each time period, divided by the number of hours in each time period. This shows resources "smoothed out" over time periods longer than their specific assignment. **T̲otal** was used for the resource profile in Figure 6–9.

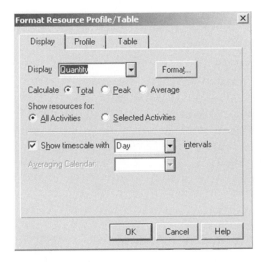

Figure 6–10   Format Resource Profile/Table Dialog Box—Display

Show resources for If **S̲elected activities** is selected, only the selected resources for the selected activities appear. If **A̲ll activities** is selected, the resource total appears and the table does not change as you scroll through the activities.

S̲how timescale with In Figure 6–10, the timescale is turned on. Click this checkbox off to remove the timescale from the resource profile. This allows you to see more resources in the table at once.

In̲terval This controls the length of the time period for which *SureTrak* calculates resources for the resource profile. Options for viewing are **Hours, Days, Weeks, Months, Quarters,** and **Years.** The default is whatever units the onscreen bar chart is configured to. To change the default, click on the down arrow, and then click on the option you prefer.

A̲veraging Calendar This field is only active if the **Average** option is chosen in the **Calculate** section on the **Display** tab of the **Format Resource Table Profile** dialog box. It lets you choose the base calendar for the averaging calculations.

The **Profile** tab (Figure 6–11) has the following options.

Sho̲w histogram Figure 6–11 is an example of the settings for a histogram showing a bar for each day. The **B̲ars** option is selected. If the **Ar̲eas** option had been chosen, the distinction between each day would not be shown—the total areas would show. Sometimes, this makes the graphic easier to understand. The **Emphasi̲ze overload with color** and **Draw l̲imits** options are discussed later in this chapter.

Show total cu̲mulative curve If this box is checked, a cumulative resource curve is superimposed on the profile histogram. The total cumulation curve was shown in Figure 6–9. The vertical scale at the far right of the histogram is the scale for the cumulative resource curve.

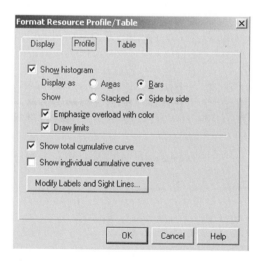

Figure 6–11   Format Resource Profile/Table Dialog Box—Profile

Show individual cumulative curves If this box is checked, individual cumulative resource curves for each resource are superimposed on the profile histogram.

Modify Labels and Sight Lines Click this button to display additional options (Figure 6–12), including **Resource profile Title** and **Maximum value for Y-axis.**

Show horizontal sight lines Mark this checkbox if you want to visually track divisions of the y-axis across the profile. *Note: In Figure 6–9, the horizontal sight lines are set for every two units on the vertical (y) axis. In this case there is a horizontal sight line between every two activities.* The spacing, color, and line type can be changed to make the interpretation of the resource profile easier.

The **Table** tab (Figure 6–13) will be discussed in the resource table section of this chapter. If you want to see all individual crafts or resources, rather than total requirements, click on the down arrow on the far left field of the resource profile (see Figure 6–9). The **Select Resources** (Figure 6–9) dialog box appears. This dialog box enables you to make a distinction between the resources in the resource profile by assigning a different color and/or pattern to each (Figure 6–14).

To look at the resource profile for a selected resource rather than all resources, use the down arrow on the far left field of the resource profile to select the resource. In Figure 6–15, CARP 1 is selected. Notice the difference in the profile when only the CARP 1 is selected rather than the total requirements. Compare Figure 6–15 with only CARP 1 selected to Figure 6–9 with Resource totals selected. The maximum requirements for CARP 1 (2) happen on April 5th, the same day as the maximum for all crafts (9).

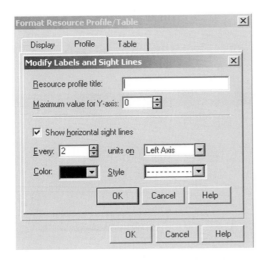

Figure 6–12   Modify Labels and Sight Lines

Figure 6–13   Format Resource Profile/Table Dialog Box—Table

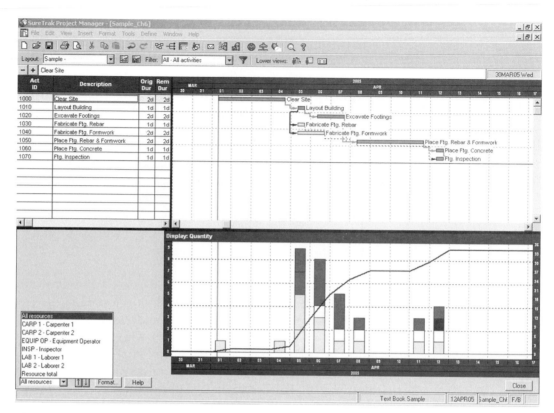

Figure 6–14 Resource Profile-All Resources

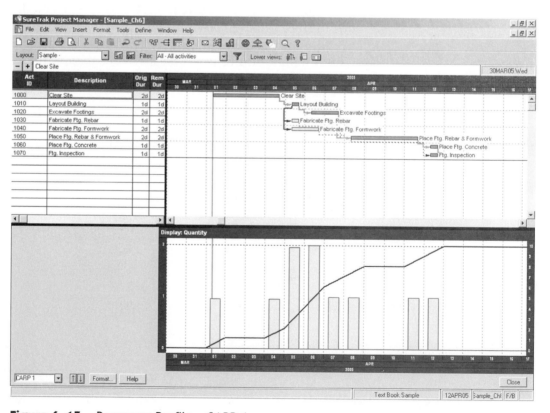

Figure 6–15 Resource Profile—CARP 1

RESOURCE TABLES

To access resource tables, click on **View** main pull-down menu (Figure 6–16), and then click on the **Resource Table.** The Resource Table appears at the bottom of the screen (Figure 6–17).

Craft Numerical Totals

The resource table in Figure 6–17 provides a numerical total for each day by craft or resource. By selecting **All resources** from the resource selection pull-down menu, all used resources appears. All used carpenters, laborers, and equipment operators as defined by the resources dictionary and specified in the resource dialog box of the individual activities are shown on the table. Any single resource can also be selected.

Labor Totals

If you want totals, rather than the breakdown by craft, choose **Resource total** from the **resource** selection pull-down menu (Figure 6–18). The peak manpower requirement is on April 5th. As you can see from Figure 6–17, this represents 2 CARP 1, 3 CARP 2, 2 LAB 1, and 2 LAB 2.

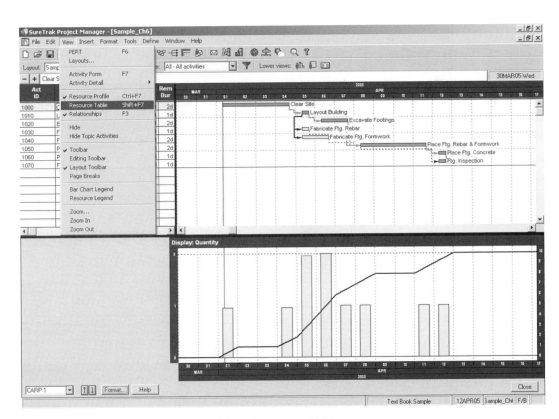

Figure 6–16   View Pull-Down Table—Resource Table

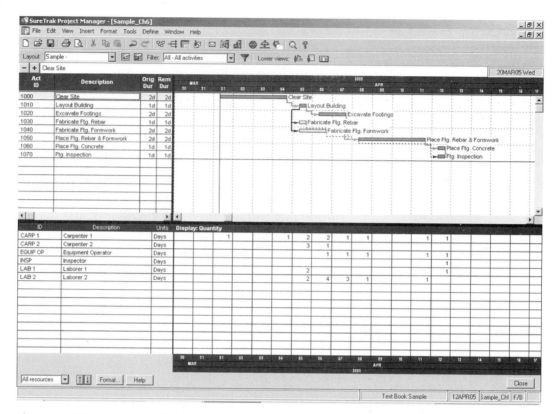

Figure 6–17   Resource Table—Detailed Resource Table

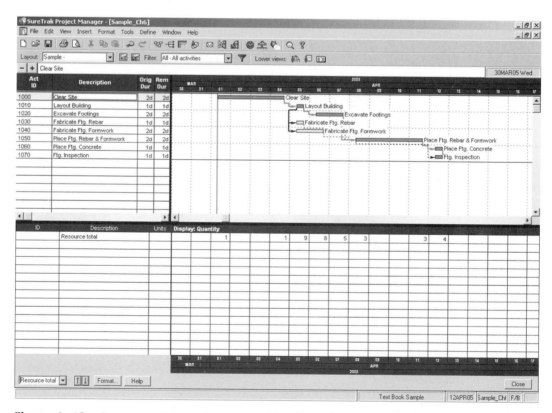

Figure 6–18   Resource Table—Resources Total

Modify Appearance

You can modify the appearance of the resource table using the **Format Resource Profile/Table** dialog box. Click on the **Format** button and the **Format Resource Profile Display Options** dialog box appears (refer to Figure 6–10). Click on the **Table** tab to view the following options (Figure 6–13):

Font Click on the down arrow to select the font type, style, and size of the resource table.

Decimals Allows you to specify the number of decimal places that will be shown in the resource table.

Show column totals When this option is selected, the column totals are shown in the resource table.

Show row totals When this option is selected, the row totals are shown in the resource table.

DRIVING RESOURCES

A *driving resource* is a resource that determines the duration of the activity to which it is assigned. Remember, you can have more than one driving resource assigned to an activity. *SureTrak* automatically calculates the activity duration based on the quantity to complete (amount of work) and the units per time period (productivity rate) of the driving resource. If an activity has more than one resource, the driving resource may need to be established. You can establish the driving resource by selecting **Yes** for the **Driving** field from the **Resources** dialog box (Figure 6–19).

SET RESOURCE LIMITS

The bottom portion of the **Define Resources** dictionary dialog box is used to set limits on the availability of a resource (see Figure 6–20). There are two columns under **Limits**:

Limit Specify the maximum or highest availability of the resource at one time.

Through date Specify time intervals for entering cut off dates.

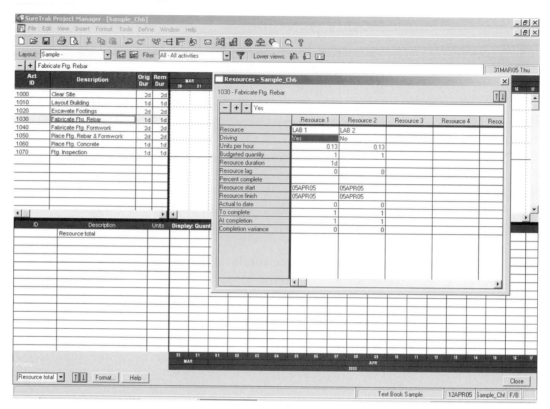

Figure 6–19 Resources Dialog Box—Driving

Limit per Unit of Measure

Limits relate to the units used to define the resource. Use a meaningful code of up to four characters, such as HRS (hours), EA (each), or CY (cubic yards). The unit of measure for LAB 1 is DAYS. Because the project being scheduled is on a daily basis, a better choice would have been EA (each). Entering .12 (1 LAB 1 divided by 8 hours per day or .12 per hour) for the limit means that only one LAB 1 is normally available to the project. Any

Figure 6–20 Define Resources Dialog Box—Driving

combination of activity requirements requiring more than one LAB 1 per day (or .12 per hour) would be an overload and a possible problem.

Overload

Resources that go over the limit are easy to identify in *SureTrak*. When the resource table is engaged (click on the **Resource Table** from the **View** main pull-down menu), the resource that is over the limit appears in a different color (red) from the other resources (black) (Figure 6–21). This color change only happens when the resource table time interval (here it is days) matches the project planning unit (also days and set when the project was created). Activity 1020, Excavate Footings, and Activity 1030, Fabricate Ftg Rebar, both require a LAB 1. Both activities are scheduled for April 05. This requires the services of two LAB 1s. Because we have established a limit of one, this is an overload.

Another View

Another way to view the requirements for a resource is to view the resource profile. Click on **Resource Profile** from the **View** main pull-down menu (Figure 6–22). Select LAB 1 by clicking the down arrow on the resource selection pull-down menu. To reconfigure the profile, click on **Format**. The **Format Resource Profile/Table** dialog box appears (Figure 6–22). Click on the **Profile** tab and check the **Draw limits** check-box. As Figure 6–22 shows, the limits are shown at one.

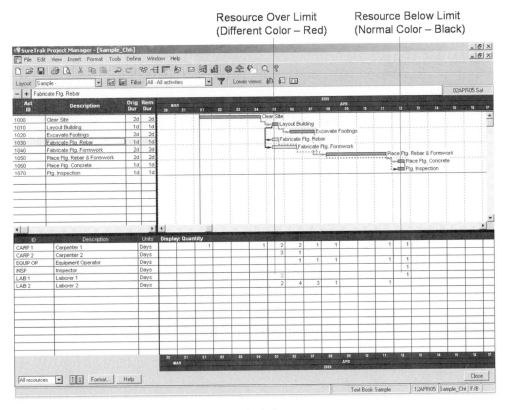

Figure 6–21   Resource Table—Over Limit Resource

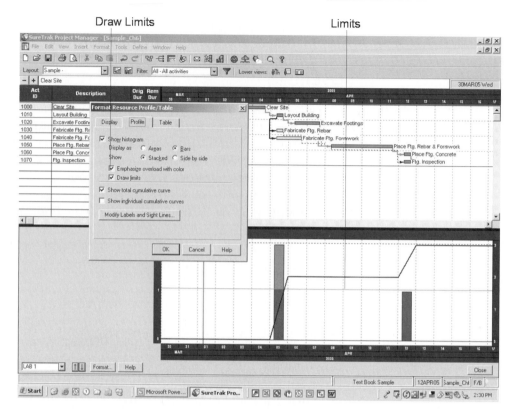

Figure 6–22 Resource Profile—Resources Limits

LEVEL RESOURCES

In Figure 6–22, the normal limit for LAB 1 is exceeded. Rearranging priorities of the project so that the resource limit will not be exceeded is called *resource leveling*. Resource leveling means trying to take the peaks out of the profile to lower the overall need for a resource, thus lessening the problems associated with mobilizing and demobilizing personnel. Of course, the first option in trying to level resources is to use positive floats to rearrange activities without prolonging project duration.

Leveling Resources To begin *SureTrak's* leveling process, click on the **Level** option on the **Tools** main pull-down menu (Figure 6–23) to open the **Level** dialog box (Figure 6–24).

Dialog Box for Resource Leveling

The following options appear in the Level dialog box:

Automatic resource leveling Click on this box if you want *SureTrak* to level resources every time it calculates the schedule.

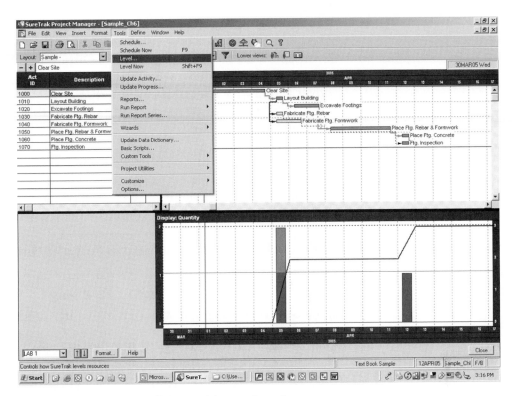

Figure 6–23   Tools Pull-Down Menu—Level

Level can extend project finish Click on this box if you want to level resources even if it delays the schedule. If you leave it unchecked, *SureTrak* will level resources only within the current calculated project finish date.

Prioritization The **Data Item** identifies priorities in the leveling process. The default is Total Float and Late Start, but you may change these priorities. Under the **Order** column, if **Ascent** is selected, the most critical activities are leveled first. If **Descend** is selected, the least critical activities get resources first.

Figure 6–24   Level Dialog Box

You assign priority to activities to tell *SureTrak* how to resolve resource conflicts when you level resources. For example, suppose you use Total Float and Remaining Duration as the priority-defining items. Activities with the least amount of total float will take priority over those with a lot of float. Activities with equal total float will get resources in order of remaining duration (those scheduled to finish first get resources first).

<u>R</u>esources If this button is clicked, the **Define Resources** dialog box appears (see Figure 6–3) and you can change which resources are levelable or nonlevelable. If only certain of the resources are to be leveled, you can select those resources.

OK Click on the **OK** button to initiate the leveling process by *SureTrak*.

Results

Figure 6–25 displays the results of the leveling process by *SureTrak*. Compare Figure 6–25 with Figure 6–23 to see the changes in the resource leveling exercise. Notice that of the activities requiring LAB 1, Activity 1030 was the only activity moved, and that the project duration was not modified. Activity 1030 had only one day of float (given our original logic). On a larger project with more activities, relationships, and floats, a lot of manipulation can take place with or without extending project duration.

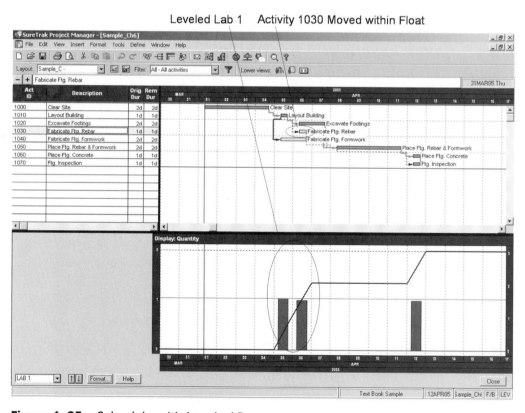

Figure 6–25 Schedule with Leveled Resources

EXAMPLE PROBLEM

Table 6–2 is an Activity List with labor resources for a house as an example for your use (see wood frame house drawings in the Appendix). The Wood Frame House—Example Problem was first shown in Figure 1–3 of Chapter 1. The activity descriptions and IDs were provided in Table 2–1 of Chapter 2. The durations and relationships were provided in Table 4–2 of Chapter 4.

Figure 6–26 shows all activity resource requirements for the Wood Frame House—Example Problem. The onscreen total resource profile (Figure 6–27) was constructed using the tabular list of twenty-eight activities (activity ID, description, and labor classification). Figure 6–28 is a resource profile for the CARPENTR craft. Figure 6–29 is a detailed resource table and Figure 6–30 is a total resource table.

Activity ID	Activity Description	Labor Classification
1000	Clear Site	Sub
1010	Building Layout	2-CARPENTR, 1-CRPN FOR, 1-LAB CL 1, 1-LAB CL2
1020	Form/Pour Footings	2-CARPENTR, 1-CRPN FOR, 1-LAB CL 1, 1-LAB CL2
1030	Pier Masonry	2-MASON, 1-LAB CL 1, 2-LAB CL2
1040	Wood Floor System	2-CARPENTR, 1-CRPN FOR, 2-CRPN HLP
1050	Rough Framing Walls	2-CARPENTR, 1-CRPN FOR, 2-CRPN HLP
1060	Rough Framing Roof	2-CARPENTR, 1-CRPN FOR, 2-CRPN HLP
1070	Doors & Windows	2-CARPENTR, 1-CRPN FOR, 2-CRPN HLP
1080	Ext Wall Board	2-CARPENTR, 1-CRPN FOR, 2-CRPN HLP
1090	Ext Wall Insulation	Sub
1100	Rough Plumbing	Sub
1110	Rough HVAC	Sub
1120	Rough Elect	Sub
1130	Shingles	Sub
1140	Ext Siding	Sub
1150	Ext Finish Carpentry	2-CARPENTR, 1-CRPN FOR, 2-CRPN HLP
1160	Hang Drywall	Sub
1170	Finish Drywall	Sub
1180	Cabinets	Sub
1190	Ext Paint	Sub
1200	Int Finish Carpentry	2-CARPENTR, 1-CRPN FOR, 2-CRPN HLP
1210	Int Paint	Sub
1220	Finish Plumbing	Sub
1230	Finish HVAC	Sub
1240	Finish Elect	Sub
1250	Flooring	Sub
1260	Garding & Landscaping	Sub
1270	Punch List	

Table 6–2 Activity List with Labor Resources—Wood Frame House

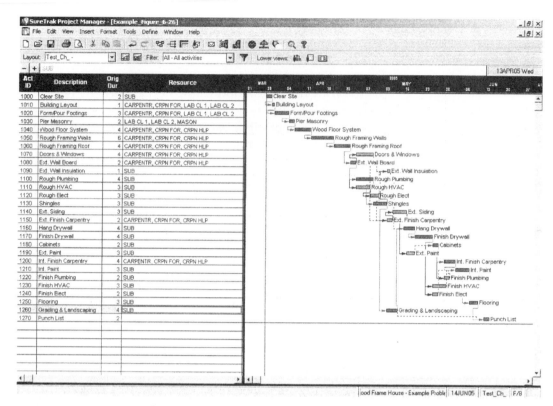

Figure 6–26 Schedule with Resources—Wood Frame House

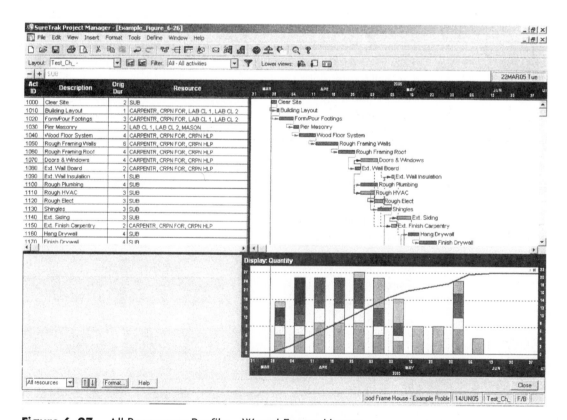

Figure 6–27 All Resources Profile—Wood Frame House

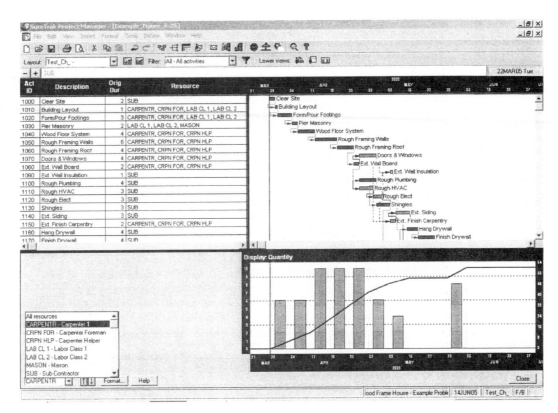

Figure 6–28   CARPENTR—Wood Frame House

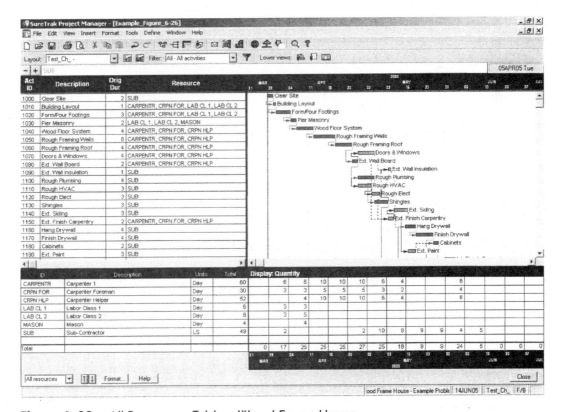

Figure 6–29   All Resources Table—Wood Frame House

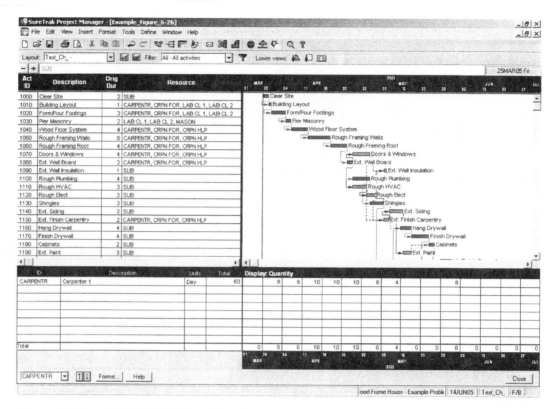

Figure 6–30   CARPENTR Table—Wood Frame House

SUMMARY

The contractor must be able to control and expend resources to complete construction projects. This chapter provided an overview of using the schedule and *SureTrak* to control resources.

Construction resources can be labor, material, or equipment. These resource definitions must be defined to *SureTrak* before they can be associated to a particular activity. This defining of a resource can include its description, unit of measurement, cost, limits, and whether it is revenue producing, driving, and leveled.

After the resources have been defined to *SureTrak*, the next step is to determine and assign the resource requirements of the activities. An activity can have multiple resources assigned to it with each resource defined by units per hour, budgeted quantity, resource duration, percent complete, resource start, resource finish, actual to date, to complete, and a completion variance.

SureTrak resource profiles and tables are used to analyze and evaluate the project resource requirements. Resource profiles provide a graphical representation whereas tables provide the same information in tabular format.

Defining which activity resources are driving, setting resource limits, and then leveling the schedule based upon resource requirements are the scheduler's tools in allocating and controlling the use of resources and time to complete the activities.

Again, any construction project involves the use of resources in the placement of resources. Having the capability to tie resources to the schedule and analyze this information for maximum efficiency is one of the major benefits of using *SureTrak*.

EXERCISES

1. **Describing the Need for Controlling Resources**
Prepare a summary describing the necessity of controlling resources.

2. **Defining Resources**
Open the project modified in Exercise 5 of Chapter 5 and define the labor listed in Table 6–3.

Resource	Description	Unit
LAB 1	Laborer 1	Days
LAB 2	Laborer 2	Days
CARP 1	Carpenter 1	Days
ELEC 1	Electrician 1	Days
PLUMB 1	Plumber 1	Days
FORE 1	Foreman	Days

Table 6–3 Labor for the Small Commercial Building in the South

3. Assigning Resources to Activities

Assign the labor to the activities of the project modified in Exercise 5 of Chapter 5 as listed in Table 6–4.

Activity ID	Description	Resource	Units per Day
1000	A	LAB 1	1
1010	B	LAB 1	2
		CARP 1	1
		FORE	1
1020	C	LAB 2	3
		ELEC1	1
		FORE	1
1030	D	PLUMB 1	2
1040	E	LAB1	1
		FORE	1
1050	F	LAB 1	1

Table 6–4 Resource List for the Small Commercial Building in the South

4. Displaying Resource Profile and Table

Display the resource profile and table of All Resources. Display the resources profile of only Laborer 1.

5. Modifying the Appearance of Resource Profiles and Tables

Display the profile of All resources with bars Stacked and then bars Side by side.

Complete a manual resource profile and resource table for each precedence diagram for Exercises 6 through 11. The labor resources provided in these diagrams is generic.

6.

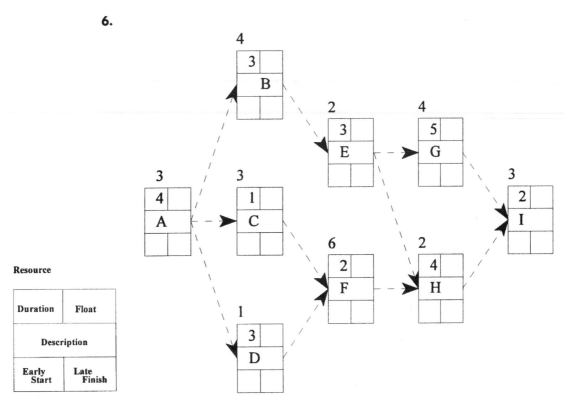

Figure 6–31   Exercise 6—Precedence Diagram with Labor Resource

7.

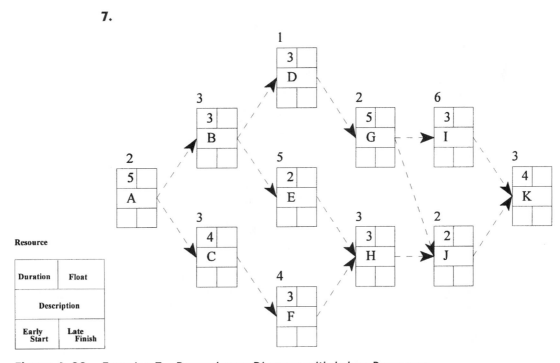

Figure 6–32   Exercise 7—Precedence Diagram with Labor Resource

8.

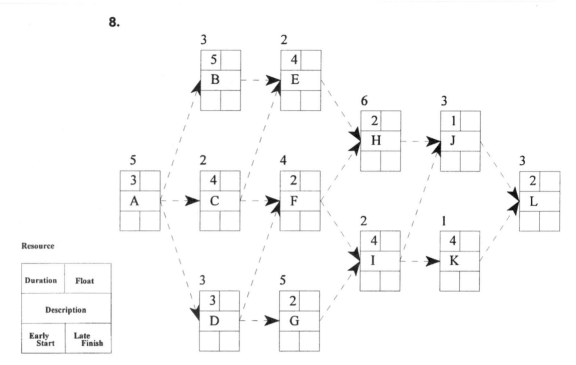

Figure 6–33   Exercise 8—Precedence Diagram with Labor Resource

9.

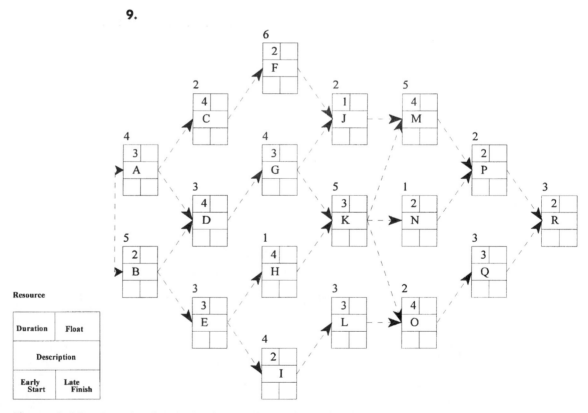

Figure 6–34   Exercise 9—Precedence Diagram with Labor Resource

10.

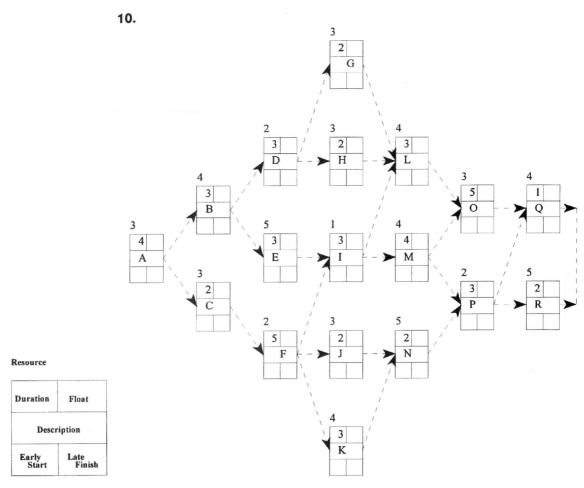

Figure 6–35   Exercise 10—Precedence Diagram with Labor Resource

11.

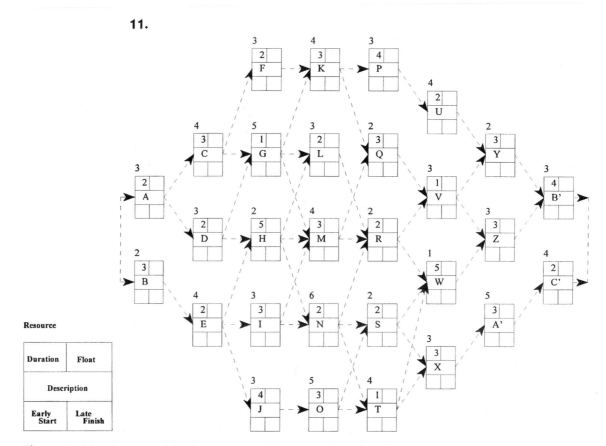

Figure 6-36 Exercise 11—Precedence Diagram with Labor Resource

12. Small Commercial Concrete Block Building—Resources
Prepare an onscreen bar chart for the small commercial concrete block building located in the Appendix. This exercise should include the following steps:
   1. Prepare a list of labor resources required by the activity.
   2. Create the onscreen resource profiles.
   3. Create the onscreen resource tables.

13. Large Commercial Building—Resources
Prepare an onscreen bar chart for the large commercial building located in the Appendix. This exercise should include the following steps:
   1. Prepare a list of labor resources required by the activity.
   2. Create the onscreen resource profiles.
   3. Create the onscreen resource tables.

7 Costs

Objectives

Upon completion of this chapter, you should be able to:

- Describe the necessity of controlling costs
- Define costs
- Display cost tables
- Assign other costs to an activity
- Change the order of the resources in the cost table
- Display cost profiles
- Display the cumulative cost curve

NECESSITY OF CONTROLLING COSTS

The most important resource needed for building is money. Scheduling and controlling the expenditure of funds is critical to the building process.

Financing

The contractor must finance the project. Having the amount necessary at the right time is a tricky business. Ultimately, the owner pays for a project through a contractor, but the contractor must make sure that money is available to finance interim periods until the monthly payment from the owner is available. With a good estimate, accurate cash projections, and intelligent distribution of funds, the contractor can finance the project without having to borrow funds. If the contractor has to borrow funds, a potential source of profit is lost in interest payments to a banker.

Conflict. Assume a contractor's bid to an owner for a new project is $10 million. The contractor never has that much money in the bank available to completely finance the project. The contractor depends on progress payments to recoup costs as the project is being put in place. Every contractor wants to finance the project completely with the owner's money; every owner wants the contractor to have some money at financial risk in the project.

Cash Needs. The contractor must be able to finance not only that particular project, but other projects he is building, and must have funds to finance the home office. Money has to be available to meet payroll costs each week for the craftspersons who work directly for the contractor. Money has to be available each month to pay for materials and supplies used at the job site and for project subcontractors. The contractor also must finance the construction equipment used at the job site, whether it is owned by the company or rented.

Timing the Expenditures.   The contractor needs to maintain cash flow to finance projects without having to borrow funds or take cash from accounts where it is earning interest. The goal is to have each project stand on its own. This requires:

- Distributing costs by activity
- Tracking costs by activity
- Controlling payment requests to owner
- Controlling labor productivity and costs
- Controlling material and supplies costs
- Controlling subcontractors
- Controlling overhead costs
- Controlling payment of funds

Owner's Requirements

The owner is as concerned about controlling the expenditure of funds as controlling time. Most project contracts call for the contractor to be paid monthly as the project progresses. The owner needs a way to make sure that the project not only is progressing toward a satisfactory completion, but also that funds are being properly spent and the contractor is not being overpaid according to the progress accomplished on the project.

Measure Progress.   The primary reason for the owner requiring a schedule is to have a document to evaluate the progress of the contractor. Usually, within a short period of time after the contract is signed, the contractor must turn over two documents to the owner. The first is the schedule and the second is the schedule of values. The *schedule of values* is a breakdown of cost by category or phase of work. When the contractor turns in a pay request at the end of the month, a judgment is made as to the percent completion of each category. Because this judgment is usually subjective, there is room for argument. Many owners now require an activity cost-loaded schedule. An activity cost-loaded schedule provides a more detailed breakdown of cost. The process of cost-loading the activities and coming to an agreement at the beginning as to their value reduces potential conflict. There is less argument at each pay period as to the percent complete because the completion of a particular activity is much easier to judge than progress in broad phases of cost.

This chapter addresses the planning of cash needs and expenditures, and the timing of the expenditures so that the contractor is in a constant positive cash flow position.

Manual Calculation. Table 7–1 shows the total cost input for each activity and the cost per day for each activity in the sample schedule. For Activity 1000, Clear Site, the cost per day is the Total Cost, $340, divided by the Duration, 2. The cost per day for this activity is $170. The total cost for the entire project is $4,519, or the summation of all activities.

Table 7–2 shows the costs according to a timescale and how the money is expected to be spent. The planned $340 expenditure for Activity 1000 is to be spent on the first ($170) and second days ($170) of the project, or April 01 and 04. According to our plan, $170 will be spent on the second day of the project, but we will have spent a cumulative total of $340 by the end of the day. This includes the $170 spent on day 1, and the $170 spent on day 2. So, we know how much will be spent at the end of each day toward the total cost of the project of $4,519. We also know that we will need $3,325 to finance the project through the end of the sixth day, April 08.

DEFINE COSTS

To define the cost within *SureTrak*, click on **Resources** from the **Define** main pull-down menu (Figure 7–1). Notice that this is the same **Define Resources** dialog box you used in Chapter 6 to define resource quantities. Now you will use it to define resource costs (Figure 7–2). As in Chapter 6, you can use the **Transfer** button to copy the resources and their associated cost from another project (Figure 7–3). If another *SureTrak* schedule has been completed and you are pleased with the way the resources/costs are defined, use the **Transfer** function. Many companies use the same or similar resources definitions for every project. It is much faster to transfer these resources to each new project than to reenter them. The **Transfer Resources** dialog box can be used to copy resources from any other project to the current project. The Resources Dictionary for the selected project completely replaces the dictionary for the current project.

Notice that in Figure 7–2, the **Cost** field for each resource is 0.00 because the cost for each resource has not been input. In Figure 7–4, the **Cost** for each resource is completed. **CARP 1** has a cost of 120.00 input or $120 per 8 hour day or $15.00 per hour. Because *SureTrak* calculates everything in hours, we need to figure out the units per hour rate. One carpenter divided by 8 hours per day is 0.12 units per hour. Again, for the purposes of this book, we chose to look at production per day. The **Units** of **Days** were chosen in the **Define Resources** dialog box (Figure 7–2). Keeping it in a simple hour format is the way many *SureTrak* users use the system.

COST TABLE

To select the cost table, click on **View** from the main pull-down menu (Figure 7–5) and click on the **Resource Table** option, to open the cost

Act. ID	Description	Duration	Total Cost	Cost/Day
1000	Clear Site	2	$340	$170
1010	Layout Building	1	$482	$482
1020	Excavate Footings	2	$1,152	$576
1030	Fabricate Ftg. Rebar	1	$188	$188
1040	Fabricate Ftg. Formwork	2	$714	$357
1050	Place Ftg. Rebar & Formwork	2	$898	$449
1060	Place Ftg. Conc	1	$505	$505
1070	Ftg. Inspection	1	$240	$240
			Total	$4,519

Table 7-1 Sample Schedule—Cost/Day

Act. ID	Description	Duration	Day 1 April 01	Weekend April 02	April 03	Day 2 April 04	Day 3 April 05	Day 4 April 06	Day 5 April 07	Day 6 April 08	Weekend April 09	April 10	Day 7 April 11	Day 8 April 12
1000	Clear Site	2	$170			$170								
1010	Layout Building	1					$482							
1020	Excavate Footings	2						$576	$576					
1030	Fabricate Ftg. Rebar	1					$188							
1040	Fabricate Ftg. Formwork	2					$357	$357						
1050	Place Ftg. Rebar & Formwork	2								$449			$449	
1060	Place Ftg. Conc	1												$505
1070	Ftg. Inspection	1												$240
	Cost/Day		$170		$0	$170	$1,027	$933	$576	$449		$0	$449	$745
	Cumulative Cost/Day		$170		$170	$340	$1,367	$2,300	$2,876	$3,325		$3,325	$3,774	$4,519

Table 7-2 Sample Schedule—Cumulative Cost

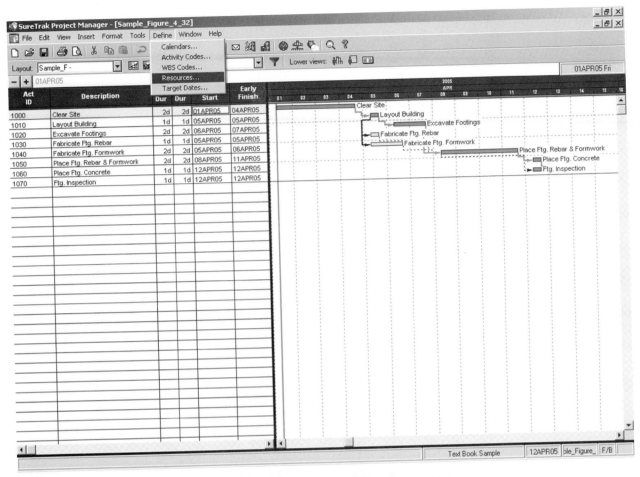

Figure 7–1   Sample Schedule—Onscreen Bar Chart—Define Resources

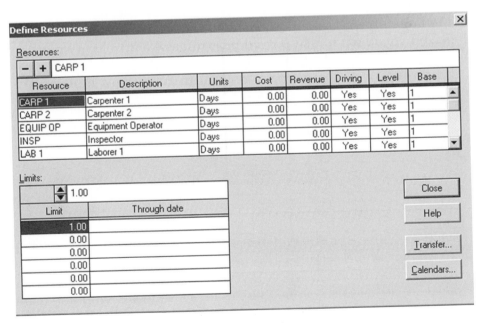

Figure 7–2   Define Resources Dialog Box—Resources

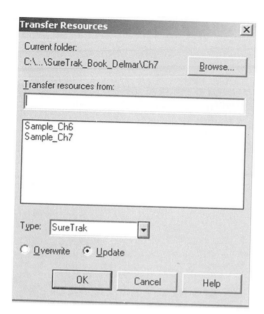

Figure 7–3 Transfer Resources Dialog Box

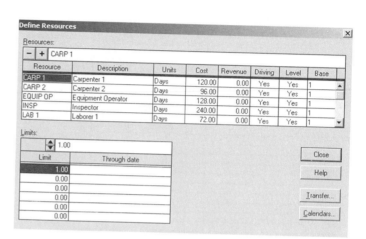

Figure 7–4 Define Resources Dialog Box—Cost

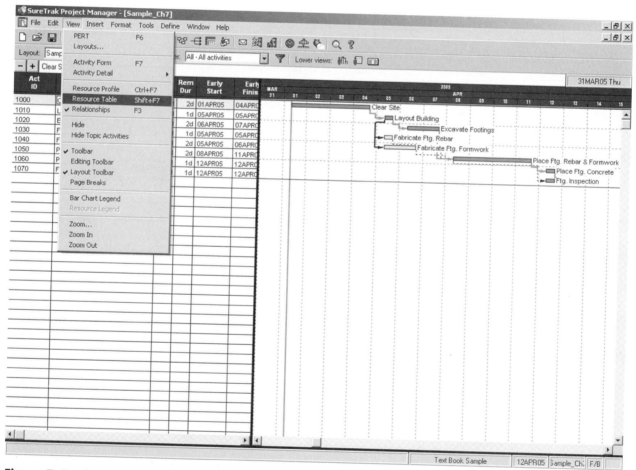

Figure 7–5 Sample Schedule—View Resource Table

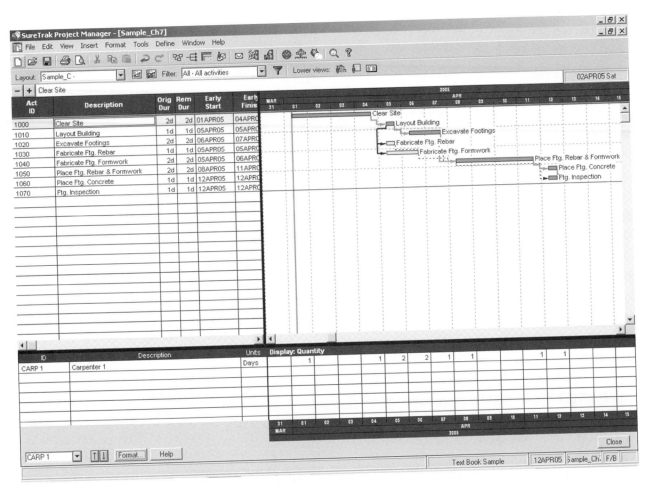

Figure 7–6 Resource Table—Quantity—CARP 1

table (Figure 7–6). You can also select the **Resource Table** button from the toolbar. The resource table in Figure 7–6 must be reconfigured because it is still formatted for quantity of resources as used in Chapter 6. Click on **Format** and the **Format Resources Profile/Table** dialog box appears (Figure 7–7). Under **Display,** select **Costs.**

Figure 7–8 is the cost table for **CARP 1.** *Note: The number of carpenters required was defined in Chapter 6 so the amount of the labor resource needed has already been inputted.* Here we are defining what the cost of the resource is per time unit. Note that the daily rate of $125 does not agree with the $120 rate input in Figure 7–4. This again is because we entered a daily cost amount and *SureTrak* is designed to calculate by the hour. The units per hour entered was 0.12 (refer to Figure 6–6) or 1 day/8 hours per day or 0.125. But *SureTrak* only has two decimal places available, hence the rounding to $125. If this rounding bothers you, you can eliminate it by entering the **Costs** dialog box for the identified activity and changing the **Budgeted cost** and **At completion** fields to 120. Figure 7–9 has **All resources** selected from the resource selection pull-down menu.

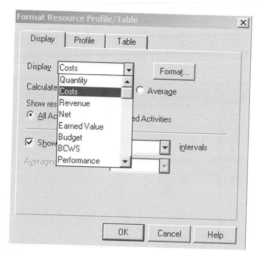

Figure 7–7 Format Resource Profile/Table—Costs

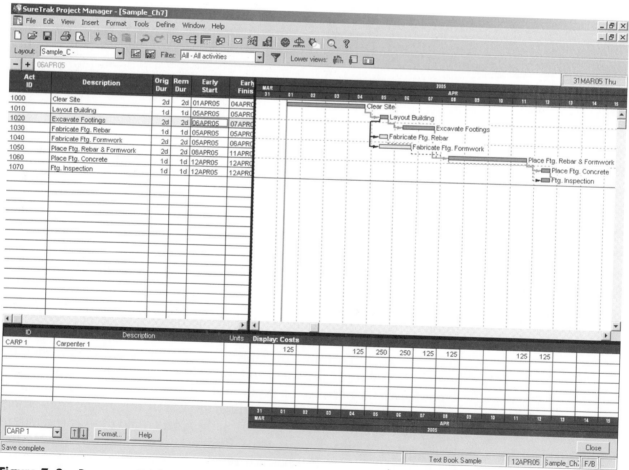

Figure 7–8 Resource Table—Costs—CARP 1

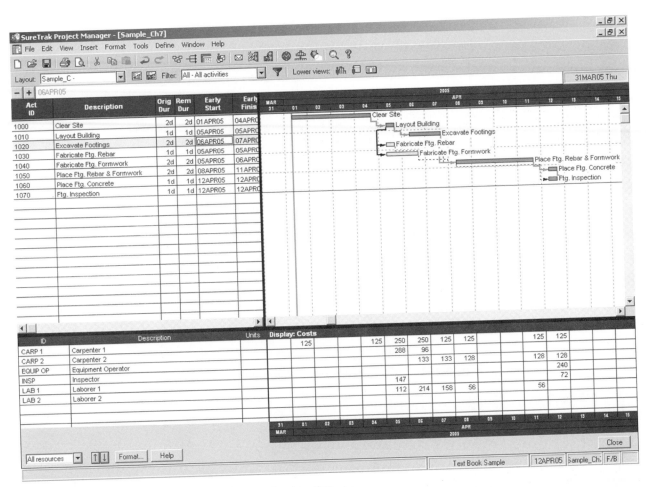

Figure 7-9 Resource Table—Costs—All Resources

ASSIGN OTHER COSTS

As you can see from Figure 7–9, only the costs for labor resources have been input into the sample schedule so far. For total project cost/schedule control, other costs—such as materials and distributable costs (project overhead and/or profit)—may need to be defined. First, these new cost resources have to be defined in the **Define Resources** dialog box. Select **Resources** from the **Define** main pull-down menu. Note in Figure 7–10 that **DIST (Distributable Costs)** and **MATL (Material Cost)** were added. The unit selected for both these new resources is **LS** or Lump Sum. The **Cost** remains at 0.00 because the budgets will be input by activity.

Now let's input costs for the two new resources by activity. In Figure 7–11, select Activity 1000, Clear Site, and right-click. Select **Activity Detail,** and then **Costs.** The **Costs** dialog box appears (Figure 7–12). In Figure 7–13, the **DIST** (Distributable cost) with a **Budgeted Cost** of $40.00 and the **MAT** (Material Cost) with a **Budgeted Cost** of $50.00 were added to the **CARP 1** resource with a **Budgeted Cost** of $250.00. When the **Budgeted Cost** and the **At Completion** cost are input, *SureTrak* automatically puts

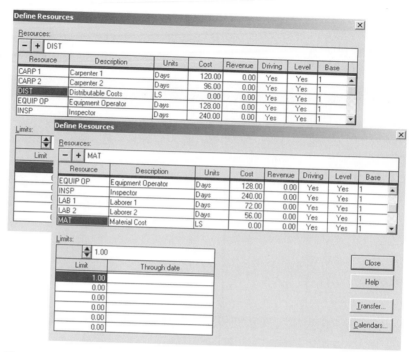

Figure 7–10   Define Resources Dialog Box—Cost

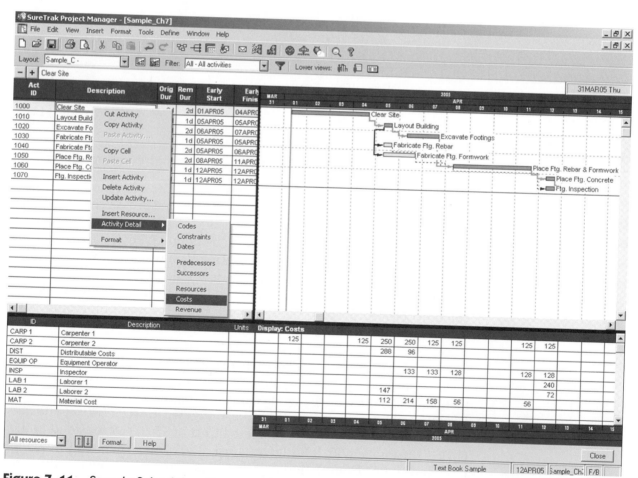

Figure 7–11   Sample Schedule—Onscreen Bar Chart—Activity Detail—Costs

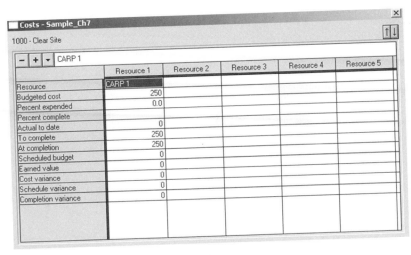

Figure 7–12   Cost Dialog Box—Resource

the same value in the **To complete** field when the information is accepted. In Figure 7–14, the MAT (Material Cost) with a **Budgeted Cost** of $60 and **DIST** (Distributable Cost) with a **Budgeted Cost** of $30 were added to the **CARP 1, CARP 2,** and **LAB 1** requirements. Compare Figure 7–15 with the new materials and distributable costs as per Table 7–3 to Figure 7–9 with only labor costs.

To change the order of the resources in the cost table, click the **Format** button, and the **Display** tab. Then click on the **Format** button and the **Resources** tab (Figure 7–16). Select the resource, and then click in the Order field. Use the up/down arrow to establish the order of the resources. For Figure 7–17, the new established order is **CARP 1** - 1, **CARP 2** - 2, **LAB 1** - 3, **LAB 2** - 4, **EQUIP OP** - 5, **INSP** - 6, **MAT** - 7, and **DIST** - 8.

Next, the timescale may need to be adjusted (Figure 7–18) so that the numbers in the table will be large enough to read or reduced to show

Costs - Sample_Ch7

1000 - Clear Site

	Resource 1	Resource 2	Resource 3	Resource 4	Resource 5
Resource	CARP 1	DIST	MAT		
Budgeted cost	250	40	50		
Percent expended	0.0	0.0	0.0		
Percent complete					
Actual to date	0	0	0		
To complete	250	40	50		
At completion	250	40	50		
Scheduled budget	0	0	0		
Earned value	0	0	0		
Cost variance	0	0	0		
Schedule variance	0	0	0		
Completion variance	0	0	0		

Figure 7–13   Cost Dialog Box—Budgeted Cost

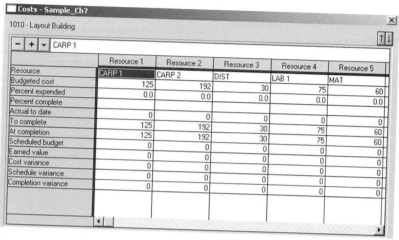

Figure 7–14   Cost Dialog Box—To Complete

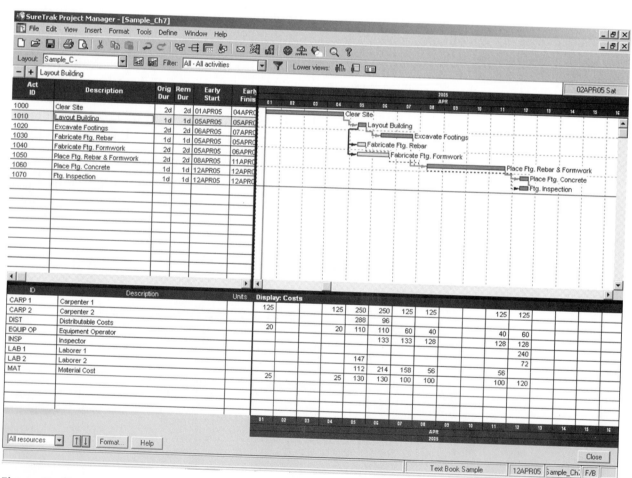

Figure 7–15   Resource Table—Costs—All Resources

Act. ID	Description	Duration	Labor Cost	Distributable Cost	Material Cost	Total Cost	Cost/Day
1000	Clear Site	2	$250	$40	$50	$340	$170
1010	Layout Building	1	$392	$30	$60	$482	$482
1020	Excavate Footings	2	$832	$120	$200	$1,152	$576
1030	Fabricate Ftg. Rebar	1	$128	$30	$30	$188	$188
1040	Fabricate Ftg. Formwork	2	$554	$100	$60	$714	$357
1050	Place Ftg. Rebar & Formwork	2	$618	$80	$200	$898	$449
1060	Place Ftg. Conc	1	$325	$60	$120	$505	$505
1070	Ftg. Inspection	1	$240	$0	$0	$240	$240
					Total	$4,519	

Table 7–3   Sample Schedule—Activity Cost

more of the schedule on the screen at one time. Double-click anywhere on the top two rows of the timescale above the bar chart to open the **Timescale** dialog box. The **Timescale density** slide bar controls the timescale spacing of days or time units on the onscreen bar chart and cost table. Click and hold the button on the slide bar and move it to the left or right.

COST PROFILES

To view the cost profile, click on the **View** main pull-down menu (Figure 7–19) and select the **Resource Profile** option. The resource profile appears at the bottom of the screen (Figure 7–20).

Configurations.   If when you view the resource profile, no usable information is included, you must reconfigure the table graphic. Click

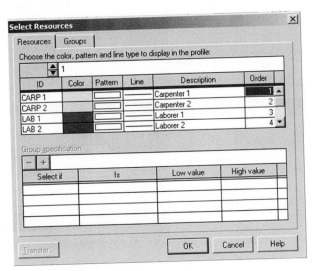

Figure 7–16   Select Resources Dialog Box

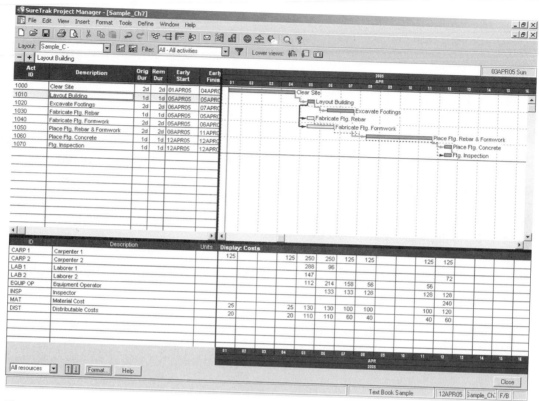

Figure 7–17   Resource Table—Costs—All Resources

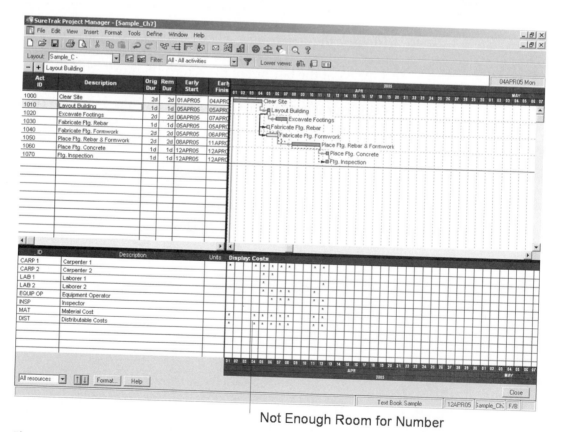

Not Enough Room for Number

Figure 7–18   Sample Schedule—Onscreen Bar Chart—View—Resource Table

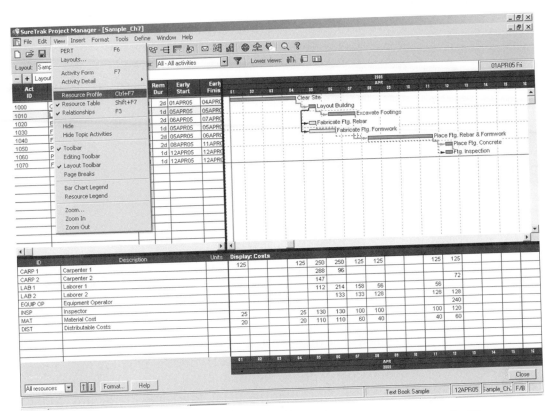

Figure 7–19   Sample Schedule—Onscreen Bar Chart—View—Resource Profile

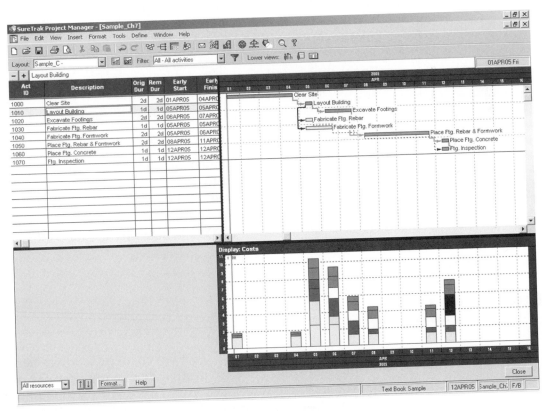

Figure 7–20   Resource Profile—Costs—All Resources

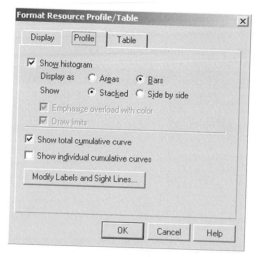

Figure 7–21   Format Resource Profile/Table—Cumulative Curve

on the **Format** button and the **Format Resource Profile/Table** dialog box (Figure 7–21) appears. Under the **Display** pull-down menu, click on the **Costs** option. You may recall that in Chapter 6 we viewed resources and selected **Quantity** using this same dialog box. Make sure the **Show histogram** option box (under the **Profile** tab) is checked. If the **Show histogram** option box is not checked, either no cost or only cumulative cost information will appear on the cost profile. Whether you use the **Areas** or **Bars** options under **Show histogram** depends on your personal preference in its appearance on the cost profile. When finished, click on **OK**.

Resources.   If the specific resource information you want still does not appear on the cost profile, use the drop-down menu on the bottom-left corner of the resource profile to change the selection of resources.

Axis.   The horizontal axis on the Cost Profile is *calendar days*. The vertical axis is *dollars*. The shaded area within the histogram represents the cost/day that is planned to be expended on your project, based on the predecessor/successor logic and the estimated cost per activity. The peak cost expended per day looks to be $1,000 (10 × 100), from Figure 7–20. The cost profile is useful, but a table with the exact dollar values per day may be even more useful.

CUMULATIVE COSTS

Cumulative Curve.   From the **View** main pull-down menu, click on the **Resource Profile**. The cost profile, displaying cost per day, appears at the bottom of the screen (Figure 7–20). Click on the **Format** button and then the **Profile** tab of the **Format Resource Profile/Table** dialog box.

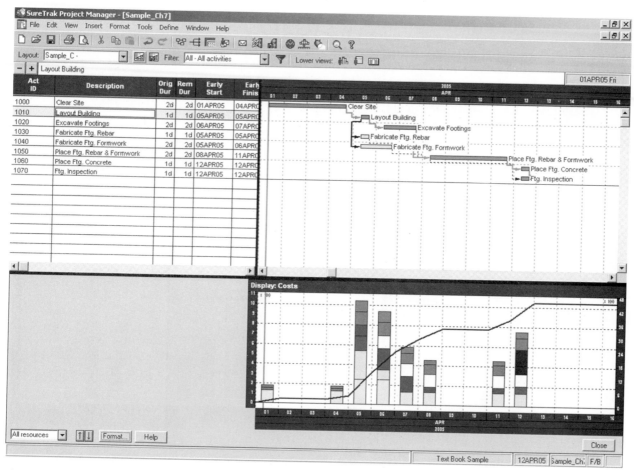

Figure 7-22   Resource Profile—Costs—All Resources—Cumulative Curve

The only change to make to the dialog box is the checkbox for **Show total cumulative curve.** As soon as the box is checked and the **OK** button is clicked, the cumulative curve appears in the cost profile screen (Figure 7–22).

Area Under the Curve.   The area under the curve represents the cumulative amount of money spent on the project to date. The vertical scale on the right of the cost profile represents the total cumulative cost for the project. The total monies anticipated to be spent (cash flow) for the project is $4,519. No money is spent until the beginning of the day on April 1. The last activity for which money is expended takes place on April 12. The anticipated monies to be spent on the project will be spent between April 1 and 12. The height of the curve shows the monies to be spent as of any particular day. Figure 7–23 is a cost profile with the cumulative curve showing for only the **DIST** or **Distributable Costs.** This change can be made by using the resource selection pull-down menu.

View.   The example project has few activities and a short duration. Obviously, the typical construction project would have many more

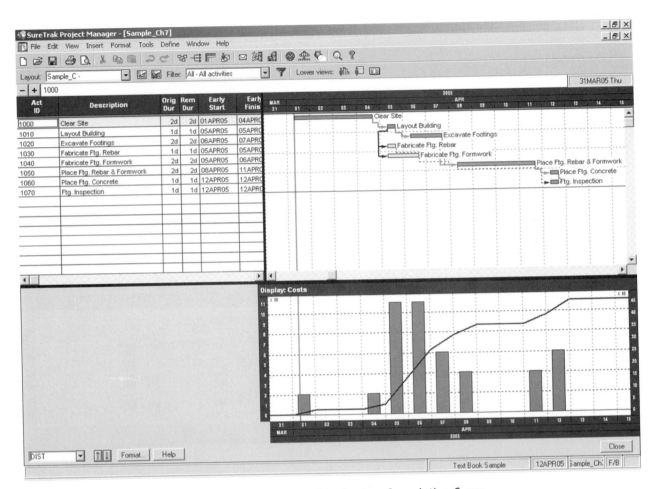

Figure 7–23 Resource Profile—Costs—Distributable Costs—Cumulative Curve

activities and a longer duration. The entire cumulative cost curve can still be viewed on a single screen. Double-click on the screen bar chart timescale. The **Timescale** dialog box appears. Observe what happens when the slide bar button is moved to the left an inch or so, and the **OK** button is clicked. Not only is the onscreen bar chart timescale condensed, but the cost profile is also modified. If the project is of a longer duration, say a year, the cash requirements for the entire project (by day and cumulative) can be observed on a single screen by adjusting the timescale.

EXAMPLE PROBLEM

Table 7–4 is an Activity List with costs and Table 7–5 is a list of resources with their associated costs for a house put together as an example for student use (see the wood frame house drawings in the Appendix). Figure 7–24 through Figure 7–28 provide additional resource information.

Act. ID	Activity Description	Total Cost
1000	Clear Site	$1,680
1010	Layout Building	$2,375
1020	Form/Pour Footings	$3,045
1030	Pier Masonry	$1,260
1040	Wood Floor System	$2,540
1050	Rough Framing Walls	$3,560
1060	Rough Framing Roof	$2,840
1070	Doors & Windows	$2,791
1080	Ext Wall Board	$1,420
1090	Ext Wall Insulation	$585
1100	Rough Plumbing	$1,000
1110	Rough HVAC	$1,568
1120	Rough Elect	$1,240
1130	Shingles	$1,441
1140	Ext Siding	$2,010
1150	Ext Finish Carpentry	$1,920
1160	Hang Drywall	$2,244
1170	Finish Drywall	$1,191
1180	Cabinets	$2,018
1190	Ext Paint	$1,300
1200	Int Finish Carpentry	$2,840
1210	Int Paint	$5,725
1220	Finish Plumbing	$4,000
1230	Finish HVAC	$4,506
1240	Finish Elect	$1,910
1250	Flooring	$2,083
1260	Grading & Landscaping	$900
1270	Punch List	$0

Table 7–4   Activity List with Costs—Wood Frame House

Resource	Description	Units	Cost
CARPENTR	Carpenter 1	Day	$120
CRPN FOR	Carpenter Foreman	Day	$150
CRPN HLP	Carpenter Helper	Day	$60
LAB CL 1	Labor Class 1	Day	$60
LAB CL 2	Labor Class 2	Day	$65
MASON	Mason	Day	$120
SUB	Sub-Contractor	LS	—
MAT	Material	LS	—
EQUIP	Equipment	LS	—
OV-PR	Overhead and Profit	LS	—

Table 7–5   Resources Cost

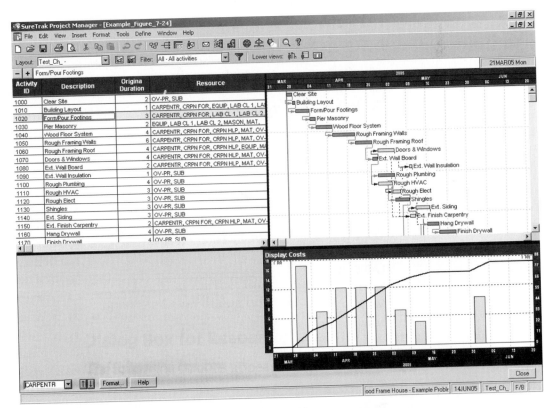

Figure 7–24 Cost Dialog Box—Wood Frame House

The onscreen **Costs** dialog box for activity 1020 (Figure 7–24) was prepared as an example of resource input. The cost profiles (detailed for
CARPENTR, Figure 7–25, and **All resources,** Figure 7–26) were prepared using the tabular list of twenty-eight activities (activity ID, description, and total activity costs). The cost tables (detailed, Figure 7–27, and
total, Figure 7–28) are also provided.

Figure 7–25 CARPENTER Cost Profile—Wood Frame House

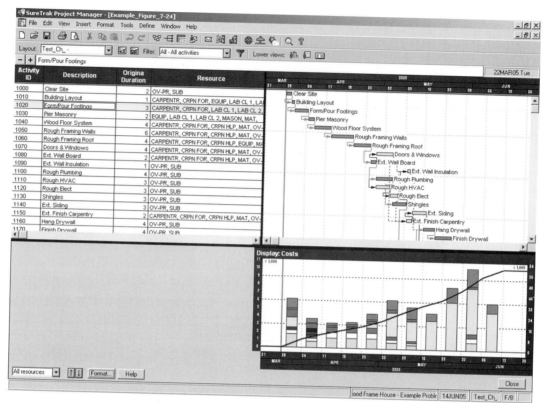

Figure 7–26   All Resources Cost Profile—Wood Frame House

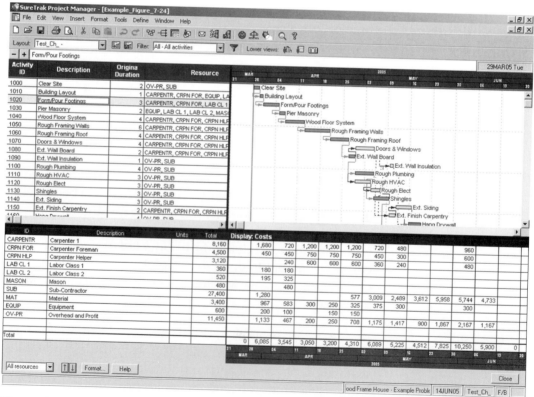

Figure 7–27   All Resources Cost Table—Wood Frame House

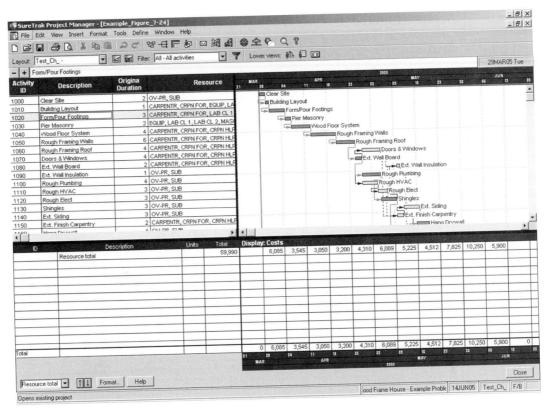

Figure 7–28 Total Cost Table—Wood Frame House

SUMMARY

The most important resource needed for construction is money. Scheduling and controlling the expenditure of funds is critical to the building process. This chapter provided an overview of using the schedule and *SureTrak* to control costs.

Cost must be defined to *SureTrak* as a resource before it can be associated to a particular activity. After the cost has been defined to *SureTrak*, the next step is to determine and assign the cost requirements of the activities. An activity can have multiple cost elements assigned to it with each element defined by budgeted cost, percent expended, percent complete, actual to date, to complete, at completion, scheduled budget, earned value, cost variance, schedule variance, and a completion variance.

SureTrak cost profiles and tables are used to analyze and evaluate the project resource requirements. The cost profiles and tables can be used to evaluate both period and cumulative cost. Cost profiles provide a graphical representation where the tables provide basically the same information in tabular format.

The analysis of cost to create cash flow requirements is critical to the management of any construction project from the point of view of both the contractor and owner.

EXERCISES

1. **Describing the Necessity of Controlling Cost**
   Prepare a summary describing the necessity of controlling cost.
2. **Defining Cost**
   Open the project modified in Exercise 5 of Chapter 6 and assign cost to the labor as show in Table 7–6.

Resource	Description	Unit	Cost
LAB 1	Laborer 1	Days	$80
LAB 2	Laborer 2	Days	$90
CARP 1	Carpenter 1	Days	$200
ELEC 1	Electrician 1	Days	$220
PLUMB 1	Plumber 1	Days	$210
FORE	Foreman	Days	$230

Table 7–6 Labor Cost for the Small Commercial Building in the South

3. **Displaying Cost Profile and Table**
   Display the cost profile and table of All Resources. Display the cost profile of only Laborer 1.
4. **Assigning Other Costs to Activities**
   Define the resource Other—Other costs—LS and assign it to activities 1000–1020 with a cost of $200 and to activities 1030–1050 with a cost of $300.
5. **Changing the Order of the Resources in the Cost Table**
   Order the resources as shown in Table 7–7.

Resource	Description	Order
LAB 1	Laborer 1	1
LAB 2	Laborer 2	2
CARP 1	Carpenter 1	3
PLUMB 1	Plumber 1	4
ELEC 1	Electrician 1	5
FORE	Foreman	6
Other	Other Costs	7

Table 7–7 Alternative Order of the Resources for the Small Commercial Building in the South

Complete a manual cost/day and cumulative cost table similar to Table 7–1 and Table 7–2 for Exercises 6 through 11. Also plot the cumulative curves.

6.

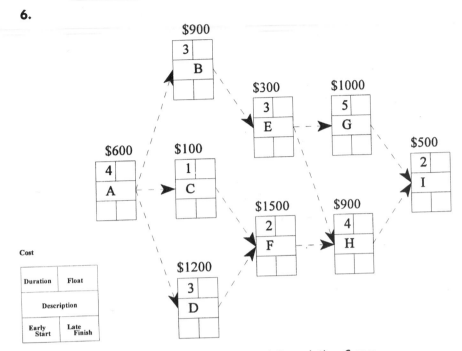

Figure 7–29 Exercise 6—Cost Table and Cumulative Curve

7.

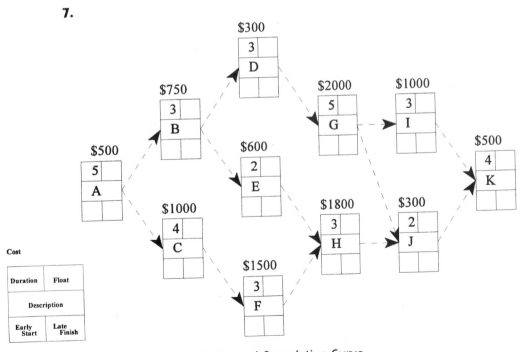

Figure 7–30 Exercise 7—Cost Table and Cumulative Curve

8.

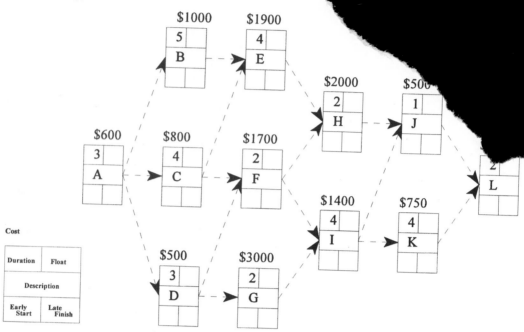

Figure 7–31 Exercise 8—Cost Table and Cumulative Curve

9.

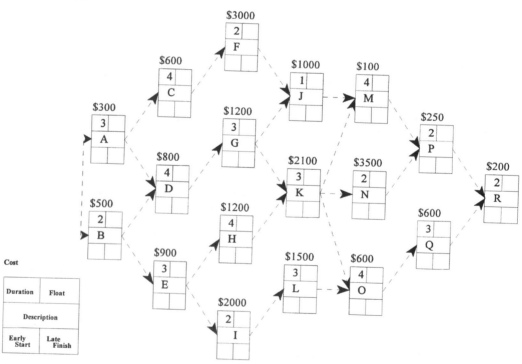

Figure 7–32 Exercise 9—Cost Table and Cumulative Curve

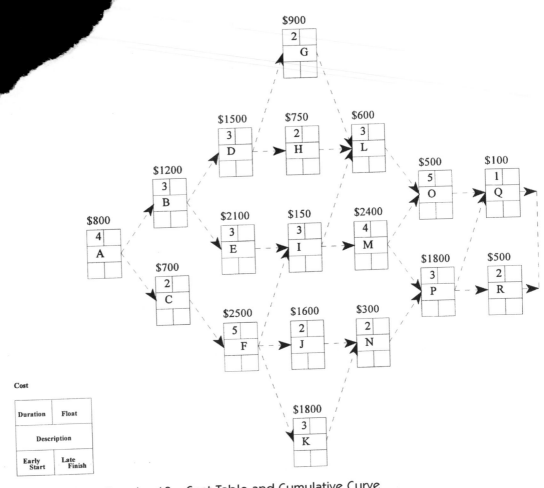

Figure 7–33 Exercise 10—Cost Table and Cumulative Curve

11.

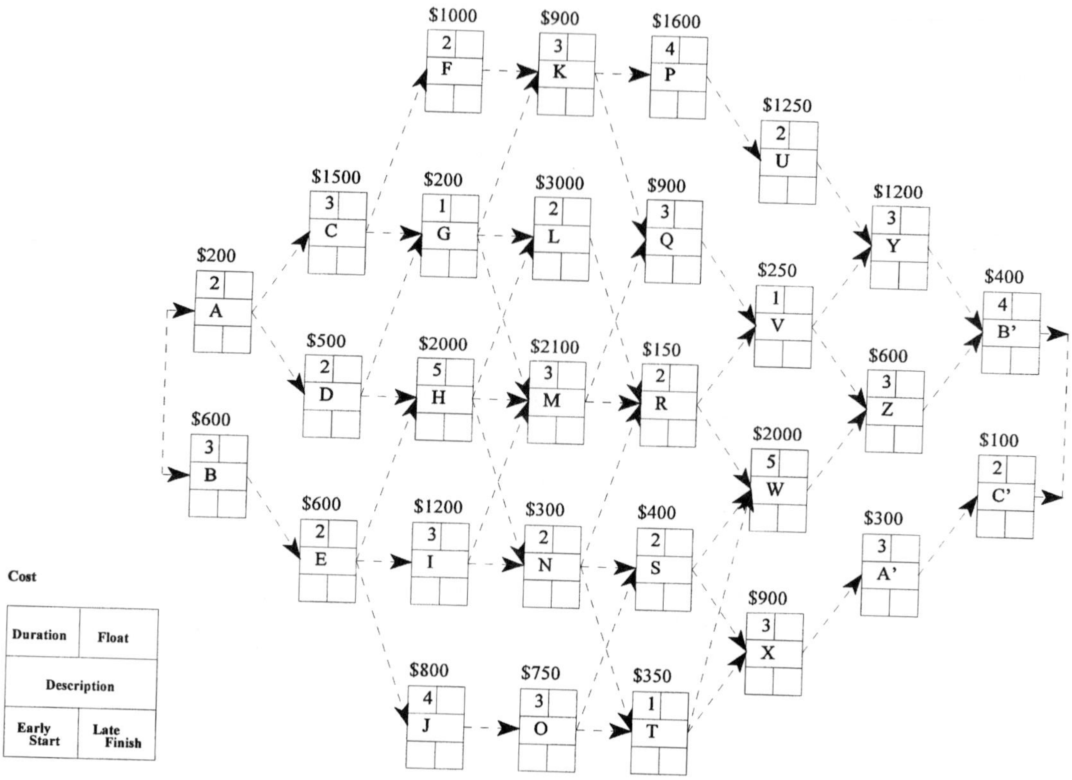

Figure 7–34   Exercise 11—Cost Table and Cumulative Curve

12. Small Commercial Concrete Block Building—Costs

Prepare an onscreen cost profile and cost table for the small commercial concrete block building located in the Appendix. This exercise should include the following steps:

1. Assign costs per activity by type of cost.
2. Create the onscreen cost profile.
3. Create the onscreen cost table.

13. Large Commercial Building—Costs

Prepare an onscreen cost profile and cost table bar chart for the following large commercial building located in the Appendix. This exercise should include the following steps:

1. Assign costs per activity by type of cost.
2. Create the onscreen cost profile.
3. Create the onscreen cost table.

8

Bar Chart Hard Copy Prints

Objectives

Upon completion of this chapter, you should be able to:

- Describe the necessity for good presentations
- Describe graphic reports
- Preview a bar chart
- Zoom in and out of bar chart print preview
- Include resource profiles/tables in the bar chart
- Print all pages or specific pages of a bar chart
- Modify the page setup
- Print bar charts in portrait or landscape orientation
- Change the timescale date span
- Modify header and footer

NECESSITY FOR GOOD PRESENTATIONS

Obviously putting a schedule together at the beginning of a project is valuable. The process forces you to plan, organize, sequence, and show the interrelationships among different activities or functions of the project. This information is useless, however, unless it is communicated to all relevant parties on the project. Good scheduling involves disseminating information to and getting feedback from all key parties regarding the original plan, updates, and progress toward completion of the project. At present, the easiest and most efficient way to disseminate the computerized *SureTrak* schedule is through hard copy (paper) graphical or tabular reports. Using e-mail with attached files or the Internet to transfer graphical information is moving construction companies toward a paperless office and scheduling function. This chapter deals with hard copy prints (printouts) of schedule information.

U.S.	Letter	8 1/2 × 11 in
	Legal	8 1/2 × 14 in
ANSI	B	11 × 17 in
	C	17 × 22 in
	D	22 × 34 in
	E	34 × 44 in
Architectural	A	9 × 12 in
	B	12 × 18 in
	C	18 × 24 in
	D	24 × 36 in
	E	36 × 48 in
ISO	A4	210 × 297 mm
	A3	297 × 420 mm
	A2	420 × 594 mm
	A1	594 × 841 mm
	A0	849 × 1189 mm

Table 8–1 Printer/Plotter Paper Sizes

GRAPHIC REPORTS

A graphical report is a printout or plot depicting information about the schedule. A plotter produces large drawings (Table 8–1). Plotters offer multiple colors, line types, line widths, and shading capabilities for producing professional presentations. Graphical printers are usually of the laser jet or ink jet variety. Multiple colors are an option with these printers and the paper is usually letter size (8.5" × 11") or legal size (8.5" × 14"). Whether you choose a printer or a plotter depends primarily on the size of the hard copy you want to produce.

BAR CHART PRESENTATION

As mentioned earlier, the primary advantage of the bar chart is its overall simplicity. It is very easy to read and interpret, and therefore can be an effective communications tool. The primary disadvantage is that it does not normally show interrelationships among project activities. By selecting **Relationships** from the **View** main pull-down menu, you can view the logic interrelationships onscreen. If an activity is delayed and the relationship lines are not shown, it is difficult to determine the impact on the rest of the schedule. The bar chart is undoubtedly the most commonly used print form of *SureTrak*.

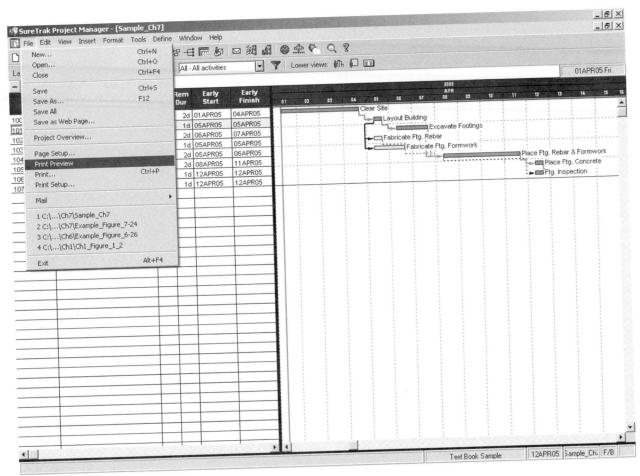

Figure 8–1　Bar Chart—File Main Pull-Down Menu—Print Preview

A hard copy of the *SureTrak* bar chart can be produced in one of two ways. It can be accessed either from the **File** or the **Tools** main pull-down menus through **Reports**. Because only the **File** main pull-down menu option offers the **Print Preview,** we will cover this option in detail.

Accessing the bar chart print from the **File** main pull-down menu (Figure 8–1) is easier than accessing it from the **Tools** main pull-down menu. The disadvantage is that the customized reports are not available. The **File** option gives essentially a copy of the onscreen bar chart. As you can see from Figure 8–1, four bar chart print functions are available from this menu. The functions are **Page Setup, Print Preview, Print,** and **Print Setup.**

PRINT PREVIEW OPTIONS

These toolbar options are customizable by the user. In this chapter, we show the *SureTrak* defaults. Click on the **Print Preview** option and the

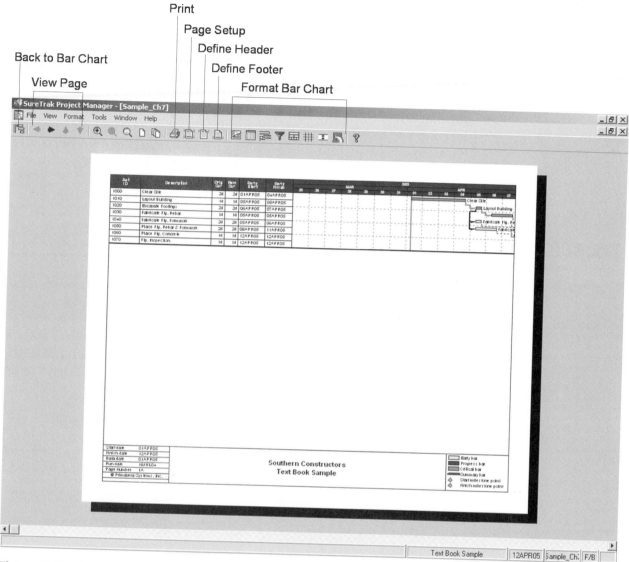

Figure 8–2   Bar Chart—Print Preview

project bar chart appears on the screen (Figure 8–2). It is an onscreen image of the hard copy print of the *SureTrak* onscreen bar chart. You can manipulate the onscreen image of the print using the toolbar at the top of the screen.

BAR CHART BUTTON

The first button on the toolbar (Figure 8–2) starting from the left side is the **Bar Chart** button that closes print preview and returns to the onscreen bar chart.

VIEW PAGE BUTTONS

The next group of four buttons (Figure 8–2) to the right of the Bar Chart on the toolbar screen, are arrows for selecting page views. If the schedule print is composed of multiple pages, use the horizontal arrows to scroll the pages horizontally and the vertical arrows to scroll pages vertically. In Figure 8–2, three of the four arrows are "grayed out," meaning the hard copy print of the sample project is composed of two horizontal pages. Click on the second arrow (darkened right arrow) to view the second horizontal page.

ZOOM BUTTONS

The third set of buttons on the toolbar relate to the zoom viewing functions (Figure 8–3).

Zoom In The **Zoom In** button (Figure 8–3) is identified by a magnifying glass with a plus sign inside. When the cursor is outside the boundaries of the reproduction of the hard copy page, it appears as an

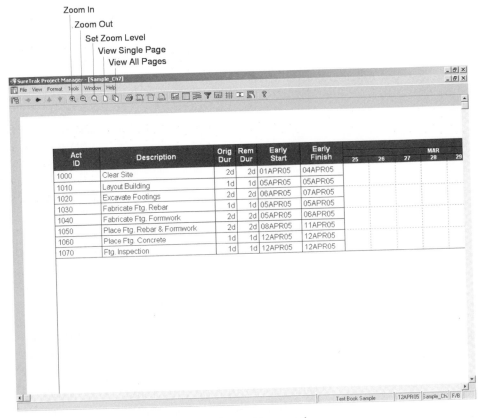

Figure 8–3   Bar Chart—Print Preview—Zoomed

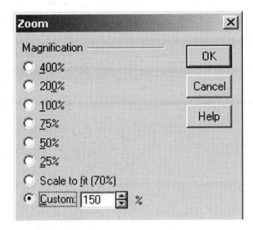

Figure 8–4   Zoom Dialog Box

arrow (normal appearance). When the cursor is inside the boundaries of the onscreen hard copy page, it appears as a magnifying glass. Move the cursor (magnifying glass) to the location to be zoomed (or magnified), and then click the mouse. *SureTrak* automatically sizes the print to fit the screen. When **Zoom In** is selected, *SureTrak* zooms in (makes larger) to the next standard zoom level (75%, 100%, 200%, or 400%). Figure 8–3 is a copy of Figure 8–2 zoomed in twice (150%) with the mouse cursor pointed to the lower-left corner of the screen (the section of the screen that we want zoomed in on).

Zoom Out The **Zoom Out** button (Figure 8–3) is identified by a magnifying glass with a minus sign inside. By clicking on this button, *SureTrak* automatically zooms out (makes smaller) to its next standard zoom level.

Set Zoom Level The **Set Zoom Level** button (Figure 8–3) is a plain magnifying glass. By clicking on this button, you can customize the zoom setting, either out or in. Figure 8–4 shows the **Magnification** choices. *SureTrak*'s default is the **Scale to fit,** in this case 70%. This is the size magnification required to fit the entire page of the hard copy print on a single screen. The choices of magnification zoom in or outs are 400%, 200%, 100%, 75%, 50%, and 25%. Also available is the **Custom** zoom sizing which can be anything between 25 and 400%.

View Single Page The **View Single Page** button (refer to Figure 8–3) is identified by a page with a bent upper-right corner. This button is used in conjunction with the next button, **View All Pages.** If the schedule to print has a number of pages, the **View Single Page** button displays only one page. **View All Pages** arranges and displays all the pages of the hard copy print.

View All Pages The **View All Pages** button (refer to Figure 8–3) is a picture of three pages with bent right corners. Viewing multiple pages on the screen simultaneously makes reading the prints very difficult,

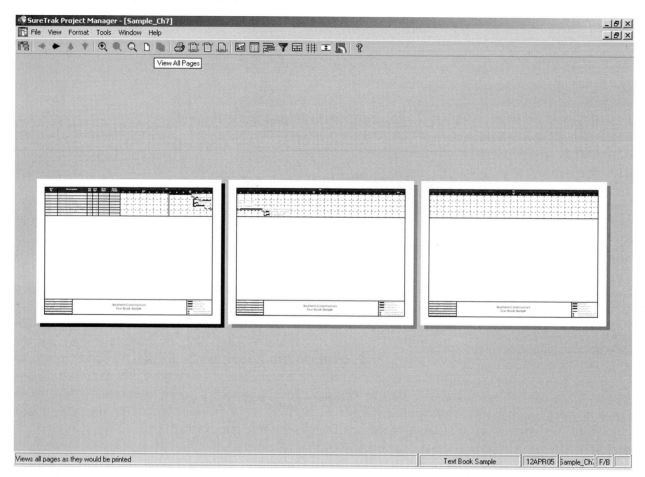

Figure 8–5 Bar Chart—Print Preview—View All Pages

but the ability to view this display makes it a great formatting tool. Figure 8–5 shows the sample schedule with the **View All Pages** button selected. To return to the single page view, simply select that button.

FORMAT PRINT BUTTONS

The next set of four buttons (refer to Figure 8–2) at the center of the toolbar, is used to execute the hard copy print or to format the print before the hard copy print is made (to avoid wasting paper). These buttons allow you to modify print options in **Print Preview** so that all changes can be proofed before the actual hard copy is executed.

Print

The **Print** button (refer to Figure 8–2) is used to actually execute the hard copy print of the **Print Preview** screen. Click on the **Print** button to produce the **Print** dialog box (Figure 8–6).

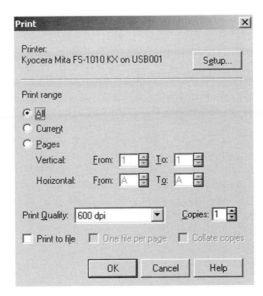

Figure 8–6   Print Dialog Box

Print range This function is used to select a limited number of pages rather than printing the entire schedule. Suppose there is a schedule whose hard copy is six pages long, three pages horizontally (numbers) and two pages vertically (alpha characters). The *SureTrak* numbering scheme for this combination is:

1A   1B   1C
2A   2B   2C

By understanding this numbering scheme, if the hard copy prints for all six pages were not necessary, only certain pages may be printed.

Print Quality Printers have varying levels of print quality. If the copy you are printing is not a final print to be used for a presentation, you may be interested in speed only. By requesting a lesser quality print, the print is much quicker.

Print to file For any number of reasons, you may want to print to a file rather than have the hard copy printed. You may not have a printer connected to your machine, you may want to copy the print file to a floppy and take it to a machine with a printer, or you may want to send the print file as an e-mail attachment to a remote site. By copying the file to a portable disk, you expand the optional use of other printers. You can use any file extension you like or none.

Copies This option is used if you want multiple copies of each page of the hard copy print.

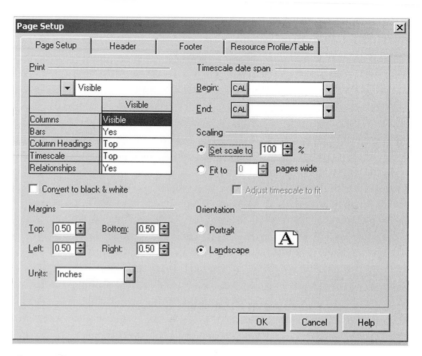

Figure 8-7 Page Setup Dialog Box

Page Setup

The second of the print format buttons (refer to Figure 8–2) is the **Page Setup** button. This is used to modify the page setup of the hard copy print of the **Print Preview** screen. Page setup of the hard copy can be accessed through **Print Preview** or directly from the **File** main pull-down menu. Both of these options lead to the same dialog box. From **Print Preview,** the **Page Setup** button is a picture of a single page of a schedule, with header and footer, and red arrows at the top and bottom. Click on the **Page Setup** button to produce the **Page Setup** dialog box (Figure 8–7).

> **Print** The <u>P</u>rint options control the information printed in the actual bar chart section of the hard copy print.
>
> **Columns** The three options available under the **Columns** are **All, Visible,** and **None.** The **Visible** option has all the columns exposed on the onscreen bar chart before the **Print Preview** selection was made. The **All** option prints all columns whether exposed on the onscreen bar chart or hidden from view. Figure 8–8 shows the sample schedule with all columns exposed. All defined columns will print. The **None** option will print no columns. This option is necessary when designing a timescale precedence diagram. See Figure 8–9 for labeling the <u>P</u>rint field options.
>
> **Bars** The two options are **Yes** and **No.** Use this option to make the activity list a tabular listing rather than graphical by deleting the bars.

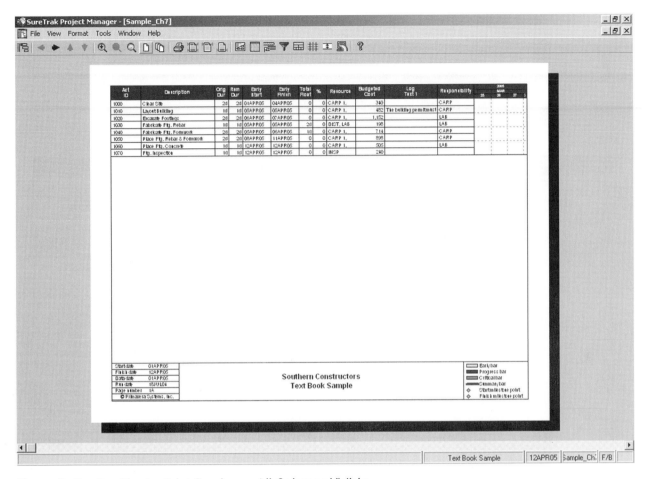

Figure 8-8   Bar Chart—Print Preview—All Columns Visible

Column Headings The four options for **Column Headings** placement are **Top, Bottom, Both,** and **None.**

Timescale The four options for **Timescale** placement are **Top, Bottom, Both,** and **None.**

Relationships The two options are **Yes** and **No.** This option is necessary when designing a timescale precedence diagram, or just simply showing the relationships on the diagram. Figure 8-9 shows the sample schedule with the **Yes** option selected for **Relationships.**

Convert to black & white Sometimes, even if you have a color printer, you want the hard copy print to be black and white. Construction can be a very conservative industry, and some construction professionals are used to looking at their schedules in black and white. Color prints can be a distraction to these people. Click this checkbox and the hard copy print will be black and white.

Margins The **Top, Bottom, Left,** and **Right** margins can be individually controlled by the **Units** field. The units options are **Inches, Centimeters,** and **Points.**

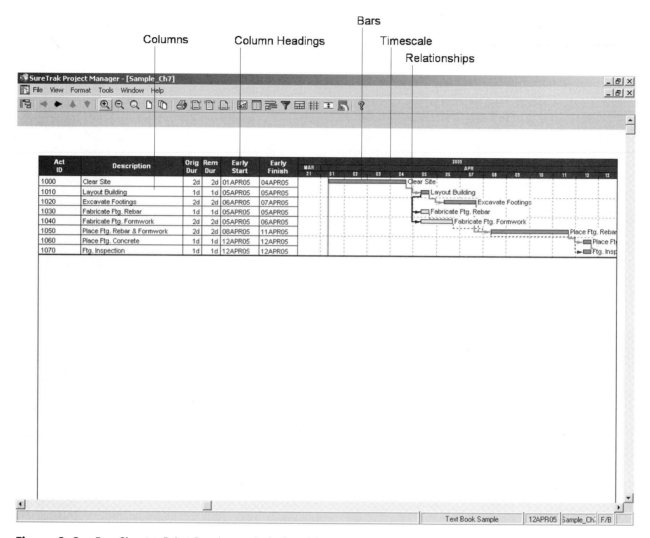

Figure 8–9 Bar Chart—Print Preview—Relationships

Orientation The two orientation options (Figure 8–7) are **Portrait** and **Landscape.** Figure 8–10 is the sample schedule in portrait format. Figure 8–11 shows landscape format.

Time Scale Date Span The **Begin** date option gives you the ability to vary the beginning date of the hard copy print, not the beginning date of the schedule. This is obviously handy when you are three months into a six-month project. You would want to start the hard copy print as of the third month. Click on the **CAL** at the left of the **Begin** field (Figure 8–12). From the menu, you can see four possibilities. The default is **Calendar Date.** This option is used to pick a specific date to start the hard copy print using the pop-up calendar that is accessed by clicking on the down arrow to the right of the **Begin** field (Figure 8–13). Here Wed, April 6 is selected. The other three options under the **CAL** pull-down menu are **Start date +, Data date +,** and **Finish date +.** Here a specific number of days or time units is added to the chosen option to determine the start date of the hard copy print.

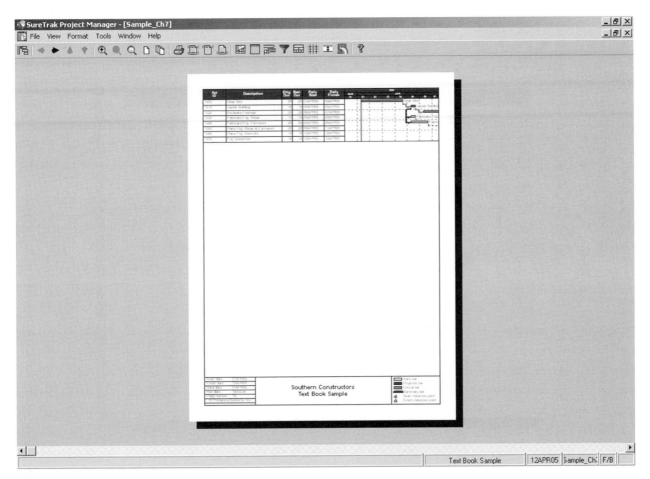

Figure 8–10   Bar Chart—Print Preview—Portrait

The **End** date option gives you the ability to vary the ending date of the hard copy print, not the ending date of the schedule. The options for this field are the same as the **Begin** field.

Scaling The **Set scale to** option provides a certain percentage (see Figure 8–7) of the hard copy print. Figure 8–14 shows the **Print Preview** of the sample schedule with a 75% scaling. Compare Figure 8–11 at 100% to Figure 8–14 at 60%. The other option under **Scaling** is the **Fit to __ pages wide.** This option is handy when the hard copy print is slightly over a single page wide. By limiting the horizontal number count of pages to one page wide, the scaling percentage to create a document one page wide is automatic.

Define Header

The third of the print format buttons (refer to Figure 8–2) is the **Define Header** button, which is used to modify the header or the title block at the top of the hard copy print. The **Define Header** and **Define Footer** options operate exactly the same way. The header is at the top of the hard copy print and the footer is at the bottom. We discuss the options of the **Define Footer** button next.

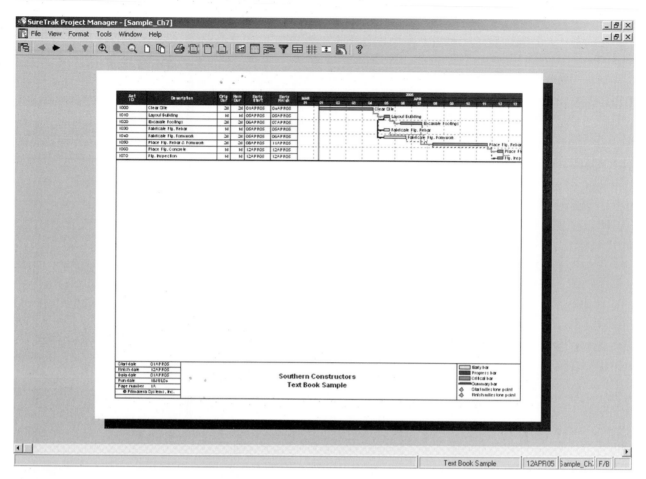

Figure 8–11 Bar Chart—Print Preview—Landscape

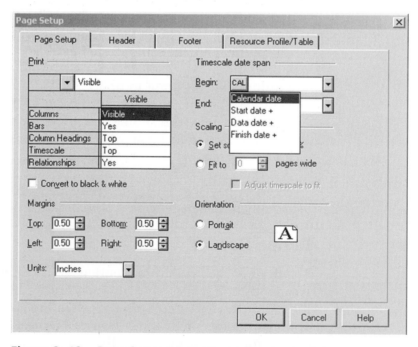

Figure 8–12 Page Setup Dialog Box—Timescale date span

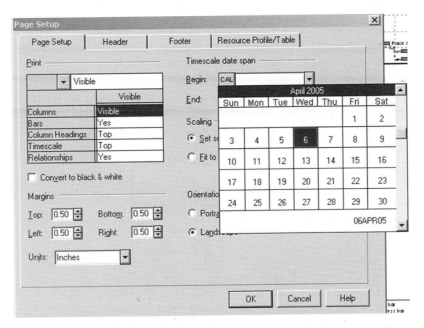

Figure 8–13   Page Setup Dialog Box—Calendar Date

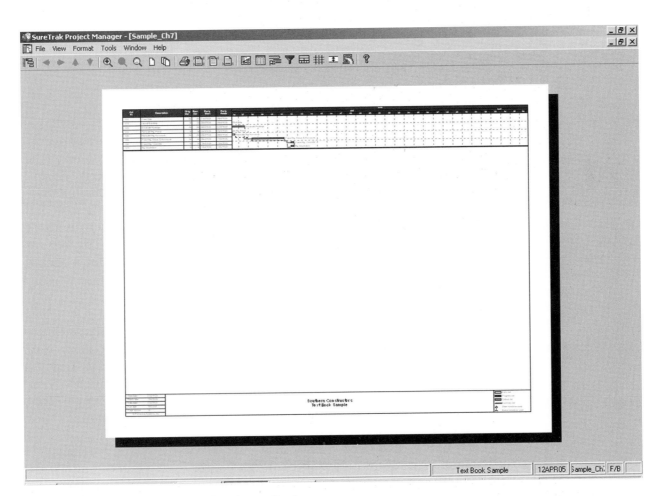

Figure 8–14   Bar Chart—Print Preview—Scale

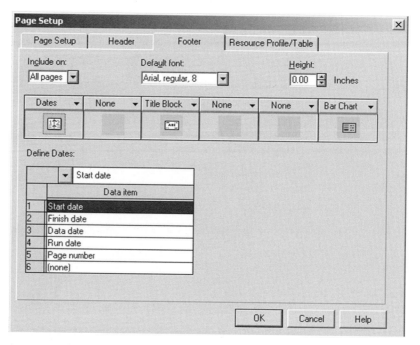

Figure 8–15   Footer Dialog Box—Dates

Define Footer

The **Define Footer** button is used to modify the footer or the title block at the bottom of the hard copy print. Select this button to produce the **Footer** dialog box in Figure 8–15.

Notice that there are six footer section fields. The first two fields appear at the left of the footer, the second two fields appear at the center of the footer, and the last two fields appear at the right of the footer. When you click on any of the footer fields, the predetermined footer/header alternatives appear (Figure 8–16). The footer in Figure 8–2 is the result of footer configuration in Figure 8–16. **Dates** appears in field 1, **Title Block** in field 3, and **Bar Chart** in field 6. The predetermined footer/header alternatives are **Title Block, Logo, Revision Box, Bar Chart Legend, Resource Legend, Dates, Project Comments,** and **None.** These alternatives are discussed further in the following section.

> **Title block** The first of the eight predetermined alternatives in the **Footer** dialog box is the <u>**Title block.**</u> Notice the center portion of the footer in Figure 8–2. Now look at Figure 8–17 for the location of the **Title block** button and dialog box options. The options available for this dialog box are the ability to vary the letter size and font from the other footer/header blocks. Note the bold, large letters of the **Title block** in Figure 8–2.

> **Logo** The second alternative in the **Footer** dialog box is **Logo.** This portion of the footer/header is intended for printing a company logo. See Figure 8–18 for the location of the **Logo** button and dialog box options. Note that the **Bar Chart Legend** was moved to section **5,** and

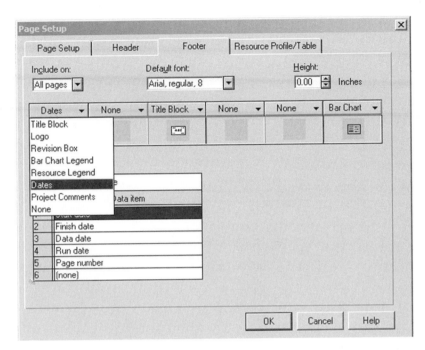

Figure 8–16 Footer Dialog Box—Predefined Footer/Header Alternatives

the **Logo** block was put in section **6.** By obtaining the logo in an electronic format, the clip art can be placed in this field. For example, in Figure 8–19, the metafiles in the *C:stwinclipart* files were accessed. *Note: with the file selected, the view field of the file to the right of the dialog box automatically shows a graphic representation of the file.* Figure 8–20 shows the result of clicking on the **OK** button to accept the new

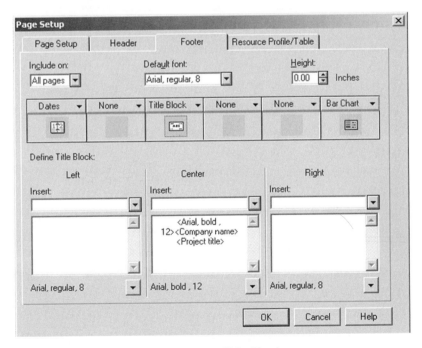

Figure 8–17 Footer Dialog Box—Title Block

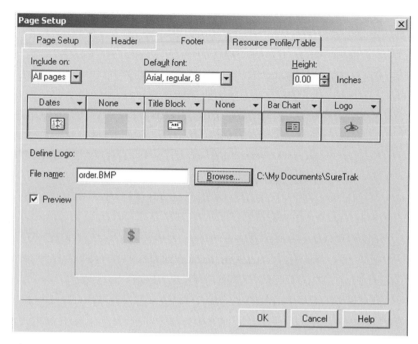

Figure 8–18   Footer Dialog Box—Logo

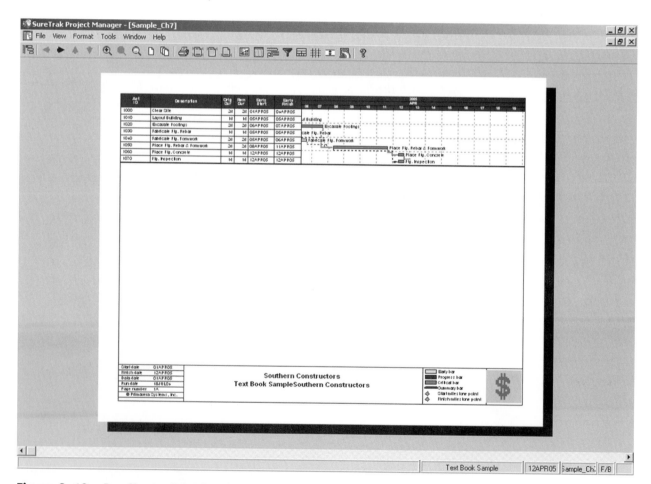

Figure 8–19   Bar Chart—Print Preview—Logo

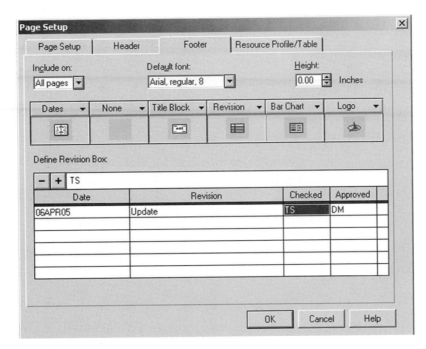

Figure 8–20   Footer Dialog Box—Revision Box

configuration. Note the new footer with the **Bar Chart Legend** moved and the dollar **Logo** in place (see Figure 8–19).

Revision Box The third alternative in the **Footer** dialog box is **Revision Box.** It is intended as a documentation of the approval cycle for a project schedule. See Figure 8–20 for the location of the **Revision Box** button and dialog box options.

Bar Chart Legend The fourth alternative in the **Footer** dialog box is **Bar Chart Legend.** Notice the right portion of the footer in Figure 8–2. Now look at Figure 8–21 for the location of the **Bar Chart Legend** button and dialog box options. The legend itself is determined by the configuration of the bar in the **Print Preview.** The format of the bar can be changed by using the **Format Bars** button from the **Print Preview** toolbar. The only possible option available in this dialog box is the **Number of Columns** that the legend is printed in. The legend in Figure 8–2 is printed in one column.

Resource Legend The fifth alternative in the **Footer** dialog box is **Resource Legend.** Figure 8–22 shows the location of the **Resource Legend** dialog box options. The legend itself is determined by the configuration of the bar in the **Print Preview.** The only option available in this dialog box is the **Number of Columns** that the legend is printed in.

Dates The sixth of seven predefined alternatives of the **Footer** dialog box is **Dates.** Notice the left portion of the footer in Figure 8–2. See Figure 8–23 for the location of the **Dates** button and dialog box options. There are six lines available for printed information in the

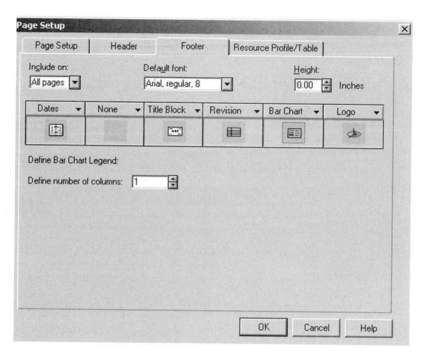

Figure 8–21 Footer Dialog Box—Bar Chart Legend

dialog box. Click on the down arrow and select the information fields for each line. The options are **Company name, Data date, Filter, Finish date, Layout, Must finish date, Number/Version, Page count, Page number, Percent complete, Percent Expended, Project name, Project title, Project type, Report ID, Report name, Run date, Start date,** and **Target finish date.**

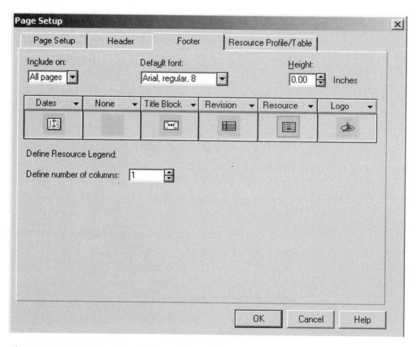

Figure 8–22 Footer Dialog Box—Resource Legend

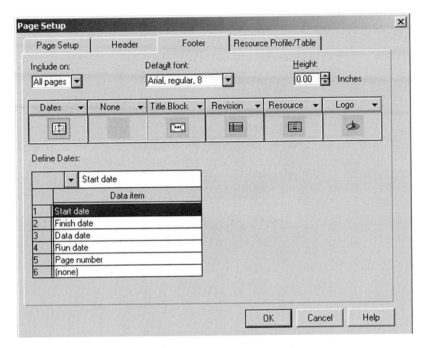

Figure 8–23   Footer Dialog Box—Dates

Project Comments The seventh alternative in the **Footer** dialog box is **Project Comments**. This portion of the footer/header is a documentation field for writing any pertinent information that you want to appear on the hard copy print. See Figure 8–24 for the location of the **Project Comments** button and dialog box options.

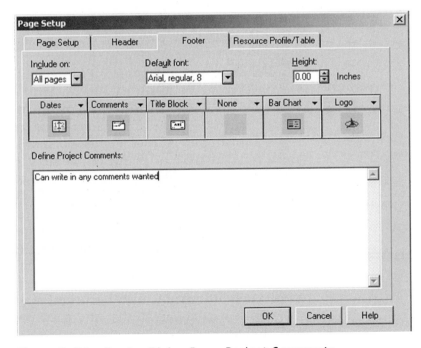

Figure 8–24   Footer Dialog Box—Project Comments

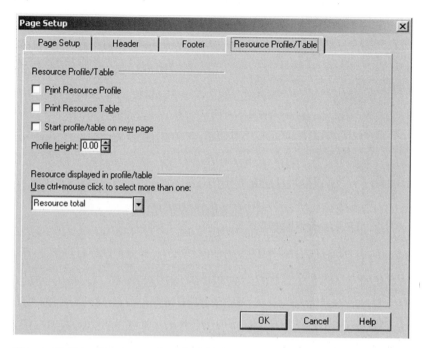

Figure 8–25 Resource Profile/Table Dialog Box

None The last option in the **Footer** dialog box is the **None** button. If one of the other blocks has been chosen for a section and you want it removed, click on the section to select it, and then click on the **None** button. The section will then go blank and the other block will be removed.

Resource Profile/Table Options Even if the resource table and/or resource profile are visible on the onscreen bar chart, they do not appear on the print preview of the hard copy print unless this option is used. Click on the **Resource Profile/Table** Tab in the Page Setup Dialog Box and the **Resource Profile/Table** dialog box appears (Figure 8–25). Click on **Print Resource Profile** or **Print Resource Table** to choose one of the two print options. **Start table/profile on new page** is good to simplify presentations. Sometimes, having the bar chart and a resource table or profile on the same page looks cluttered and confusing. Click on this option and the resource table or profile will start on a new page. **Resource Displayed in Profile/Table** allows you to choose individual resources or a composite of all resources. Figure 8–26 is the resource profile of the sample schedule with the bar chart and the profile printed on the same page. Stacking the individual resources and the color combination is defined on the onscreen bar chart with the resource table/profile exposed using the **Select** button. **Profile height** is used to vary the height of the resource table/profile. By controlling the height, the hard copy print looks more balanced and proportional.

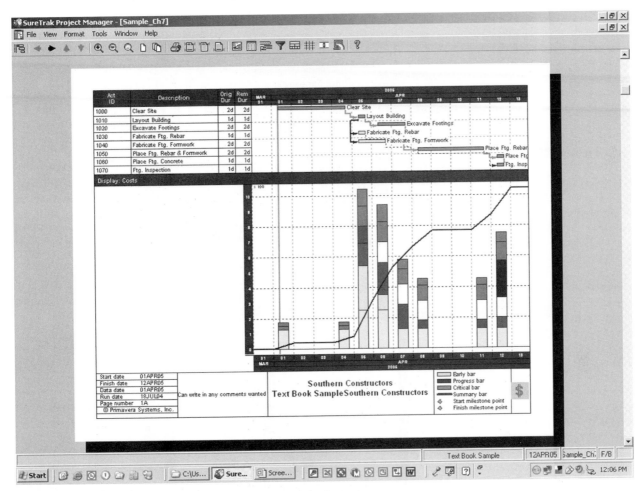

Figure 8–26   Bar Chart—Print Preview—Resource Profile

FORMAT BAR CHART BUTTONS

Refer to Figure 8–2 for the format bar chart buttons. These eight buttons (Figure 8–27) at the right of the toolbar are used to format the bar chart before the hard copy print is made. These buttons allow you to make and proof formatting changes in **Print Preview** before making the actual hard copy print of the bar chart.

The first button is the **Layouts** button (Figure 8–27).

Format Columns

The second button is **Format Columns** (Figure 8–27), which is used to change the format of the bar chart columns appearing in the **Print Preview** screen. Click on the **Format Columns** to produce the **Columns** dialog box (Figure 8–28).

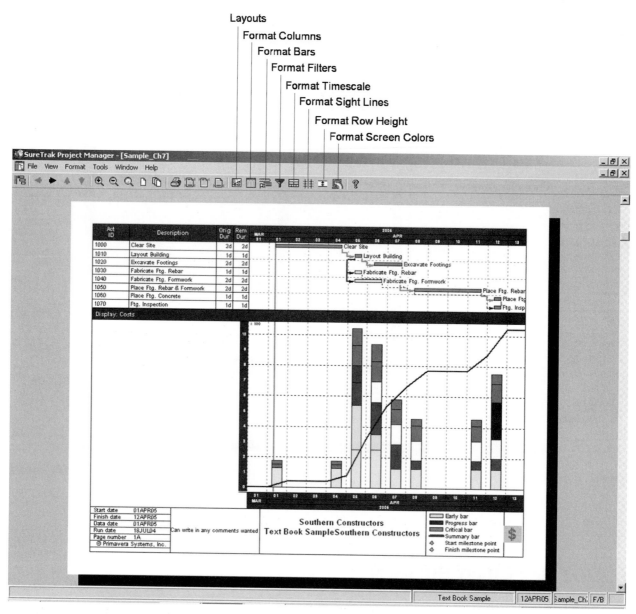

Figure 8–27 Bar Chart—Print Preview—Bar Chart Format Buttons

Columns Dialog Box The function of this dialog box is to format the bar chart columns as they appear in the **Print Preview** screen and the hard copy print. The **Columns** dialog box for formatting the bar chart print (Figure 8–28) is the same for formatting the onscreen bar chart (Chapter 5, Figure 5–6). The **Columns** dialog box, accessed from the onscreen bar chart, was reached by clicking on the **Format** main pull-down menu, and then selecting **Columns.** Note that if you change the dialog box (from either the **Print Preview** screen or the onscreen bar chart), the onscreen bar chart, the **Print Preview** screen, and the hard copy print are *all* changed.

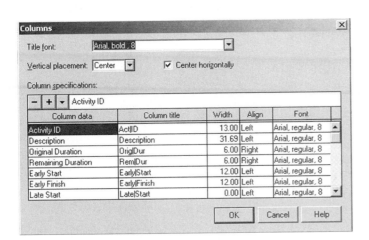

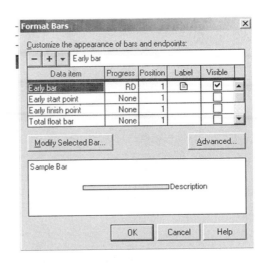

Figure 8–28   Bar Chart Format Buttons—Format Columns

Figure 8–29   Bar Chart Format Buttons—Format Bars

Format Bars

The **Format Bars** button (Figure 8–27) is used to change the format of the bar chart bars appearing in the **Print Preview** screen. Click on the **Format Bars** button to produce the **Format Bars** dialog box (Figure 8–29).

Format Bars Dialog Box The function of this dialog box is to format the bar chart bars as they appear on the **Print Preview** screen and the hard copy print. Note that this dialog box and its use are the same as the **Bars** dialog box (as discussed in Chapter 5), which was accessed from the onscreen bar chart. You can open the **Format Bars** dialog box by clicking on the **Format** main pull-down menu, and then selecting **Bars.** Note again that if you change the dialog box (from either the **Print Preview** screen or the onscreen bar chart), the onscreen bar chart, the **Print Preview** screen, and the hard copy print are *all* changed.

Format Filters

The **Format Filters** button (refer to Figure 8–27) is used for filtering activities to change the format of the bar chart bars appearing in the **Print Preview** screen. A filter temporarily limits the activities that appear in the onscreen or **Print Preview** screen, according to criteria you establish. You may only want to look at a particular subcontractor's work, or at critical activities, or at activities scheduled for the next month. Click on **Format Filters** to produce the **Filter** dialog box (Figure 8–30).

Filter Dialog Box The function of this dialog box is to limit the activities that appear in the **Print Preview** screen and the hard copy print. Note that this dialog box and its use are the same as the **Filter** dialog box that was accessed from the onscreen bar chart. To access the **Filter** dialog box, click on the **Format** main pull-down menu, and then select

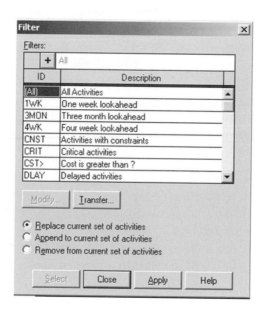

Figure 8–30 Bar Chart Format Buttons—Format Filters

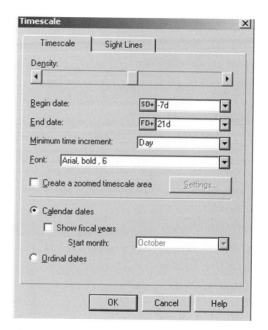

Figure 8–31 Bar Chart Format Buttons—Format Timescale

Filter, or click the **Format Filters** button from the onscreen bar chart toolbar. The **Format Filters** button on the **Print Preview** toolbar and the onscreen bar chart toolbar are identical. If you change the dialog box (from either the **Print Preview** screen or the onscreen bar chart), the onscreen bar chart, the **Print Preview** screen, and the hard copy print are *all* changed.

Format Timescale

The **Format Timescale** button (refer to Figure 8–27) is used to change the timescale of the bar chart bars appearing in the **Print Preview** screen. Click on the **Format Timescale** button to produce the **Timescale** dialog box (Figure 8–31).

Timescale Dialog Box The function of this dialog box is to modify the amount of the bar chart that appears in the **Print Preview** screen and the hard copy print by modifying the timescale. Note that this dialog box and its use are the same as the **Timescale** dialog box that was discussed in Chapter 4. To access the **Timescale** dialog box, click on the **Format** main pull-down menu, and then select **Timescale,** or double-click on one of the two top rows of the timescale above the onscreen bar chart. The **Format Timescale** options on the **Print Preview** toolbar and the onscreen options, as with the other **Print Preview** buttons, are identical. If you change the dialog box (from either the **Print Preview** screen or the onscreen bar chart), the onscreen bar chart, the **Print Preview** screen, and the hard copy print are *all* changed.

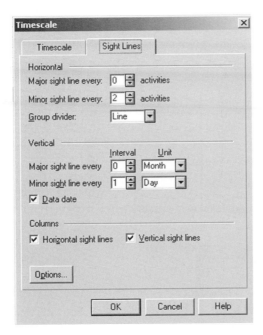

Figure 8–32   Bar Chart Format Buttons—Format Sight Lines

Format Sight Lines

The **Format Sight Lines** button (refer to Figure 8–27) is used to change the sight lines of the bar chart bars appearing in the **Print Preview** screen. Click on the **Format Sight Lines** button to open the **Sight Lines** tab (Figure 8–32).

> **Sight Lines Tab** The function of this tab is to modify the sight line spacing of the bar chart that appears in the **Print Preview** screen and the hard copy print. Note that this tab and its use are the same as the **Sight Lines** dialog box that was discussed in Chapter 5. Access the **Sight Lines** dialog box by clicking on the **Format** main pull-down menu and selecting **Sight Lines.** The **Format Sight Lines** options on the **Print Preview** toolbar and the onscreen options, as with the other **Print Preview** buttons, are identical. If you change the dialog box (from either the **Print Preview** screen or the onscreen bar chart), the onscreen bar chart, the **Print Preview** screen, and the hard copy print are *all* changed.

Format Row Height

The **Format Row Height** button (refer to Figure 8–27), is used to change the row heights of the bar chart bars appearing in the **Print Preview** screen. Click on the **Format Row Height** button to produce the **Row Height** dialog box (Figure 8–33).

> **Row Height Dialog Box** The function of this dialog box is to modify the row height spacing of the bar chart that appears in the **Print Preview** screen and the hard copy print. Note that this dialog box and its

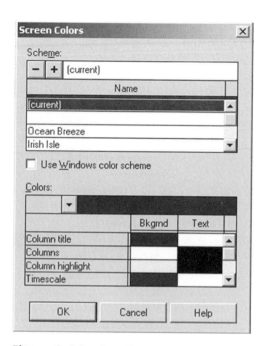

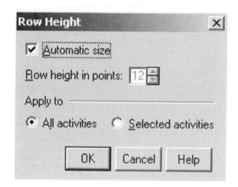

Figure 8–33   Bar Chart Format
Buttons—Format Row Heights

Figure 8–34   Bar Chart Format
Buttons—Format Screen Colors

use are the same as the **Row Height** dialog box that was discussed in Chapter 5. You access the **Row Height** dialog box by clicking on the **Format** main pull-down menu and selecting **Row Height.** The **Format Row Height** options on the **Print Preview** toolbar and the onscreen options, as with the other **Print Preview** buttons, are identical. If you change the dialog box (from either the **Print Preview** screen or the onscreen bar chart), the onscreen bar chart, the **Print Preview** screen, and the hard copy print are *all* changed.

Format Screen Colors

The **Format Screen Colors** button (refer to Figure 8–27), is used to change the colors of the bar chart bar component parts appearing in the **Print Preview** screen. Click on the **Format Screen Colors** button to produce the **Screen Colors** dialog box (Figure 8–34).

Screen Colors Dialog Box The function of this dialog box is to modify the colors of the bar chart component parts that appear in the **Print Preview** screen and the hard copy print. Note that this dialog box and its use are the same as the **Screen Colors** dialog box discussed in Chapter 5. Access the **Screen Colors** dialog box by clicking on the **Format** main pull-down menu and selecting **Screen Colors.** The **Format Screen Colors** options on the **Print Preview** toolbar and the onscreen options, as with the other **Print Preview** buttons, are identical. If you change the dialog box (from either the **Print Preview** screen or the onscreen bar chart), the onscreen bar chart, the **Print Preview** screen, and the hard copy print are *all* changed.

LAYOUT BUTTON OPTIONS

The first of the bar chart format buttons is the **Layouts** button (refer to Figure 8–27). We cover it last because the multiple options need to be discussed individually. The purpose of **Layouts** is to save the onscreen and hard copy print layout so that the next time a screen or print is requested, it will be the same as when it was saved even if the same report or print is used for another schedule (file) with a different format. Click on the **Layouts** button to produce the **Layouts** dialog box (Figure 8–35).

Layouts Dialog Box

The function of this dialog box is two-fold. The first is to name and save the layout for a schedule. Under **Layouts** at the top portion of the dialog box, the layout is named. The layout for the **Sample Schedule** is **(Sample-C).** The second function of the **Layouts** dialog box is to provide additional button functions not provided on the **Print Preview** toolbar. To see these additional functions, click on the **Format Selected Layout** button (Figure 8–36). The five options not available on the **Print Preview** toolbar are **Organize, Relationship Lines, Resource Profile/Table, Summary Bars,** and **Summarize All.** Note that this dialog box and its use are the same as the **Layouts** dialog box discussed in Chapter 5. Access the **Layouts** dialog box from the onscreen bar chart by clicking on the **View** main pull-down menu and selecting **Layouts.** If you change the dialog box (from either the **Print Preview** screen or the onscreen bar chart), the onscreen bar chart, the **Print Preview** screen, and the hard copy print are *all* changed. The **Layouts** dialog box for formatting bar chart print (Figure 8–35) is the same as formatting the onscreen bar chart (Chapter 5, Figure 5–2).

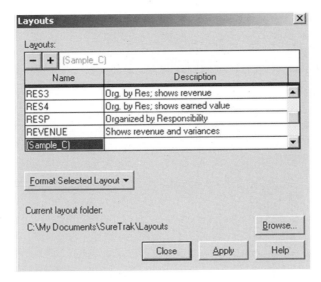

Figure 8–35   Bar Chart Format Buttons—Layouts

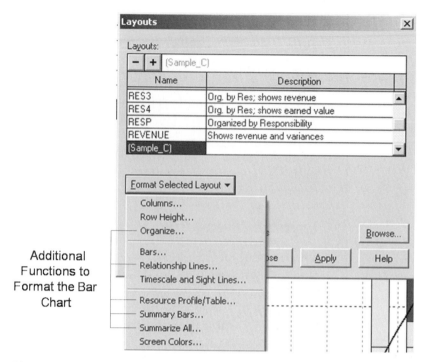

Additional
Functions to
Format the Bar
Chart

Figure 8–36   Bar Chart Format Buttons—Additional Functions

Organize

Click on the **Organize** button (refer to Figure 8–35) to produce the **Organize** dialog box (Figure 8–37).

Organize Dialog Box   The function of this button is to limit by some criteria the activities that will appear on the **Print Preview** screen and the hard copy print. The **Organize** dialog box for formatting the bar chart print (Figure 8–37) is the same as formatting the onscreen bar

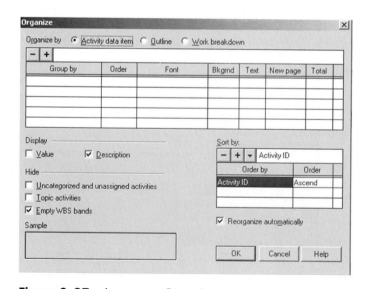

Figure 8–37   Layouts—Organize

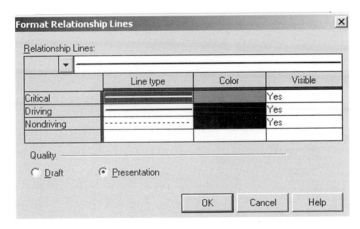

Figure 8–38   Layouts—Format Relationship Lines

chart. Access the **Organize** dialog box by clicking on the **Format** main pull-down menu and selecting **Organize.** If you change the dialog box (from either the **Print Preview** screen or the onscreen bar chart), the onscreen bar chart, the **Print Preview** screen, and the hard copy print are *all* changed.

Relationship Lines

Click on the **Relationship Lines** option (refer to Figure 8–36) to produce the **Format Relationship Lines** dialog box (Figure 8–38).

> **Relationship Lines Dialog Box** The function of this dialog box is to format the activity relationship lines, if they are requested, that will appear on the **Print Preview** screen and the hard copy print. This dialog box and its use is the same as the **Format Relationship Lines** dialog box as discussed in Chapter 5. Access the **Format Relationship Lines** dialog box by clicking on the **Format** main pull-down menu and selecting **Relationship Lines.** To actually get the relationship lines to appear on the onscreen bar chart, select **Relationships** from the **View** main pull-down menu. The relationship lines will now appear on the onscreen bar chart and **Print Preview** screen and the hard copy print. If you change the dialog box (from either the **Print Preview** screen or the onscreen bar chart), the onscreen bar chart, the **Print Preview** screen, and the hard copy print are *all* changed in format.

Resource Profile/Table

The **Format Selected Layout** button (Figure 8–35) is used to change the format of the resource profile and resource table as they appear in the **Print Preview** screen. Click on the **Format Selected Layout** button to produce the **Format Resource Profile/Table** dialog box (Figure 8–39).

> **Format Resource Profile/Table Dialog Box** The function of this dialog box is to format the resource profile and resource table as they

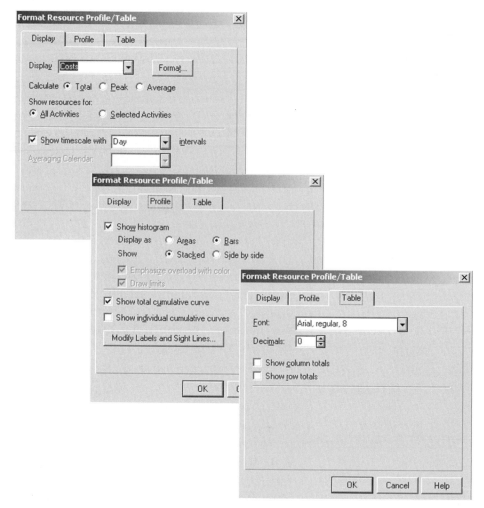

Figure 8-39 Layouts—Resource Profile/Table

appear in the **Print Preview** screen and the hard copy print. Note that this dialog box and its use are the same as the **Format Resource Profile/Table** dialog box discussed in Chapter 5. Access the **Format Resource Profile/Table** dialog box by clicking on the **Format** main pull-down menu and selecting **Resource Profile/Table.** If you change the dialog box (from either the **Print Preview** screen or the onscreen bar chart), the onscreen bar chart, the **Print Preview** screen, and the hard copy print are *all* changed.

Summary Bars

Click on the **Summary Bars** option (refer to Figure 8–36) to produce the **Summary Bars** dialog box (Figure 8–40).

Summary Bars Dialog Box The function of this dialog box is to summarize, by some criteria, the activities that will appear on the **Print Preview** screen and the hard copy print. This dialog box and its use is the same as the **Summary Bars** dialog box as discussed in Chapter 5.

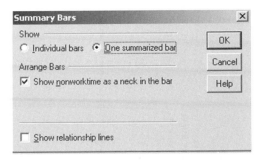

Figure 8–40   Layouts—Summary Bars

Figure 8–41   Layouts—Summarize All

Access the **Summary Bars** dialog box by clicking on the **Format** main pull-down menu, selecting **Summarization,** and then clicking **Format Summary Bars.** If you change the dialog box (from either the **Print Preview** screen or the onscreen bar chart), the onscreen bar chart, the **Print Preview** screen, and the hard copy print are *all* changed.

Summarize All

Click on the **Summarize All** option (refer to Figure 8–36) to produce the **Summarize All** dialog box (Figure 8–41).

Summarize All Dialog Box   The function of this dialog box is to provide another level of summarization beyond the summarize bar function. This dialog box and its use are the same as the **Summarize All** dialog box as discussed in Chapter 5. Access the **Summarize All** dialog box by clicking on the **Format** main pull-down menu, selecting **Summarization,** and then clicking **Summarize All.** If you change the dialog box (from either the **Print Preview** screen the onscreen bar chart), the onscreen bar chart, the **Print Preview** screen, and the hard copy print are *all* changed.

EXAMPLE PROBLEM: BAR CHART PRINT

Figure 8–42 shows the bar chart hard copy prints for a house put together as an example for student use (see the wood frame house drawings in the Appendix).

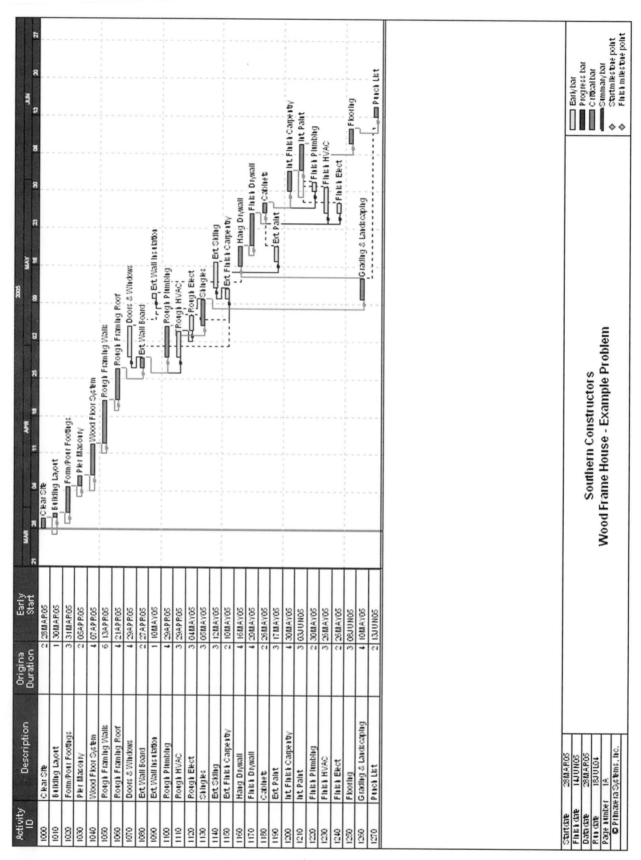

Figure 8–42 Bar Chart Hard Copy Prints

SUMMARY

The process of putting a schedule together is useless unless it is communicated to all relevant parties on the project. Good scheduling involves disseminating information to and getting feedback from all key parties regarding the original plan, updates, and progress toward completion of the project. The hard copy (paper) graphical or tabular reports can be used to disseminate the *SureTrak* schedule to all parties.

The primary advantage of the bar chart is its overall simplicity, which makes it an effective communications tool. Modifying the hard copy print to maximize the communication is a real asset for the scheduler. The Print Preview options are used to customize the *SureTrak* hard copy printouts.

The Zoom Print Preview button is used to enlarge or reduce the onscreen version of the hard copy printout. This makes evaluating and changing the prints possible before the print is actually executed.

The Format Print buttons—Print, Page Setup, Define Header, and Define Footer—are used to choose the printer, modify the paper size, define the columns to be printed, choose beginning/ending dates, and select paper orientation, footer/header fields, logos, title blocks, and legends.

The Format Bar Chart buttons—Layouts, Columns, Bars, Filters, Timescale, Sight Lines, Row Height, Screen Colors, and Layouts—can be used to modify the print preview of the hard copy printout. All of these *SureTrak* options can be used within print preview without having to return to the onscreen schedule.

EXERCISES

1. **Describing the Necessity for Good Presentations**
   Prepare a summary describing the necessity for good presentations.
2. **Previewing a Bar Chart**
   Open the project modified in Exercise 5 of Chapter 7 and print preview the hard copy of the bar chart.
3. **Zooming In and Out of the Preview**
   Zoom out of the preview to 25% and then zoom in to 350%.
4. **Including the Resource Profile in the Bar Chart**
   Include the resources profile for All Resources and set the profile width to 1.5".
5. **Printing the Bar Chart with Resource Profile**
   Print Exercise 4 to fit in one page.
6. Prepare bar chart prints for Figure 3–13 (Exercise 4, Chapter 4).
7. Prepare bar chart prints for Figure 3–14 (Exercise 5, Chapter 4).

8. Prepare bar chart prints for Figure 3–15 (Exercise 6, Chapter 4).
9. Prepare bar chart prints for Figure 3–16 (Exercise 7, Chapter 4).
10. Prepare bar chart prints for Figure 3–17 (Exercise 8, Chapter 4).
11. Prepare bar chart prints for Figure 3–18 (Exercise 9, Chapter 4).
12. **Small Commercial Concrete Block Building—Bar Chart Print**
Prepare bar chart printouts for the small commercial concrete block building (Exercise 12, Chapter 4) located in the Appendix. Prepare a hard copy print using the File main pull-down menu option.
13. **Large Commercial Building—Bar Chart Print**
Prepare bar chart printouts for the large commercial building (Exercise 13, Chapter 4) located in the Appendix. Prepare a hard copy print using the File main pull-down menu option.

9

PERT Diagrams

Objectives

Upon completion of this chapter, you should be able to:

- Describe the PERT diagram characteristics
- Insert activities using the PERT diagram view
- Move activities in a PERT diagram
- Zoom in and out of a PERT diagram
- View a PERT diagram using the Trace Logic and Cosmic View options
- Format a PERT diagram
- Organize a PERT diagram
- Filter a PERT diagram
- Preview a PERT diagram
- Print a PERT diagram

VIEW THE PERT DIAGRAM

The PERT diagram (activity-on-node logic diagram) is a visual representation of activities that shows the relationships between the activities. To see the PERT view of the sample schedule, click on **View** from the main pull-down menu, then select **PERT** (Figure 9–1). The onscreen view changes from the bar chart format (Figure 9–1) to the PERT format (Figure 9–2).

The arrows in this view show the relationships (predecessor/successor) between the activities. Activity 1000, Clear Site, is a predecessor to Activity 1010, Layout Building; therefore, Clear Site must be completed before Layout Building can begin. Activity 1010, Layout Building, is a successor to Activity 1000, Clear Site. Showing the relationships between activities on the screen makes them easier to see and understand. Each of the activities or rectangular shapes in Figure 9–2 is called a *node* and contains information about the activity.

Activity Form

You can edit an activity in the PERT view within *SureTrak* the same way as in the bar chart format. The easiest way is to access the **Activity Form,**

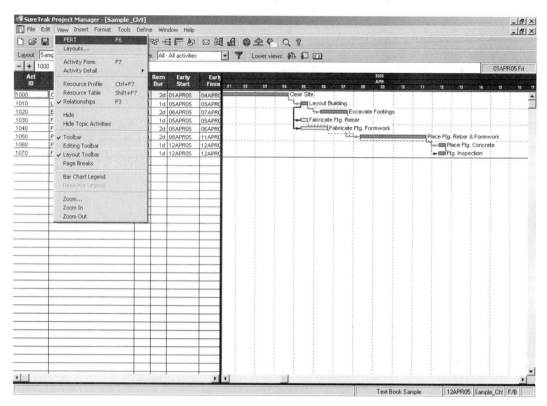

Figure 9–1   Bar Chart—View Main Pull-Down Menu—PERT

shown at the bottom of Figure 9–3. This form is accessed in one of two ways, either by clicking on the **View** main pull-down menu and selecting **Activity Form** or by selecting **Edit Activity** from the **Edit** main pull-down menu.

With the **Activity Form** selected, click on the activity to be edited. Figure 9–3 shows Activity 1050, Place Ftg Rebar & Formwork with a bold box around it, meaning that it has been selected. The information appearing in the **Activity Form** relates to the selected activity. You can edit any of the **Activity Form** fields other than the **ID** by simply clicking on the field and modifying the information.

Activity Detail

The resource and cost functions work the same way with the PERT view as with the onscreen bar chart. The **Define Resources** dialog box also works the same way (select **Resources** from the **Define** main pull-down menu) as the onscreen bar chart. To edit specific activity resource or cost information, the same three options as the bar chart are also available. Select the activity, click the right mouse button, click **Activity Detail,** and then select **Resources** or **Costs** depending on which dialog box you want. The second way is to select **Activity Detail** from the **View** main pull-down menu. The third way is to use the **F7** function key. Figure 9–4 shows the **Resources** dialog box selected.

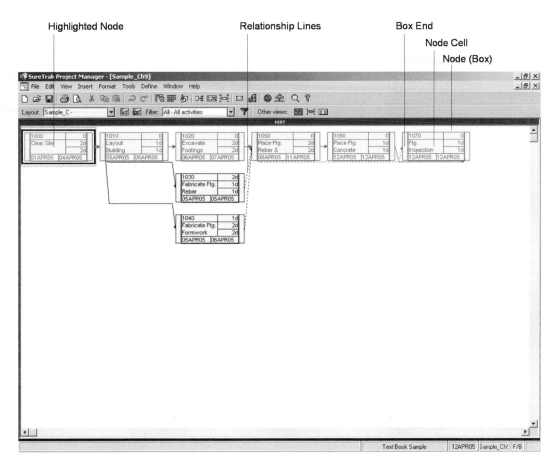

Figure 9–2   PERT

Insert Activity

There are advantages to adding a new activity using the PERT view rather than the bar chart view. The primary advantage is in actually seeing the placement of the activity and its relationship to the other activities. To insert an activity, click on the spot or placement on the PERT screen where the new activity will be placed. An empty box (Figure 9–5) will appear in the location of the new activity. Next, select **Activity** from the **Insert** main pull-down menu (Figure 9–5).

In Figure 9–6, the new activity was inserted in the empty box location. Note that the Activity Form automatically comes up at the bottom on the screen. In Figure 9–7, the activity **ID** field is input at XX. *SureTrak* automatically assigns the number X, but that number does not relate to placement in our diagram. Because we set the interval at 10 for each new activity, there is room between established activities to insert a new activity. The new **ID** is set at 1045. Input the activity description (Order Concrete) field and the **Duration** (1 day) next. When you click on the **OK** button, the information from the Activity Form is entered in the cells of the new activity box (Figure 9–8).

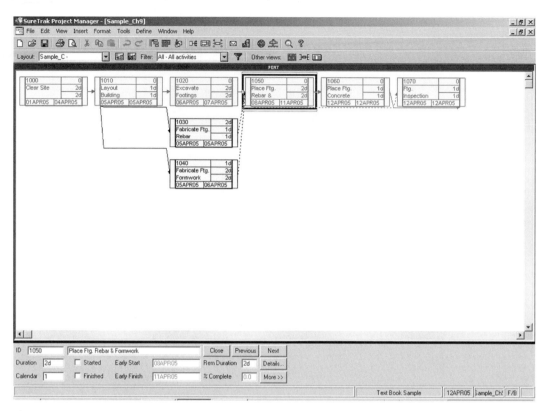

Figure 9–3 PERT—Activity Form

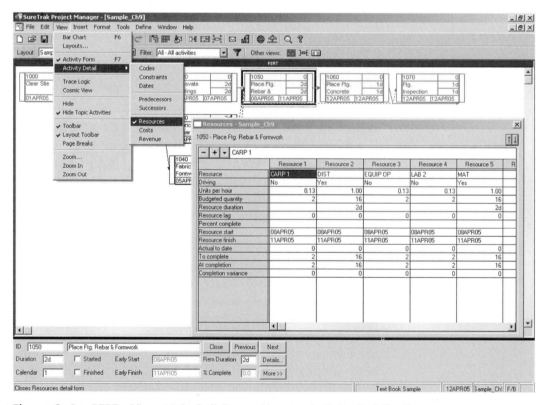

Figure 9–4 PERT—View Main Pull-Down Menu—Activity Detail—Resources

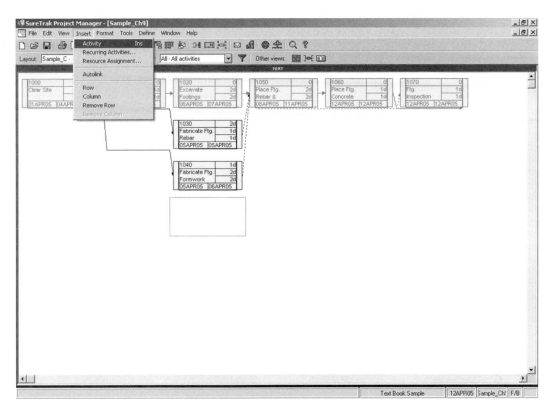

Figure 9–5 PERT—Insert Main Pull-Down Menu—Activity

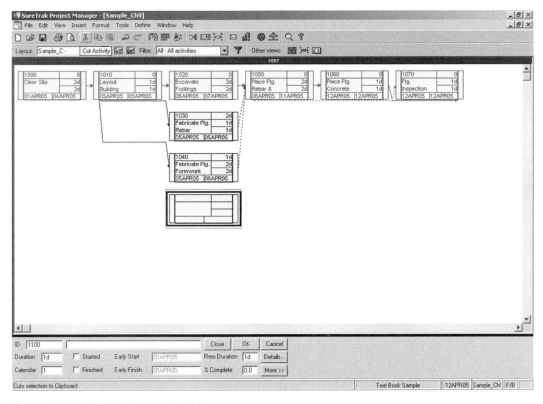

Figure 9–6 PERT—Insert Activity

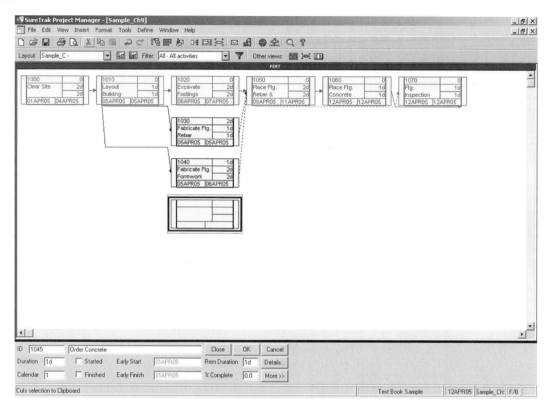

Figure 9–7 PERT—Activity Form

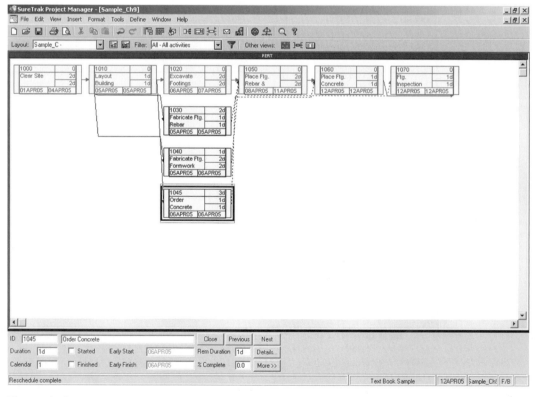

Figure 9–8 PERT—Relationships

The only relationship, so far, between the inserted activity and the rest of the diagram is the sequencing of the activity **ID** number. The relationships (predecessor/successor activities) between the inserted activity and the other activity need to be established. Activity 1010, Layout Building, needs to be established as a predecessor activity. Activity 1060, Place Ftg. Conc., needs to be established as a successor activity. To establish a SF relationship between Activity 1010 and the new activity (1045), place the mouse pointer in the box end of Activity 1010. The mouse pointer changes from an arrow to a pitchfork to indicate that a relationship is being established. Hold the mouse button down, drag it to the end box of Activity 1045, and release the mouse button. The new relationship line between the two activities has been established. Following the same procedure, establish the FS relationship between Activity 1045 and Activity 1060. Now the relationships between the inserted activity and the rest of the diagram have been established. Figure 9–8 shows the inserted activity with the new established relationships.

Move Activity

Many times, the activity placement in the view automatically created by *SureTrak,* or the view that results when an activity is inserted, needs to be rearranged to be more meaningful. Moving activities to rearrange the *SureTrak* PERT view is simple. Click on the activity to be moved and drag it to the new location. In Figure 9–9, the newly created Activity 1045 is selected (it has a dark border around it). Hold down the left mouse button and drag the box to the new location. Figure 9–9 shows the new activity location indicated by a double border around an empty box. This shows the location that the box was dragged to. When the mouse button is released, the activity moves to the new location (Figure 9–10).

Trace Logic

The **Trace Logic** feature from the **View** main pull-down menu is a valuable tool when there are logic problems with a diagram. (This view is not available with the bar chart format.) Click on the activity you want (Activity 1010 in Figure 9–11) and **Trace Logic** displays all predecessors and successors to the selected activity. On the **Trace Logic** screen in the lower half of Figure 9–11, Activity 1010 has a box around it, designating it as the selected activity. Activity 1000 is shown as a predecessor activity and Activities 1020, 1030, 1040, and 1045 are shown as successor activities.

Zoom

The zoom functions (**Zoom, Zoom In,** and **Zoom Out**) on the PERT view are the same as the onscreen bar chart. Figure 9–12 shows the **Zoom** dialog box available by selecting **Zoom** from the **View** main pull-down menu. The PERT diagram shown in Figure 9–12 is zoomed in at 65% of its original size. Compare Figure 9–12 with Figure 9–2. Obviously, being able to place many activities on the same screen and still being able to read all pertinent information is a very valuable tool.

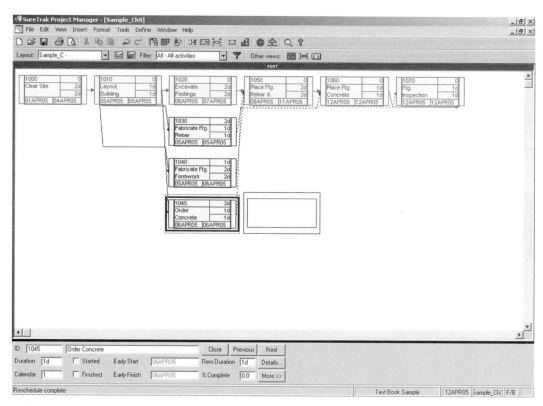

Figure 9–9　PERT—Move Activity

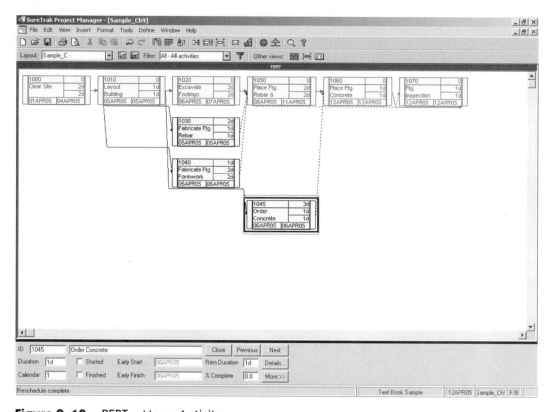

Figure 9–10　PERT—Move Activity

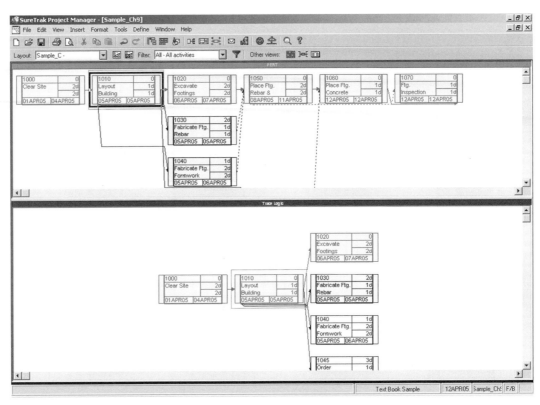

Figure 9–11   PERT—Trace Logic

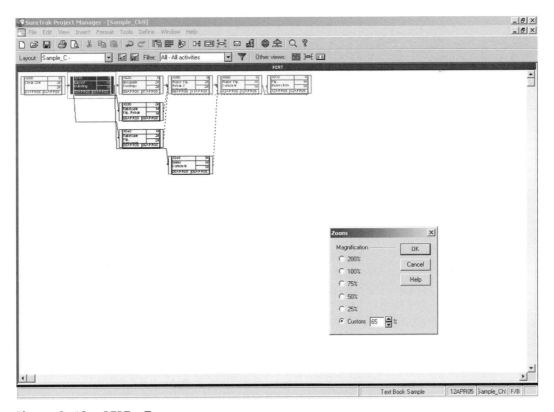

Figure 9–12   PERT—Zoom

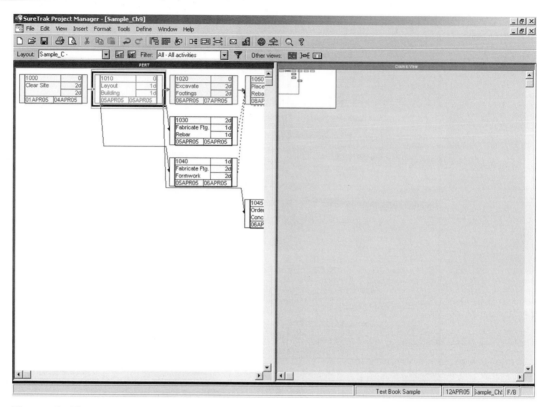

Figure 9–13 PERT—Cosmic View

Cosmic View

This feature of the PERT view is also not available with the bar chart format. The **Cosmic View** is available by selecting the **View** main pull-down menu. Figure 9–13 is a **Cosmic View** of the sample schedule. The purpose of this option is to move around a large schedule quickly to access the part of the schedule onscreen that you want to see. In the **Cosmic View** at the right side of the screen, the white portion of the **Cosmic View** represents the full screen before the **Cosmic View** was requested. The darkened box inside the white portion represents the **Cosmic View** of the left portion of the screen in Figure 9–13. Notice that Activity 1010 is highlighted (has the dark border around it). In the **Cosmic View,** this activity is darkened so that it can be recognized. By clicking on any portion of the **Cosmic View** (Figure 9–14), that portion of the schedule is shown in larger scale on the left portion of the screen. Then by clicking on or selecting a specific activity, that activity is darkened in the **Cosmic View.** This tool may not seem so valuable for a schedule like the sample with only eight activities, but when thousands of activities are involved, the **Cosmic View** is a valuable tool for quickly accessing the part of the schedule you want to move to.

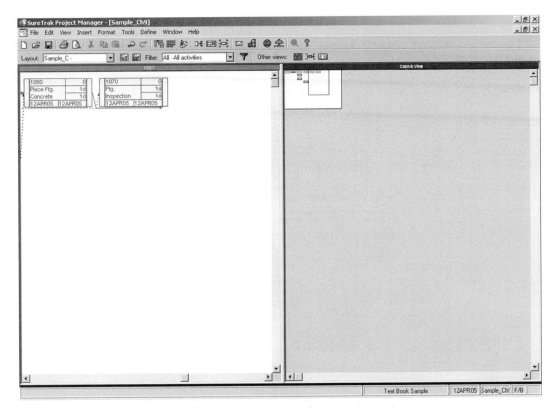

Figure 9–14   PERT—Cosmic View

FORMAT THE PERT DIAGRAM

Many times for clarity, you may want to change the information in the node (Figure 9–2) for the onscreen and the hard copy print of the PERT diagram.

Activity Box Configuration

To change the configuration of the node, select **Activity Box Configuration** from the **Format** main pull-down menu (Figure 9–15). The **Activity Box Configuration** dialog box appears (Figure 9–16). *Note: in the Activity box templates field there are many preconfigured templates for node format.* The **Standard Schedule** template is highlighted in Figure 9–16. This means the nodes for the PERT view for the sample schedule are in this format. The **Sample,** or preview field, at the right of the dialog box shows the information and configuration of the information contained in the **Activity box templates** selection. Using multiple configurations, each for a particular need, is the primary purpose of this dialog box. The preconfigured templates are:

Cosmic View   Log Text 1
Durations   Logic Review

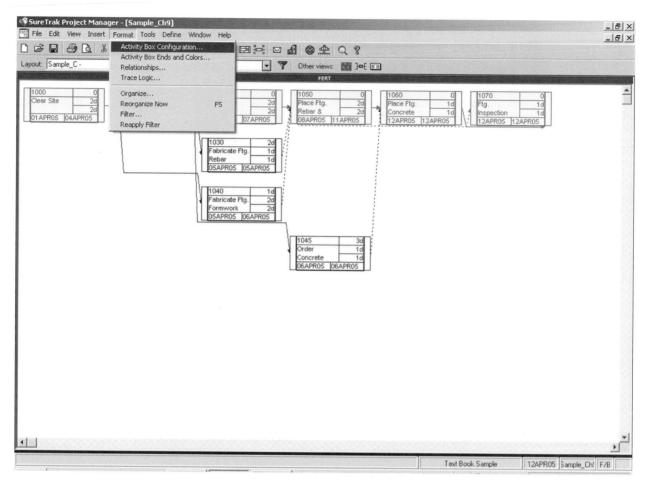

Figure 9–15 PERT—Format Main Pull-Down Menu—Activity Box Configuration

Logic Review–Early Dates   Update by Activity ID
Multiple Calendar ID   Update by WBS
New Activity   Update with Activity Codes
Resource Costs   WBS Code
Resource Quantities   WBS Code and Title
Standard Schedule   WBS Standard Schedule
Standard with Activity Type

The **Activity Box Configuration** dialog box also allows you to modify preconfigured templates to contain the fields and information you desire. Click on the **Modify template** button to produce the **Modify Template** dialog box (Figure 9–17). Note that in the **Name:** field, the **Standard Schedule** template is being modified. By clicking on the down arrow, you can select and modify any of the preconfigured templates.

The window that defines the template is divided into **Activity box cells** (see Figure 9–2). Use the **Insert row** or **Delete row** buttons to add or reduce the number of vertical fields. Use the **Split cell** or **Delete cell** buttons to add or reduce the number of horizontal fields. The field in the upper-left

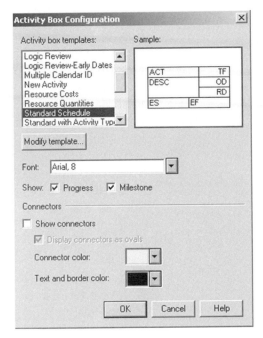

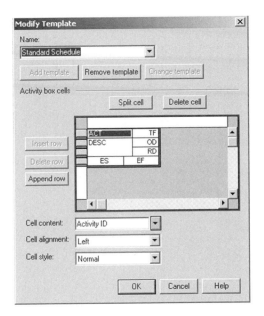

Figure 9–16   Activity Box Configuration
Dialog Box

Figure 9–17   Modify Template
Dialog Box

corner of the window is darkened or selected. The three fields at the bottom of the dialog box (**Cell content, Cell alignment,** and **Cell style**) are used to configure each of the cells. The highlighted field with the abbreviated **ACT** in it has the following information: **Cell content** is **Activity ID, Cell alignment** is **Left** justified, and **Cell style** is **Normal.** The **Cell content** options can be put into any of the seven fields that have been defined in this window for the PERT node box. The **Cell content** options are:

None	Cost Variance
Activity ID	Constraint Finish
RESP–Responsibility	Constraint Finish Type
AREA–Area	Constraint Start
PHAS–Phase	Constraint Start Type
MAIL–Mail codes	Cost at Completion
SUBP–Project ID	Cost to Complete
Log Record	Description
Activity Type	Early Finish
Actual Duration	Early Start
Actual Finish	Earned Value Cost
Actual Start	Earned Value Quantity
Actual Cost to Date	Finish Variance
Actual Quantity to Date	Free Float
Budgeted Cost	Late Finish
Budgeted Quantity	Late Start
Calendar	Level Delay
Completion Variance Cost	Log Text 1
Completion Variance Quantity	Log Text 2

Log Text 3
Log Text 4
Log Text 5
Log Text 6
Log Text 7
Log Text 8
Log Text 9
Log Text 10
Net at Completion
Object
Original Duration
Outline Code
Percent Complete
Percent Earned
Percent Expended
Predecessors
Priority
Units per Hour
Quantity at Completion
Quantity to Complete

Remaining Duration
Resource
Resource Description
Resource Designation/Cost
   Account
Resume Date
Revenue at Completion
Revenue to Date
Scheduled Budget Cost
Scheduled Budget Quantity
Schedule Variance
Successors
Suspend Date
Target Finish
Target Start
Total Float
Unleveled Finish
Unleveled Start
WBS Code

The **Cell alignment** options are:

Left
Center
Right

The **Cell style** option is Normal.

Activity Box Ends and Colors

Again, for clarity, you may want to change the activity box ends (see Figure 9–2) and/or colors for the onscreen and the hard copy print of the PERT diagram. You may change either all the boxes or select specific activities. Segregating the activities for identification purposes makes the PERT diagram quicker to use for identification purposes. In Figure 9–18, Activities 1000, 1010, and 1040 have been selected for box changes. These activities were selected by holding the Ctrl key down and clicking on each of the activities. All three of the activities are highlighted (have the dark border around them).

To change the configuration of the box shape and/or color, select **Activity Box Ends and Colors** from the **Format** main pull-down menu (refer to Figure 9–15). The **Activity Box Ends and Colors** dialog box appears (Figure 9–19). The **New Activities (default)** tab is used to change the default setting for the box appearance and the **Selected Activities** tab is used for changing specific selected activities. Because we have identified three activities for change, this is the tab to use. In Figure 9–20, the down arrow for the **Left end shape** has been clicked, exposing the choice of

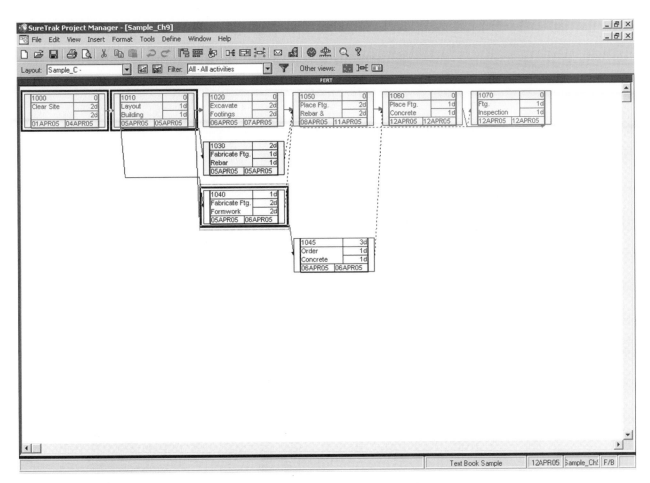

Figure 9–18   PERT—Select Activities

possible shapes. In Figure 9–21, the **Left end shape** and the **Right end shape** fields have been changed to rounded ends rather than boxed. Compare the box ends in the preview screen in Figure 9–20 to Figure 9–21. Next the **Left end fill** and the **Right end fill** were changed to diagonal lines. Finally, the **Activity box color** was changed. Look at the three selected activities in Figure 9–22 with the changed box ends and colors. The change is dramatic and makes identifying these activities very easy.

Figure 9–19   Activity Box Ends and Colors Dialog Box

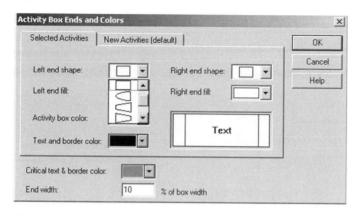

Figure 9–20   Activity Box Ends and Colors Dialog Box

Relationship Lines

As we discussed earlier in the chapter, the major advantage of the PERT diagram over the bar chart is that the focus is on the relationships between activities rather than the activities' bar length against a timescale. The dialog boxes for showing relationships are different for formatting the relationship lines for PERT diagrams. The default for the PERT diagrams is to show relationship lines (see Figure 9–2). To access the **Relationships** dialog box (Figure 9–23), select **Relationship Lines** from the **Format** main pull-down menu (see Figure 9–15). Under **Display,** the **Driving relationships** (activities on the critical path or having 0 total float) can be configured differently from **Nondriving relationships** (activities not on the critical path or not having 0 total float). Notice that in Figure 9–23, the **Driving relationships** has a **Line style** as a solid line. In Figure 9–22, the relationship line between Activity 1020 and 1050 is a solid line because both activities have a 0 total float (upper-right corner cell). The **Nondriving relationships** has a **Line style** as a dashed line. In Figure 9–22, the relationship line between Activity 1030 and 1050 is a dashed line because Activity 1030 has 1 day of total float.

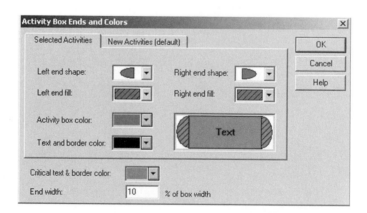

Figure 9–21   Activity Box Ends and Colors Dialog Box

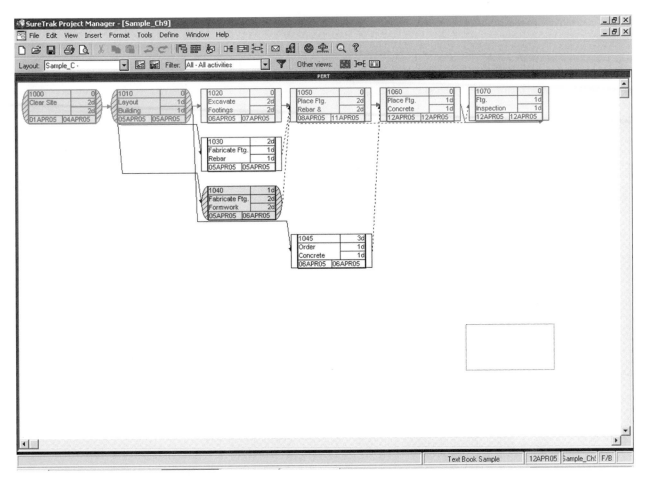

Figure 9–22   PERT—Modified Box Activities

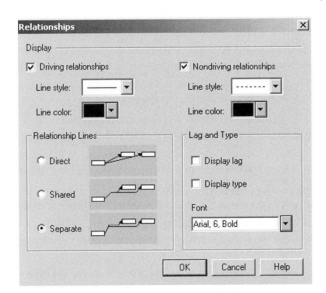

Figure 9–23   Relationships Dialog Box

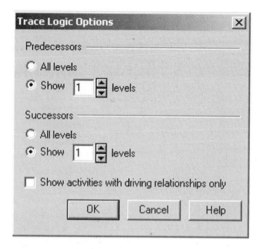

Figure 9–24 Trace Logic Options Dialog Box

Trace Logic

Figure 9–11 earlier in this chapter shows the **Trace Logic** screen accessed from the **View** main pull-down menu. To format the **Trace Logic** screen, select **Trace Logic,** from the **Format** main pull-down menu (refer to Figure 9–15). The **Trace Logic Options** dialog box appears (Figure 9–24). Figure 9–11 was generated with the default shown in Figure 9–24. If **Show_levels** is set at 2 for **Predecessors,** an activity's predecessor's predecessor, or the second level predecessor is also shown. The other primary option is the **Show activities with driving relationships only** checkbox. Checking this box produces only driving or critical (0 total float) relationships in the **Trace Logic** screen.

Organize

You may sometimes want to change the organization of the activities (see Figure 9–2) for the onscreen and the hard copy print of the PERT diagram. You may want to group similar activities together, sorted by some criteria. For communication purposes, you may want to group together all the activities of a particular subcontractor, crew, location, or responsibility. To change the screen activity organization, select **Organize,** from the **Format** main pull-down menu (refer to Figure 9–15). The **Organize** dialog box appears (Figure 9–25). The **Grouping** is defined in the **Group by** pull-down menu. Figure 9–2 was executed with **None** in the **Group by** field. The **Group by** field options are accessed by clicking the down arrow on the **Organize** dialog box. The options for the sample schedule appear in the pull-down menu in Figure 9–26. The majority of the options were defined using the **Activity Codes** dialog box, which you access by selecting the **Define** main pull-down menu, and then clicking on **Activity Codes.**

Figure 9–27 is an onscreen PERT view of the sample schedule organized by responsibility/activity code (the configuration requested in Figure 9–26). Compare Figure 9–27, with organization changes to Figure 9–2, with no organization formatting.

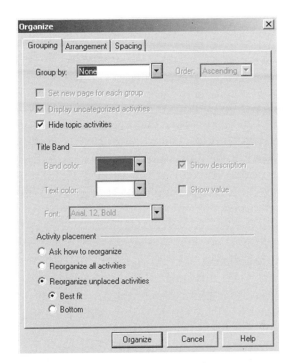

Figure 9–25   Organize Dialog Box—Grouping

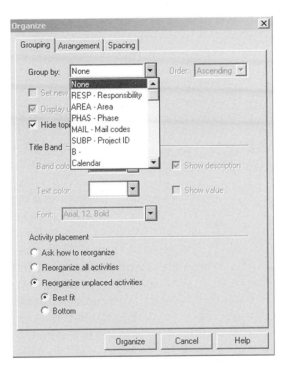

Figure 9–26   Organize Dialog Box—Group By

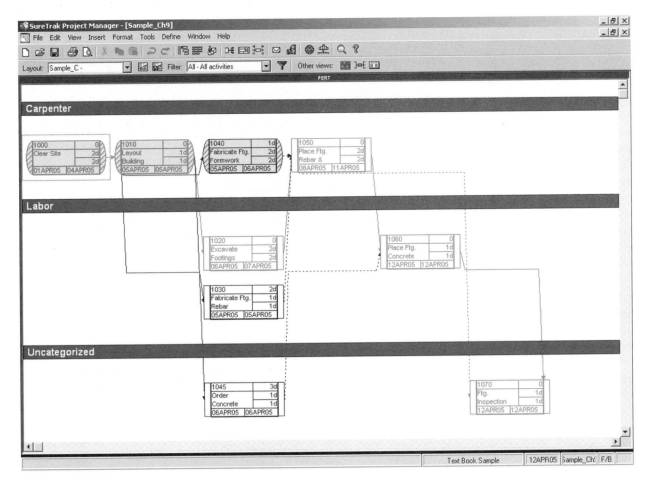

Figure 9–27   PERT—Organize—Group By

In addition to the grouping options (Figure 9–25), the **Organize** dialog box also has the arrangements and spacing options. Figure 9–28 shows the arrangements options, which you can access by clicking on the **Arrangement** tab.

By default, the PERT diagram is shown using the **Standard PERT layout.** By clicking on **PERT layout with timescale** and selecting an interval for the timescale (Figure 9–29), time information is added to the PERT diagram. Notice that when you click the **Organize** button in addition to having a timescale added, the nodes (boxes) are arranged to match the timescale. Compare Figure 9–30 with arrangement changes to Figure 9–22, with no arrangement changes.

Figure 9–31 shows the spacing options, which you can access by clicking on the **Spacing** tab of the **Organize** dialog box. By default, the spacing between nodes (boxes) is .25 inches vertically and .15 inches horizontally. You can modify these default values by clicking any of the arrows or typing a different number in any of the boxes. Figure 9–32 shows the **Spacing** dialog box with the vertical spacing modified from .25 inches to .95 inches. When you click the **Organize** button, the spacing is modified in the PERT diagram. Figure 9–33 shows the PERT diagram with nodes (boxes) vertically spaced .95 inches rather than .25 inches as shown in Figure 9–30.

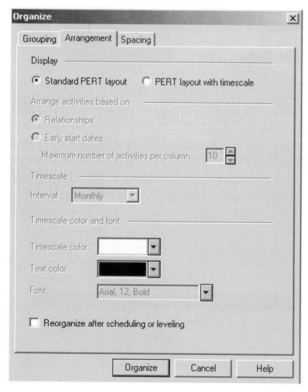

Figure 9–28 Organize Dialog Box— Arrangement

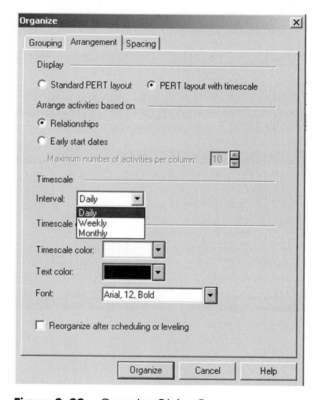

Figure 9–29 Organize Dialog Box— Arrangement

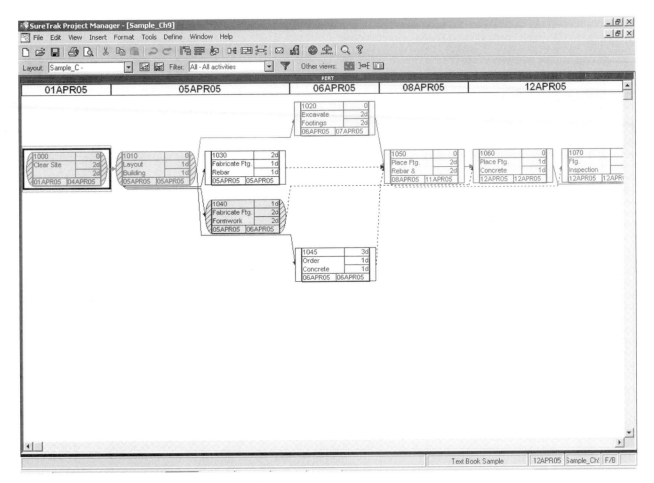

Figure 9–30   PERT—Arrange—Timescale

Filter

You may occasionally want to filter or limit the activities (refer to Figure 9–2) that appear on the onscreen and the hard copy print of the PERT diagram. You may want to show only the activities filtered by some criterion. To filter the screen activity selection, select **Filter,** from the **Format** main pull-down menu (refer to Figure 9–15). The **Filter** dialog box appears (Figure 9–34). The preconfigured filter options are:

All	All Activities
1 WK	One week lookahead
3 MON	Three month lookahead
4 WK	Four week lookahead
CNST	Activities with constraints
CRIT	Critical activities
CST>	Cost is greater than?
DLAY	Delayed activities
DONE	Completed activities
INDP	Independent activities
MEET	Meeting activities

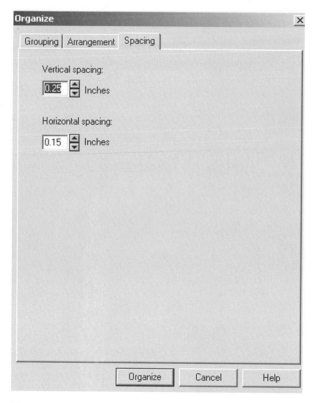

Figure 9–31   Organize Dialog Box—Spacing

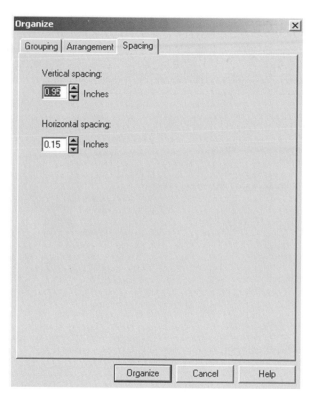

Figure 9–32   Organize Dialog Box—Spacing—
Vertical and Horizontal Spacing

MILE	Milestone activities
NEAR	Near critical activities
NET-	Negative net income
NET+	Positive net income
NONE	Activities with no progress
OVER	Overbudget activities
PROG	Work in progress
RES	Activities for resource?
STRT	Activities that start between ? and ?
TASK	Task activities
TODO	Activities underway or with no progress
TOPX	Topic activities
XTOP	Nontopic activities

Figure 9–34 shows **CRIT, Critical activities** selected. With the **Replace current set of activities** option selected, click on the **Apply** button. Figure 9–35 is the result. Because not all the activities in the sample schedule are critical (0 total float), the noncritical activities (1030, 1040, and 1045) were filtered.

As an example of filtering activities, Figure 9–36 shows the **TOPX, Topic activities** option selected from the **Filter** dialog box. Here "topic activities" can be filtered or selected and the nonselected activities hidden. The

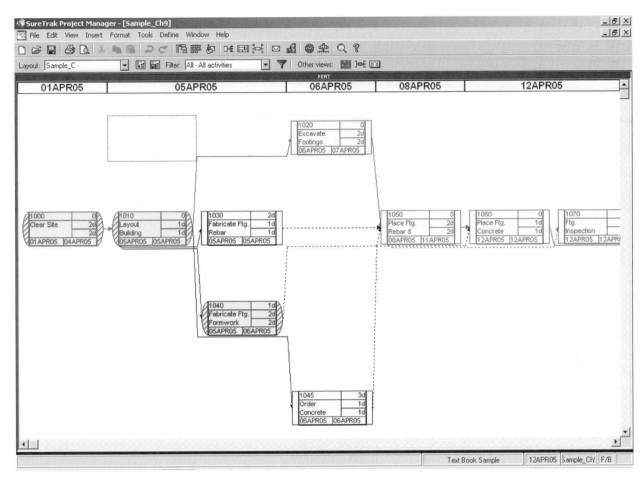

Figure 9–33   PERT—Arrange—Spacing

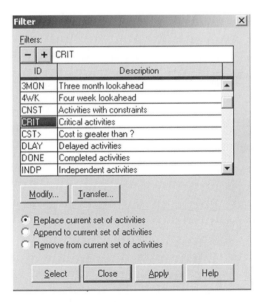

Figure 9–34   Filter Dialog Box

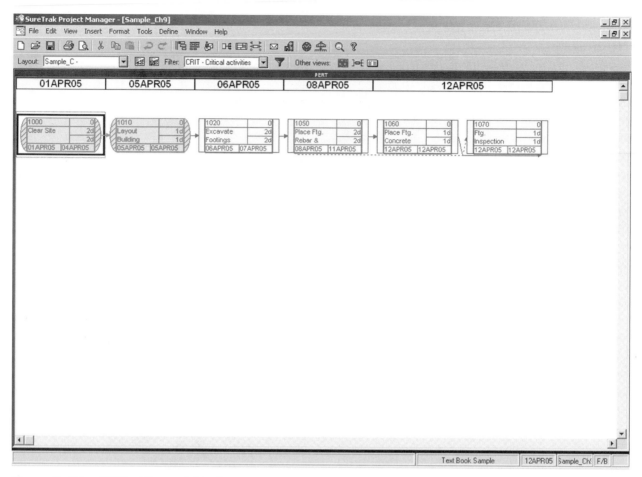

Figure 9–35 PERT—Filter—Critical Activities

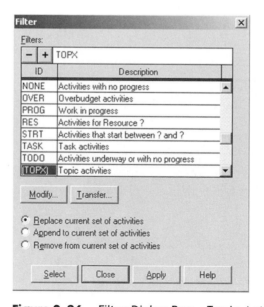

Figure 9–36 Filter Dialog Box—Topic Activities

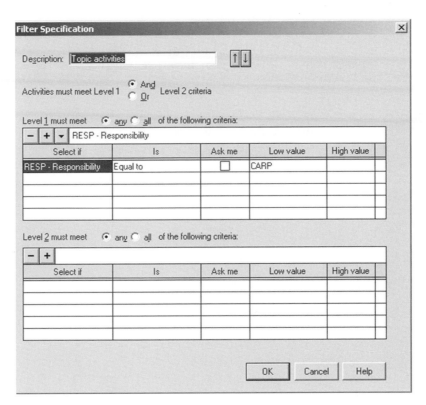

Figure 9–37   Filter Specification Dialog Box

TOPX filter is a *SureTrak* predefined filter used primarily to define topic activities under *SureTrak*'s Outline feature; however, it can also be used for defining selected or topic activities. First you have to define what the "topic activities" are. Select the **Modify** button to identify the topic activities to be filtered. This brings up the **Filter Specification** dialog box (Figure 9–37). The **Select if** column is used to define the filter limitation. Click on the + button to add a new filter specification. This highlights the first cell in the **Select if** column. Next, click on the down arrow to define the topic activities. The options are the same as the **Organize** dialog box (Figure 9–26). The majority of these options were defined using the **Activity Codes** dialog box from the **Define** main pull-down menu. Figure 9–38 is the onscreen PERT view of the sample schedule filtered to show only the carpentry crew activities (the configuration requested in Figure 9–37).

PRINT PREVIEW

The print options for the PERT view are very similar to the bar chart view discussed in Chapter 8. The **Print Preview** for the PERT view is accessed from the **File** main pull-down menu. Figure 9–39 is the **Print Preview** screen for the sample schedule. Compare the PERT **Print Preview** screen (Figure 9–39) to the bar chart view in Chapter 8 (see Figure 8–2).

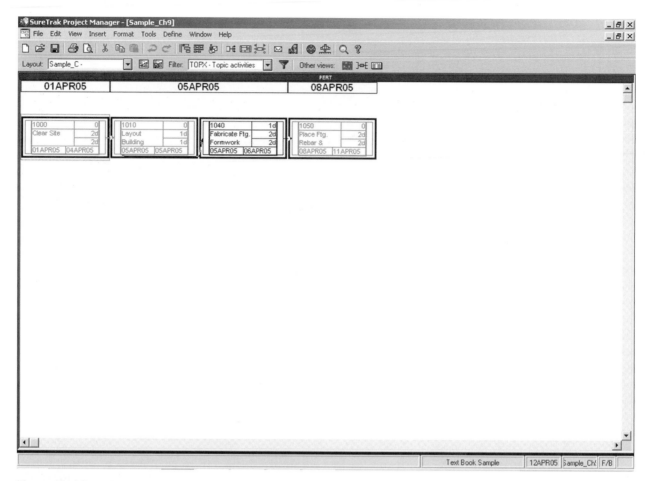

Figure 9–38   PERT—Filter—Topic Activities—RESP

Toolbar

Note that all the toolbar options for the two **Print Preview** screens are essentially the same and have the same functions except for the **Format PERT Diagram** and the **Format Bar Chart** buttons. The **Activity Box Configuration, Format Relationships, Format Filters,** and **Layouts** all call up format dialog boxes that were discussed earlier in this chapter. Again, when the onscreen PERT view is changed, the same change is carried over to the **Print Preview** screen and to the hard copy print. The print preview options for the PERT view operate essentially the same way as discussed in detail in Chapter 8.

Print

Figure 9–40 is a hard copy print of the PERT view of the sample schedule.

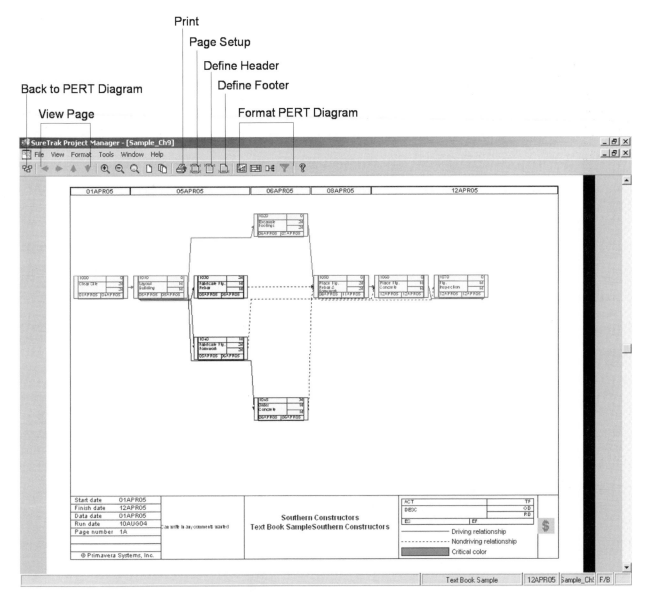

Figure 9–39   PERT—Filter—Topic Activities—RESP

EXAMPLE PROBLEM

Figures 9–41a, b, c, and d show four PERT hard copy prints for a house put together as an example for student use (see the wood frame house drawings in the Appendix).

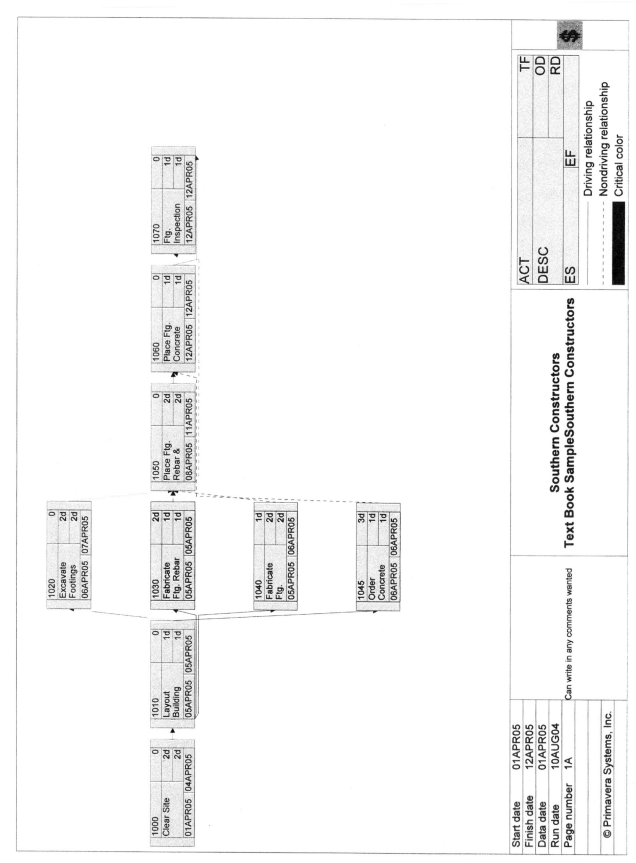

Figure 9–40 PERT Print—Sample Schedule

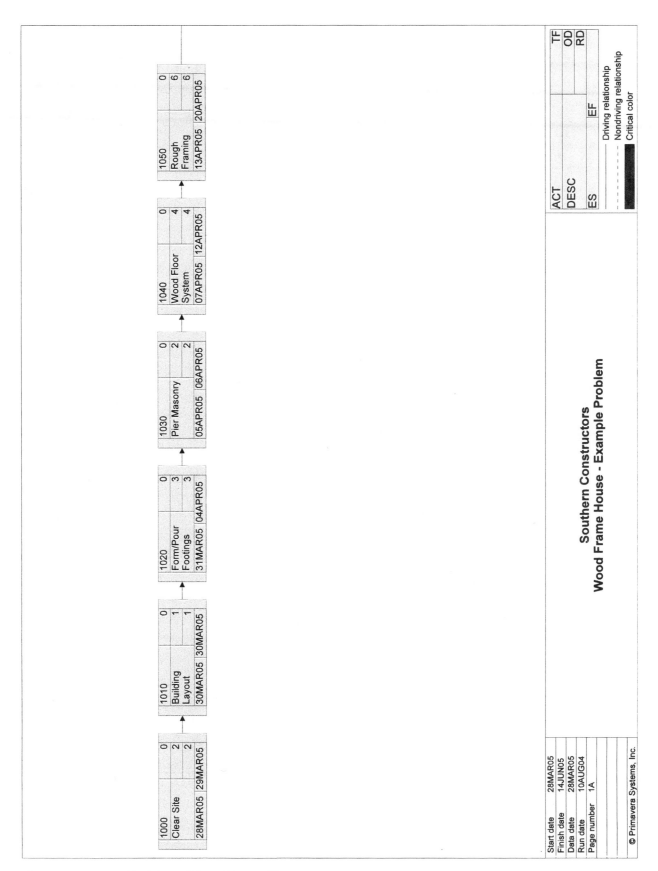

Figure 9–41a   PERT Print—Wood Frame House

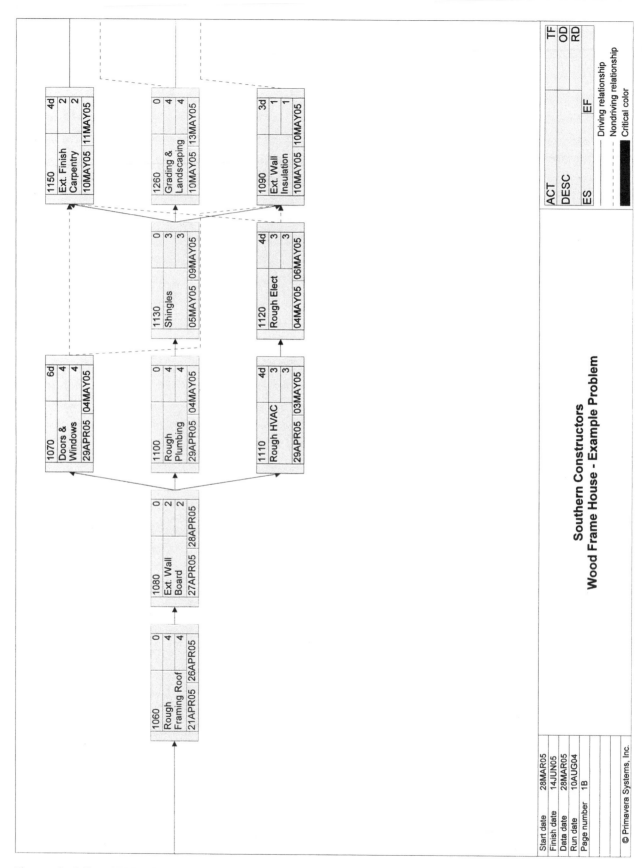

Figure 9–41b PERT Print—Wood Frame House

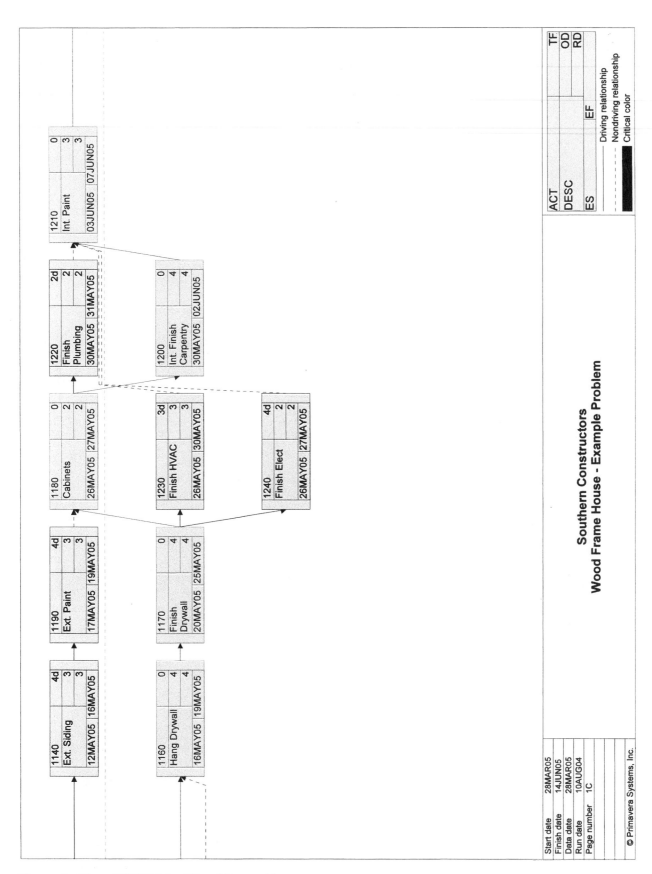

Figure 9–41c   PERT Print—Wood Frame House

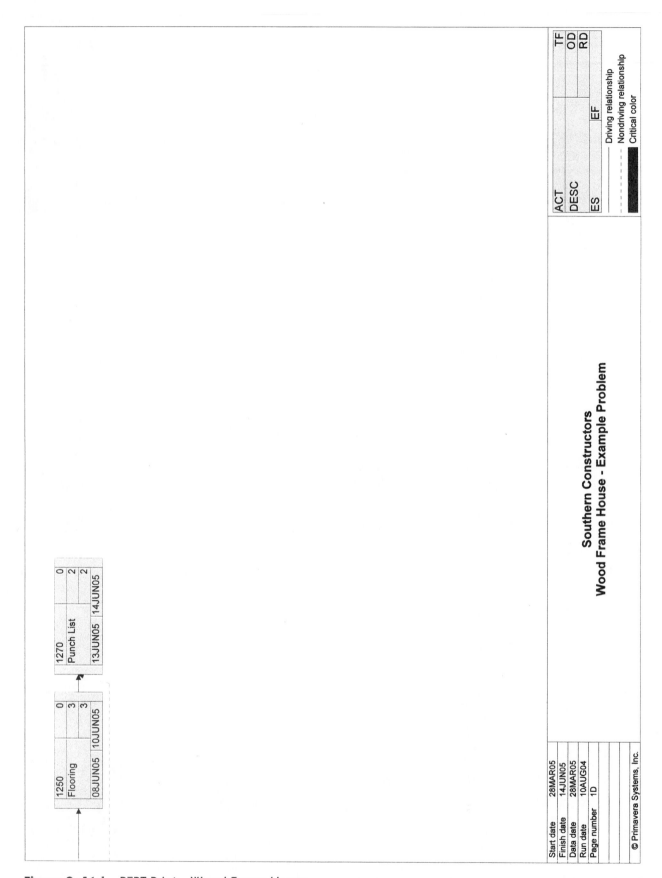

Figure 9–41d PERT Print—Wood Frame House

SUMMARY

The PERT diagram (activity-on-node logic diagram) is a visual representation of the relationships between activities. The arrows in this view show predecessor/successor relationships. Because the logic or relationships between the activities form the basis of the schedule and the basis for impacts when the schedule is updated, many schedulers like to use the PERT diagram for creating the original schedule rather than the method described in this book. This chapter provides an overview of formatting and printing the PERT diagram using *SureTrak*.

The primary advantage of the PERT diagram is the graphical representation of logic relationships. This chapter explained the process of adding and viewing activities using *SureTrak*'s **Activity Form, Trace Logic, Zoom,** and **Cosmic View** options.

This chapter also described the process of formatting the PERT view by changing the activity box configurations and relationship lines. The trace logic, organize, and filter PERT formatting capabilities were used.

SureTrak's print options for the PERT view are very similar to the bar chart options. The chapter showed examples of *SureTrak*'s PERT hard copy prints.

EXERCISES

1. **Describing the PERT Diagram Characteristics**
   Prepare a summary describing the characteristics of a PERT diagram.
2. **Inserting Activities Using the PERT Diagram View**
   Open the project modified in Exercise 5 of Chapter 7 and switch to the PERT diagram view. Insert a new activity as follows:

Activity ID:	900
Description:	X
Duration:	2 days
Successor:	1000

3. **Filtering Activities In and Out of the Preview**
   Filter the project from the previous exercise to show only critical path activities.
4. **Modifying Vertical and Horizontal Spacing Between Nodes**
   Modify the vertical and horizontal spacing to .5 inch.
5. **Printing the PERT Diagram**
   Print the Exercise 4 to fit on one page.
6. Prepare PERT diagram hard copy prints for Figure 3–12 (Exercise 4, Chapter 4).

7. Prepare PERT diagram hard copy prints for Figure 3–13 (Exercise 5, Chapter 4).

8. Prepare PERT diagram hard copy prints for Figure 3–14 (Exercise 6, Chapter 4).

9. Prepare PERT diagram hard copy prints for Figure 3–15 (Exercise 7, Chapter 4).

10. Prepare PERT diagram hard copy prints for Figure 3–16 (Exercise 8, Chapter 4).

11. Prepare PERT diagram hard copy prints for Figure 3–17 (Exercise 9, Chapter 4).

12. **Small Commercial Concrete Block Building**
Prepare PERT diagram hard copy prints for the small commercial concrete block building located (Exercise 12, Chapter 4) in the Appendix.

13. **Large Commercial Building**
Prepare PERT diagram hard copy prints for the large commercial building located (Exercise 13, Chapter 4) in the Appendix.

10 | List Reports

Objectives

Upon completion of this chapter, you should be able to:

- Describe the use of a list report
- Use a predefined list report
- Modify the columns of a list report
- Organize a list report
- Set up a list report for printing
- Preview a list report diagram
- Print a list report diagram

DAILY SCHEDULE LIST REPORTS

Although there are many styles of *SureTrak* formatted tabular reports, we have chosen the list tabular report as a demonstration example. List reports are very useful as a communication tool.

This form is particularly useful as a communication tool when it is necessary to give a craft or subcontractor only the information pertinent to the subcontractor's work. A sort by the activities pertinent to the work, with only necessary information about relevant activities, will reduce confusion and should increase efficiency. List reports can be used in conjunction with diagrams or used as stand-alone reports. They are also useful as worksheets when updating the schedule.

The list reports are accessed through **Reports** from the **Tools** main pull-down menu (Figure 10–1). The **Reports** dialog box appears (Figure 10–2). This dialog box lists all reports defined for this project (Sample), as well as those that come with *SureTrak*. As with previous encounters with this dialog box, you can either delete or modify the specifications to existing reports, or add a new report title and specification. The top half of the dialog box is used as a selection box for choosing preconfigured standardized **Reports.** You can also delete, add, name, format, and save new reports by clicking on the – or + buttons. Both the onscreen window and the hard copy prints are controlled by a report. This means that the onscreen version and the hard copy print are both changed when modifications are

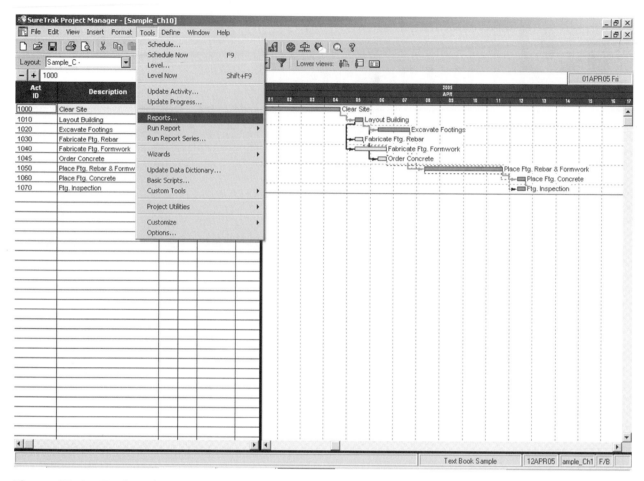

Figure 10–1   Tools Main Pull-Down Menu—Reports

Figure 10–2   Reports Dialog Box—List Reports

made in the **Reports** dialog box. The formatted report is a combination of layout, filter, view, and page options for the specified report within the specified project.

Within the **Reports** dialog box, click on **LST1 - Daily to-do list** to select it. The list (tabular) reports within *SureTrak* begin with **LST** in the **ID** field. The **Daily to-do list** is the report **Description**. The hard copy print will show the listed columns as selected. To change the new onscreen format back to the way it was before being reformatted, click on the **(DEFAULT)** report **Description** and then on the **Apply** button. The onscreen view will be returned to the **Current** view before the screen is reformatted.

The **Layout** and **Filter** options within the **Reports** dialog box display the layout and filter selected for the reports. Use the **Transfer** button if you have defined reports in another *SureTrak* project that you want to use on this project.

Figure 10–3 is an example of a schedule (Figure 10–1 in **Bar** chart format) in list or tabular format. This report is a hard copy print of selected information from the *SureTrak* activity table data in list or tabular form.

Figure 10–3 is a copy of the hard copy print on the **LST1** report with no modifications. Note that the activities are sorted by day (Early Start) and have the same columns as appear on the onscreen version if it is uncovered. Column selection and formatting was discussed in Chapter 5.

LAYOUTS

The icon buttons at the bottom of the **Reports** dialog box (see Figure 10–2) are used to reformat the selected report. With the **LST1** report selected, click on the **Modify** button to modify it. The **Layouts** dialog box can be selected for report configuration (Figure 10–4 and Figure 10–5). The **DAY - Organized by Day** format is selected under **Layouts** in the top half of the dialog box. Note that you are changing the *layout* associated with this report, not the actual report. This is important because you could have this layout assigned to multiple reports and you would therefore be changing the appearance of more than one report. The **Layouts** dialog box for formatting list reports (Figure 10–5) is the same as for formatting the onscreen bar chart (see Chapter 5, Figure 5–2) and the bar chart print (see Chapter 8, Figure 8–35).

COLUMNS

The columns specification function under the list reports works the same way as with the bar chart and PERT hard copy prints discussed in Chapter 8 and Chapter 9. By changing the columns selected for the list report,

Report: Daily to-do list **Layout:** Organized by Day **Filter:** All Activities			SureTrak Project Manager Text Book Sample				Southern Constructors Report Date: 11AUG04 Page 1A of 1A		

Act ID	Activity Description	Orig Dur	Rem Dur	Early Start	Early Finish	Total Float	%	Resource	Budgeted Cost
Text Book Sample									
01APR05									
1000	Clear Site	2d	2d	01APR05	04APR05	0	0	CARP 1,	340
05APR05									
1010	Layout Building	1d	1d	05APR05	05APR05	0	0	CARP 1,	482
1030	Fabricate Ftg. Rebar	1d	1d	05APR05	05APR05	2d	0	DIST, LAB	198
1040	Fabricate Ftg. Formwork	2d	2d	05APR05	06APR05	1d	0	CARP 1,	714
06APR05									
1020	Excavate Footings	2d	2d	06APR05	07APR05	0	0	CARP 1,	1,152
1045	Order Concrete	1d	1d	06APR05	06APR05	3d	0		0
08APR05									
1050	Place Ftg. Rebar & Formwork	2d	2d	08APR05	11APR05	0	0	CARP 1,	898
12APR05									
1060	Place Ftg. Concrete	1d	1d	12APR05	12APR05	0	0	CARP 1,	505
1070	Ftg. Inspection	1d	1d	12APR05	12APR05	0	0	INSP	240
		8d	8d	01APR05	12APR05	0	0		4,528

			Date	Revision		Checked	Approved
Data date	01APR05						
Start date	01APR05						
Finish date	12APR05						
Must finish date							
Target finish date							
© Primavera Systems, Inc.							

Figure 10–3 List Report

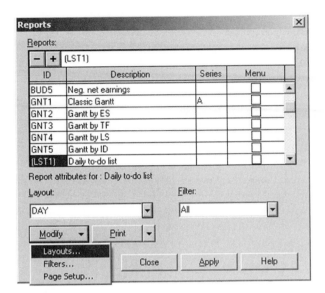

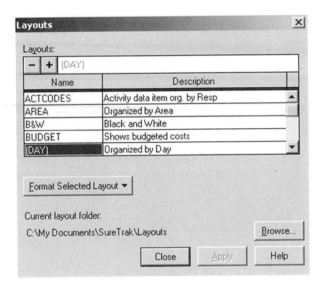

Figure 10–4 Reports Dialog Box—Modify

Figure 10–5 Layouts Dialog Box

the report can be changed to a more traditional scheduling report, or to a resource or cost report. To change the columns selected for print in the **LST1 - Daily to-do list** report, within the **Layouts** dialog box, click on the **Format Selected Layout** button, and select **Columns** (Figure 10–6). The **Columns** dialog box appears (Figure 10–7).

To change the list report in Figure 10–3 to a more traditional scheduling report, move the Percent Complete column after Rem Dur (remaining duration). Add a Late Start column after Early Start, add an Early Finish

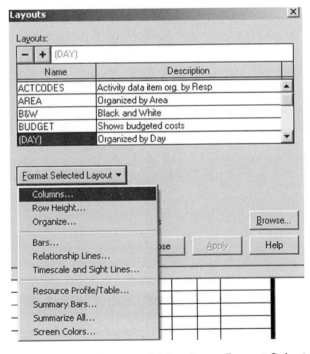

Figure 10–6 Layouts Dialog Box—Format Selected Layout

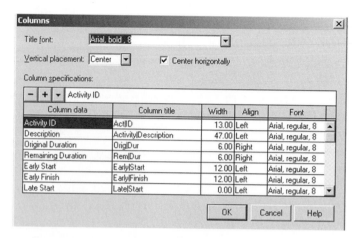

Figure 10–7   Columns Dialog Box

column before the Late Finish and delete the Resource and Budget columns. To delete the Percent Complete from its present location, use the scroll bar under **Column specifications,** find the **Column data** entitled **Percent Complete,** select it, and click on the – button. To add it in the new location, click on the **Column data** that will come after the new column **Early Start,** and click on the + button. A blank line for a new column will appear. Click on the down arrow to bring up the **Column data** menu. The **Columns** dialog box for formatting list reports (Figure 10–7) uses the same formatting as the onscreen bar chart (Chapter 5, Figure 5–6) and the bar chart print (Chapter 8, Figure 8–28). Figure 10–8 is the hard copy of the list report with the changes just described. You can modify the column title that appears on the hard copy print by changing the **Column title** field in the **Columns** dialog box (Figure 10–7). Use the **Width, Align,** and **Font** fields to change the appearance of the hard copy print (the change here will also affect the onscreen column appearance).

ORGANIZE

The organize function under the list reports works the same way as with the bar chart and PERT hard copy prints discussed in Chapter 5 and Chapter 9. Click on the **Layouts** button of the **Reports** dialog box (Figure 10–4). Click on the **Format Selected Layout** button and select **Organize** (Figure 10–9). The **Organize** dialog box appears (Figure 10–10). The **Organize** dialog box for formatting list reports (Figure 10–10) is the same as the onscreen bar chart (Chapter 5, Figure 5–30) and the bar chart print (Chapter 8, Figure 8–37). To reorganize the hard copy print in Figure 10–8 to sort by responsibility, click on the first field (the field after the inserted organize criterion) under **Group by,** under **Organized by.** We want to change from **Project** in Figure 10–10 to **RESP - Responsibility** in Figure 10–11. Figure 10–12 shows a hard print of the *SureTrak* list report reorganized by responsibility.

Report: Daily to-do list Layout: Organized by Day Filter: All Activities		SureTrak Project Manager Text Book Sample					Southern Constructors Report Date: 11AUG04 Page 1A of 1A		

Act ID	Activity Description	Orig Dur	Rem Dur	Percent Complete	Early Start	Late Start	Early Finish	Total Float	
Text Book Sample									
01APR05									
1000	Clear Site	2d	2d	0	01APR05	01APR05	04APR05	0	
05APR05									
1010	Layout Building	1d	1d	0	05APR05	05APR05	05APR05	0	
1030	Fabricate Ftg. Rebar	1d	1d	0	05APR05	07APR05	05APR05	2d	
1040	Fabricate Ftg. Formwork	2d	2d	0	05APR05	06APR05	06APR05	1d	
06APR05									
1020	Excavate Footings	2d	2d	0	06APR05	06APR05	07APR05	0	
1045	Order Concrete	1d	1d	0	06APR05	11APR05	06APR05	3d	
08APR05									
1050	Place Ftg. Rebar & Formwork	2d	2d	0	08APR05	08APR05	11APR05	0	
12APR05									
1060	Place Ftg. Concrete	1d	1d	0	12APR05	12APR05	12APR05	0	
1070	Ftg. Inspection	1d	1d	0	12APR05	12APR05	12APR05	0	
		8d	8d	0	01APR05	01APR05	12APR05	0	

Data date	01APR05		Date	Revision	Checked	Approved
Start date	01APR05					
Finish date	12APR05					
Must finish date						
Target finish date						
© Primavera Systems, Inc.						

Figure 10–8   List Report—Columns Changed

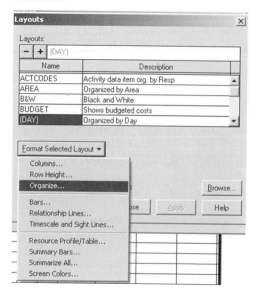

Figure 10–9   Layouts Dialog Box—Format Selected Layout

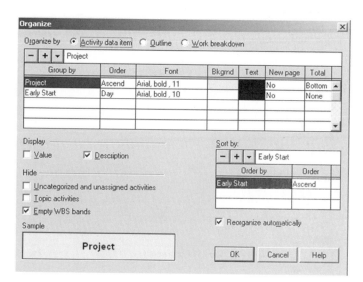

Figure 10–10   Organize Dialog Box

PAGE SETUP

The Page Setup function under the list reports also works the same way as the bar chart and PERT hard copy prints. From the **Reports** dialog box (Figure 10–12), click on **Modify,** then select the **Page Setup** button, and the **Page Setup** dialog box (Figure 10–13) appears. We want to modify the hard copy print in Figure 10–12 to include a title block and a logo. Click on the **Footer** button of the **Page Setup** dialog box (Figure 10–13) to open the **Footer** dialog box (Figure 10–14). Click on the third field from the left, labeled **None,** and then select the **Title Block** footer option from the drop-down menu that appears (Figure 10–15). In the **Define Title**

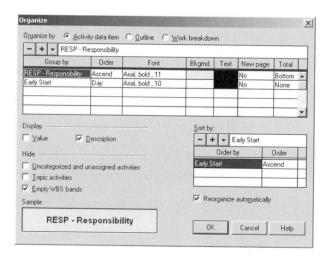

Figure 10–11   Organize Dialog Box—RESP

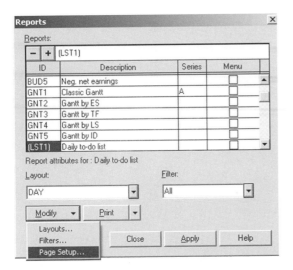

Figure 10–12   Reports Dialog Box—Page Setup

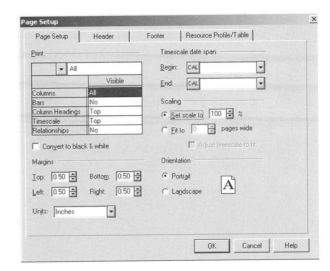

Figure 10–13   Page Setup Dialog Box

Block field at the bottom of the dialog box (Figure 10–15), type the title block information you want.

To include the logo in the **Footer** dialog box, click on the last field to the right (or the field you want the title logo in), then select the **Logo** footer option from the drop-down menu that appears (Figure 10–16). Under the **Define Logo** field, we selected the order.BMP file. Figure 10–17 shows the resulting new title block and logo added to the hard copy *SureTrak* list report.

WEEKLY SCHEDULE LIST REPORTS

On a larger project, the daily list (**LST1**) report can become too "busy." The weekly (**LST2**) or monthly (**LST3**) can be a more informative communications tool (refer to Figure 10–2). Figure 10–18 is a hard copy print of the

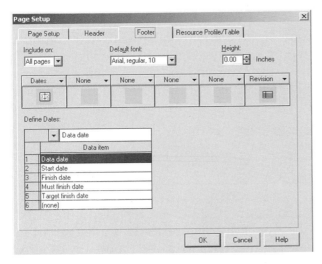

Figure 10–14   Page Setup Dialog Box—Footer

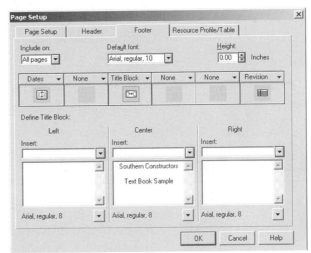

Figure 10–15   Page Setup Dialog Box—Title Block

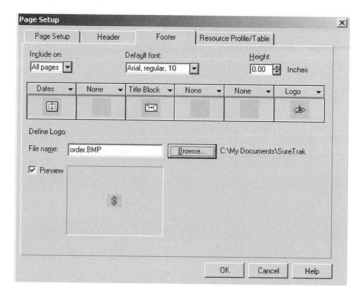

Figure 10–16 Page Setup Dialog Box—Logo

Weekly to-do list (LST2) of the sample schedule. Compare Figure 10–18 to Figure 10–3, the same list report executed in the daily format (**LST1**).

RESOURCE AND COST LIST REPORTS

The columns of the list reports can be reconfigured to make them resource and cost reports. Figure 10–19 is the hard copy of the daily to-do list with new columns specified to make it a resource report. Compare Figure 10–19 to Figure 10–17. The Early Start, Late Start, Early Finish, and Late Finish columns have been deleted. The Budgeted Quantity, Actual Quantity to Date, and Completion Variance Quantity columns have been added. This tabular list report is a valuable tool for analyzing and updating resource usage.

Figure 10–20 is the hard copy of the daily to-do list with new columns specified to make it a cost report. Compare Figure 10–20 to Figure 10–17. The Budgeted Cost, Actual Cost to Date, and the Cost Variance columns have been added. This tabular list report is a valuable tool for analyzing and updating cash flow.

EXAMPLE PROBLEM

Figure 10–21, Figure 10–22, and Figure 10–23 are list reports for the wood frame house drawings in the Appendix. Figure 10–21 is a list report in the classic schedule format. Figure 10–22 is a list report in resource format. Figure 10–23 is a list report in cost format.

Report: Daily to-do list
Layout: Organized by Day
Filter: All Activities

SureTrak Project Manager
Text Book Sample

Southern Constructors
Report Date: 11AUG04
Page 1A of 1A

Act ID	Activity Description	Orig Dur	Rem Dur	Percent Complete	Early Start	Late Start	Early Finish	Total Float	
Carpenter									
01APR05									
1000	Clear Site	2d	2d	0	01APR05	01APR05	04APR05	0	
05APR05									
1010	Layout Building	1d	1d	0	05APR05	05APR05	05APR05	0	
1040	Fabricate Ftg. Formwork	2d	2d	0	05APR05	06APR05	06APR05	1d	
08APR05									
1050	Place Ftg. Rebar & Formwork	2d	2d	0	08APR05	08APR05	11APR05	0	
		7d	7d	0	01APR05	01APR05	11APR05	0	
Labor									
05APR05									
1030	Fabricate Ftg. Rebar	1d	1d	0	05APR05	07APR05	05APR05	2d	
06APR05									
1020	Excavate Footings	2d	2d	0	06APR05	06APR05	07APR05	0	
12APR05									
1060	Place Ftg. Concrete	1d	1d	0	12APR05	12APR05	12APR05	0	
		6d	6d	0	05APR05	06APR05	12APR05	0	
06APR05									
1045	Order Concrete	1d	1d	0	06APR05	11APR05	06APR05	3d	
12APR05									
1070	Ftg. Inspection	1d	1d	0	12APR05	12APR05	12APR05	0	
		5d	5d	0	06APR05	11APR05	12APR05	0	

Data date	01APR05		
Start date	01APR05	Southern Constructors	
Finish date	12APR05		
Must finish date		Text Book Sample	
Target finish date			
© Primavera Systems, Inc.			

Figure 10–17   List Report with Title Block and Logo

Report: Weekly to-do list Layout: Organized by Week Filter: All Activities		SureTrak Project Manager Text Book Sample						Southern Constructors Report Date: 11AUG04 Page 1A of 1A	

Act ID	Activity Description	Orig Dur	Rem Dur	Early Start	Early Finish	Total Float	%	Resource	Budgeted Cost
Text Book Sample									
28MAR05									
1000	Clear Site	2d	2d	01APR05	04APR05	0	0	CARP 1,	340
04APR05									
1010	Layout Building	1d	1d	05APR05	05APR05	0	0	CARP 1,	482
1030	Fabricate Ftg. Rebar	1d	1d	05APR05	05APR05	2d	0	DIST, LAB	198
1040	Fabricate Ftg. Formwork	2d	2d	05APR05	06APR05	1d	0	CARP 1,	714
1020	Excavate Footings	2d	2d	06APR05	07APR05	0	0	CARP 1,	1,152
1045	Order Concrete	1d	1d	06APR05	06APR05	3d	0		0
1050	Place Ftg. Rebar & Formwork	2d	2d	08APR05	11APR05	0	0	CARP 1,	898
11APR05									
1060	Place Ftg. Concrete	1d	1d	12APR05	12APR05	0	0	CARP 1,	505
1070	Ftg. Inspection	1d	1d	12APR05	12APR05	0	0	INSP	240
		8d	8d	01APR05	12APR05	0	0		4,528

Data date	01APR05		Date	Revision	Checked	Approved
Start date	01APR05					
Finish date	12APR05					
Must finish date						
Target finish date						
© Primavera Systems, Inc.						

Figure 10–18 Weekly List Report

Report: Daily to-do list
Layout: Organized by Day
Filter: All Activities

SureTrak Project Manager
Text Book Sample

Act ID	Activity Description	Orig Dur	Rem Dur	Percent Complete	Budgeted Quantity	Actual Quantity to Date	Completion Variance Quantity	Total Float
Carpenter								
01APR05								
1000	Clear Site	2d	2d	0	34	0	0	0
05APR05								
1010	Layout Building	1d	1d	0	20	0	0	0
1040	Fabricate Ftg. Formwork	2d	2d	0	38	0	0	1d
08APR05								
1050	Place Ftg. Rebar & Formwork	2d	2d	0	38	0	0	0
		7d	7d	0	130	0	0	0
Labor								
05APR05								
1030	Fabricate Ftg. Rebar	1d	1d	0	18	0	0	2d
06APR05								
1020	Excavate Footings	2d	2d	0	42	0	0	0
12APR05								
1060	Place Ftg. Concrete	1d	1d	0	19	0	0	0
		6d	6d	0	79	0	0	0
06APR05								
1045	Order Concrete	1d	1d	0	0	0	0	3d
12APR05								
1070	Ftg. Inspection	1d	1d	0	1	0	0	0
		5d	5d	0	1	0	0	0

Data date	01APR05	
Start date	01APR05	Southern Constructors
Finish date	12APR05	
Must finish date		Text Book Sample
Target finish date		
© Primavera Systems, Inc.		

Figure 10–19 List Resource Report

| Report: Daily to-do list
Layout: Organized by Day
Filter: All Activities | SureTrak Project Manager
Text Book Sample | Southern Constructors
Report Date: 11AUG04
Page 1A of 1A |

Act ID	Activity Description	Orig Dur	Rem Dur	Percent Complete	Budgeted Cost	Actual Cost to Date	Cost Variance	Total Float
Carpenter								
01APR05								
1000	Clear Site	2d	2d	0	340	0	0	0
05APR05								
1010	Layout Building	1d	1d	0	482	0	0	0
1040	Fabricate Ftg. Formwork	2d	2d	0	714	0	0	1d
08APR05								
1050	Place Ftg. Rebar & Formwork	2d	2d	0	898	0	0	0
		7d	7d	0	2,432	0	0	0
Labor								
05APR05								
1030	Fabricate Ftg. Rebar	1d	1d	0	198	0	0	2d
06APR05								
1020	Excavate Footings	2d	2d	0	1,152	0	0	0
12APR05								
1060	Place Ftg. Concrete	1d	1d	0	505	0	0	0
		6d	6d	0	1,855	0	0	0
06APR05								
1045	Order Concrete	1d	1d	0	0	0	0	3d
12APR05								
1070	Ftg. Inspection	1d	1d	0	240	0	0	0
		5d	5d	0	240	0	0	0

Data date	01APR05
Start date	01APR05
Finish date	12APR05
Must finish date	
Target finish date	
© Primavera Systems, Inc.	

Southern Constructors

Text Book Sample

Figure 10–20 List Cost Report

Report: Daily to-do list
Layout: Organized by Day
Filter: All Activities

SureTrak Project Manager
Wood Frame House - Example Problem

Southern Constructors
Report Date: 11AUG04
Page 1A of 1A

Act ID	Activity Description	Orig Dur	Rem Dur	Percent Complete	Early Start	Late Start	Early Finish	Late Finish	Total Float	
28MAR05									0	
1000	Clear Site	2	2	0	28MAR05	28MAR05	29MAR05	29MAR05		
30MAR05									0	
1010	Building Layout	1	1	0	30MAR05	30MAR05	30MAR05	30MAR05		
31MAR05									0	
1020	Form/Pour Footings	3	3	0	31MAR05	31MAR05	04APR05	04APR05		
05APR05									0	
1030	Pier Masonry	2	2	0	05APR05	05APR05	06APR05	06APR05		
07APR05									0	
1040	Wood Floor System	4	4	0	07APR05	07APR05	12APR05	12APR05		
13APR05									0	
1050	Rough Framing Walls	6	6	0	13APR05	13APR05	20APR05	20APR05		
21APR05									0	
1060	Rough Framing Roof	4	4	0	21APR05	21APR05	26APR05	26APR05		
27APR05									0	
1080	Ext. Wall Board	2	2	0	27APR05	27APR05	28APR05	28APR05		
29APR05										
1070	Doors & Windows	4	4	0	29APR05	09MAY05	04MAY05	12MAY05	6d	
1100	Rough Plumbing	4	4	0	29APR05	29APR05	04MAY05	04MAY05	0	
1110	Rough HVAC	3	3	0	29APR05	05MAY05	03MAY05	09MAY05	4d	
04MAY05										
1120	Rough Elect	3	3	0	04MAY05	10MAY05	06MAY05	12MAY05	4d	
05MAY05										
1130	Shingles	3	3	0	05MAY05	05MAY05	09MAY05	09MAY05	0	
10MAY05										
1090	Ext. Wall Insulation	1	1	0	10MAY05	13MAY05	10MAY05	13MAY05	3d	
1150	Ext. Finish Carpentry	2	2	0	10MAY05	16MAY05	11MAY05	17MAY05	4d	
1260	Grading & Landscaping	4	4	0	10MAY05	10MAY05	13MAY05	13MAY05	0	
12MAY05										
1140	Ext. Siding	3	3	0	12MAY05	18MAY05	16MAY05	20MAY05	4d	
16MAY05										
1160	Hang Drywall	4	4	0	16MAY05	16MAY05	19MAY05	19MAY05	0	
17MAY05										
1190	Ext. Paint	3	3	0	17MAY05	23MAY05	19MAY05	25MAY05	4d	
20MAY05										
1170	Finish Drywall	4	4	0	20MAY05	20MAY05	25MAY05	25MAY05	0	
26MAY05										
1180	Cabinets	2	2	0	26MAY05	26MAY05	27MAY05	27MAY05	0	
1230	Finish HVAC	3	3	0	26MAY05	31MAY05	30MAY05	02JUN05	3d	
1240	Finish Elect	2	2	0	26MAY05	01JUN05	27MAY05	02JUN05	4d	
30MAY05										
1200	Int. Finish Carpentry	4	4	0	30MAY05	30MAY05	02JUN05	02JUN05	0	
1220	Finish Plumbing	2	2	0	30MAY05	01JUN05	31MAY05	02JUN05	2d	
03JUN05										
1210	Int. Paint	3	3	0	03JUN05	03JUN05	07JUN05	07JUN05	0	
08JUN05										
1250	Flooring	3	3	0	08JUN05	08JUN05	10JUN05	10JUN05	0	
13JUN05										
1270	Punch List	2	2	0	13JUN05	13JUN05	14JUN05	14JUN05	0	
		57	57	0	28MAR05	28MAR05	14JUN05	14JUN05	0	

			Date	Revision	Checked	Approved
Data date	28MAR05					
Start date	28MAR05					
Finish date	14JUN05					
Must finish date						
Target finish date						
© Primavera Systems, Inc.						

Figure 10–21　List Report—Schedule—Wood Frame House

Report: Daily to-do list Layout: Organized by Day Filter: All Activities			SureTrak Project Manager Wood Frame House - Example Problem			Southern Constructors Report Date: 11AUG04 Page 1A of 1A	

Act ID	Activity Description	Orig Dur	Rem Dur	Percent Complete	Budgeted Quantity	Actual Quantity to Date	Completion Variance Quantity	Total Float
28MAR05								
1000	Clear Site	2	2	0	18	0	0	0
30MAR05								
1010	Building Layout	1	1	0	37	0	0	0
31MAR05								
1020	Form/Pour Footings	3	3	0	63	0	0	0
05APR05								
1030	Pier Masonry	2	2	0	58	0	0	0
07APR05								
1040	Wood Floor System	4	4	0	84	0	0	0
13APR05								
1050	Rough Framing Walls	6	6	0	126	0	0	0
21APR05								
1060	Rough Framing Roof	4	4	0	116	0	0	0
27APR05								
1080	Ext. Wall Board	2	2	0	42	0	0	0
29APR05								
1070	Doors & Windows	4	4	0	84	0	0	6d
1100	Rough Plumbing	4	4	0	36	0	0	0
1110	Rough HVAC	3	3	0	27	0	0	4d
04MAY05								
1120	Rough Elect	3	3	0	27	0	0	4d
05MAY05								
1130	Shingles	3	3	0	27	0	0	0
10MAY05								
1090	Ext. Wall Insulation	1	1	0	9	0	0	3d
1150	Ext. Finish Carpentry	2	2	0	42	0	0	4d
1260	Grading & Landscaping	4	4	0	36	0	0	0
12MAY05								
1140	Ext. Siding	3	3	0	27	0	0	4d
16MAY05								
1160	Hang Drywall	4	4	0	36	0	0	0
17MAY05								
1190	Ext. Paint	3	3	0	27	0	0	4d
20MAY05								
1170	Finish Drywall	4	4	0	36	0	0	0
26MAY05								
1180	Cabinets	2	2	0	18	0	0	0
1230	Finish HVAC	3	3	0	27	0	0	3d
1240	Finish Elect	2	2	0	18	0	0	4d
30MAY05								
1200	Int. Finish Carpentry	4	4	0	84	0	0	0
1220	Finish Plumbing	2	2	0	18	0	0	2d
03JUN05								
1210	Int. Paint	3	3	0	27	0	0	
08JUN05								
1250	Flooring	3	3	0	27	0	0	0
13JUN05								
1270	Punch List	2	2	0	0	0	0	0
		57	57	0	1,177	0	0	0

			Date	Revision	Checked	Approved
Data date	28MAR05					
Start date	28MAR05					
Finish date	14JUN05					
Must finish date						
Target finish date						
© Primavera Systems, Inc.						

Figure 10–22   List Report—Resource Control—Wood Frame House

| Report: Daily to-do list
Layout: Organized by Day
Filter: All Activities | | SureTrak Project Manager
Wood Frame House - Example Problem | | | | | | Southern Constructors
Report Date: 11AUG04
Page 1A of 1A | |

Act ID	Activity Description	Orig Dur	Rem Dur	Percent Complete	Budgeted Cost	Actual Cost to Date	Cost Variance	Total Float	
28MAR05									
1000	Clear Site	2	2	0	1,680	0	0	0	
30MAR05									
1010	Building Layout	1	1	0	2,375	0	0	0	
31MAR05									
1020	Form/Pour Footings	3	3	0	3,045	0	0	0	
05APR05									
1030	Pier Masonry	2	2	0	1,260	0	0	0	
07APR05									
1040	Wood Floor System	4	4	0	2,540	0	0	0	
13APR05									
1050	Rough Framing Walls	6	6	0	3,560	0	0	0	
21APR05									
1060	Rough Framing Roof	4	4	0	2,840	0	0	0	
27APR05									
1080	Ext. Wall Board	2	2	0	1,420	0	0	0	
29APR05									
1070	Doors & Windows	4	4	0	2,790	0	0	6d	
1100	Rough Plumbing	4	4	0	1,000	0	0	0	
1110	Rough HVAC	3	3	0	1,568	0	0	4d	
04MAY05									
1120	Rough Elect	3	3	0	1,240	0	0	4d	
05MAY05									
1130	Shingles	3	3	0	1,441	0	0	0	
10MAY05									
1090	Ext. Wall Insulation	1	1	0	585	0	0	3d	
1150	Ext. Finish Carpentry	2	2	0	1,920	0	0	4d	
1260	Grading & Landscaping	4	4	0	900	0	0	0	
12MAY05									
1140	Ext. Siding	3	3	0	2,010	0	0	4d	
16MAY05									
1160	Hang Drywall	4	4	0	2,244	0	0	0	
17MAY05									
1190	Ext. Paint	3	3	0	1,300	0	0	4d	
20MAY05									
1170	Finish Drywall	4	4	0	1,190	0	0	0	
26MAY05									
1180	Cabinets	2	2	0	2,018	0	0	0	
1230	Finish HVAC	3	3	0	4,506	0	0	3d	
1240	Finish Elect	2	2	0	1,910	0	0	4d	
30MAY05									
1200	Int. Finish Carpentry	4	4	0	2,840	0	0	0	
1220	Finish Plumbing	2	2	0	4,000	0	0	2d	
03JUN05									
1210	Int. Paint	3	3	0	5,725	0	0	0	
08JUN05									
1250	Flooring	3	3	0	2,083	0	0	0	
13JUN05									
1270	Punch List	2	2	0	0	0	0	0	
		57	57	0	59,990	0	0	0	

			Date	Revision	Checked	Approved
Data date	28MAR05					
Start date	28MAR05					
Finish date	14JUN05					
Must finish date						
Target finish date						
© Primavera Systems, Inc.						

Figure 10–23 List Report—Cost Control—Wood Frame House

SUMMARY

SureTrak's list reports provide project information in a tabular format. Sometimes a tabular report is a more convenient format for transferring, analyzing, or updating data. This chapter provided an overview of formatting and printing *SureTrak's* list reports.

This chapter showed examples of the list reports shown in a daily, weekly, and monthly format.

As with other *SureTrak* screens and reports, **Layouts** is a convenient method for modifying the list reports columns and bars. You also saw how to use Organize within Layouts to sort project data.

SureTrak's print options for list reports are very similar to those for reports discussed earlier in the book.

EXERCISES

1. **Describing the Use of a List Report**
   Prepare a summary describing the uses of a list report.
2. **Using a Predefined List Report**
   Open the project modified in Exercise 5 of Chapter 7 and use the report (LST1) Daily-to-do list.
3. **Modifying the Columns of a List Report**
   Include the percentage complete column following the activity description.
4. **Printing the PERT Diagram**
   Print Exercise 3.
5. Prepare a classic schedule list hard copy print for Figure 3–13 (Exercise 4, Chapter 4).
6. Prepare a classic schedule list hard copy print for Figure 3–14 (Exercise 5, Chapter 4).
7. Prepare a classic schedule list hard copy print for Figure 3–15 (Exercise 6, Chapter 4).
8. Prepare a classic schedule list hard copy print for Figure 3–16 (Exercise 7, Chapter 4).
9. Prepare a classic schedule list hard copy print for Figure 3–17 (Exercise 8, Chapter 4).
10. Prepare a classic schedule list hard copy print for Figure 3–18 (Exercise 9, Chapter 4).
11. **Small Commercial Concrete Block Building—List Reports**
    Prepare list reports for the small commercial concrete block building located (Exercise 12, Chapter 4) in the Appendix.
12. **Large Commercial Building—List Reports**
    Prepare list reports for the large commercial building located (Exercise 13, Chapter 4) in the Appendix.

Section 3

Controlling

11 | Updating the Schedule

Objectives

Upon completion of this chapter, you should be able to:

- Describe the importance of updating the schedule
- Create a copy before updating a schedule
- Establish target dates (or baselines)
- Establish a new data date
- Update activities
- Calculate the updated schedule
- Input progress through the activity table
- Display a target bar
- Display a PERT diagram
- Change activity duration and logic relationships
- Add a new activity to an updated schedule

COMMUNICATING CURRENT INFORMATION

Why update a project? Very few projects ever go exactly according to the original plan/schedule. Among the influences likely to change the original plan are weather, acts of God, better or worse productivity than anticipated, delivery problems, labor problems, changes in the scope of the project, interferences between crafts or subcontractors, and mismanagement in the flow of work. After a project is underway, it is necessary to modify the schedule to keep it current.

When you update a project, you show that time has passed and progress has been made on the project. *SureTrak* calculates the effects of this progress on the remainder of the project.

Making Changes

For a schedule to remain viable, the project changes must continually be incorporated into the current schedule. If the update shows that an activity was delayed, extended, interrupted, or accelerated, *SureTrak* shows the effects of that change on the successors to the activity, and all their successors, all the way to project completion. Before starting the update, you need to save the project baselines. The target dates are the schedule baseline; the budget amounts are the cost baseline. It is also a good idea to copy your project before updating for documentation purposes, in case any later problems develop. Generally, a project is updated at the same time each month. The *data date* for recording progress is usually at the beginning of the new month. The current schedule is thus the current plan for project completion. An update should result in a *SureTrak* project that accurately reflects the current status of the project.

Tracking Physical Progress

Keeping the schedule current involves two phases.

Monitoring Progress.   The first phase is monitoring all activities to determine their status. Each activity is either complete, partially complete, or no work has been accomplished. The activities for which there is no physical progress keep their original activity relationships and durations. Where there is partial physical progress, the original activity relationships are kept and the remaining duration is calculated either as number of days or percent complete. After all progress to date is recorded for each activity, the schedule is recalculated. Individual activities and the ensemble are determined to be either on, ahead of, or behind schedule.

Documenting Changes.   The second phase is documenting changes. Very seldom is a project built in exactly the sequence planned. Either more detail is needed, or with more time for analysis, better plans are developed. When the project scope changes, change orders must be incorporated. As the approach to building the project changes, the contractor needs to modify the schedule to show the changes.

GETTING STARTED

SureTrak offers tools to aid the scheduler in keeping the schedule current. The first step should be to copy the previous schedule. This step is not only in line with good practice, but should be considered an absolute requirement. As good practice, one of the advantages of *SureTrak* is to play "what if" games; therefore, for whatever reason, it can be valuable to have an original, "uncontaminated" version of the schedule that you can recopy and restart with any intended changes.

Using the Save As Dialog Box

To make a copy of the schedule, click on the **File** main pull-down menu and select **Save As** (Figure 11–1). The **Save As** dialog box appears (Figure 11–2). In the default configuration, project files are saved in the **C:\My Documents\SureTrak** directory. Under **Project name**, type in the new name for the copied project. We chose the name **Sample_Ch11_Update_04_04_05.** Click on **OK** to execute the copy of the original

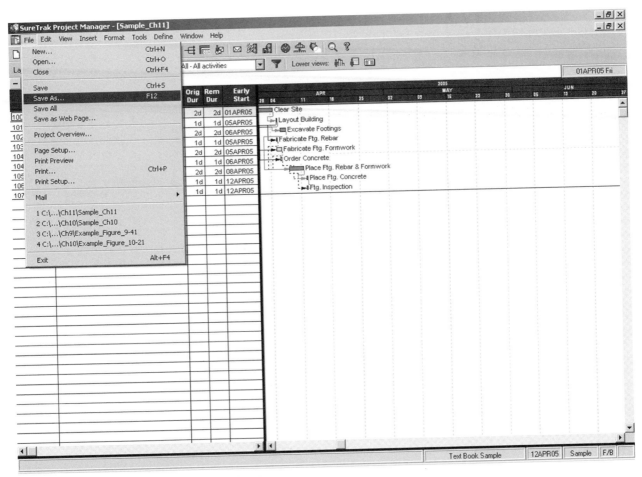

Figure 11-1   Sample Schedule—File Main Pull-Down Menu

schedule. For each schedule, there are approximately twenty-five *SureTrak* files created, so it is easier to copy projects using the **Save As** dialog box than by other means.

Establishing Target Dates

Before starting the actual update, you need to save the target dates or project baselines. The target dates represent the baseline or "original" schedule to use for comparison purposes. Click on the **Define** main pull-down menu, and then select **Target Dates**. The **Target Dates** dialog box appears (Figure 11–3). For the sample schedule, we assigned **Early dates** to **All activities**. Click on the **OK** button to save the target dates as the project baseline.

Establishing a New Data Date

In Figure 11–4, the data date is 07APR05. This data date was established when the schedule was originally created. To change the data date, click on and hold down the mouse on the existing data date line (Figure 11–4). The cursor pointer arrow will turn into a double-headed arrow with the letters DD (data date) inside. Note in Figure 11–4 that *SureTrak* creates a **New data date** box to help you establish the new data date. While holding

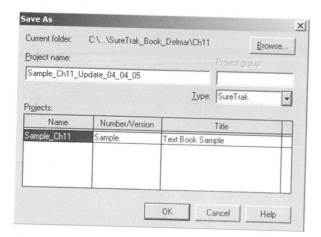

Figure 11-2   Save As Dialog Box

Figure 11-3   Target Dates Dialog Box

down the mouse button, drag it to the right. Notice in Figure 11–5 that a highlighted field is created. The activities that fall within the highlighted time period are also highlighted in the **Act ID** and **Description** fields. When you release the mouse, you establish the new data date. In

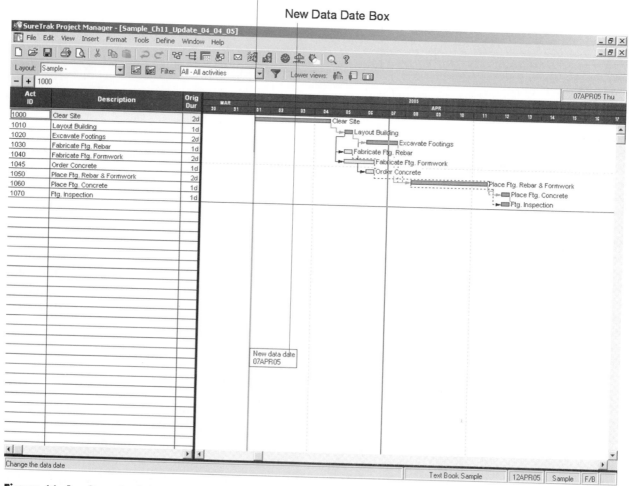

Figure 11-4   Sample Schedule—Data Date

Activities Updated   Highlighted Update Period

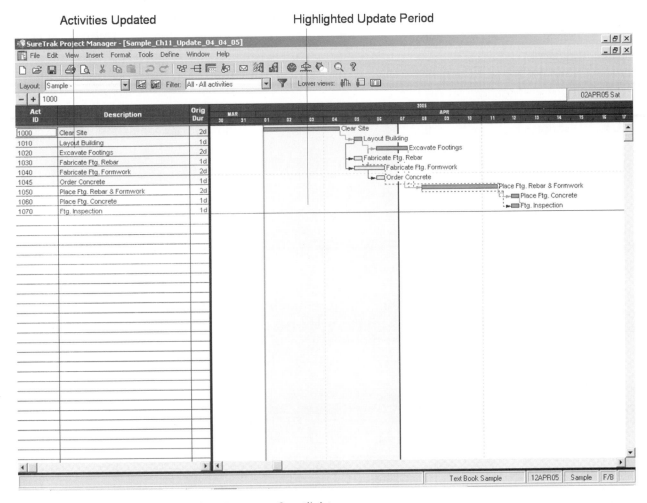

Figure 11–5   Sample Schedule—Progress Spotlight

Figure 11–5, the new data date is 07APR05. The update period goes from the beginning of the workweek of Friday morning (April 01) to Thursday morning (April 07). Our project update is for the work accomplished during this week, or activities 1000, 1010, 1020, 1030, 1040, and 1045.

RECORD PROGRESS

Progress must be measured before it can be recorded. Physical progress is determined either from daily reports or a weekly mark-up of a list schedule report by whomever is responsible for updating the schedule. Table 11–1 is a report showing the actual physical progress as determined in the field for activities 1000, 1010, 1020, 1030, and 1040.

Update Activity Dialog Box

The next step is to record physical progress in *SureTrak*. Select the first activity for recording progress.

Act. ID	Description	Original Duration	Actual Duration	Remaining Duration	% Complete	Actual Start	Actual Finish	Early Finish
1000	Clear Site	2	3	0	100	01APR05	05APR05	
1010	Layout Building	1	1	0	100	06APR05	06APR05	
1020	Excavate Footings	2						
1030	Fabricate Ftg. Rebar	1		1	50	06APR05		07APR05
1040	Fabricate Ftg. Formwork	2		1	50	06APR05		07APR05
1045	Order Concrete	1		1				
1050	Place Ftg. Rebar & Formwork	2		2				
1060	Place Ftg. Conc	1		1				
1070	Ftg. Inspection	1		1				

Table 11–1 Sample Schedule—Update Date

Activity 1000. Choose Activity 1000 - Clear Site (Figure 11–6). Access the **Update Activity** dialog box in one of two ways. One method is to click the right mouse button to bring up the Edit Activity menu, and then select **Update Activity.** The other is to click on the **Tools** main

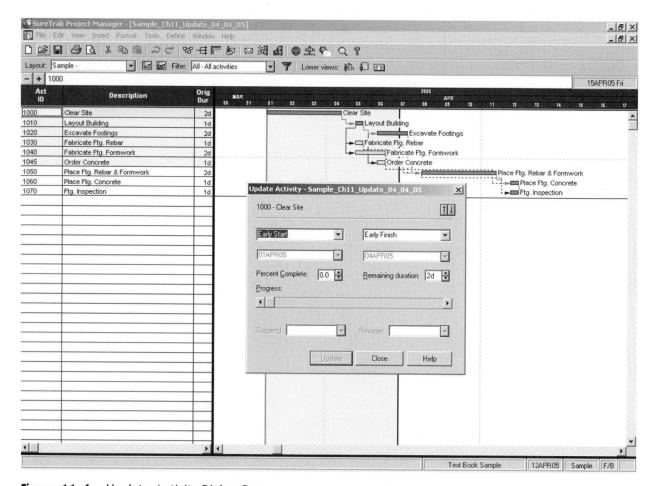

Figure 11–6 Update Activity Dialog Box

Automatic
Schedule
Calculation

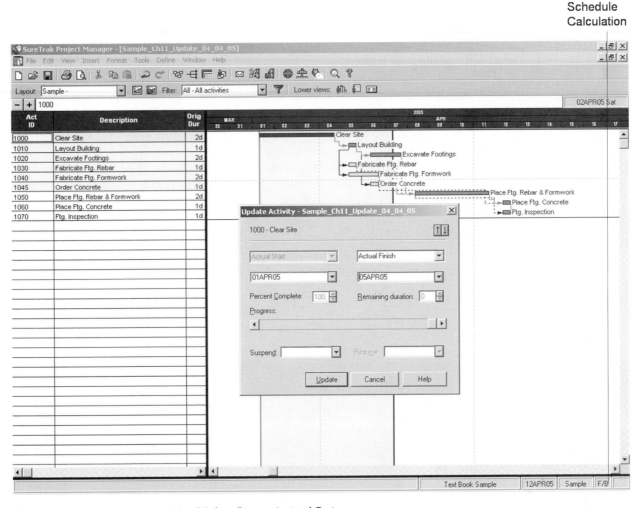

Figure 11-7   Update Activity Dialog Box—Actual Dates

pull-down menu, and then select **Update Activity.** The **Update Activity** dialog box appears as shown in Figure 11–6.

Note that Activity 1000 is highlighted, and the dialog box itself has an identifier to show which activity is being updated. When you click **Early Start,** the **Actual Start** option is shown (Figure 11–7). The actual start of Activity 1000 is 01APR05 according to Table 11–1. Because the planned **Early Start** (Figure 11–6) and the **Actual Start** (Figure 11–7) are the same date, 01APR05, no change is necessary.

When you click the **Early Finish** down arrow (Figure 11–6), the **Actual Finish** option is shown (Figure 11–7). The actual finish of Activity 1000 is 05APR05 according to Table 11–1. Because the planned **Early Finish** (Figure 11–6) and the **Actual Finish** (Figure 11–7) are not the same date, use the down arrow to expose the pull-down calendar to make the change. The original duration was two days. The actual duration is three days.

Note that when the **Actual Finish** is selected in Figure 11–7, the **Remaining Duration** and the Percent **Complete** fields become grayed out or nonfunctional. This is because when we say the activity is complete, *SureTrak* automatically sets the Remaining Duration at 0, and the Percent Complete at 100%.

Notice the activity bar color of Activity 1000 in Figure 11–7 has changed to show progress, and the bar is longer to show the three-day duration rather than the original two-day duration. Note also that even though the Automatic Schedule Calculation is turned on (Figure 11–7), the impact of changing the duration of Activity 1000 did not affect the rest of the activities. The only impact is that Activity 1000 is shown with a three-day duration rather than two.

Activity 1010.   Choose Activity 1010-Layout Building (Figure 11–8). With the **Update Activity** dialog box still active, either click on the next activity in the **Description** field to select it and place its information in

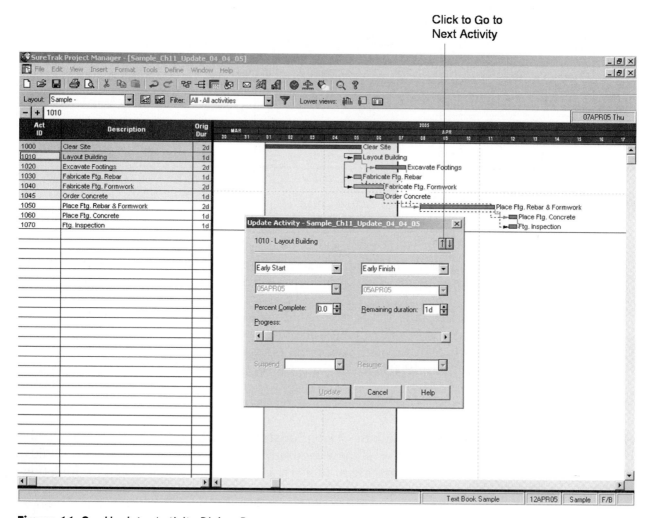

Figure 11–8   Update Activity Dialog Box

the **Update Activity** dialog box or click on the down arrow in the **Update Activity** dialog box to take you to the next activity (Figure 11–8).

According to Table 11–1, Activity 1010 has an actual duration of 1 day, is 100% complete, and it both started and finished on 06APR05. Because the actual start and finish are not the same as planned, simply change the **Early Start** and **Early Finish** to the **Actual Start** and **Actual Finish** and select 06APR05 as the starting and ending dates.

Activity 1030.   Choose Activity 1030, Fabricate Ftg. Rebar (Figure 11–9). With the **Update Activity** dialog box still active, click on Activity 1030. According to Table 11–1, Activity 1030, started on 06APR05, is 50% complete, has 1 day remaining duration, and has a planned early finish of 07APR05.

Because the activity has been started but not finished, select only the **Actual Start** of 06APR05. The **Remaining duration** is set at 1 day, and **Percent Complete** is set at 50%. If you try inputting this combination into *SureTrak* and it does not accept it, there is a reason. Under the **Tools**

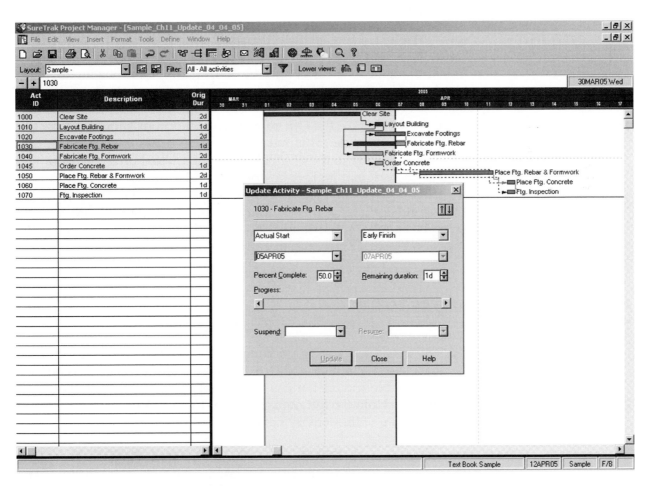

Figure 11–9   Update Activity Dialog Box

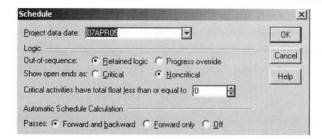

Figure 11–10 Schedule Dialog Box

main pull-down menu, select **Options**, and then the **Resource** tab. Under **Autocost Rules**, if **Link remaining duration and schedule percent complete** is checked, these two updating actions are linked. If you change either **Percent Complete** or **Remaining duration,** *SureTrak* recalculates the other item in proportion to the one you changed. To input the information for Activity 1030, this box needs to be unchecked.

Activity 1040 was also updated according to the information in Table 11–1.

Schedule

Now we have updated all the activities within the shaded update period (Figure 11–9) and/or on which we have physical progress. The next step is to calculate the schedule. From the **Tools** main pull-down menu, select **Schedule.** The **Schedule** dialog box appears (Figure 11–10). The new data date of 07APR05 has already been established. Under the **Logic** portion of the dialog box, the **Out-of-sequence** checkbox should be checked for **Progress override.** This means that if progress is attained out-of-sequence according to the original logic, your actual progress input will override the original logic. Click on **OK** and *SureTrak* will perform the forward and backward passes with all calculations. Figure 11–11 is the result of our update on the Sample Schedule. Compare Figure 11–11 to Figure 11–1. We are essentially one day behind schedule; finishing the project on 13APR05 rather than 12APR05.

ACTIVITY TABLE PROGRESS INPUT

Another convenient way to update physical progress is using the activity table. Place the mouse arrow on the heavy dark vertical line between the activity table and the bar chart. The mouse arrow becomes two short vertical lines. Hold the left mouse button down to adjust the screen and to expose more of the activity table (Figure 11–12).

Note that Figure 11–12 is the original nonupdated schedule. It is being used to compare the schedule updated earlier in this chapter. The

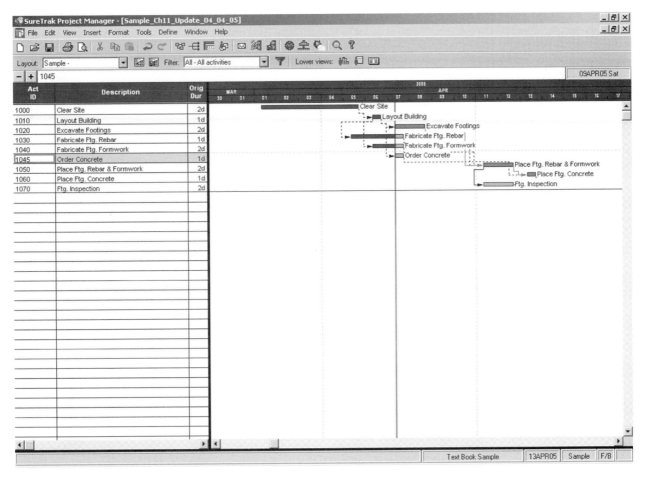

Figure 11–11   Sample Schedule—Updated

columns that show on this activity table are all dependent on the current column format.

Compare Figure 11–12 to Figure 11–11. The Actual Start and Actual Finish columns were added to the table. The **Actual Start, Actual Finish, Rem Dur** (Remaining Duration), and % (Percentage Complete) can be used for determining physical progress. Notice that in Figure 11–12, the **Early Start** for Activity 1000 is listed as 01ARP05. To enter the **Actual Start** date in the activity table, click in the **Actual Start** field, and then click on the down arrow on the menu bar to expose the pull-down calendar (Figure 11–13). Double-click on the date you want to accept.

TARGET COMPARISON

Visually comparing the updated schedule to the target dates makes analysis much easier. Click on the **Format** main pull-down menu and select **Bars.** The **Format Bars** dialog box appears (Figure 11–14). Use the

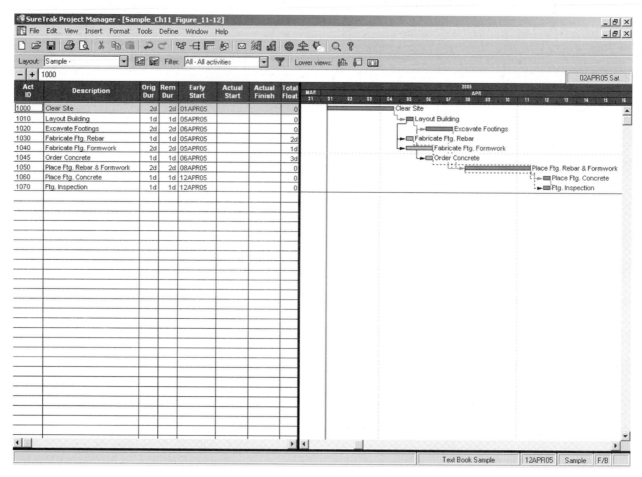

Figure 11–12   Reformatted Columns

button slide bar to scroll down to **Target bar,** and then click on the **Visible** column. Note the view screen at the bottom of the dialog box now shows two bars: the early start/actual duration bar and the target bar. Compare Figure 11–1 (before the update) to Figure 11–11 (after the update). Then look at Figure 11–15, which shows both of these schedules on the same screen. For each activity, the top dark bar represents the current (updated) schedule, and the bottom light bar represents the target or original schedule. When you look at Figure 11–15, it is easy to tell where we fell behind.

PERT VIEW

To see the result of the update on the *SureTrak* PERT view, click on the **View** main pull-down menu and select **PERT.** The onscreen bar chart will convert to the PERT view (Figure 11–16). Notice that the activities that are 100% complete (Activity 1000 and Activity 1010) have two diagonal

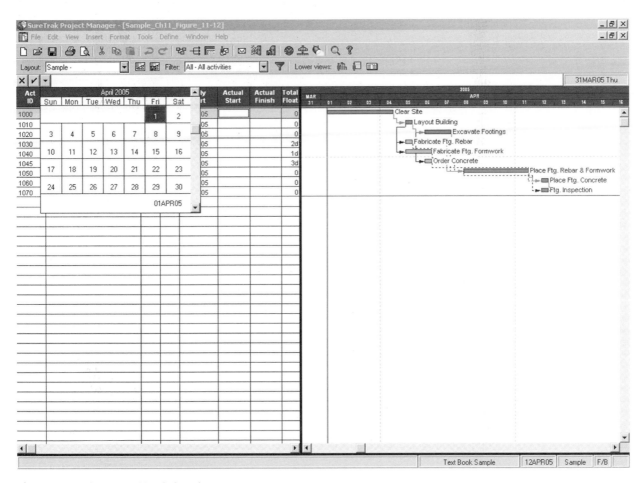

Figure 11–13   Pop-Up Calendar

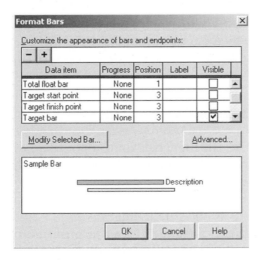

Figure 11–14   Format Bar Dialog Box

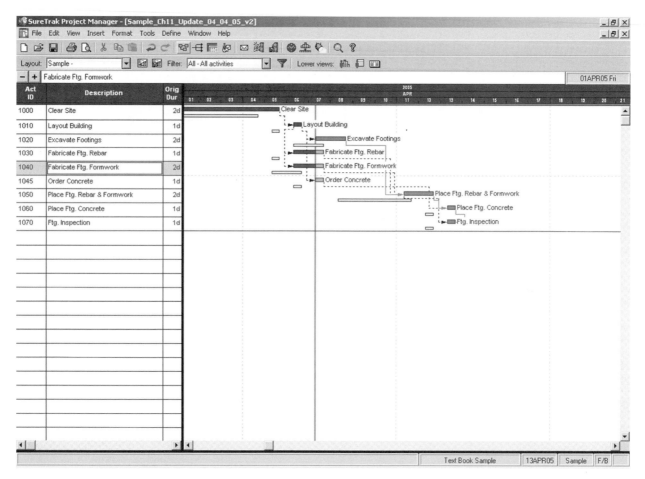

Figure 11-15 Sample Schedule—Target Bar

crossing lines within the box denoting completion. The activities that are partially complete (Activity 1030 and Activity 1040) have a single diagonal line denoting partial progress. These diagonal lines makes it easy to look at the PERT diagram and quickly determine which activities are finished, which have partial progress, and which have not been started.

REPORTS

Figure 11–17 is the bar chart hard copy print with the target bar showing. When the target bar shows onscreen, it also appears when making the hard copy print.

Figure 11–18 is the PERT diagram hard copy print with the progress showing. When progress has been input, it automatically shows on a hard copy print of the PERT diagram.

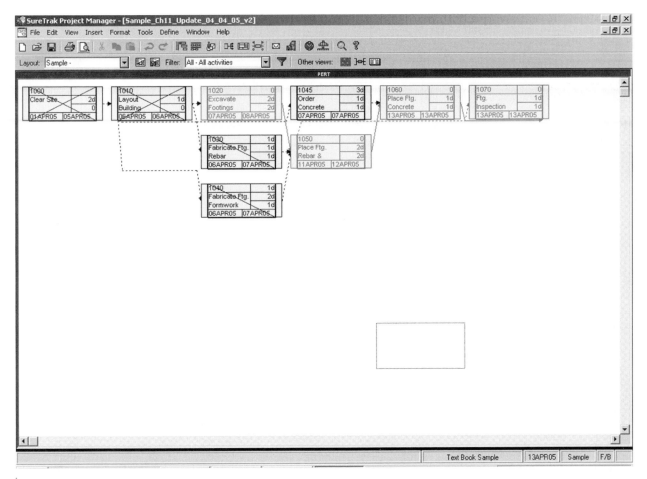

Figure 11–16 PERT View—Updated

Figure 11–19 is the list report showing progress. Here the **Target Start, Target Finish**, and **Finish Variance** columns were added. Finish Variance is handy for analyzing progress.

DOCUMENTING CHANGES

As the approach to building the project changes, the schedule should be modified to show the changes. There are many kinds of changes. Activity durations may change if crew sizes are increased to make up for lost time. The logic of activity sequences and interrelationships may be changed to more accurately reflect interferences. Activities may be added or deleted to reflect changes in the scope of the project. Activities may be added for greater detail. As the project proceeds, usually more detailed planning, such as the 10 days look-ahead, is required. As better data become available with actual activity on which to base projections

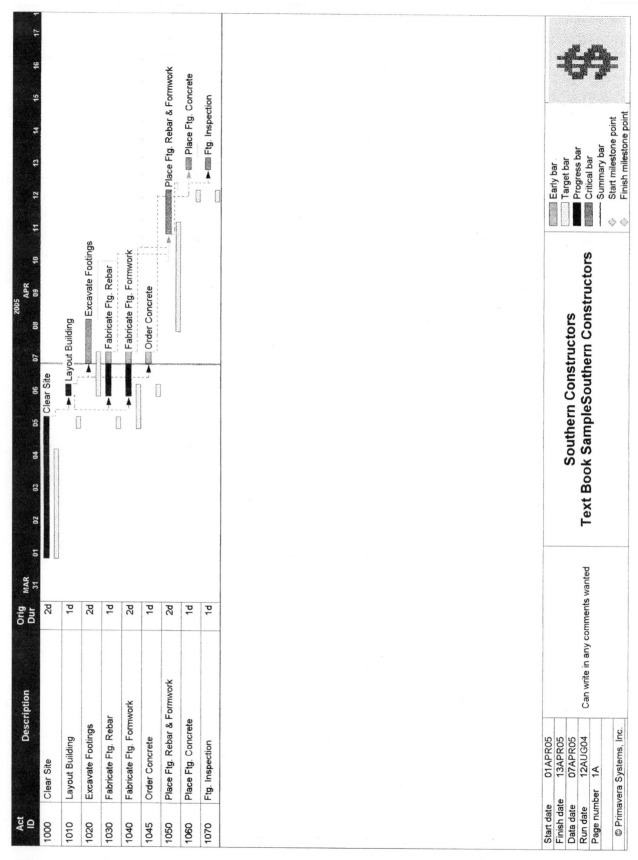

Figure 11–17 Bar Chart Print —Updated with Target Bar

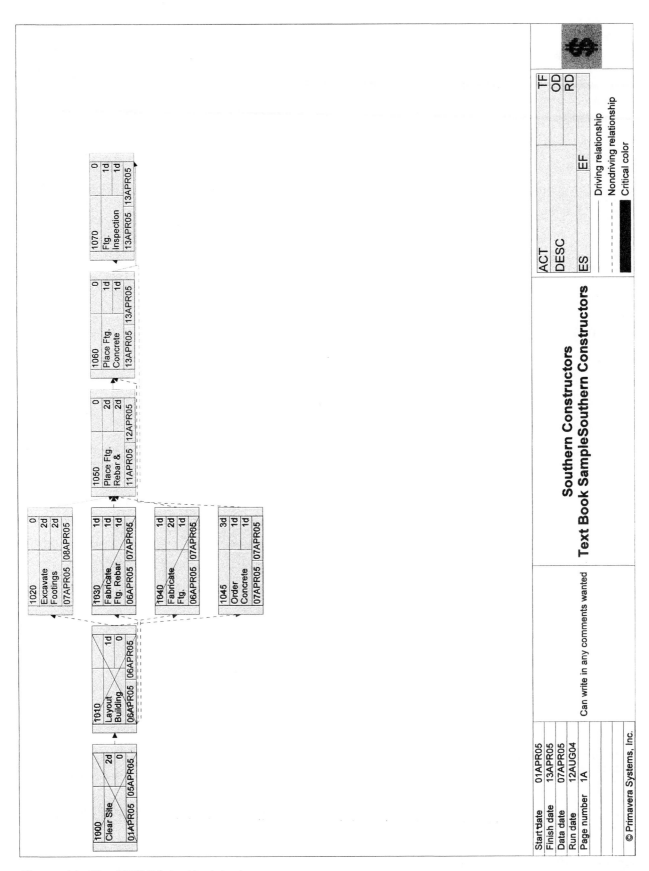

Figure 11–18   PERT Print—Updated

Report: Daily to-do list
Layout: Organized by Day
Filter: All Activities

SureTrak Project Manager
Text Book Sample

Southern Constructors
Report Date: 12AUG04
Page 1A of 1A

Act ID	Activity Description	Orig Dur	Rem Dur	Percent Complete	Early Start	Target Start	Late Finish	Target Finish	Finish Variance
Carpenter									
01APR05									
1000	Clear Site	2d	0	100	01APR05 A	01APR05	05APR05 A	04APR05	-1d
06APR05									
1010	Layout Building	1d	0	100	06APR05 A	05APR05	06APR05 A	05APR05	-1d
1040	Fabricate Ftg. Formwork	2d	1d	50	06APR05 A	05APR05	08APR05	06APR05	-1d
11APR05									
1050	Place Ftg. Rebar & Formwork	2d	2d	0	11APR05	08APR05	12APR05	11APR05	-1d
		8d	4d	57	01APR05 A	01APR05	12APR05	11APR05	-1d
Labor									
06APR05									
1030	Fabricate Ftg. Rebar	1d	1d	0	06APR05 A	05APR05	08APR05	05APR05	-2d
07APR05									
1020	Excavate Footings	2d	2d	0	07APR05	06APR05	08APR05	07APR05	-1d
13APR05									
1060	Place Ftg. Concrete	1d	1d	0	13APR05	12APR05	13APR05	12APR05	-1d
		6d	5d	0	06APR05 A	05APR05	13APR05	12APR05	-1d
07APR05									
1045	Order Concrete	1d	1d	0	07APR05	06APR05	12APR05	06APR05	-1d
13APR05									
1070	Ftg. Inspection	1d	1d	0	13APR05	12APR05	13APR05	12APR05	-1d
		5d	5d	0	07APR05	06APR05	13APR05	12APR05	-1d

Data date	07APR05
Start date	01APR05
Finish date	13APR05
Must finish date	
Target finish date	12APR05
© Primavera Systems, Inc.	

Southern Constructors

Text Book Sample

Figure 11–19 List Report—Updated

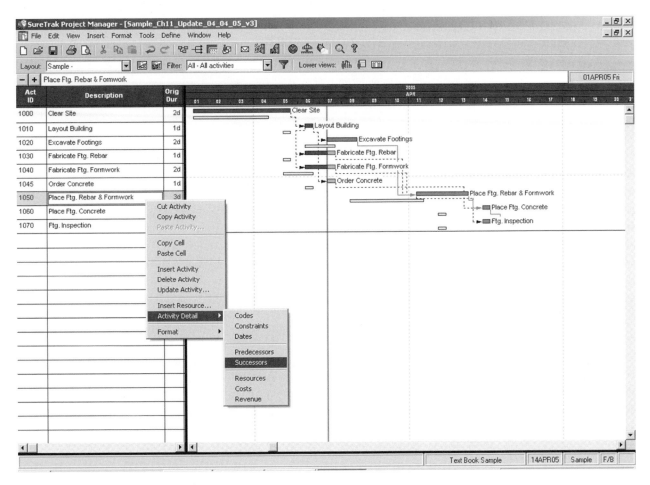

Figure 11–20   Activity Detail Menu

more details arise. Whatever the source, changes need to be incorporated into the schedule.

Duration Changes

A change in duration may occur because similar activities are performed repeatedly within the schedule. An example is pouring a multistory concrete building with each floor divided into several pours. By the time several floors have been placed, very good empirical data on the time required for each pour are available. This information is used to adjust projections of activities remaining to be completed.

The original duration of Activity 1050, Place Ftg Rebar & Formwork, was 2 days (refer to Figure 11–17). In Figure 11–20, it was changed to 3 days. This change was made by changing the number in the **Orig Dur** field of the activity table. Notice that because the **Automatic Schedule Calculation** function is on, the impact of this is immediately shown on its successors, Activity 1060 and Activity 1070. The project duration has been extended from the end of April 13 to April 14.

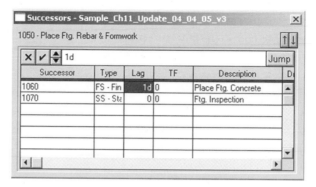

Figure 11–21   Successors Dialog Box

Logic Changes

The original logic of the relationships between activities may change for any number of reasons. The following is an example of a logic change and how to incorporate it into the schedule. It is necessary for inspection purposes to include one day of lag between Activity 1050 and 1060. Select Activity 1050 by clicking the mouse arrow on any of the activities in the activity table fields. Press the right mouse button, and select **Activity Detail**. The **Activity Detail** selection menu (Figure 11–20) appears. Select **Successors** and the **Successors** dialog box appears (Figure 11–21). Activity 1060 is a **Successor** to Activity 1050. Select Activity 1060 for changes, click on the **Lag** field, and enter 1. Compare the calculated schedule with this change incorporated (Figure 11–22) to the schedule before the change (Figure 11–20). In the schedule before the change, Activity 1050 ends at the end of the day, April 13. Activity 1060 ends at the end of the day, April 14. With the addition of the lag day in Activity 1050, it ends at the end of the day, April 13. April 14 is the lag day. Activity 1060 now begins and ends on April 15.

Added Activities

Activity Table.   Sometimes it is necessary to either add or delete activities from the schedule. Adding activities to the schedule within *SureTrak* is not difficult. Place the mouse cursor on the activity after which the new activity will be placed, or click in the empty space after the last activity, and then click on the **+** button on the button bar (Figure 11–23). Note that in Figure 11–24, you can input the activity **Description** and **Org Dur** directly from the onscreen bar chart columns, without activating the Activity Form. If you feel more comfortable having the Activity Form in place when adding activities, select **View,** and then **Activity Form.** Another option is to have the activity form immediately activate when a new activity is added. To have *SureTrak* do this, select **Options** from the **Tools** main pull-down menu. The **Options** dialog box appears.

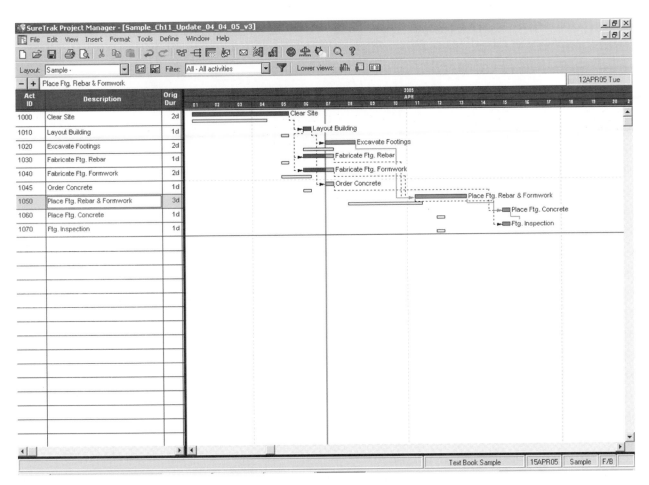

Figure 11–22   Sample Schedule—Duration Change

Select the **Defaults** tab, and then click on the **Display activity form when adding activities in the bar view** checkbox.

In Figure 11–25, the new activity, 1080 - Strip Clean & Oil Forms, with an original duration of 1 day, is placed after 1070 - Ftg Inspection. In Figure 11–24, *SureTrak* places the new activity after the data date. This is because we have not yet established relationships with the other activities, and this is the earliest possible time that the activity could be started as the time before the data date has been expended. Also notice that because this activity was not included in the original (baseline schedule), there is no target bar included for the new activity.

Relationships.   Relationships need to be defined next. Because Activity 1080 is the last activity of the schedule and will have no successors, predecessors need to be established. Highlight Activity 1080, click the right mouse button, select **Activity Detail**, and then select **Predecessors.** The **Predecessors** dialog box appears (Figure 11–25). In Figure 11–25,

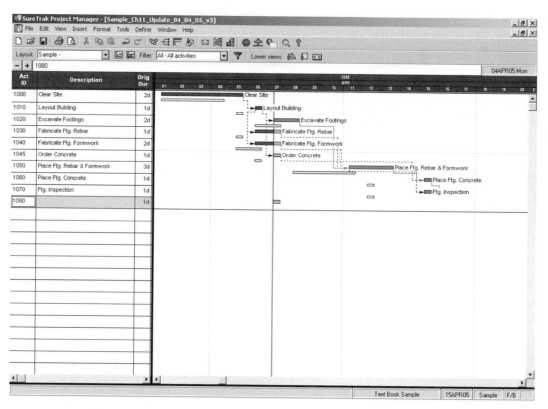

Figure 11–23   Sample Schedule—New Activity

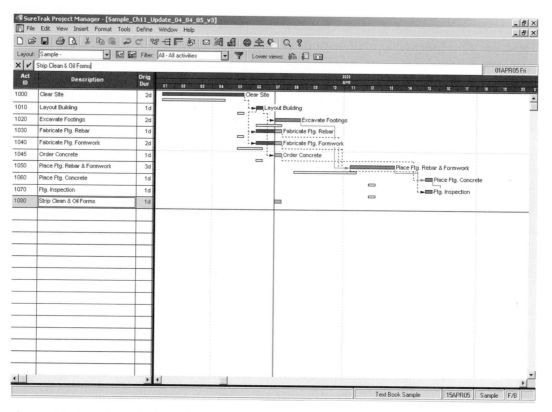

Figure 11–24   Sample Schedule—New Activity

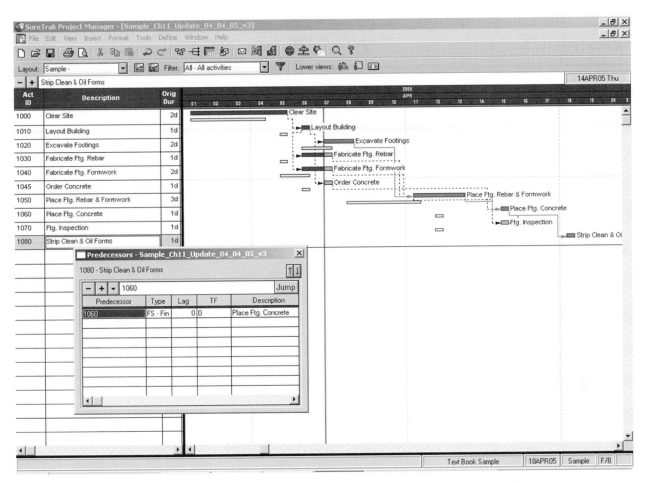

Figure 11-25   Sample Schedule—New Activity—Predecessors

Activity 1060, Place Ftg Conc, is made a predecessor to Activity 1080, Strip Clean & Oil Forms. The default relationship **Type** of Finish-to-Start **(FS)**, and 0 **Lag** time is accepted. Because the **Automatic Schedule Calculation** function is on in our sample schedule, Figure 11-25 is the schedule calculated with the new activity. The new planned project finish date is April 18.

EXAMPLE PROBLEM

Figure 11-26 is a hard copy print of the updated *SureTrak* list report. Figure 11-27 is a hard copy print of the updated bar chart comparison report. Table 11-2 was used to update the schedule for the house put together as an example for student use (see the wood frame house drawings in the Appendix).

Report: Daily to-do list
Layout: Organized by Day
Filter: All Activities

SureTrak Project Manager
Wood Frame House - Example Problem

Southern Constructors
Report Date: 12AUG04
Page 1A of 1A

Act ID	Activity Description	Orig Dur	Rem Dur	Percent Complete	Early Start	Target Start	Late Finish	Target Finish	Finish Variance
28MAR05									
1000	Clear Site	2	0	100	28MAR05 A	28MAR05	31MAR05 A	29MAR05	-2
30MAR05									
1010	Building Layout	1	0	100	30MAR05 A	30MAR05	30MAR05 A	30MAR05	0
31MAR05									
1020	Form/Pour Footings	3	0	100	31MAR05 A	31MAR05	04APR05 A	04APR05	0
05APR05									
1030	Pier Masonry	2	0	100	05APR05 A	05APR05	05APR05 A	06APR05	1
07APR05									
1040	Wood Floor System	4	0	100	07APR05 A	07APR05	11APR05 A	12APR05	1
13APR05									
1050	Rough Framing Walls	6	3	20	13APR05 A	13APR05	18APR05	20APR05	2
18APR05									
1060	Rough Framing Roof	4	4	0	18APR05	21APR05	22APR05	26APR05	2
22APR05									
1080	Ext. Wall Board	2	2	0	22APR05	27APR05	26APR05	28APR05	2
26APR05									
1070	Doors & Windows	4	4	0	26APR05	29APR05	10MAY05	04MAY05	2
1100	Rough Plumbing	4	4	0	26APR05	29APR05	02MAY05	04MAY05	2
1110	Rough HVAC	3	3	0	26APR05	29APR05	05MAY05	03MAY05	2
29APR05									
1120	Rough Elect	3	3	0	29APR05	04MAY05	10MAY05	06MAY05	2
02MAY05									
1130	Shingles	3	3	0	02MAY05	05MAY05	05MAY05	09MAY05	2
05MAY05									
1090	Ext. Wall Insulation	1	1	0	05MAY05	10MAY05	11MAY05	10MAY05	2
1150	Ext. Finish Carpentry	2	2	0	05MAY05	10MAY05	13MAY05	11MAY05	2
1260	Grading & Landscaping	4	4	0	05MAY05	10MAY05	11MAY05	13MAY05	2
09MAY05									
1140	Ext. Siding	3	3	0	09MAY05	12MAY05	18MAY05	16MAY05	2
11MAY05									
1160	Hang Drywall	4	4	0	11MAY05	16MAY05	17MAY05	19MAY05	2
12MAY05									
1190	Ext. Paint	3	3	0	12MAY05	17MAY05	23MAY05	19MAY05	2
17MAY05									
1170	Finish Drywall	4	4	0	17MAY05	20MAY05	23MAY05	25MAY05	2
23MAY05									
1180	Cabinets	2	2	0	23MAY05	26MAY05	25MAY05	27MAY05	2
1230	Finish HVAC	3	3	0	23MAY05	26MAY05	31MAY05	30MAY05	2
1240	Finish Elect	2	2	0	23MAY05	26MAY05	31MAY05	27MAY05	2
25MAY05									
1200	Int. Finish Carpentry	4	4	0	25MAY05	30MAY05	31MAY05	02JUN05	2
1220	Finish Plumbing	2	2	0	25MAY05	30MAY05	31MAY05	31MAY05	2
31MAY05									
1210	Int. Paint	3	3	0	31MAY05	03JUN05	03JUN05	07JUN05	2
03JUN05									
1250	Flooring	3	3	0	03JUN05	08JUN05	08JUN05	10JUN05	2
08JUN05									
1270	Punch List	2	2	0	08JUN05	13JUN05	10JUN05	14JUN05	2
		55	42	18	28MAR05 A	28MAR05	10JUN05	14JUN05	2

Data date	14APR05		
Start date	28MAR05		
Finish date	10JUN05		
Must finish date			
Target finish date	14JUN05		

© Primavera Systems, Inc.

Date	Revision	Checked	Approved

Figure 11–26 Wood Frame House—Updated List Report

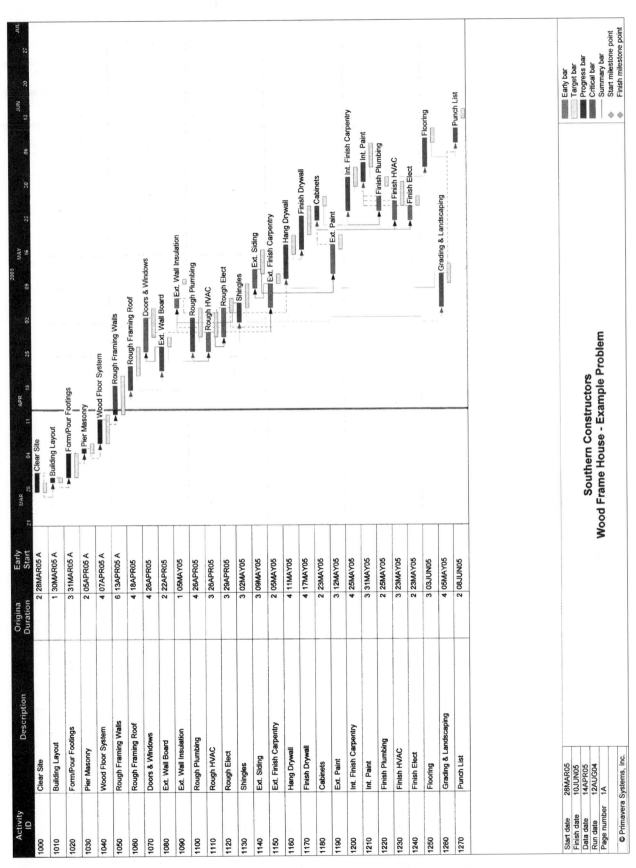

Southern Constructors
Wood Frame House - Example Problem

		Early bar
		Target bar
		Progress bar
		Critical bar
		Summary bar
		Start milestone point
		Finish milestone point

Activity ID	Description	Original Duration	Early Start
1000	Clear Site	2	28MAR05 A
1010	Building Layout	1	30MAR05 A
1020	Form/Pour Footings	3	31MAR05 A
1030	Pier Masonry	2	05APR05 A
1040	Wood Floor System	4	07APR05 A
1050	Rough Framing Walls	6	13APR05 A
1060	Rough Framing Roof	4	18APR05
1070	Doors & Windows	4	26APR05
1080	Ext. Wall Board	2	22APR05
1090	Ext. Wall Insulation	1	05MAY05
1100	Rough Plumbing	4	26APR05
1110	Rough HVAC	3	26APR05
1120	Rough Elect	3	29APR05
1130	Shingles	3	02MAY05
1140	Ext. Siding	3	09MAY05
1150	Ext. Finish Carpentry	2	05MAY05
1160	Hang Drywall	4	11MAY05
1170	Finish Drywall	4	17MAY05
1180	Cabinets	2	23MAY05
1190	Ext. Paint	3	12MAY05
1200	Int. Finish Carpentry	4	26MAY05
1210	Int. Paint	3	31MAY05
1220	Finish Plumbing	2	25MAY05
1230	Finish HVAC	3	23MAY05
1240	Finish Elect	3	23MAY05
1250	Flooring	3	03JUN05
1260	Grading & Landscaping	4	05MAY05
1270	Punch List	2	08JUN05

Start date	28MAR05
Finish date	10JUN05
Data date	14APR05
Run date	12AUG04
Page number	1A

© Primavera Systems, Inc.

Figure 11–27   Wood Frame House—Updated Bar Chart Comparison Report

Act. ID	Description	Original Duration	Actual Duration	Remaining Duration	% Complete	Actual Start	Actual Finish
1000	Clear Site	2	4	0	100	28MAR05	31MAR05
1010	Building Layout	1	1	0	100	30MAR05	30MAR05
1020	Form/Pour Footings	3	3	0	100	31MAR05	04APR05
1030	Pier Masonry	2	1	0	100	05APR05	05APR05
1040	Wood Floor System	4	3	0	100	07APR05	11APR05
1050	Rough Framing Walls	6		4	20	13APR05	
1060	Rough Framing Roof	4					
1070	Doors & Windows	4					
1080	Ext. Wall Board	2					
1090	Ext. Wall Insulation	1					
1100	Rough Plumbing	4					
1110	Rough HVAC	3					
1120	Rough Elect	3					
1130	Shingles	3					
1140	Ext. Siding	3					
1150	Ext. Finish Carpentry	2					
1160	Hang Drywall	4					
1170	Finish Drywall	4					
1180	Cabinets	2					
1190	Ext. Paint	3					
1200	Int. Finish Carpentry	4					
1210	Int. Paint	3					
1220	Finish Plumbing	2					
1230	Finish HVAC	3					
1240	Finish Elect	2					
1250	Flooring	3					
1260	Grading & Landscaping	4					
1270	Punch List	2					

Table 11–2   Wood Frame House—Update Date

SUMMARY

Very few projects ever go exactly according to the original plan/schedule. Once a project is under way, it is necessary to update it to show progress, document changes, and make modifications to improve the plan. This chapter showed examples of an updated schedule.

Target dates are used within *SureTrak* to establish the target or baseline schedule to which the current or updated schedule is compared to establish progress. The data date is used to establish the date upon which progress is determined and is the date of the current schedule.

Actual progress is recorded by updating the activity with actual start, actual or planned finish, and percent complete information. You can input this information in several ways. When all information on activities that have progress has been input, the schedule is calculated to show the impact.

Formatting bars can be used to display both the current and target schedules onscreen at the same time. This is a great help in progress analysis and communication.

Progress is shown on the PERT view by diagonal lines appearing in the activity box.

As the approach to building a project changes, the schedule should be modified and used as a tool to communicate these changes. There are many kinds of changes; for example, the scope of the project may be changed; the builder may be ordered to accelerate the project; activity durations may be changed if crew sizes are increased or decreased; the logic of activity sequences and interrelationships may be changed to more accurately reflect interferences; or activities may be added for greater detail. A current schedule should be modified to communicate and document these changes.

EXERCISES

1. **Describing the Importance of Updating a Schedule**
   Prepare a summary describing the importance of updating a schedule.
2. **Creating a Copy Before Updating a Schedule**
   Open the project modified in Exercise 5 of Chapter 7 and save it with a different name.
3. **Establishing Target Dates (or Baselines)**
   Establish that target date for Exercise 2.
4. **Establishing a New Data Date**
   Establish a new data date that is one week later than the original data date.
5. **Update Activities**
   Update the activities that were highlighted upon establishing the new data date in Exercise 4.
6. **Calculate the Updated Schedule**
   Calculate the updated schedule of Exercise 5.

7. Update Figure 11–28 (Exercise 6, Chapter 4) manually, and then update it onscreen using the *SureTrak* bar chart.

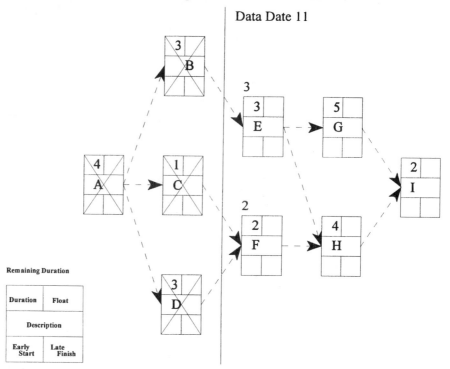

Figure 11–28   Exercise #1

8. Update Figure 11–29 (Exercise 7, Chapter 4) manually, and then update it onscreen using the *SureTrak* bar chart.

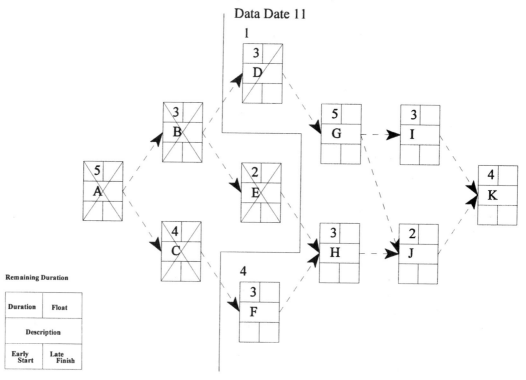

Figure 11–29   Exercise #2

9. Update Figure 11–30 (Exercise 8, Chapter 4) manually, and then update it onscreen using the *SureTrak* bar chart.

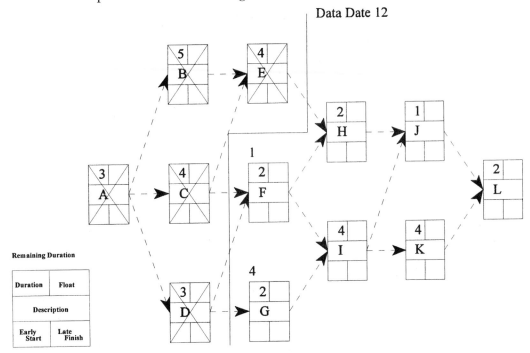

Figure 11–30   Exercise #3

10. Update Figure 11–31 (Exercise 9, Chapter 4) manually, and then update it onscreen using the *SureTrak* bar chart.

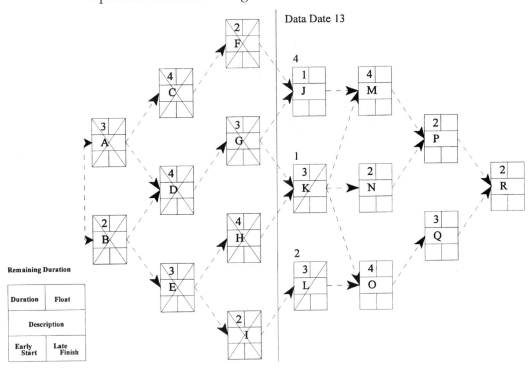

Figure 11–31   Exercise #4

11. Update Figure 11–32 (Exercise 10, Chapter 4) manually, and then update it onscreen using the *SureTrak* bar chart.

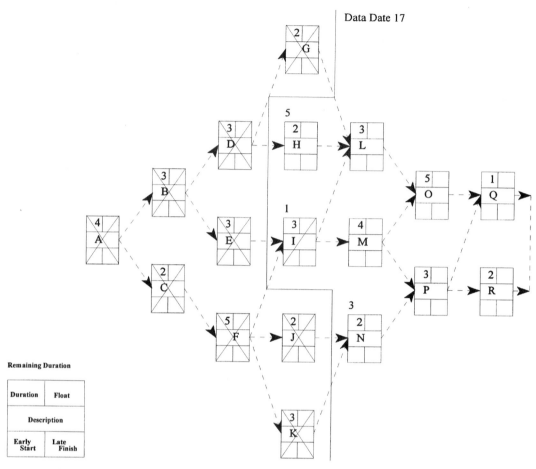

Figure 11–32 Exercise #5

12. Update Figure 11–33 (Exercise 11, Chapter 4) manually, and then update it onscreen using the *SureTrak* bar chart.

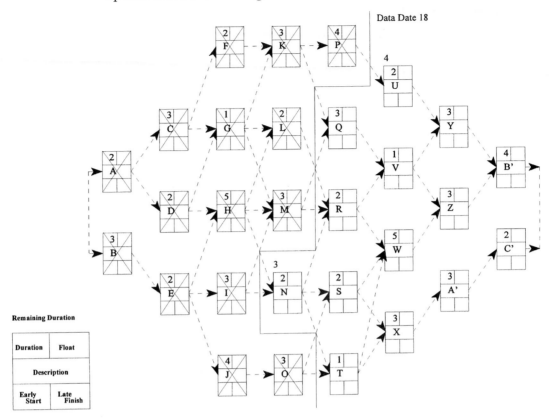

Figure 11–33   Exercise #6

13. Small Commercial Concrete Block Building—Updating
Prepare the following reports for the small commercial concrete block building located (Exercise 12, Chapter 4) in the Appendix:
- Updated bar chart

14. Large Commercial Building—Updating
Prepare the following reports for the large commercial building located (Exercise 13, Chapter 4) in the Appendix:
- Updated bar chart

Tracking Resources

Objectives

Upon completion of this chapter, you should be able to:

- Modify the autocost rules
- Update resources used
- Update resources needed to complete an activity
- Print an updated resource profile
- Print an updated resource table
- Print a tabular comparison report for tracking

TRACKING RESOURCE: SURETRAK VERSUS CONTRACTORS

SureTrak always does all schedule calculations in hours regardless of the (time) units indicated in the **Define Resource** dialog box (refer to Figure 6–3) or the **Format** selected in the **Resources Profile/Table.** However, in most cases, contractors think in terms of days as the unit of time and therefore assign resources in terms of days. For example, a carpenter is assigned to an activity called "Fabricate Ftg. Formwork" for 1 day rather than 8 hours. This chapter presents a method to input the resources in terms of days and still be compatible with *SureTrak*'s calculations in hours.

TRACKING RESOURCES: ACTUAL VERSUS PLANNED EXPENDITURES

After a project is underway, the contractor tracks (monitors) resources to compare actual to planned costs. The resources can be labor, equipment, materials, or other resources that were assigned to the activity (usually allocated from the original estimate) when the target schedule was constructed. Figure 12–1 shows the target (original) schedule with the resource table in view. Table 12–1 shows the detailed craft requirements sorted by activity by day for the target (original) schedule.

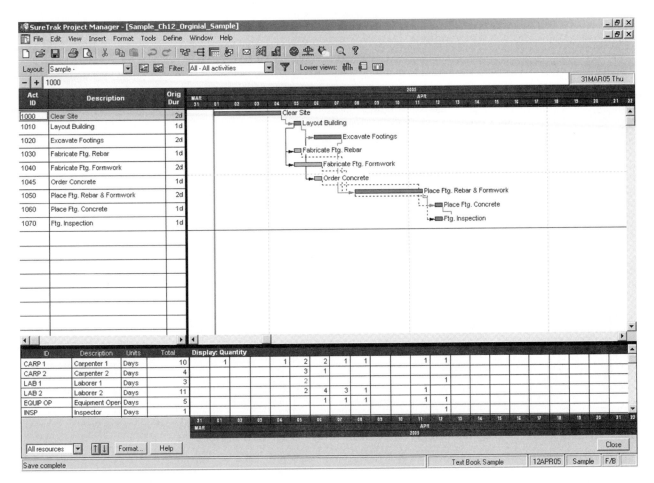

Figure 12–1   Target Schedule—Resource Table

Resource	Total	01	02	03	04	05	06	07	08	09	10	11	12	13
CARP 1	10	1000-1			1000-1	1010-1 1040-1	1020-1 1040-1	1020-1	1050-1			1050-1	1060-1	
CARP 2	4					1010-2 1040-1	1040-1							
LAB 1	3					1010-1 1030-1							1060-1	
LAB 2	11					1030-1 1040-1	1020-3 1040-1	1020-3	1050-1			1050-1		
EQUIP OP	5						1020-1	1020-1	1050-1			1050-1	1060-1	
INSP	1												1070-1	
Total	34	1			1	9	8	5	3			3	4	

Table 12–1   Craft Requirements by Activity by Day—Target Schedule

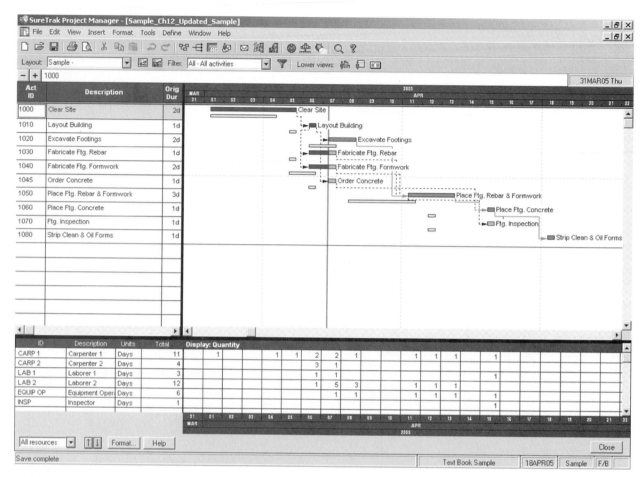

Figure 12–2 Current Schedule—Resource Table

In Chapter 11, we updated the project for current progress. This updated schedule is now called the current schedule. It can be compared to the target schedule to determine actual physical progress. Figure 12–2 shows the current schedule as updated in Chapter 11.

Only by tracking and comparing the actual usage of resources to the target budget can the contractor determine physical progress and earned value. Management must gauge the progress gained for the amount of resources expended to determine whether it is over or under the original budget. This analysis provides the information to help solve potential cost and time problems.

On a cost-plus or negotiated project, the owner has access to the contractor's cost information. The resource-loaded and then resource-monitored schedule thus gives the owner valuable information in managing her company's time and cash.

Figure 12–2 is the current schedule. The lighter bars show the target schedule as updated in Chapter 11, with progress, modified logic, and new activity input. The resource table is also shown. Figure 12–1 is the target schedule also with the resource table showing. Compare the resource table in Figure 12–1 to Figure 12–2. There are obvious differences by date. One striking difference is that if you add the total resource units required from the resource table—(all resource-days) CARP 1 (10) + CARP 2 (4) + LAB 1 (3) + LAB 2 (11) + EQUIP OP (5) + INSP (1)—they total 34 days in Figure 12–1 and 37 in Figure 12–2. These differences are explained in the following sections as we update the expended resources for each activity. Again, as with the rest of this book working along with the text and *SureTrak* at the same time will make this much easier to understand.

RESOURCE TABLE

Select the **View** main pull-down menu, and then **Resource Table.** If costs rather than resources appear when the **Resource Profile/Table** appears at the bottom of the screen (refer to Figures 12–1 and 12–2), select the **Format** button. The **Format Resource Profile/Table** dialog box appears (Figure 12–3). Select the **Table** option. Select **Quantity** from the **Display** options. For Figure 12–1 and Figure 12–2, the **Show timescale with Day intervals** was checked.

AUTOCOST RULES

It is critical to understand the Autocost Rules within *SureTrak* when updating the schedule with current resource information. To get to the **Resource** options screen (Figure 12–4), select the **Tools** main pull-down

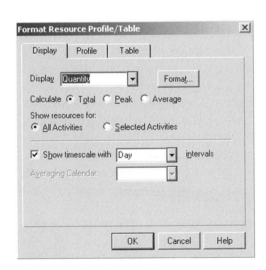

Figure 12–3   Format Resource
Profile/Table Dialog Box

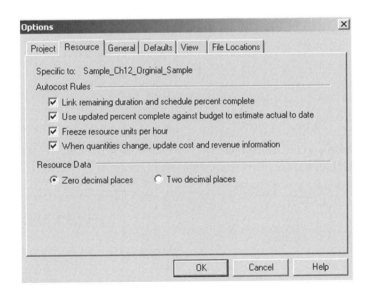

Figure 12–4   Options Dialog Box

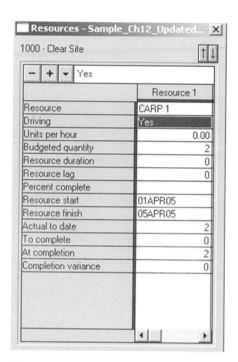

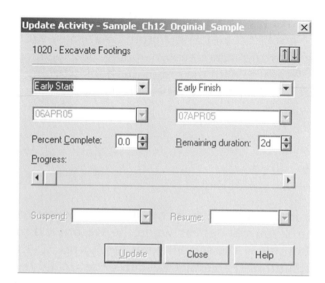

Figure 12–5 Update Activity Dialog Box

Figure 12–6 Resources Dialog Box—Activity 1000

menu, then **Options,** and then the **Resources** tab. The first set of four checkboxes relates to the **Autocost Rules.**

> **Link remaining duration and schedule percent complete** If you choose to link (have a check in this checkbox) these updating actions, you can either change **Percent Complete** (Figure 12–5) or the **Remaining duration** and *SureTrak* will calculate the other item in proportion to the one you changed. Access the **Update Activity** dialog box, Figure 12–5, by selecting an activity (in this case Activity 1020 - Excavate Footings). Click the right mouse button and select **Update Activity.** If you do not link (no check in the checkbox) the remaining duration and schedule percent complete, a change in one will not affect the other value; you can update them separately. If you link the remaining duration and percent complete, you cannot edit percent complete for activities to which **Driving** (Figure 12–6) resources are assigned. Access the **Resources** dialog box by selecting an activity (in this case Activity 1000 - Clear Site), then clicking the right mouse button, selecting **Activity Detail,** and then selecting **Resources.** For this chapter, the **Link remaining duration and schedule percent complete** checkbox is left unchecked.

> **Use updated percent complete against budget to estimate actual to date** Mark this box (see Figure 12–4) to tell *SureTrak* to multiply the updated percent complete by the budgets to estimate actual to-date resource quantities when either the percent complete or a budget changes. Leave it unchecked if you want to track expenditures rather than use estimates made by *SureTrak.* For this chapter, the **Use**

updated percent complete against budget to estimate actual to-date checkbox is left unchecked.

Freeze resource units per hour If you mark this checkbox (see Figure 12–4), *SureTrak* recalculates the quantity to complete when the remaining duration of an activity or actual to-date resource use changes, but will not change the units per hour. *SureTrak* uses the equation: Quantity to Complete = Units per Hour/Remaining Duration. Leave it unchecked if you want *SureTrak* to recalculate units per hour when the remaining duration or actual to-date quantity changes. *SureTrak* uses the equation: Units per Hour = Quantity to Complete/Remaining Duration. For this chapter, the **Freeze resource units per hour** checkbox is left unchecked.

When quantities change, update cost and revenue information This checkbox (see Figure 12–4) will be discussed in Chapter 13 when cost information is updated.

Resource Data Also included under the **Resources** tab of the **Options** dialog box (see Figure 12–4) is **Resource Data.** Figure 12–1 and Figure 12–2 were executed with the **Zero decimal places** option selected. With this option selected, *SureTrak* rounds to the nearest whole number. For this chapter, the **Two decimal places** option was chosen.

Figure 12–7 shows the **Options** dialog box with the **Resource** tab selected. This has been modified for use in this chapter. Figure 12–8 shows the sample schedule recalculated with the preceding changes made to the **Options** dialog box. Note the changes in days 6 and 7 from Figure 12–2 to Figure 12–8. The numbers went from 1 to .69. This difference will be explained when we update activity 1000 - Clear Site in the next section.

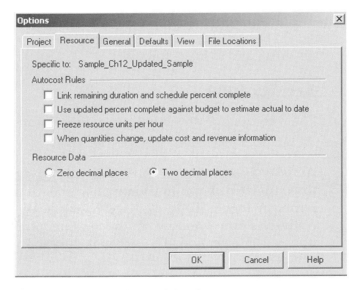

Figure 12–7   Options Dialog Box

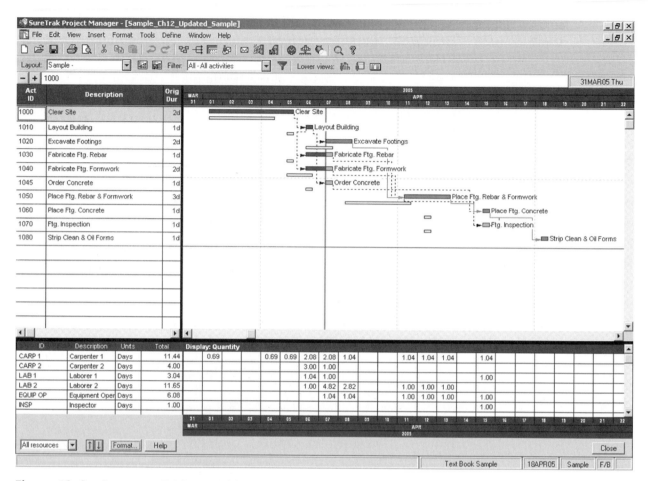

Figure 12–8 Resource Table—Activity 1000

RECORD EXPENDITURES

The labor resources that were input in Chapter 6 are tracked in this section. The schedule will be updated to show actual expenditures of resources and to make forecasts.

Activity 1000

The recording of the actual expenditure of resources for Activity 1000 will be input in the following sections.

Activity Form.   On the *SureTrak* bar chart screen, click on the activity to be updated. From the **View** main pull-down menu, select **Activity Form**. In Figure 12–9, Activity 1000 has been selected. Look at the **Activity Form** window at the bottom of the screen to check the updated information. The original duration is 2d (days), the percent complete is 100%, the actual start is 01APR05, and the actual finish is 05APR05.

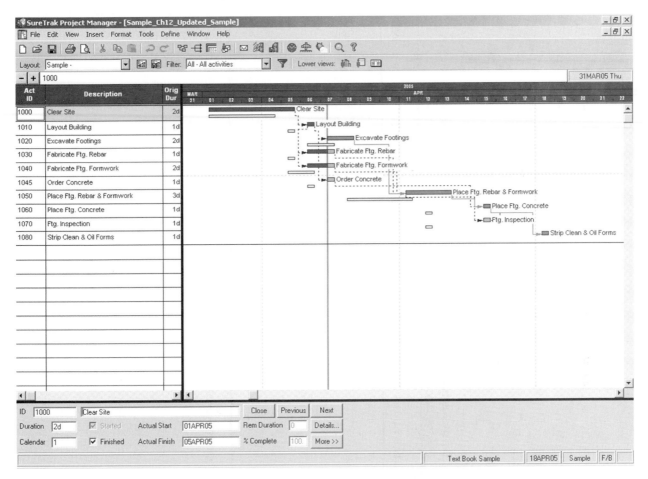

Figure 12–9   Activity Form—Activity 1000

Resources Dialog Box.   Now return to the resource table window. With Activity 1000 still selected, click the right mouse button to expose the **Activity Detail** selection box. Select **Resources** to open the **Resources** dialog box for Activity 1000 (refer to Figure 12–6). Notice that the **Budgeted quantity, Actual to date,** and **At completion** fields are all showing a quantity of 2 person-days. This target budget was put in when the resources were originally loaded into the schedule (in this case, Chapter 6).

In Figure 12–1, the target schedule showed a 2-day duration, with one CARP 1 for APR 01 and one CARP 1 for APR 04. When the schedule was updated in Chapter 11, the actual duration for Activity 1000 was 3 days (01APR05 to 05APR05). In Figure 12–8, *SureTrak* has taken the 3-day actual duration and divided it by the 2 person-days of budgeted quantity. This equals .69 person-days for 3 days ($2/3 \approx .69$).

Actual Labor Expenditures.   The actual labor expenditures and projections to completion are gathered from project time sheets (Table 12–2). For Activity 1000, 1 person for 3 days was actually used, rather than

Act. ID	Description	Percent Complete	Actual Start	Actual Expenditures to Date (Person-Days)						Projected Expenditures to Completion (Person-Days)					
				CARP 1	CARP 2	LAB 1	LAB 2	EQUIP OP	INSP	CARP 1	CARP 2	LAB 1	LAB 2	EQUIP OP	INSP
1000	Clear Site	100%	01APR05	3	—	—	—	—	—	—	—	—	—	—	—
1010	Layout Building	100%	06APR05	1	2	2	—	—	—	—	—	—	—	—	—
1020	Excavate Footings	0%		0	—	—	0	0	—	2	—	—	6	2	—
1030	Fabricate Ftg. Rebar	50%	06APR05	—	—	1	1	—	—	1	—	1	0	—	—
1040	Fabricate Ftg. Formwork	50%	06APR05	1	1	—	2	—	—	1	1	—	1	—	—
1045	Order Concrete	0%		—	—	—	—	—	—	—	—	—	—	—	—
1050	Place Ftg. Rebar & Formwork	0%		0	—	—	0	0	—	2	—	—	2	2	—
1060	Place Ftg. Conc	0%		0	—	0	—	0	—	1	—	1	—	1	—
1070	Ftg. Inspection	0%		—	—	—	—	—	0	—	—	—	—	—	1

Table 12-2 Craft Requirements—Actual Expenditures to Date/Projected Expenditures to Completion

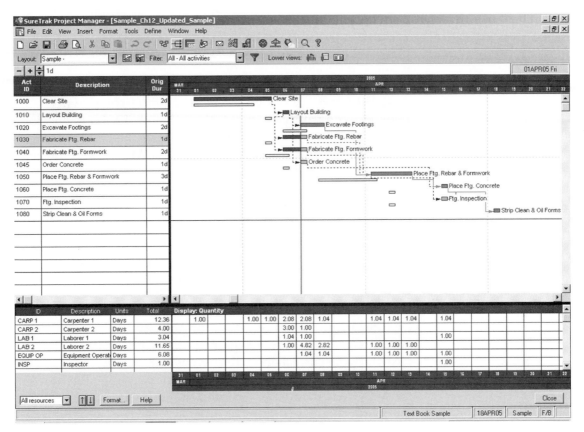

Figure 12–10   Resource Table—Activity 1000

1 person for 2 days; therefore, a total of 3 days of CARP 1 time was expended. The **Resources** dialog box for Activity 1000 (refer to Figure 12–6) must be updated to show actual progress by making the following changes (Figure 12–10):

1. **Percent complete.** Click on the **Percent complete** field and input 100 (100%).
2. **Actual to date.** Click on **Actual to date,** and input 3 (3 person-days).
3. **At completion.** Click on the **At completion** fields and input 3 (3 person-days).

When these changes are made (Figure 12–10), the **Budgeted quantity** field remains at 2.00, because the original budget has not changed. The **Completion variance** field at the bottom of the dialog box shows a −1, representing 1 person-day over the original budget (Budgeted quantity − Actual to date or 2 − 3 = −1).

Resource Table.   Compare the resource table before the resource update for Activity 1000 (refer to Figure 12–8) with that after the update (refer to Figure 12–10). The .69 person-days for 3 days has been changed to 1 person-day for the 3 days (01APR05 to 05APR05).

Activity 1010

Recording the actual expenditure of resources for Activity 1010 will be input in the following sections.

Activity Form.   Click on the next activity to be updated, Activity 1010. Look at the **Activity Form** window (Figure 12–11) to check the updated information. The original duration is 1 day, the percent complete is 100%, the actual start is beginning of the day 06APR05, and the actual finish is the end of the day 06APR05.

Resources Dialog Box.   Now return to the resource table window and the **Resources** dialog box for Activity 1010 (Figure 12–12). From the target budget, the **Budgeted quantity, Actual to date,** and **At completion** fields all show a quantity of 1 person-day each for CARP 1 AND LAB 1, and a quantity of 2 person-days for CARP 2.

Actual Labor Expenditures.   When the schedule was updated in Chapter 11, the actual duration for Activity 1010 was 1 day (06APR05 to 06APR05). The **Budgeted quantity** amounts of 1 for CARP 1 and 2 for CARP 2 were correct according to cost information received (Table 12–2), but 2 person-days for LAB 1 was expended, so the **Resources** dialog box for Activity 1010 must be updated to show actual progress by making the following changes (Figure 12–13):

ID	1010	Layout Building			Close	Previous	Next	
Duration	1d	☑ Started	Actual Start	06APR05	Rem Duration	0	Details...	
Calendar	1	☑ Finished	Actual Finish	06APR05	% Complete	100	More >>	

			Text Book Sample	18APR05	Sample	F/B

Figure 12–11   Activity Form—Activity 1010

Resources - Sample_Ch12_Updated_Sample   ×

1010 - Layout Building   ↑↓

−	+	▼	CARP 1			
		Resource 1	**Resource 2**	**Resource 3**	**Resource 4**
Resource		CARP 1	CARP 2	DIST	LAB 1
Driving		No	No	Yes	No
Units per hour		0.00	0.00	0.00	0.00
Budgeted quantity		1.04	2.00	8.00	1.04
Resource duration				0	
Resource lag		0	0	0	0
Percent complete					
Resource start		06APR05	06APR05	06APR05	06APR05
Resource finish		06APR05	06APR05	06APR05	06APR05
Actual to date		1.04	2.00	8.00	1.04
To complete		0.00	0.00	0.00	0.00
At completion		1.04	2.00	8.00	1.04
Completion variance		0.00	0.00	0.00	0.00

Figure 12–12   Resource Dialog Box—Activity 1010

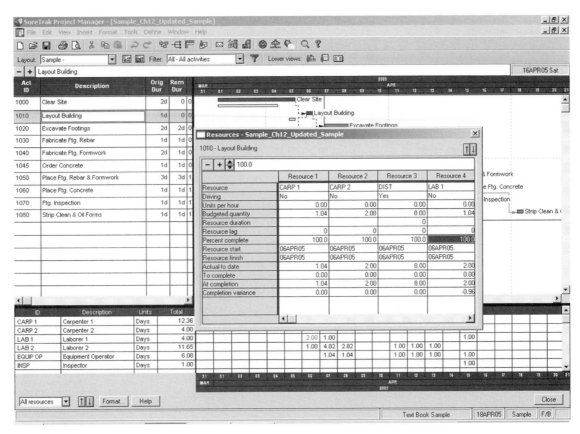

Figure 12–13   Resource Table—Activity 1010

1. **Percent complete.** Click on the **Percent complete** field and input 100 (100%) for resources CARP 1, CARP 2, and LAB 1.
2. **Actual to date.** Because the actual expenditure for CARP 1 and CARP 2 matches the original budget of 1 person-day, no change has to be made. Click on the **Actual to date** field for LAB 1, and input 2 (2 person-days).
3. **At completion.** Click on the **At completion** field and input 2 (2 person-days) for LAB 1.

Resource Table.   When the updated resource requirements for this activity are input, the new resource table window shows that requirements for LAB 1s for 06APR05 went from 1 (Figure 12–10) to 2 (Figure 12–13).

Activity 1030

Recording the actual expenditure of resources for Activity 1030 will be input in the following sections.

Activity Form.   Click on the next activity to be updated, Activity 1030. Look at the **Activity Form** window (Figure 12–14) to check the updated information. The original duration is 1 day, the percent complete is 50%, the actual start is beginning of the day 06APR05, and the early finish is the end of the day 07APR05.

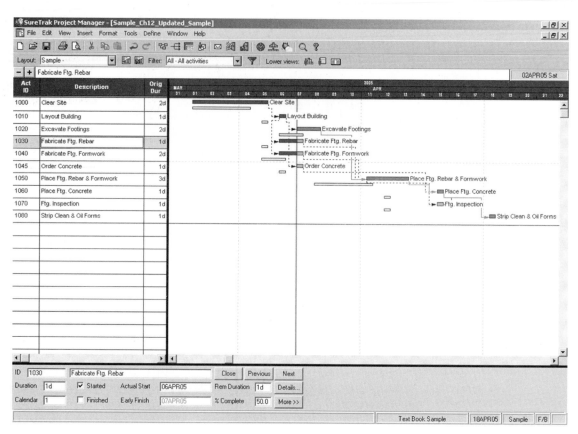

Figure 12–14   Activity Form—Activity 1030

Resources Dialog Box.   Now return to the resource table window and the **Resources** dialog box for Activity 1030 (Figure 12–15). From the target budget, the **Budgeted quantity** and **Actual to date** fields show a quantity of 1 person-day each for LAB 1 and LAB 2.

Figure 12–15   Resources Dialog Box—Activity 1030

Actual Labor Expenditures. When the schedule was updated in Chapter 11, the actual duration for Activity 1030 was 2 days (06APR05 and 07APR05). The **Budgeted quantity** of 1 person-day for LAB 1 was correct according to cost information received (refer to Table 12–2). The budget for LAB 1 is overbudget. The **Resources** dialog box for Activity 1030 must be updated to show actual progress by making the following changes (Figure 12–16):

1. **Driving.** Driving means the resource is controlling the duration of the activity to which it is assigned. Because we want to input actual durations and resources, we do not want *SureTrak* controlling this calculation. So change the driving designation to No on all three resources.
2. **Percent complete.** Click on the **Percent complete** field and input 50 (50%) of resources LAB 1 and LAB 2.
3. **Actual to date.** Click on the **Actual to date** field for LAB 1 and LAB 2, and input 1 (1 person-day).
4. **To complete.** Click on the **To Complete** field for LAB 2, and input 1 (1 person-day). Click on the **To complete** field for LAB 1 and input 1.
5. **At completion.** The **At completion** field now shows 1 person-day for LAB 2 and 2 person-days for LAB 1.
6. **Completion variance.** The **Completion variance** field now shows 0 for LAB 2 and −1 person-days (overbudget) for LAB 1.

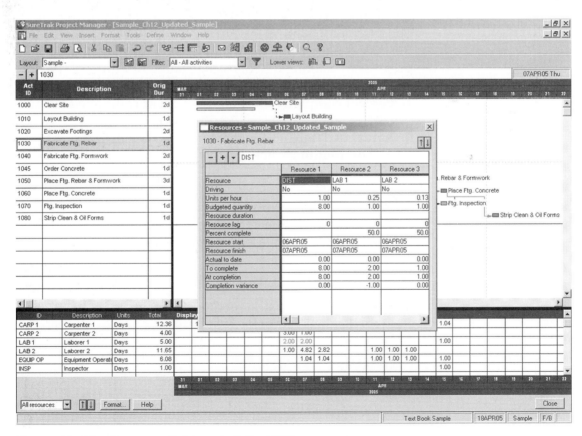

Figure 12–16   Resource Table—Activity 1030

Resource Table. When the updated resource requirements for this activity are input, the new resource table window shows that the actual to-date requirements for 07APR05 and 07APR05 for LAB 2 are now at 2 person-days (Figure 12–16).

Activity 1040

Recording the actual expenditure of resources for Activity 1040 will be input in the following sections.

Activity Form. Click on the next activity to be updated, Activity 1040. Look at the **Activity Form** window (Figure 12–17) to check the updated information. The original duration is 2 days, the percent complete is 50%, the actual start is beginning of the day 06APR05, and the actual finish is the finish of the day 07APR05.

Resources Dialog Box. Now return to the resource table window and the **Resources** dialog box for Activity 1040 (Figure 12–18). Notice that from the target (original) budget, the **Budgeted quantity,** shows a quantity of 2 and the **Actual to date** and **At completion** fields show a quantity of 1 person-day each for CARP 1, CARP 2, and LAB 2.

Actual Labor Expenditures. When the schedule was updated in Chapter 11, the actual duration for Activity 1040 was 2 days. The **Budgeted quantity** of 1 for LAB 2 was incorrect according to cost information received (Table 12–2). Three person-days will be expended, so the **Resources** dialog box for Activity 1040 must be updated to show actual progress by making the following changes (Figure 12–19):

1. **Percent complete.** Click on the **Percent complete** field and input 100 (50%) for resources LAB 1 and LAB 2.
2. **Actual to date.** Click on the **Actual to date** field for LAB 2 and input 2.
3. **At completion.** Click on the **At completion** field for LAB 2 and input 3.

Resource Table. When the updated resource requirements for this activity are input, the new resource table window shows that requirements for 07APR05 were increased by 1 person-day for LAB 2 (Figure 12–19).

Activities 1045–1080

Because no progress is claimed for activities 1045, 1050, 1060, 1070, and 1080, the **Percent complete** field is left at 0%. There are no changes.

Figure 12–17 Activity Form—Activity 1040

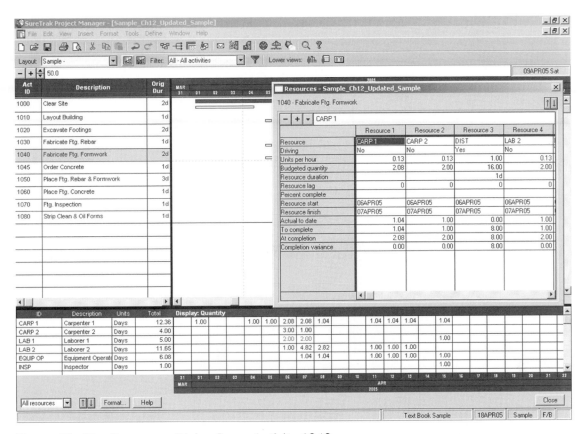

Figure 12–18 Resources Dialog Box—Activity 1040

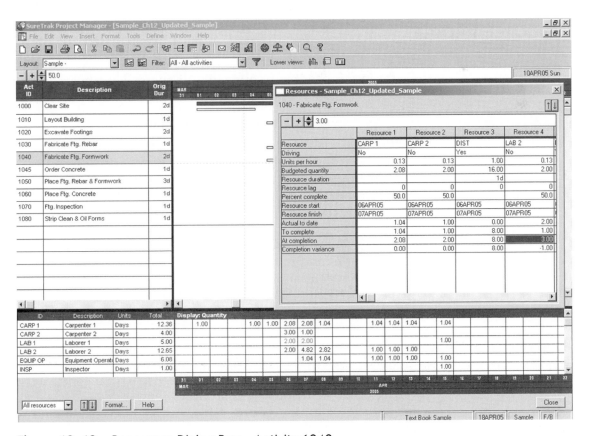

Figure 12–19 Resources Dialog Box—Activity 1040

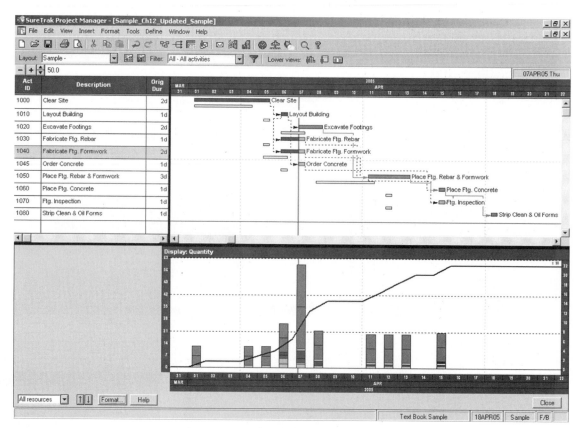

Figure 12–20   Resource Profile—Current Schedule

RESOURCE PROFILE

So far in this chapter, resource requirements have been presented in the onscreen tabular representation. Sometimes a graphical representation is more helpful.

Figure 12–20 is the resource profile of the current schedule. It includes the activity and duration update changes made in Chapter 11 and the resource update changes made in this chapter. To access this screen, select **Resource Profile** from the **View** main pull-down menu, and then select **All resources.**

Figure 12–21 is the resource profile of the target schedule as put together in Chapter 6, and relates to Table 12–1. Compare Figure 12–20 to Figure 12–21 to determine the change in labor requirements from the target to the current schedule.

RESOURCE REPORTS

Figure 12–22 is the hard copy print of the sample schedule in bar chart format showing the resource table. All resources to the left of the data

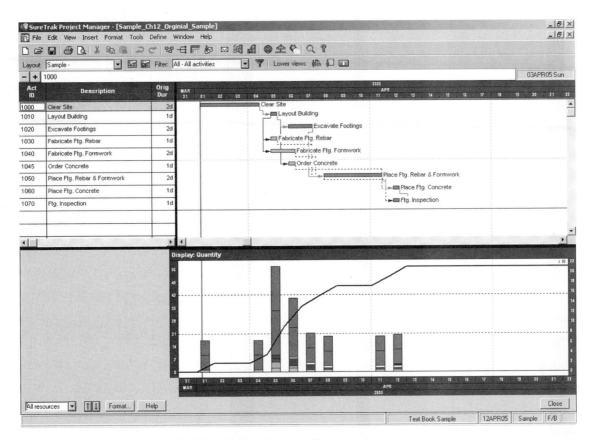

Figure 12–21   Resource Profile—Target Schedule

date are actual to-date entries. All entries to the right of the data date are to-complete estimates. Figure 12–23 provides the same information in resource profile format.

Figure 12–24 shows what can be done with *SureTrak*'s customized reports. This comparison report (current versus the target resource requirements) was created using the **Reports** selection from the **Tools** main pull-down menu. This brings up the **Reports** dialog box. Select the UP01 **Target comparison** report format. Click on the **Layouts** and then the **Columns** buttons to modify this report to meet your needs. Using the **Columns** dialog box, the columns and column widths were specified to create the report as shown in Figure 12–24.

Example Problem

Figures 12–25a, b, c, and d and Figures 12–26a and b show the updated resources of the schedule for the house put together as an example for student use (see the wood frame house drawings in the Appendix). Figure 12–25 is a hard copy print of the updated *SureTrak* bar chart print with the resource table shown. Figure 12–26 is a hard copy print of the updated *SureTrak* bar chart print with the resource profile shown.

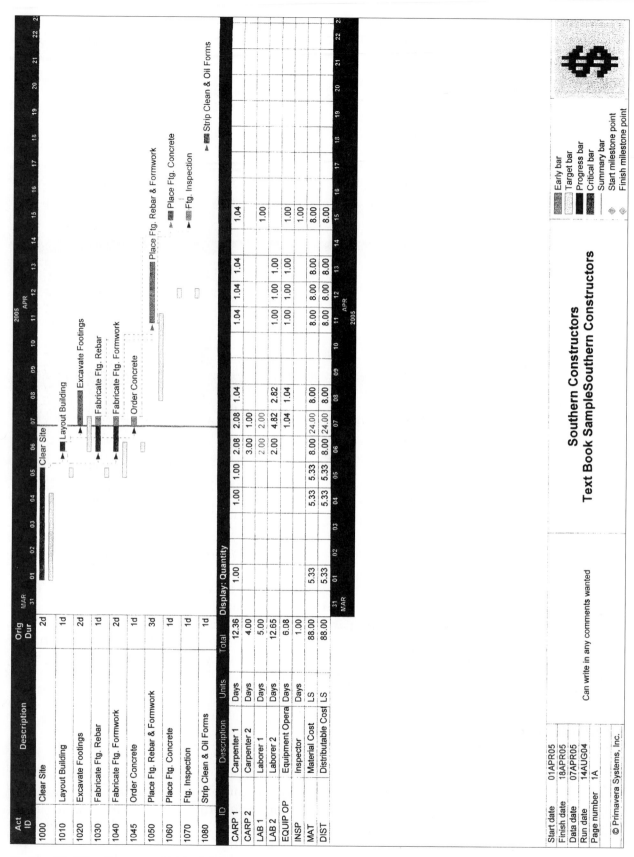

Figure 12–22 Resource Table Print

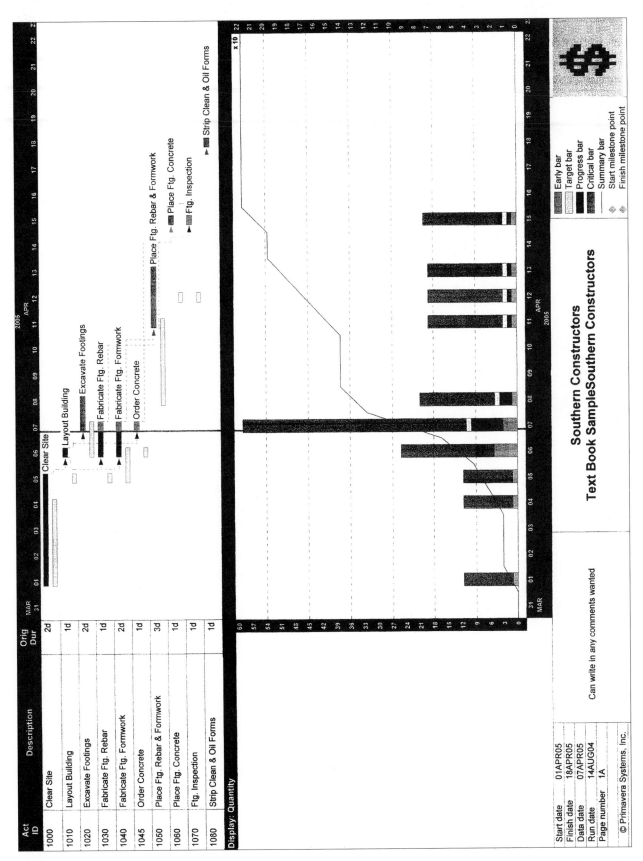

Figure 12–23   Resource Profile Print

Report: Target comparison
Layout: REPORTS
Filter: All Activities

SureTrak Project Manager
Text Book Sample

Southern Constructors
Report Date: 14AUG04
Page 1A of 1B

Act ID	Description	Orig Dur	Rem Dur	Resource	Budgeted Quantity	Actual Quantity to Date	Quantity to Complete	Quantity at Completion	Percent Complete	Percent Expended	Completion Variance Quantity
1000	Clear Site	2	0	CARP 1, DIST, MAT	34	35	0	35	100	100	-1
1010	Layout Building	1	0	CARP 1, CARP 2, DIST, LAB 1, MAT	20	21	0	21	100	100	-1
1020	Excavate Footings	2	2	CARP 1, DIST, EQUIP OP, LAB 2, MAT	42	0	42	42	0	0	0
1030	Fabricate Ftg. Rebar	1	1	DIST, LAB 1, LAB 2, MAT	18	0	19	19	50	0	-1
1040	Fabricate Ftg. Formwork	2	1	CARP 1, CARP 2, DIST, LAB 2, MAT	38	4	19	23	50	50	15
1045	Order Concrete	1	1		0	0	0	0	0	0	0
1050	Place Ftg. Rebar & Formwork	3	3	CARP 1, DIST, EQUIP OP, LAB 2, MAT	57	0	57	57	0	0	0
1060	Place Ftg. Concrete	1	1	CARP 1, DIST, EQUIP OP, LAB 1, MAT	19	0	19	19	0	0	0
1070	Ftg. Inspection	1	1	INSP	1	0	1	1	0	0	0
1080	Strip Clean & Oil Forms	1	1		0	0	0	0	0	0	0

2005 MAR 21

Data date 07APR05
Start date 01APR05
Finish date 18APR05
Must finish date
Target finish date 12APR05
© Primavera Systems, Inc.

Early bar — Summary bar
Progress bar — Start milestone point
Critical bar — Finish milestone point

Date   Revision   Checked | Approved

Figure 12–24   Target Comparison Report Print

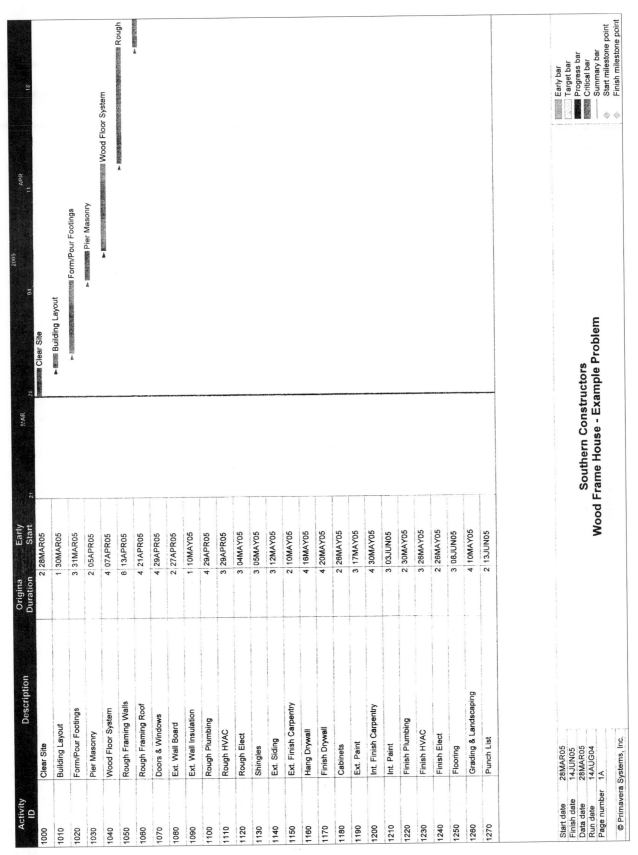

Activity ID	Description	Origina Duration	Early Start				
1000	Clear Site	2	28MAR05				
1010	Building Layout	1	30MAR05				
1020	Form/Pour Footings	3	31MAR05				
1030	Pier Masonry	2	05APR05				
1040	Wood Floor System	4	07APR05				
1050	Rough Framing Walls	6	13APR05				
1060	Rough Framing Roof	4	21APR05				
1070	Doors & Windows	4	29APR05				
1080	Ext. Wall Board	2	27APR05				
1090	Ext. Wall Insulation	1	10MAY05				
1100	Rough Plumbing	4	29APR05				
1110	Rough HVAC	3	29APR05				
1120	Rough Elect	3	04MAY05				
1130	Shingles	3	05MAY05				
1140	Ext. Siding	3	12MAY05				
1150	Ext. Finish Carpentry	2	10MAY05				
1160	Hang Drywall	4	16MAY05				
1170	Finish Drywall	4	20MAY05				
1180	Cabinets	2	26MAY05				
1190	Ext. Paint	3	17MAY05				
1200	Int. Finish Carpentry	4	30MAY05				
1210	Int. Paint	3	03JUN05				
1220	Finish Plumbing	2	30MAY05				
1230	Finish HVAC	3	26MAY05				
1240	Finish Elect	2	26MAY05				
1250	Flooring	3	08JUN05				
1260	Grading & Landscaping	4	10MAY05				
1270	Punch List	2	13JUN05				

Southern Constructors
Wood Frame House - Example Problem

Start date	28MAR05
Finish date	14JUN05
Data date	28MAR05
Run date	14AUG04
Page number	1A

© Primavera Systems, Inc.

Early bar
Target bar
Progress bar
Critical bar
Summary bar
◇ Start milestone point
◇ Finish milestone point

Figure 12–25a   Resource Table Print—Wood Frame House—Page 1

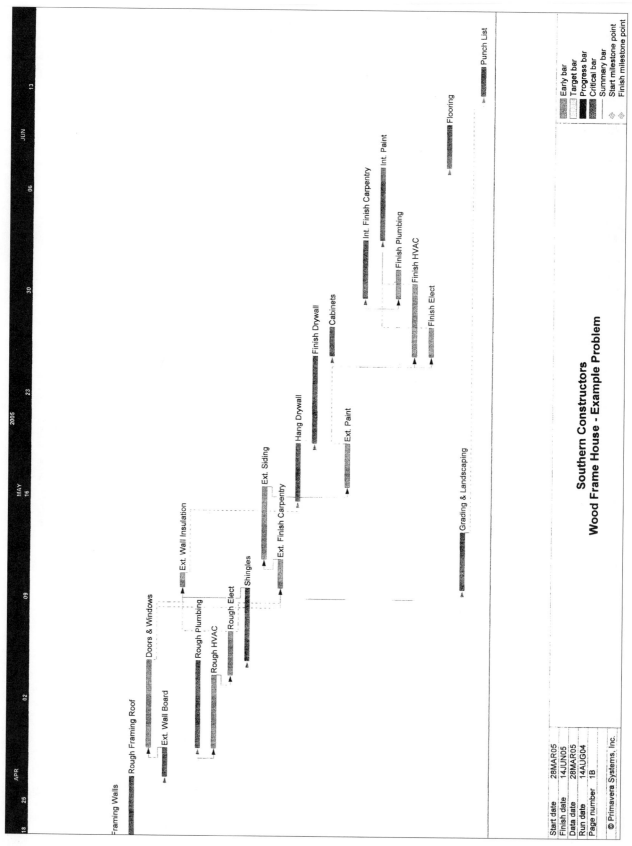

Figure 12–25b Resource Table Print—Wood Frame House—Page 2

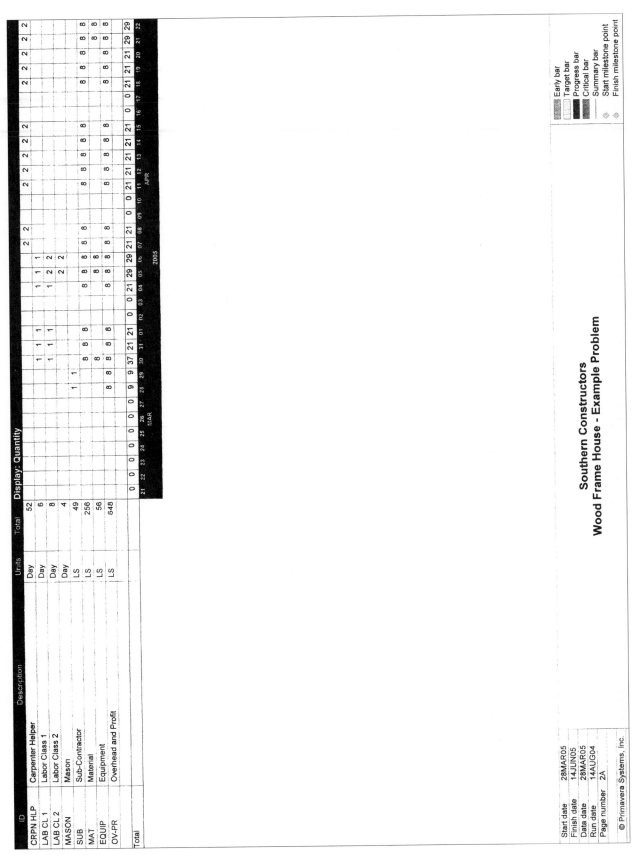

Figure 12–25c   Resource Table Print—Wood Frame House—Page 3

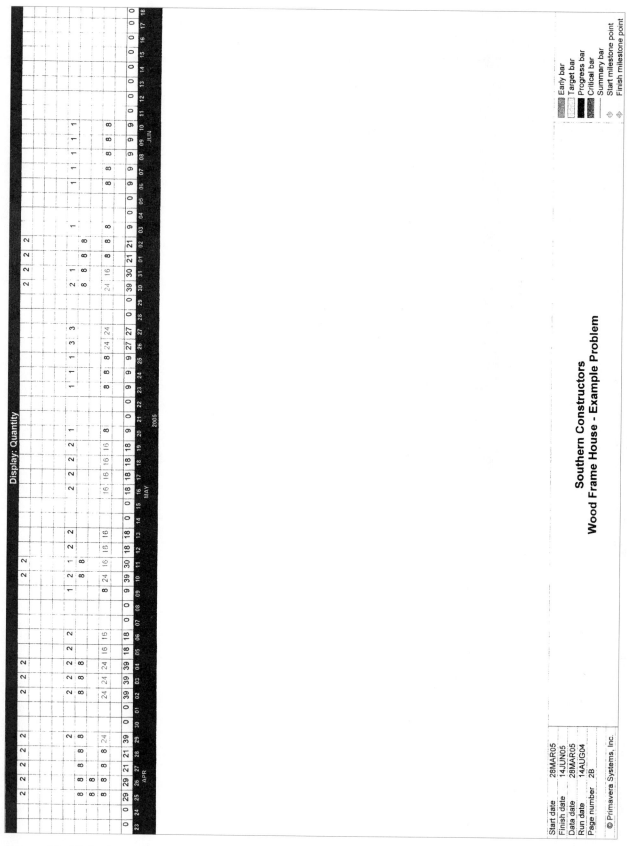

Figure 12–25d   Resource Table Print—Wood Frame House—Page 4

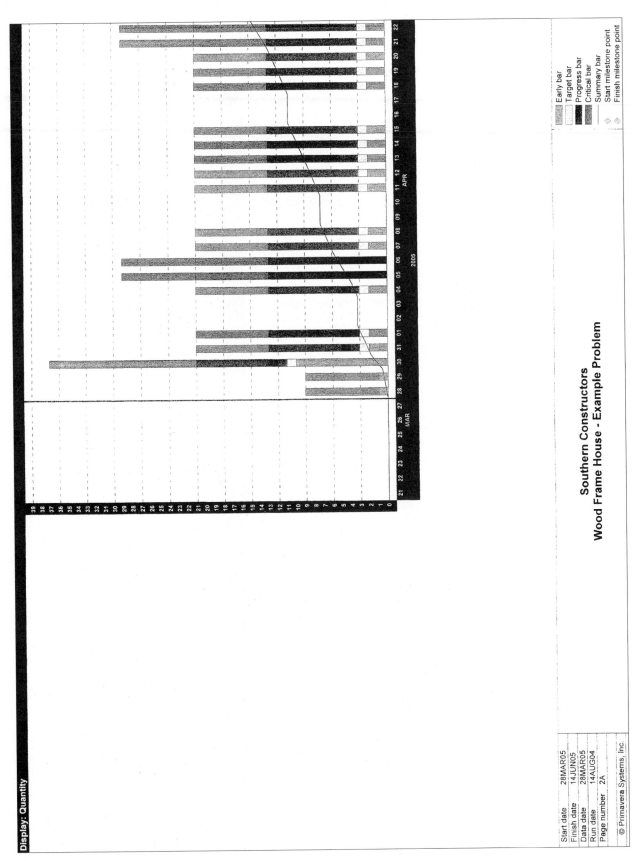

Figure 12–26a   Resource Profile Print—Wood Frame House—Page 1

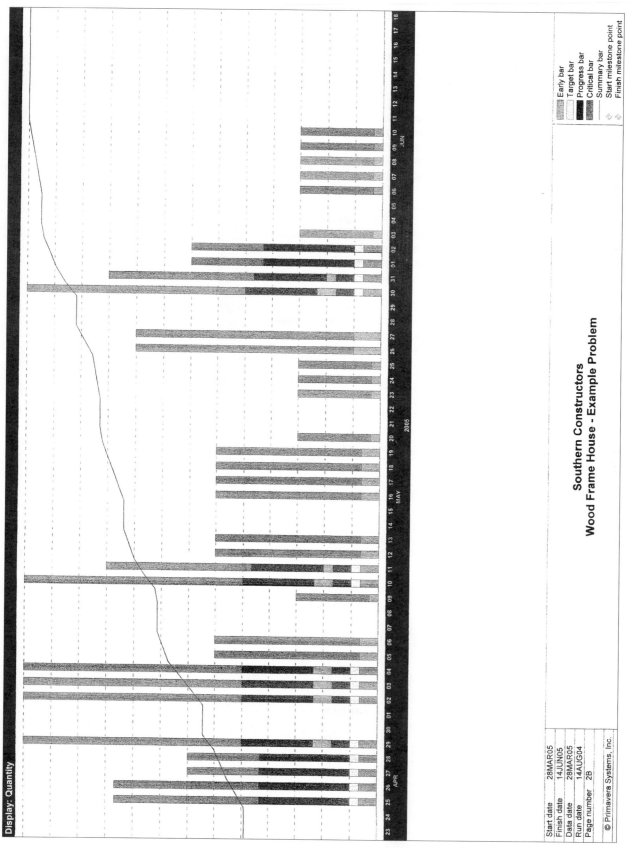

Figure 12–26b Resource Profile Print—Wood Frame House—Page 2

SUMMARY

After a project is underway, the contractor tracks (monitors) resources to compare actual to planned progress. The target (original) resources are updated to show the physical progress attained in the current (updated) schedule.

The resource table is used within *SureTrak* to show actual expenditures through the data date and then revert to planned expenditures thereafter. Activities with the actual expenditures of resources must be updated.

SureTrak's autocost rules changes the way resource updating expenditure calculations are made. The relationship between the resource duration and percent complete can be linked.

Updating resource expenditures are made using the actual to date, to complete, and at completion fields of the Resources dialog box.

The graphical representation of the resource profile helps analyze the updated resource expenditures and remaining requirements.

Resource reports, both tabular and graphical, help communicate the updated progress attained through the expenditure of resources and how these resources have been expended over time.

EXERCISES

1. **Modifying the Autocost Rules**
   Open the project modified in Exercise 6 of Chapter 11 and modify the autocost rules as described in this chapter.
2. **Updating Resources Used and Needed to Complete Activities**
   Update the resources used and needed to complete the activities in Exercise 1.
3. **Printing an Updated Resource Profile and Table**
   Print the schedule, including both the resource profile and table of Exercise 2.
4. **Printing a Tabular Comparison Report**
   Print a tabular report of Exercise 2 with the columns shown in Figure 12–24.

5. Produce a manual resource update (resource table) for Figure 12–27.

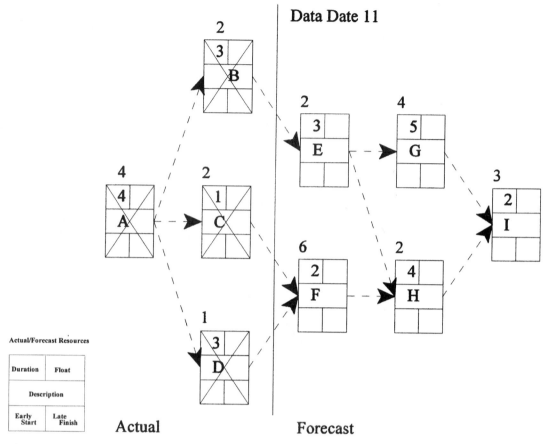

Figure 12–27   Exercise #5

6. Produce a manual resource update (resource table) for Figure 12–28.

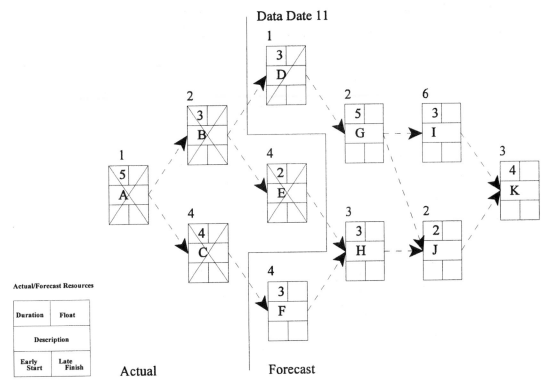

Figure 12–28 Exercise #6

7. Produce a manual resource update (resource table) for Figure 12–29.

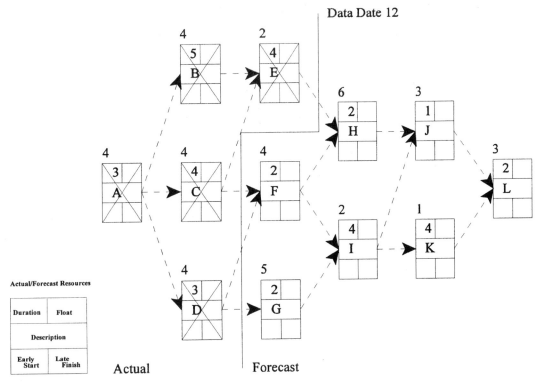

Figure 12–29 Exercise #7

8. Produce a manual resource update (resource table) for Figure 12–30.

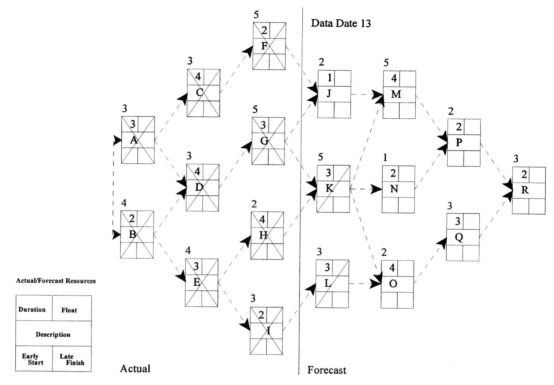

Figure 12–30 Exercise #8

9. Produce a manual resource update (resource table) for Figure 12–31.

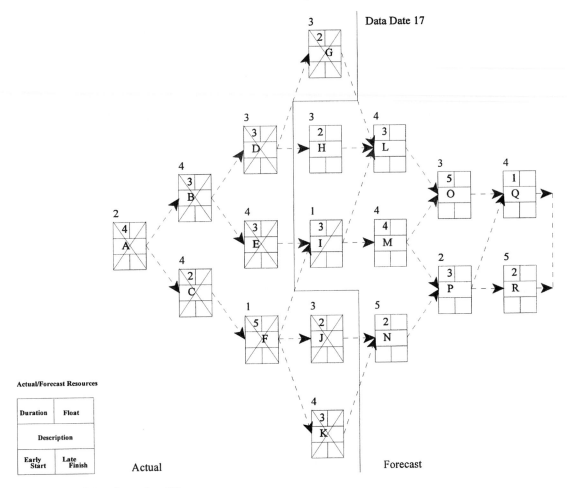

Figure 12–31   Exercise #9

10. Produce a manual resource update (resource table) for Figure 12–32.

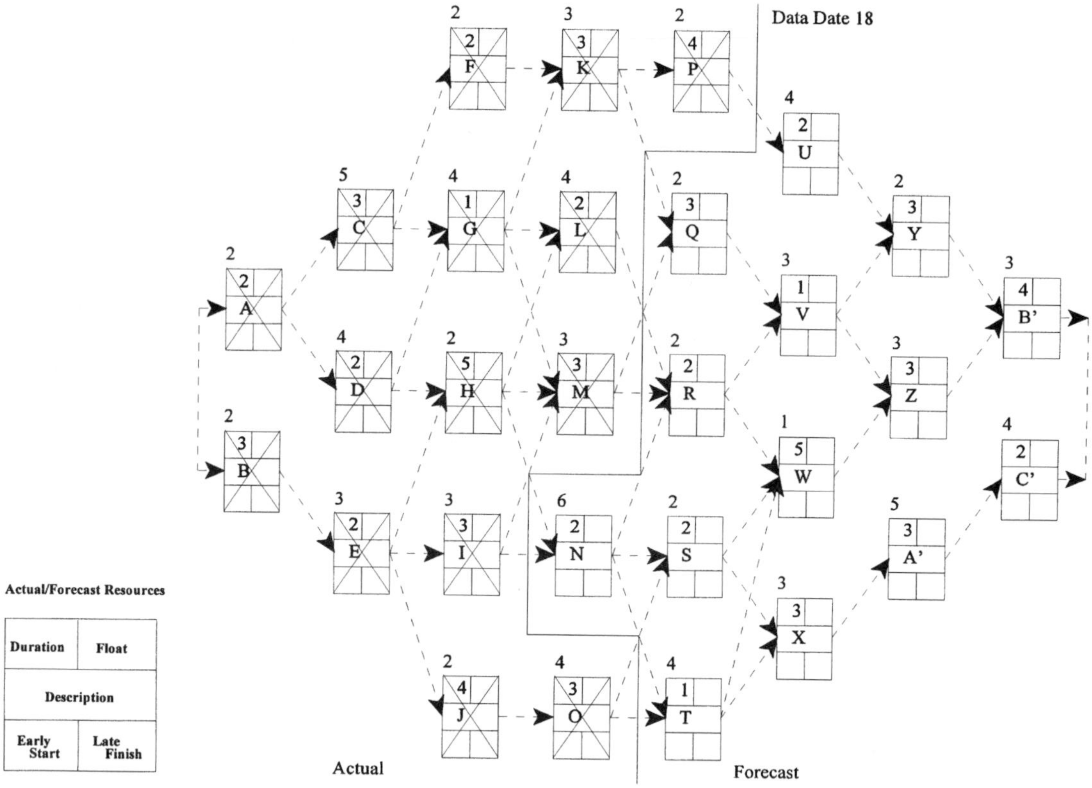

Figure 12–32 Exercise #10

11. Small Commercial Concrete Block Building—Resource Tracking
Prepare the following reports for the small commercial concrete block building located (Exercise 12, Chapter 6) in the Appendix:
- Updated resource table
- Updated resource profile

12. Large Commercial Building—Resource Tracking
Prepare the following reports for the large commercial building located (Exercise 13, Chapter 6) in the Appendix:
- Updated resource table
- Updated resource profile

13 Tracking Costs

Objectives

Upon completion of this chapter, you should be able to:

- Compare actual to planned expenditures
- Update a schedule
- Record expenditures
- Define a scheduled budget
- Define earned value
- Create a cost profile
- Print a tabular report for tracking cost

TRACKING COSTS: COMPARING ACTUAL TO PLANNED EXPENDITURES

After a project is underway, tracking costs and comparing actual to planned expenditures are tools used to control costs. The costs can either be total costs or the accumulation of the resource costs of labor, equipment, materials, and/or other resources. These costs are usually assigned to the activity (allocated from the original estimate) when the target schedule is constructed. By tracking actual costs and comparing them to the original budget, you can determine cost progress and earned value.

The construction managers must determine whether the project is making or losing money. If there are problems, he determines where the problems lie, and then passes on the information to the right people in time to do something about it.

On a cost-plus or negotiated project, the owner also can use the cost-loaded and then cost-monitored schedule as valuable information in managing cash assets. This chapter shows how to use *SureTrak* to update costs to match the updated durations, logic, and activity changes made to the schedule in Chapter 11.

COST TABLE

The cost table provides a convenient tool for tracking and analyzing cost.

Updating the Target Schedule

Figure 13–1 is the target schedule with the original costs. The original plan called for $170 to be spent on 01APR05 on Activity 1000 - Clear Site. Likewise, $170 was planned for 04APR05, $1036 for 05APR05, and so on. Figure 13–2 is the current schedule as updated in Chapter 11, with progress, modified logic, and new activities input. The current plan calls for $113 to be spent on 01APR05 on Activity 1000 - Clear Site. Thus, $113 was spent on 04APR05, $113 on 05APR05, and so on. Obviously the changes made in updating the schedule in Chapter 11 changed the cost information contained in the schedule without inputting the actual cost information. You need to understand what *SureTrak* has done to create these new cost totals. The next step is to further update the schedule showing the actual expenditure of costs and to make forecasts.

The easiest way to evaluate costs onscreen is through the resource table. Select the **View** main pull-down menu, and then select **Resource Table.** The resources appear at the bottom of the screen (Figure 13–2). If resources rather than costs appear, click the **Format** button. The **Format Resource Profile/Table** dialog box appears (Figure 13–3). Under the **Display** pull-down menu, select **Costs.**

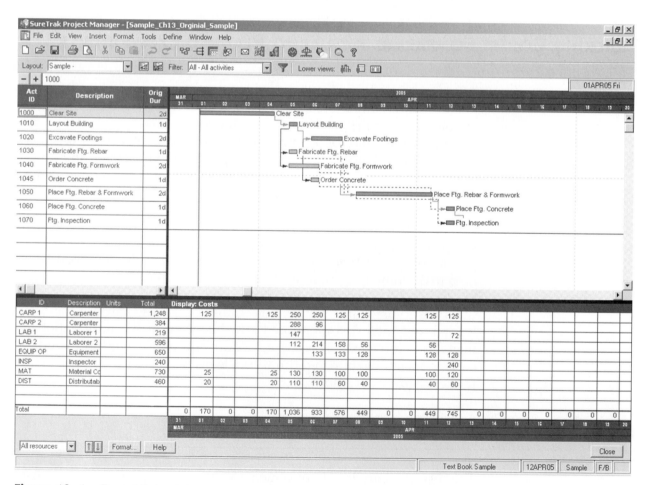

Figure 13–1 Target Schedule

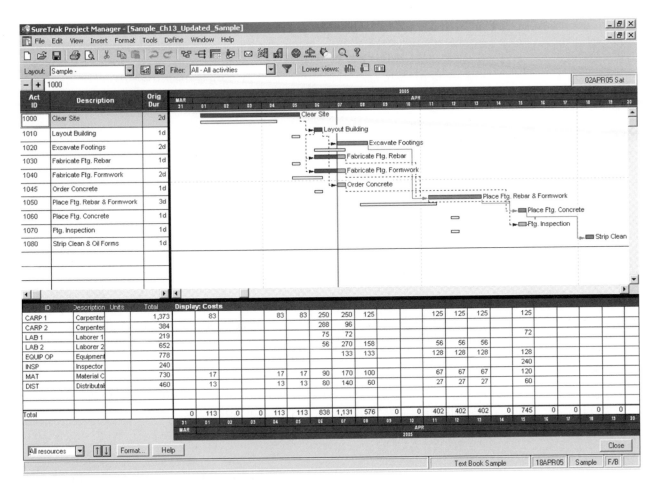

Figure 13–2   Current Schedule

Current Schedule.   Figure 13–2 is the current schedule, or the update of the target schedule as modified in Chapter 11. It contains the schedule and logic modifications made to update the actual current data date in Chapter 11. It also contains the actual input of resources made in Chapter 12. Now it is time to input the actual cost information to match the

Figure 13–3   Format Resource Profile/Table Dialog Box

actual cost to date. The comparison of the planned costs (target schedule) to the actual costs of work performed (current schedule) makes cost forecasting possible. *Cost forecasting* is the projection of the future cost based on costs to date and management's knowledge of the project.

RECORDING EXPENDITURES

Costs must be gathered by activity. This is not the way most contractors' cost accounting systems work. The usual construction cost system gathers costs by cost account code. Labor time sheets, work measurement reports, purchase orders, and all other cost accounting documents are coded by cost account code. The estimate is organized the same way. To gather costs by activity, another step must be added to the level of information gathered in the field. The documents must be coded with the cost accounting code and the activity ID for cost-gathering purposes. Another element that makes the process of gathering costs by activity difficult is that indirect costs and profit are not typically placed in any specific activity, but are prorated (or spread) over all activities. In the following section, we show how to record the actual cost expenditures for sample schedule activities where physical progress has been claimed during the updating of the target schedule.

Activity 1000

Select the **Costs** from the **Format Resource Profile/Table** dialog box (Figure 13–2). Click on the first activity to be updated with cost information, Activity 1000 - Clear Site. Click the right mouse button, select the **Activity Detail** box, and then select **Costs.** The **Cost** dialog box for Activity 1000 appears (Figure 13–4). Keep in mind that costs were originally loaded in the target schedule.

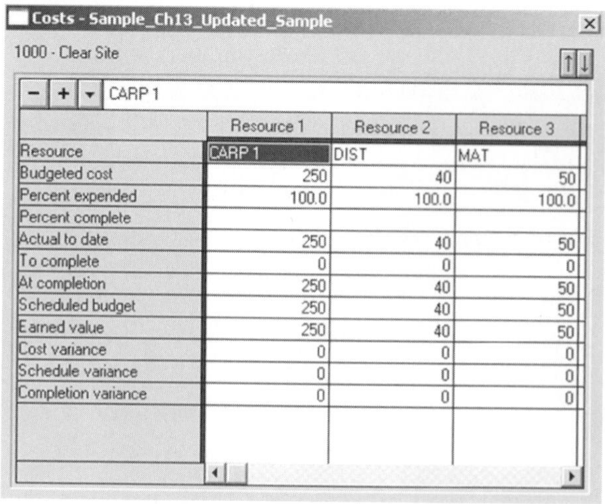

Costs - Sample_Ch13_Updated_Sample			
1000 - Clear Site			
CARP 1			
	Resource 1	Resource 2	Resource 3
Resource	CARP 1	DIST	MAT
Budgeted cost	250	40	50
Percent expended	100.0	100.0	100.0
Percent complete			
Actual to date	250	40	50
To complete	0	0	0
At completion	250	40	50
Scheduled budget	250	40	50
Earned value	250	40	50
Cost variance	0	0	0
Schedule variance	0	0	0
Completion variance	0	0	0

Figure 13–4   Cost Dialog Box—Activity 1000

Current Schedule.   From Figure 13–4, the **Budgeted cost, Actual to date, At completion,** and **Scheduled budget** fields all show a value of $250 for CARP 1, 40 for DIST, and 50 for MAT fields for a total of $340.

Budgeted cost is the amount included for the resource in the original budget. If the resource is cost bearing, as with CARP 1 at $250/2 days, *Sure-Trak* calculates the total cost of the resource assignment here. Note that for an activity that has *not* started, you can enter data in the **At completion** field and it will roll up to the **Budgeted cost** field. However, if you enter something in the **Budgeted cost** field, the **At completion** field is unaffected.

Actual to date is the amount of money spent for this resource on this activity, up to the data date. You can update the value in this field to show actual expenditures for the resource, or you can have *SureTrak* estimate progress and resource usage (choose the **Tools** main pull-down menu and then select **Update Progress**).

At completion is the current forecast cost for the resource. Before the activity is updated, *SureTrak* calculates this cost as equal to the **Budgeted cost.** After the activity is updated, *SureTrak* calculates this number as the sum of the **Actual to date** plus **To complete** fields. Note that for an activity that has *not* started, you can enter data in the **At completion** field and they will roll up to the **Budgeted cost** field. However, if you enter something in the **Budgeted cost** field, the **At completion** field is unaffected.

As you can see in the target cash flow by activity in (Table 13–1) the original duration for Activity 1000 was 2 days, or $340/2 days or $170 per day. When progress was updated to create the current schedule, the actual duration for Activity1000 was a 3-day duration. The cost update to current schedule (Table 13–2) shows the $340 budget is now $113 per day ($340/3 days).

Actual Cost Expenditures.   Note the **Percent expended** and **Percent complete** fields from Figure 13–4.

Percent expended is the portion of the budget for this activity that has been spent to date. For Activity 1000, *SureTrak* automatically placed 100% expended when the activity was updated as 100% complete.

Percent complete is the percentage of the activity that is complete. You can change it, or have *SureTrak* estimate it during a progress update.

The actual/forecast cost to date (Table 13–3) shows the actual costs for Activity 1000 as a total of $390 (not $340), or $130 per each of the three days and not $113. Figure 13–5 shows the changes you would make to update Activity 1000 with current cost information:

1. **Percent complete.** Compare Figure 13–4 to 13–5. Change the CARP 1, DIST, and MAT resources to reflect 100 **Percent complete.**

Table 13–1 Target Estimate by Activity

Act. ID	Description	Duration	Day 1 April 01	Weekend April 02	April 03	Day 2 April 04	Day 3 April 05	Day 4 April 06	Day 5 April 07	Day 6 April 08	Weekend April 09	April 10	Day 7 April 11	Day 8 April 12
1000	Clear Site	2	$170											
1010	Layout Building	1				$170								
1020	Excavate Footings	2					$482	$576						
1030	Fabricate Ftg. Rebar	1					$198		$576					
1040	Fabricate Ftg. Formwork	2					$357	$357						
1050	Place Ftg. Rebar & Formwork	2								$449			$449	
1060	Place Ftg. Conc	1												$505
1070	Ftg. Inspection	1												$240
	Cost/Day		$170	$0	$0	$170	$1,037	$933	$576	$449	$0	$0	$449	$745
	Cumulative Cost/Day		$170	$170	$170	$340	$1,377	$2,310	$2,886	$3,335	$3,335	$3,335	$3,784	$4,529

Table 13–2 Update Schedule

Act. ID	Description	Duration	Day 1 April 01	Weekend April 02	April 03	Day 2 April 04	Day 3 April 05	Day 4 April 06	Day 5 April 07	Day 6 April 08	Weekend April 09	April 10	Day 7 April 11	Day 8 April 12	Day 9 April 13	Day 10 April 14	Day 11 April 15	Weekend April 16	April 17	Day 12 April 18
1000	Clear Site	2	$113			$113														
1010	Layout Building	1					$113													
1020	Excavate Footings	2						$482	$576											
1030	Fabricate Ftg. Rebar	1								$576										
1040	Fabricate Ftg. Formwork	2						$357	$357											
1045	Order Concrete	1						$99	$99											
1050	Place Ftg. Rebar & Formwork	2											$299	$299	$299					
1060	Place Ftg. Conc	1															$505			
1070	Ftg. Inspection	1															$240			
1080	Strip Clean & Oil Forms	1															$0			
	Cost/Day		$113	$0	$0	$113	$113	$938	$1,032	$576	$0	$0	$299	$299	$299	$0	$745	$0	$0	$0
	Cumulative Cost/Day		$113	$113	$113	$227	$340	$1,278	$2,310	$2,886	$2,886	$2,886	$3,185	$3,485	$3,784	$3,784	$4,529	$4,529	$4,529	$4,529

Table 13–3 Actual/Forecast Cost Input

Act. ID	Description	Duration	Day 1 April 01	Weekend April 02	April 03	Day 2 April 04	Day 3 April 05	Day 4 April 06	Day 5 April 07	Day 6 April 08	Weekend April 09	April 10	Day 7 April 11	Day 8 April 12	Day 9 April 13	Day 10 April 14	Day 11 April 15	Weekend April 16	April 17	Day 12 April 18
1000	Clear Site	2	$130			$130														
1010	Layout Building	1					$130													
1020	Excavate Footings	2						$510	$576											
1030	Fabricate Ftg. Rebar	1								$576										
1040	Fabricate Ftg. Formwork	2						$357	$357											
1045	Order Concrete	1						$99	$99											
1050	Place Ftg. Rebar & Formwork	2											$299	$299	$299					
1060	Place Ftg. Conc	1															$505			
1070	Ftg. Inspection	1															$240			
1080	Strip Clean & Oil Forms	1															$0			
	Cost/Day		$130	$0	$0	$130	$130	$966	$1,032	$576	$0	$0	$299	$299	$299	$0	$745	$0	$0	$0
	Cumulative Cost/Day		$130	$130	$130	$260	$390	$1,356	$2,388	$2,964	$2,964	$2,964	$3,263	$3,563	$3,862	$3,862	$4,607	$4,607	$4,607	$4,607

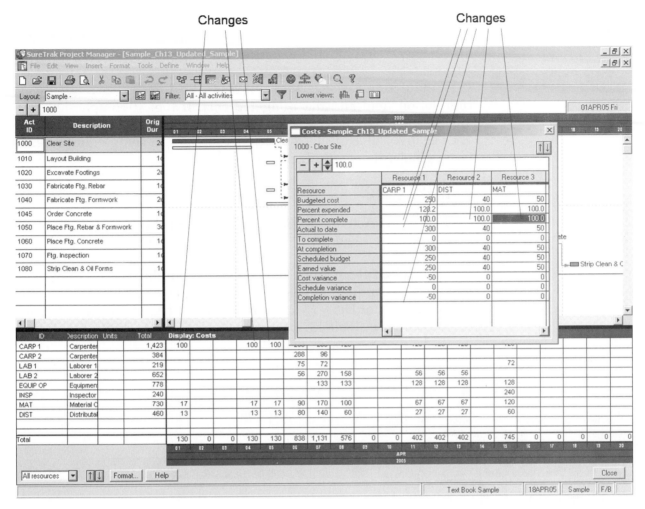

Figure 13–5 Cost Table—Activity 1000

2. **Actual to date.** Change the actual to date for CARP 1 from 250 to 300 to reflect a $50 increase in actual expenditures. The actuals to date for DIST ($40) and MAT ($50) remain unchanged.

3. **Resource table.** Look at the values for Activity 1000 in the resource table in Figure 13–5. Increase the CARP 1 costs for APR 01, 04, and 05 by $16.67/day (total $50 increase). CARP 1 now shows 3 days at $100/day. MAT shows 3 days at $17/day, and DIST shows 3 days at $13/day: $(3 \times \$100) + (3 \times \$17) + (3 \times \$13) = \$390$.

Note also from the **Costs** dialog box for Activity 1000, that when we made the preceding changes, two other fields were automatically changed. The values for **Cost variance** (0 to −50) and **Completion variance** (0 to −50) changed.

The **Cost variance** field compares the **Actual to date** costs for work performed to the **Budgeted cost** (measured in percent complete). If the

Actual to date (300) exceeds the **Budgeted cost** (250), the difference is a negative value (−50). **Cost variance** is **Earned value** (discussed later in this chapter) minus **Actual to date** costs.

The **Completion variance** field represents the difference between the **At completion** field and the **Budgeted cost**. If the **At completion** (300) exceeds the **Budgeted cost** (250), the difference is a negative value (−50). The negative value means the current estimate (**At completion**) exceeds the original estimate (**Budgeted cost**).

Activity 1010

Click next on Activity 1010. The **Cost** dialog box for Activity 1010 appears (Figure 13–6).

Current Schedule. From Figure 13–6, the **Budgeted cost, Actual to date,** and **At completion** fields all show a value of 125 for CARP 1, 192 for CARP 2, 30 for DIST, 75 for LAB 1, and 60 for MAT for a total of $482. As you can see in Table 13–1, the target cash flow by activity, the original

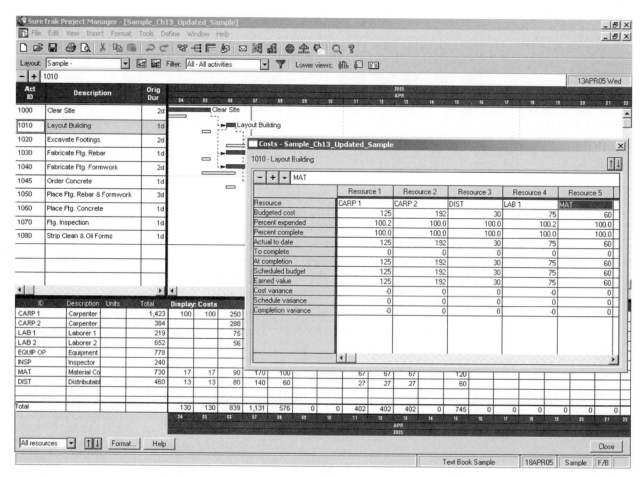

Figure 13–6 Cost Table—Activity 1010—Target Estimate

duration for activity 1010 was 1 day, or $482 per day. When progress was updated to create the current schedule, the actual duration for Activity 1010 was still a 1-day duration. From Table 13–2, the cost update to current schedule, the $482 budget is unchanged.

Actual Cost Expenditures.   From Table 13–3, actual/forecast cost to date, the actual cost for Activity 1010 is $510 (not $482). Figure 13–7 shows the following changes you would make to update Activity 1010 with current cost information:

1. **Percent complete.** Compare Figure 13–7 to 13–6. Change the CARP 1, CARP 2, DIST, LAB 1, and MAT resources to reflect 100 **Percent complete.**
2. **Actual to date.** Change the actual to date for CARP 1 from 125 to 135 to reflect a $10 increase in actual expenditures. Change the actual to date for CARP 2 from 192 to 200 to reflect an $8 increase in actual expenditures. The actual to date for DIST ($30) remains unchanged. Change the actual to date for LAB 1 from 75 to 78 to reflect a $3 increase in actual expenditures. Change the actual to date for MAT from 60 to 67 to reflect a $7 increase in actual expenditures.

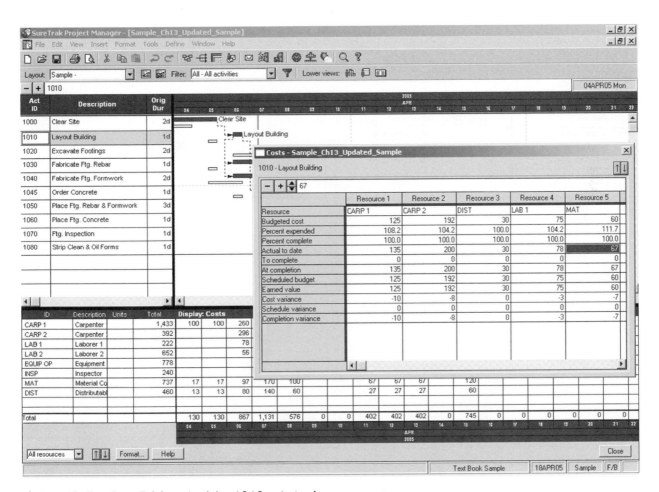

Figure 13–7   Cost Table—Activity 1010—Actual

Resource Table.   Compare the values for Activity 1010 in the resource table in Figure 13–7 to Figure 13–6 before the change. The CARP 1 costs for APR 06 have increased by $10. CARP 1 now shows a $260 total ($135 for Activity 1010 and $125 for other activities), not $250. CARP 2 now shows an $8 increase or a $296 total ($200 for Activity 1010 and $196 for other activities), not $288. LAB 1 now shows a $3 increase, or a $78 total, not $75. MAT now shows a $7 increase. DIST is unchanged at $30. The total for Activity 1010 is $135 + 200 + 30 + 78 + 67 = $510.

Activity 1030

Click on the next activity to be updated with cost information, Activity 1030 - Fabricate Ftg. Rebar. The **Cost** dialog box for Activity 1030 appears (Figure 13–8).

Current Schedule.   Refer to Figure 13–2 and Table 13–1. Activity 1030 originally had a 1-day duration with a budget of $198 but the schedule was updated to show current progress in Chapter 11. The activity was 50% complete, with completion to be achieved in 1 day (07APR05). For Activity 1030, Figure 13–8, the **Budgeted cost** is $72 for LAB 1 and $56 for LAB 2. The **At completion** value for LAB 1 is $72 and for LAB 2 is $56.

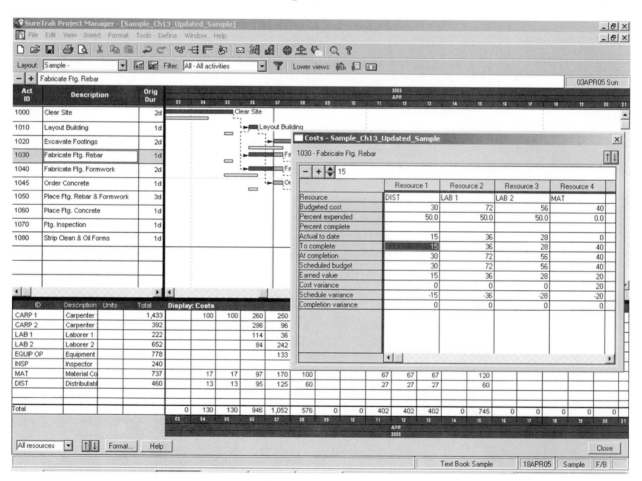

Figure 13–8   Cost Dialog Box—Activity 1030

The **To complete** field is the amount of money that remains to be spent for this resource on the current activity. *SureTrak* calculates this amount as the difference between **At completion** and the **Actual to date** costs (**At completion – Actual to date = To complete**).

Because the units for LAB 1 and LAB 2 are days, *SureTrak* multiplied the budget of 72 and 56 by the activity percent complete (50%) to obtain the 36 and 28 values for earned value of LAB 1 and LAB 2. The **Percent complete** field on the **Costs** dialog box (Figure 13–8) relates to cost percent complete and not activity percent complete. So, the $36 for LAB 1 and the $28 (plus other activities to total $242) for LAB 2 to be spent on Monday, 07APR05, is shown in the cost table, but the monies spent to attain the 50% progress is not shown until we claim cost percent complete. Note from Figure 13–8 that MAT is calculated the same way with 0 in the **Percent expended** field. This is because the units are days. The DIST resource has 50 **Percent expended** because its unit was LS or lump sum.

The monies on the cost table (Figure 13–8 and Table 13–2) relating to Activity 1020 are:

06APR05	DIST	$15	($30 × .50)
	LAB 1	$36	($72 × .50)
	LAB 2	$28	($56 × .50)
	MAT	$ 0	($40 × 0)
		$79	

07APR05	DIST	$ 15	($30 − $15)
	LAB 1	$ 36	($72 − $36)
	LAB 2	$ 28	($56 − $28)
	MAT	$ 40	($40 − 0)
		$119	

Actual Cost Expenditures.   From Table 13–3, actual/forecast cost to date, the actual cost for Activity 1030 is $198. Actual costs are $79 (15 + 36 + 28) for 06APR05 and projected costs of $119 (15 + 36 + 28 + 40) are to be spent on 07APR05. Figure 13–9 shows the following change you would make to update Activity 1030 with current cost information:

1. **Percent complete.** Compare Figure 13–8 to Figure 13–9. Change the CARP 1, DIST, and LAB 1 and LAB 2 resources to reflect 50 **Percent complete.**

Activity 1040

Click on Activity 1040. The **Cost** dialog box for Activity 1040 appears (Figure 13–10).

Current Schedule.   From Figure 13–10, the **Budgeted cost** and **At completion** fields both show a value of $250 for CARP 1, $192 for CARP 2,

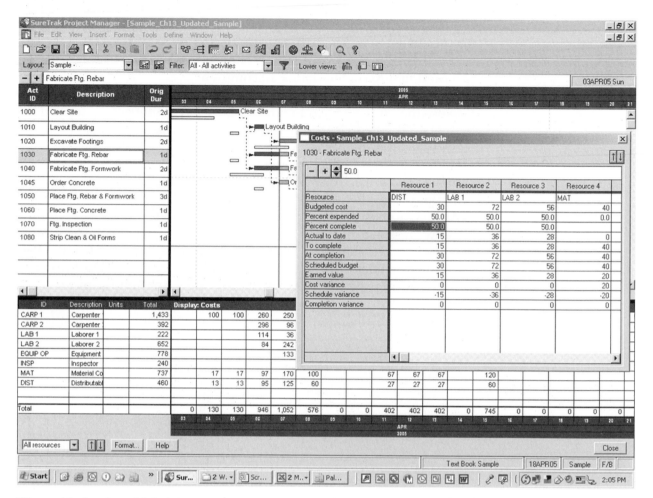

Figure 13–9   Cost Dialog Box—Activity 1030

$100 for DIST, $112 for LAB 2, and $60 for MAT. Table 13–1 shows the original duration for Activity 1040 was 2 days, or $357 per day. When progress was updated to create the current schedule, the actual duration for Activity 1030 was still a 2-day duration. From Table 13–2, the $357 budget is unchanged.

Actual Cost Expenditures.   From Table 13–3, actual/forecast cost to date, the actual costs for Activity 1040 are $357. Figure 13–11 shows the following change you would make to update Activity 1030 with current cost information:

1. **Percent complete.** Compare Figure 13–11 to Figure 13–10. Change the CARP 1, CARP 2, DIST, LAB 2, and MAT resources to reflect 50 **Percent complete.**

Resource Table.   Compare the values for Activity 1030 in the resource table in Figure 13–11 to Figure 13–9 before the change. There were no changes.

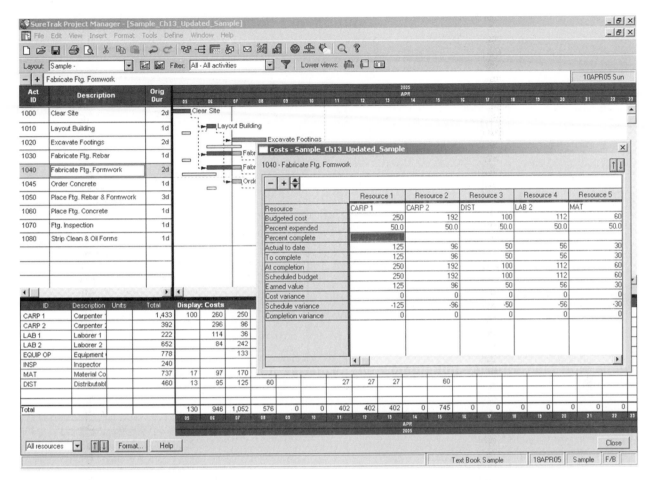

Figure 13–10   Cost Dialog Box—Activity 1040

Scheduled Budget

In Figure 13–4, the **Costs** dialog box for Activity 1000, there is a field called **Scheduled budget.** The scheduled budget (target/original schedule budget) is the budgeted cost of work scheduled (BCWS). It is the cost of the work that should have been accomplished as of the data date, according to the target or original plan. These costs show none of the duration changes, the activities added, or the actual progress input. The scheduled budget shows how costs were to be spent according to the target schedule. The numbers in the **Scheduled budget** field match the target schedule in Figure 13–1 and Table 13–1.

Figure 13–12 is a tabular budget report showing the scheduled budget as one of the columns of information provided. This is a handy tool for comparing current and original projections.

Earned Value

In Figure 13–4, the **Costs** dialog box for Activity 1000, there is a field called **Earned value.** Earned value is the budgeted cost for work performed

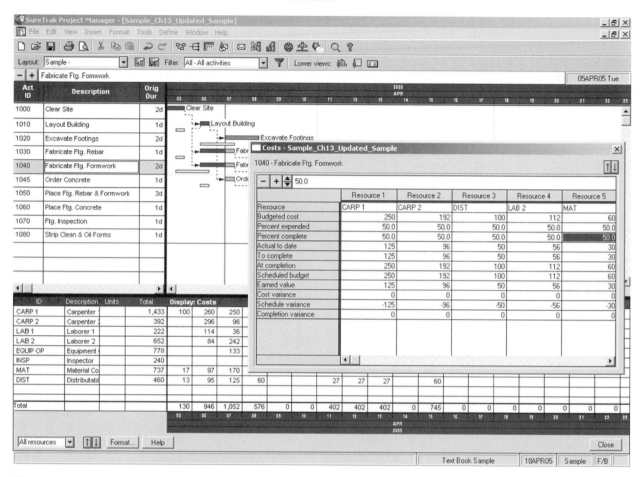

Figure 13–11   Cost Dialog Box—Activity 1040

(BCWP). The *earned value* represents the value of the work performed rather than the actual cost of work performed. *SureTrak* calculates **Earned value** as a product of **Percent complete** times **Budgeted cost.** It represents the portion of the budget that was allocated to the work that was actually accomplished.

Earned value is a measure of the costs of the actual physical progress earned according to the **Scheduled budget.** Remember the **Scheduled budget** is the cost of the work that should have been accomplished as of the data date, according to the original plan. Notice that the activities where no progress has been made show no earned value.

For Activity 1000 (refer to Figure 13–5), the **Earned value** is $250 for CARP 1. This represents the **Budgeted cost** (250) times the **Percent complete** (100%), so even though we have spent $300 on the activity, we have earned only $250.

Figure 13–13 is a tabular budget report showing earned value as one of the columns of information provided. This is a handy tool for analyzing project cost.

Act ID	Description	Orig Dur	Rem Dur	Early Start	Late Start	Early Finish	Late Finish	Total Float	Budgeted Cost	Scheduled Budget Cost	Actual Cost to Date	Cost at Completion	Completion Variance Cost
1000	Clear Site	2d	0	01APR05 A	01APR05 A	05APR05 A	05APR05 A		340	340	390	390	-50
1010	Layout Building	1d	0	06APR05 A	06APR05 A	06APR05 A	06APR05 A	0	482	482	510	510	-28
1020	Excavate Footings	2d	2d	07APR05	07APR05	08APR05	08APR05	0	1,152	576	0	1,152	0
1030	Fabricate Ftg. Rebar	1d	1d	06APR05 A	06APR05 A	07APR05	08APR05	1d	198	198	79	198	0
1040	Fabricate Ftg. Formwork	2d	1d	06APR05 A	06APR05 A	07APR05	08APR05	1d	714	714	357	714	0
1045	Order Concrete	1d	1d	07APR05	14APR05	07APR05	14APR05	5d	0	0	0	0	0
1050	Place Ftg. Rebar & Formwork	3d	3d	11APR05	11APR05	13APR05	13APR05	0	1,206	0	0	1,206	0
1060	Place Ftg. Concrete	1d	1d	15APR05	15APR05	15APR05	15APR05	0	505	0	0	505	0
1070	Ftg. Inspection	1d	1d	15APR05	18APR05	15APR05	18APR05	1d	240	0	0	240	0
1080	Strip Clean & Oil Forms	1d	1d	18APR05	18APR05	18APR05	18APR05	0	0	0	0	0	0

Start date	01APR05
Finish date	18APR05
Data date	07APR05
Run date	17AUG04
Page number	1A

Can write in any comments wanted

Southern Constructors
Text Book Sample

Early bar
Progress bar
Critical bar
Summary bar
Start milestone point
Finish milestone point

© Primavera Systems, Inc.

Figure 13–12 Budgeted Costs Tabular Report

Act ID	Description	Early Start	Early Finish	Percent Complete	Budgeted Cost	Scheduled Budget Cost	Actual Cost to Date	Earned Value Cost	Cost Variance	Schedule Variance	Completion Variance Cost
1000	Clear Site	01APR05 A	05APR05 A	100	340	340	390	340	-50	0	-50
1010	Layout Building	06APR05 A	06APR05 A	100	482	482	510	482	-28	0	-28
1020	Excavate Footings	07APR05	08APR05	0	1,152	576	0	0	0	-576	0
1030	Fabricate Ftg. Rebar	06APR05 A	07APR05	50	198	198	79	99	20	-99	0
1040	Fabricate Ftg. Formwork	06APR05 A	07APR05	50	714	714	357	357	0	-357	0
1045	Order Concrete	07APR05	07APR05	0	0	0	0	0	0	0	0
1050	Place Ftg. Rebar & Formwork	11APR05	13APR05	0	1,206	0	0	0	0	0	0
1060	Place Ftg. Concrete	15APR05	15APR05	0	505	0	0	0	0	0	0
1070	Ftg. Inspection	15APR05	15APR05	0	240	0	0	0	0	0	0
1080	Strip Clean & Oil Forms	18APR05	18APR05	0	0	0	0	0	0	0	0

Start date   01APR05
Finish date   18APR05
Data date   07APR05
Run date   17AUG04
Page number   1A
© Primavera Systems, Inc.

Can write in any comments wanted

Southern Constructors
Text Book Sample

Early bar
Progress bar
Critical bar
Summary bar
Start milestone point
Finish milestone point

Figure 13–13   Earned Value Tabular Report

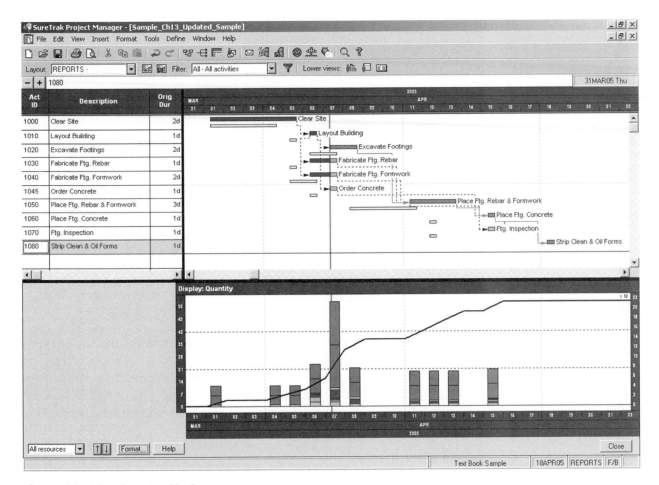

Figure 13–14   Cost Profile Screen

COST PROFILE

So far in this chapter, cost information has been presented in a tabular format in the onscreen representation. Sometimes a graphical representation is more helpful. Click on the **View** main pull-down menu. Select **Resource Profile** and the graphical representation of the cost information appears (Figure 13–14). If a cost profile does not appear on the screen, click on the **Format** button of the **Resource Profile.** The **Format Resource Profile/Table** dialog box appears (Figure 13–15).

COST REPORTS

A number of built-in *SureTrak* cost reports are useful for evaluating progress and updating information.

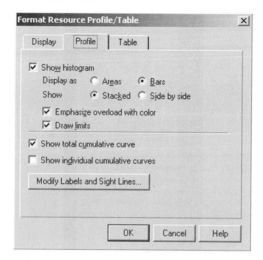

Figure 13–15 Format Resource Profile/Table Dialog Box

SureTrak's tabular reports put the cost information in a spreadsheet format for ease in evaluating numbers. The budgeted costs report (refer to Figure 13–12) is handy for comparing actual to planned expenditures and for forecasting final costs. The earned value report (refer to Figure 13–13) is great for analyzing the budgets and determining where the project stands from a cost point of view.

EXAMPLE PROBLEM

Figures 13–16a, b, c, d, e, and f and Figures 13–17a, b, and c show the cost update for the house put together as an example for student use (see the wood frame house drawings in the Appendix). Figure 13–16 is a hard copy print of the updated *SureTrak* bar chart with the costs table shown. Figure 13–17 is a hard copy print of the updated *SureTrak* cost profile.

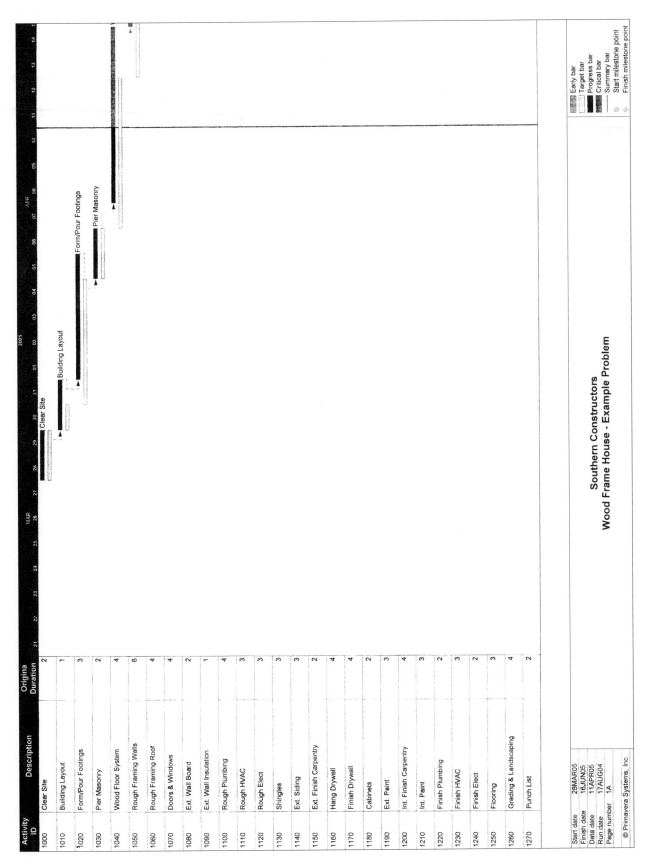

Activity ID	Description	Origina Duration
1000	Clear Site	2
1010	Building Layout	1
1020	Form/Pour Footings	3
1030	Pier Masonry	2
1040	Wood Floor System	4
1050	Rough Framing Walls	6
1060	Rough Framing Roof	4
1070	Doors & Windows	4
1080	Ext. Wall Board	2
1090	Ext. Wall Insulation	1
1100	Rough Plumbing	4
1110	Rough HVAC	3
1120	Rough Elect	3
1130	Shingles	3
1140	Ext. Siding	3
1150	Ext. Finish Carpentry	2
1160	Hang Drywall	4
1170	Finish Drywall	4
1180	Cabinets	2
1190	Ext. Paint	3
1200	Int. Finish Carpentry	4
1210	Int. Paint	3
1220	Finish Plumbing	2
1230	Finish HVAC	3
1240	Finish Elect	2
1250	Flooring	3
1260	Grading & Landscaping	4
1270	Punch List	2

Start date	28MAR05
Finish date	16JUN05
Data date	11APR05
Run date	17AUG04
Page number	1A

© Primavera Systems, Inc.

Southern Constructors
Wood Frame House - Example Problem

Early bar
Target bar
Progress bar
Critical bar
Summary bar
Start milestone point
Finish milestone point

Figure 13–16a   Cost Table Print—Wood Frame House—Page 1

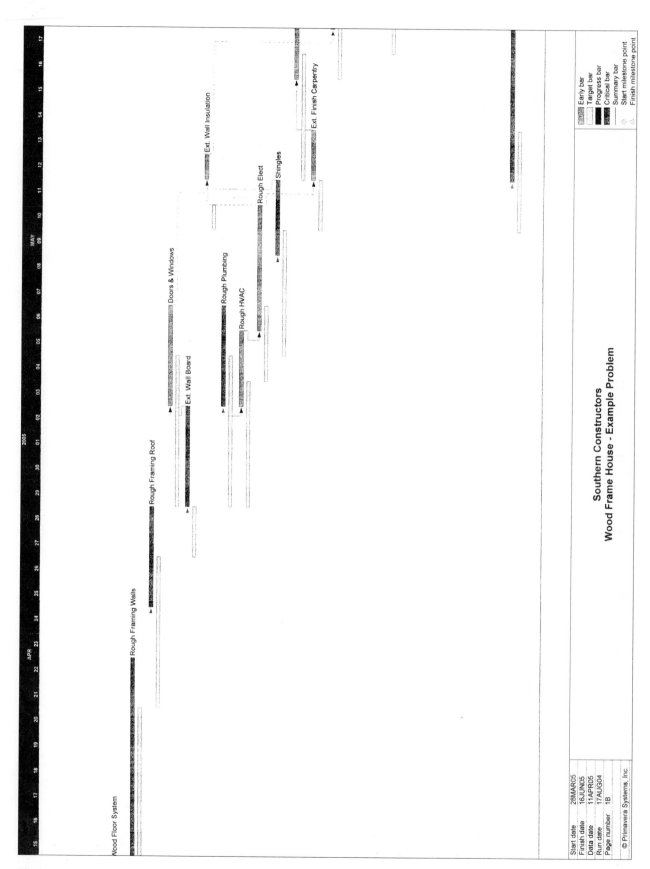

Figure 13–16b Cost Table Print—Wood Frame House—Page 2

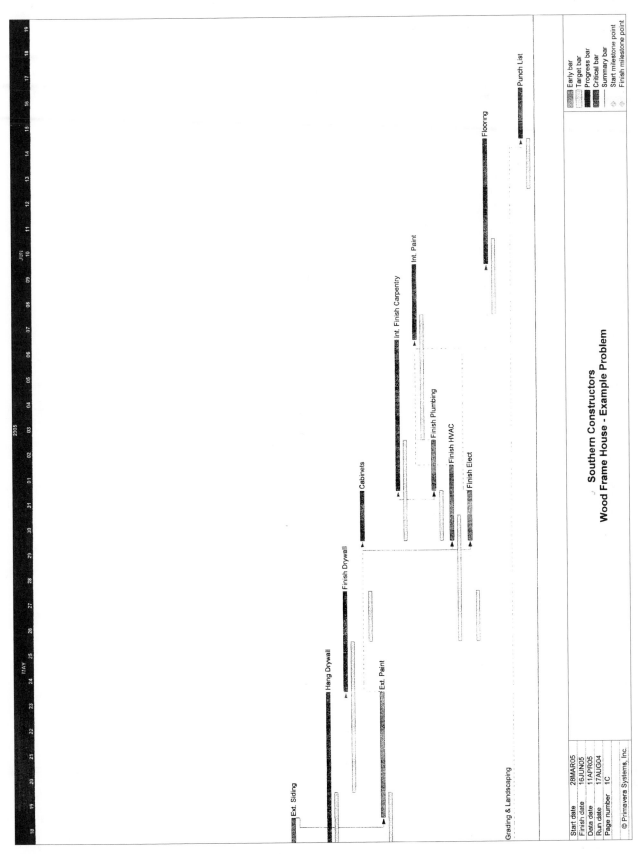

Figure 13–16c   Cost Table Print—Wood Frame House—Page 3

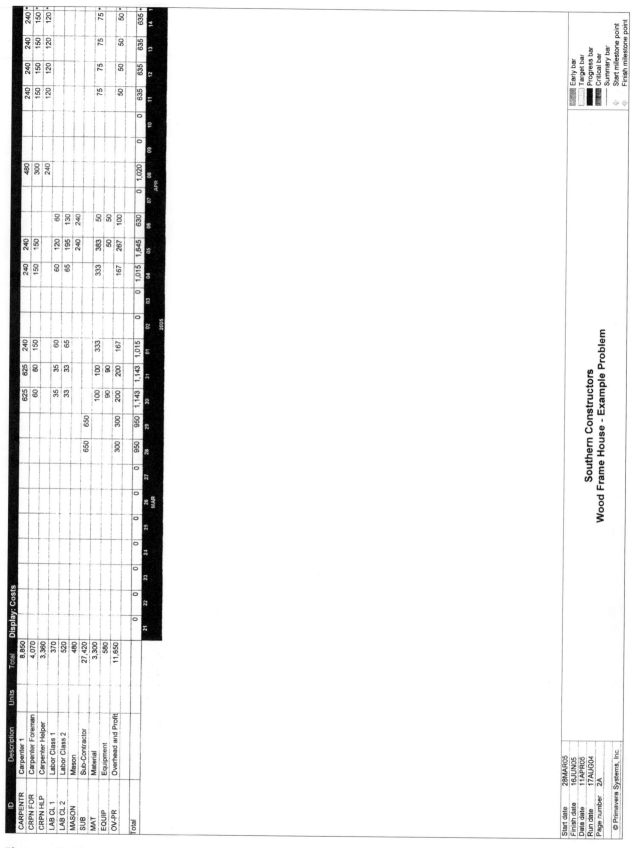

Southern Constructors
Wood Frame House - Example Problem

		Early bar
		Target bar
		Progress bar
		Critical bar
		Summary bar
◇		Start milestone point
◇		Finish milestone point

Start date	28MAR05
Finish date	16JUN05
Data date	11APR05
Run date	17AUG04
Page number	2A

© Primavera Systems, Inc.

Figure 13–16d Cost Table Print—Wood Frame House—Page 4

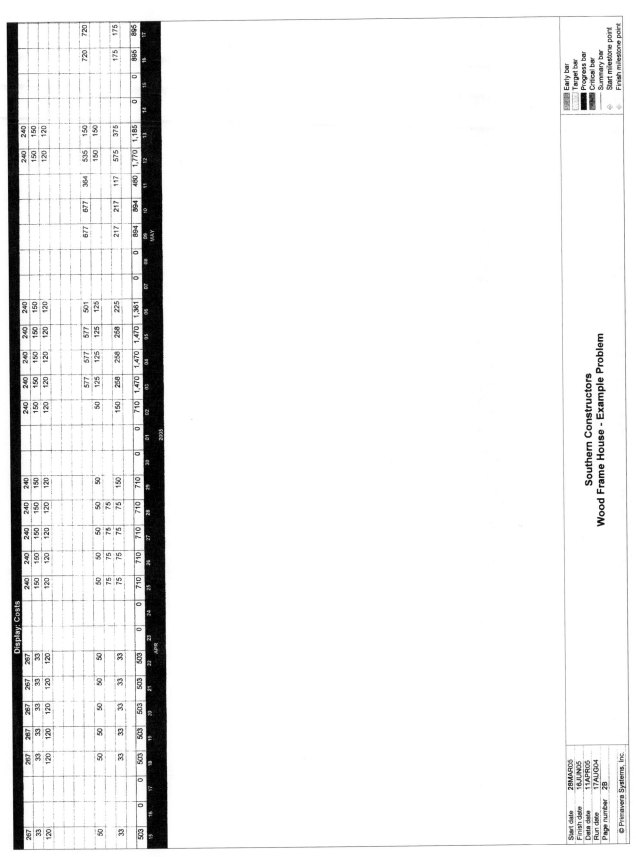

		267	267	267	267	267	267	267								240	240	240	240	240	240				240	240							720	720
267																240	240	240	240	240	240				150	150								
33		33	33	33	33	33	33	33								150	150	150	150	150	150				120	120								
120		120	120	120	120	120	120	120								120	120	120	120	120	120												175	175
													50	50	50	577	577	577	577	577	501				535	150			677	677	364			
50		50	50	50	50	50	50	50					50	50	50	125	125	125	125	125	125				150	150								
														75	75	258	258	258	258	258	225				575	375			217	217	117			
33		33	33	33	33	33	33	33					150	75	75						150													
503	0	503	503	503	503	503	503	503	0		0	710	710	710	710	1,470	1,470	1,470	1,470	1,470	1,361		894	894	1,770	1,185		0	894	894	480		895	895
15	16	17	18	19	20	21	22	23	24		25	26	27	28	29	30	01	02	03	04	05	06	07	08	09	10	11	12	13	14	15	16	17	
						APR											2005								MAY									

Southern Constructors
Wood Frame House - Example Problem

Display: Costs

Start date	28MAR05
Finish date	16JUN05
Data date	11APR05
Run date	17AUG04
Page number	2B

© Primavera Systems, Inc.

Figure 13–16e   Cost Table Print—Wood Frame House—Page 5

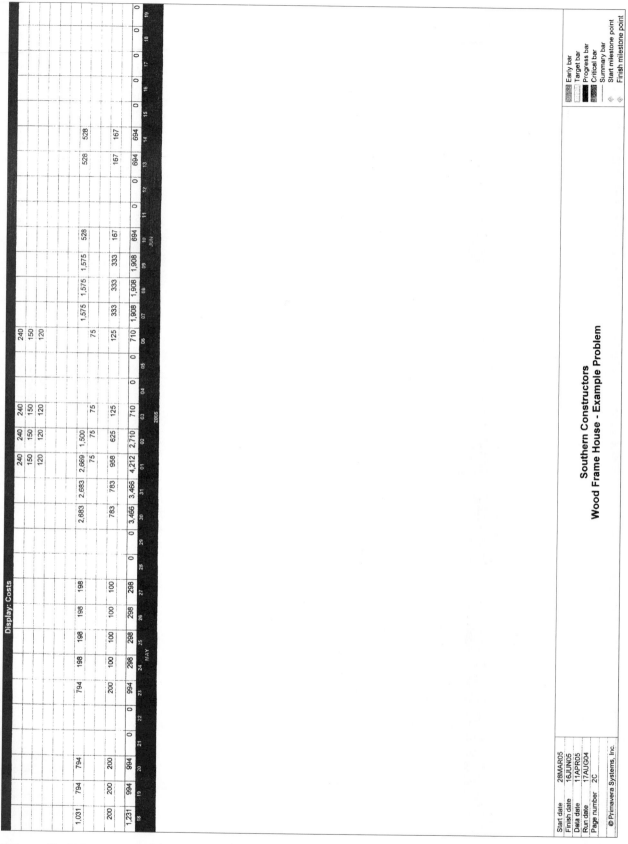

Figure 13–16f Cost Table Print—Wood Frame House—Page 6

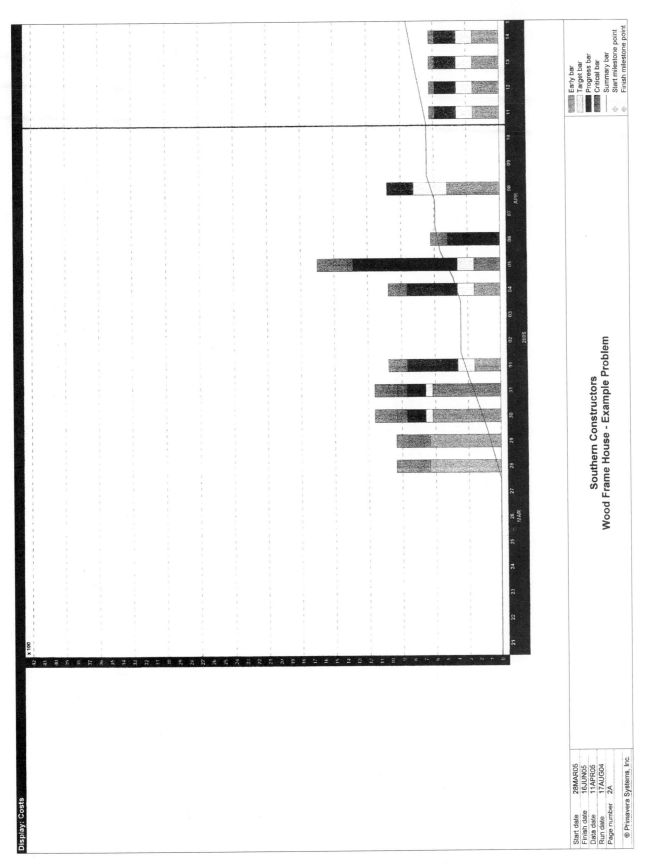

Figure 13–17a   Cost Profile Print—Wood Frame House—Page 1

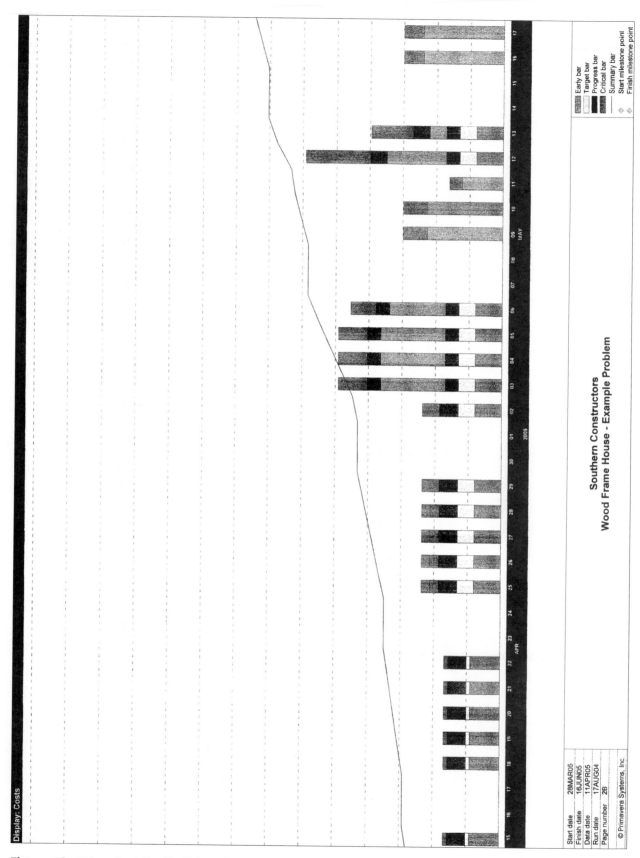

Figure 13–17b Cost Profile Print—Wood Frame House—Page 2

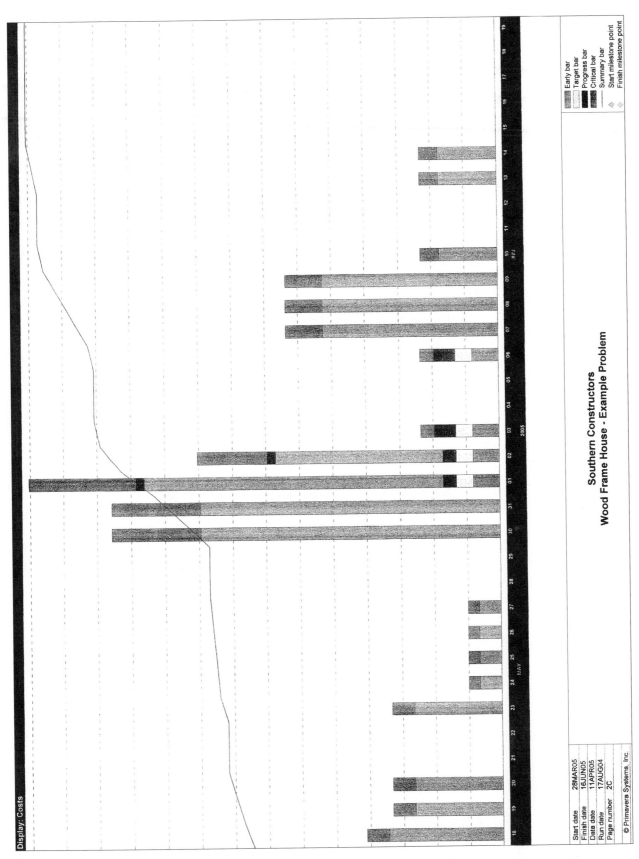

Figure 13–17c Cost Profile Print—Wood Frame House—Page 3

Summary

After a project is underway, the contractor tracks (monitors) costs to compare the actual to planned expenditure of funds. The target (original) costs are updated to show the expenditures of funds in the current (updated) schedule.

The cost table is used within *SureTrak* to show actual expenditure of funds through the data date and then reverts to planned expenditures thereafter. Updating activities with the actual expenditures of costs is required.

Updating cost expenditures are made using the actual to-date, to-complete, and at-completion fields of the **Costs** dialog box.

The graphical representation of the cost profile helps analyze the updated cost expenditures and remaining requirements.

Cost reports, both tabular and graphical, help communicate the actual project cash flow and how these expenditures have been made over time.

Exercises

1. **Comparing Actual to Planned Expenditures**
   Summarize the differences between actual and planned expenditures.
2. **Updating a Schedule and Recording Expenditures**
   Open the project modified in Exercise 2 of Chapter 12 and update and record expenditures for a few activities.
3. **Printing a Tabular Tracking Report**
   Print a tabular report of Exercise 2 with the columns shown in Figure 13–12 and Figure 13–13.

4. Produce a manual cost update (cost table) for Figure 13–18.

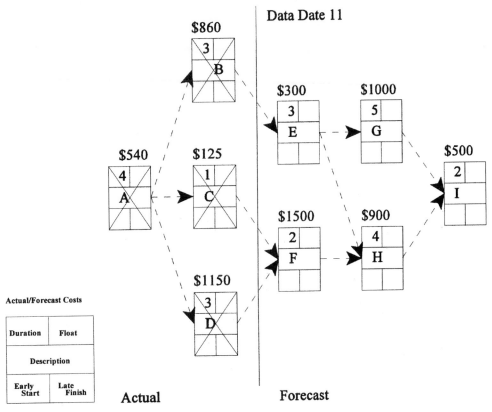

Figure 13–18   Exercise #4

5. Produce a manual cost update (cost table) for Figure 13–19.

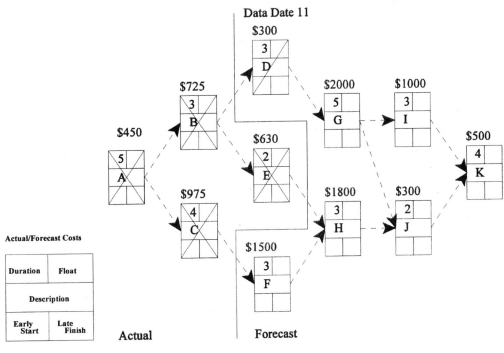

Figure 13–19   Exercise #5

6. Produce a manual cost update (cost table) for Figure 13–20.

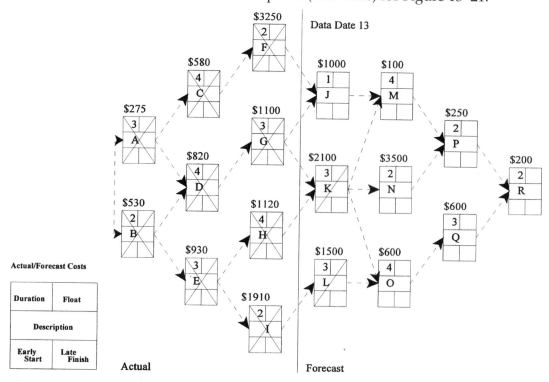

Figure 13–20 Exercise #6

7. Produce a manual cost update (cost table) for Figure 13–21.

Figure 13–21 Exercise #7

8. Produce a manual cost update (cost table) for Figure 13–22.

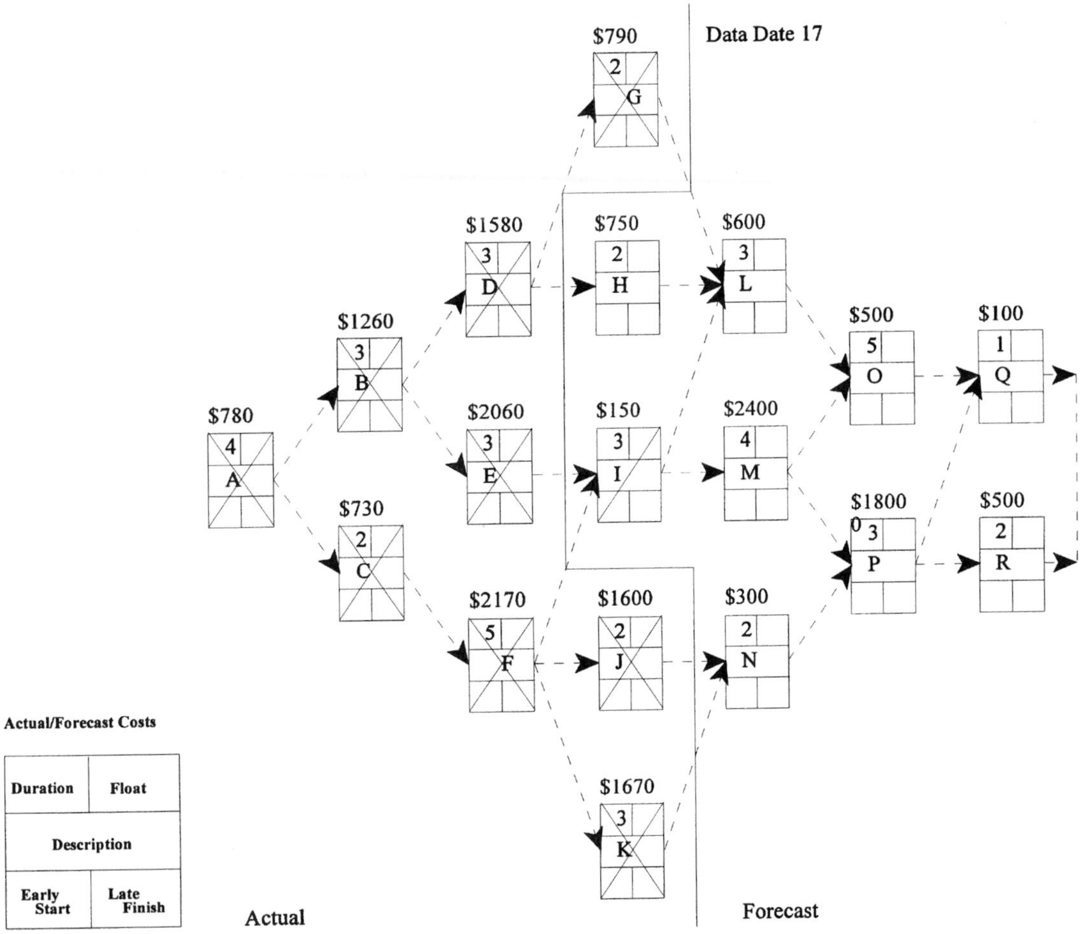

Figure 13–22 Exercise #8

9. Produce a manual cost update (cost table) for Figure 13–23.

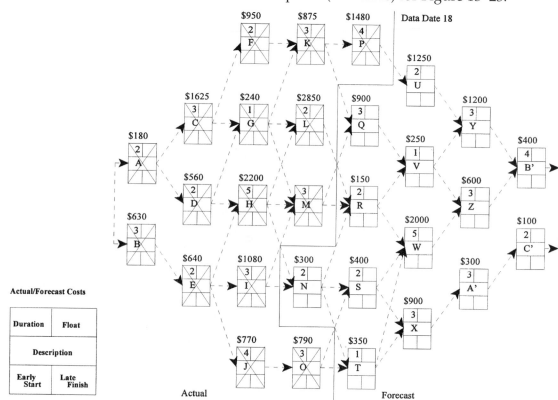

Figure 13–23 Exercise #9

10. Small Commercial Concrete Block Building—Resource Tracking
Prepare the following reports for the small commercial concrete block building located (Exercise 12, Chapter 7) in the Appendix:
- Updated cost table
- Updated cost profile

11. Large Commercial Building—Resource Tracking
Prepare the following reports for the large commercial building located (Exercise 13, Chapter 7) in the Appendix:
- Updated cost table
- Updated cost profile

Drawings for Example Problems and Exercises

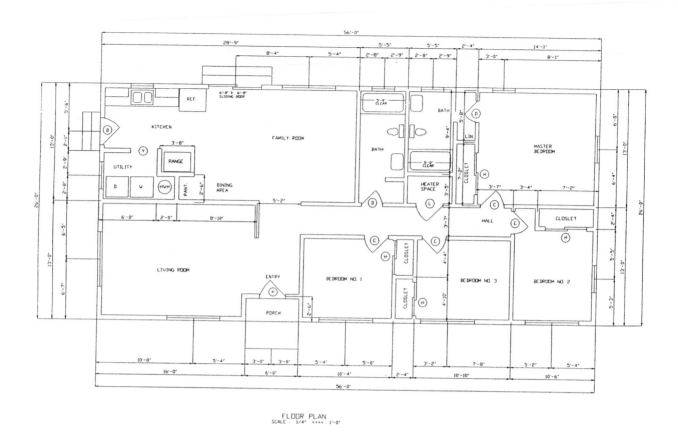

FLOOR PLAN
SCALE : 1/4" ==== 1'-0"

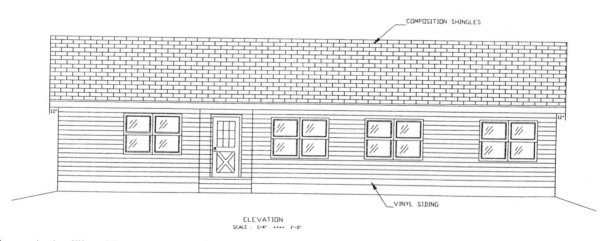

ELEVATION
SCALE : 1/4" ==== 1'-0"

Figure A-1   Wood Frame House—Floor Plan

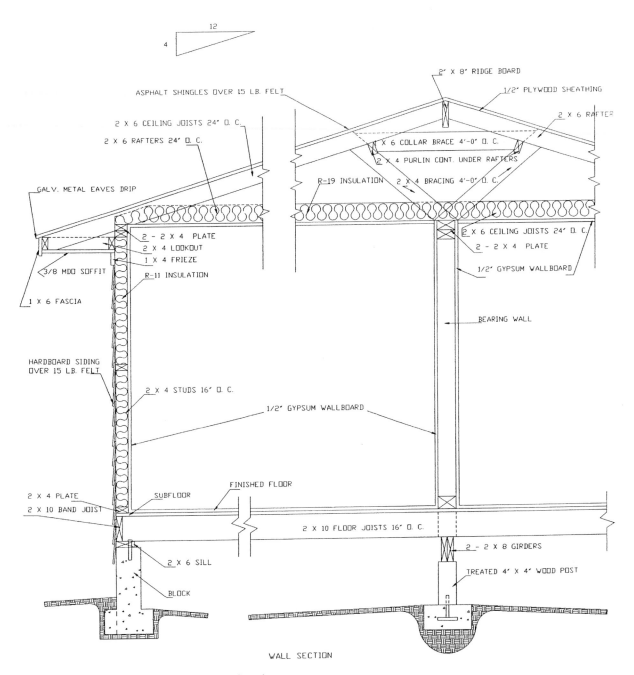

Figure A–2   Wood Frame House—Section

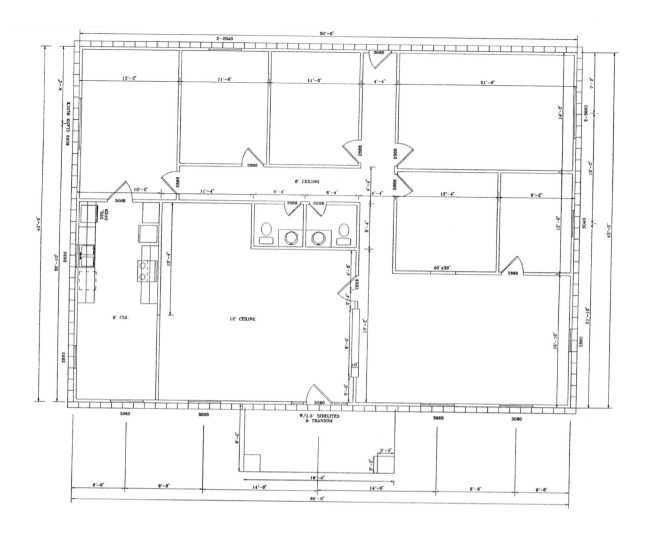

FLOOR PLAN
SCALE : 1/4" ==== 1'-0"

Figure A–3   Small Commercial Concrete Block Building—Floor Plan

FRONT ELEVATION
SCALE : 1/4" ==== 1'-0"

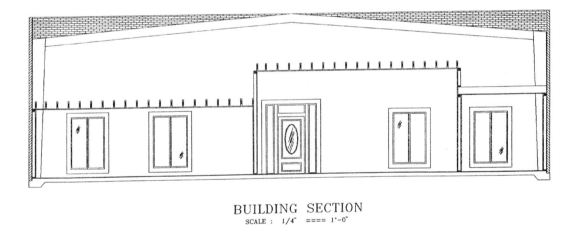

BUILDING SECTION
SCALE : 1/4" ==== 1'-0"

Figure A–4   Small Commercial Concrete Block Building—Section

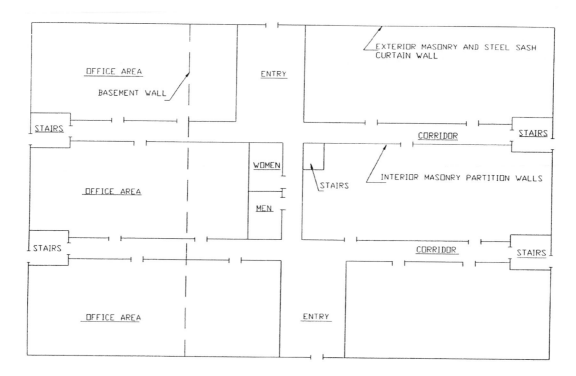

FLOOR PLAN

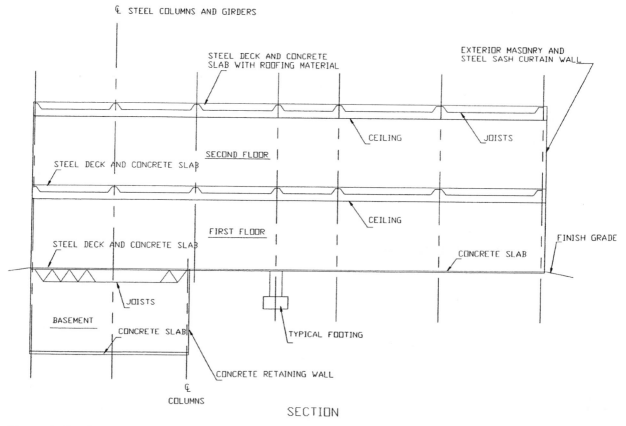

SECTION

Figure A–5 Large Commercial Building—Floor Plan

Index

Page numbers followed by a *t* or *f* indicate that the entry is included in a table or figure.

Invitation to

HOLISTIC HEALTH

A Guide to Living a Balanced Life

Charlotte Eliopoulos, PhD, MPH, RN
Executive Director
American Association for Long Term Care Nursing
Glen Arms, Maryland

JONES & BARTLETT
LEARNING

World Headquarters
Jones & Bartlett Learning
5 Wall Street
Burlington, MA 01803
978-443-5000
info@jblearning.com
www.jblearning.com

Jones & Bartlett Learning books and products are available through most bookstores and online booksellers. To contact Jones & Bartlett Learning directly, call 800-832-0034, fax 978-443-8000, or visit our website, www.jblearning.com.

Substantial discounts on bulk quantities of Jones & Bartlett Learning publications are available to corporations, professional associations, and other qualified organizations. For details and specific discount information, contact the special sales department at Jones & Bartlett Learning via the above contact information or send an email to specialsales@jblearning.com.

Production Credits
Executive Publisher: William Brottmiller
Acquisitions Editor: Amanda Harvey
Editorial Assistant: Rebecca Myrick
Associate Production Editor: Sara Fowles
Senior Marketing Manager: Jennifer Stiles
VP, Manufacturing and Inventory Control: Therese Connell
Composition: Paw Print Media
Cover Design: Kristin E. Parker
Cover Image: © Subbotina Anna/ShutterStock
Printing and Binding: Edwards Brothers Malloy
Cover Printing: Edwards Brothers Malloy

Library of Congress Cataloging-in-Publication Data
Eliopoulos, Charlotte.
Invitation to holistic health: a guide to living a balanced life / Charlotte Eliopoulos, RN, MPH, PhD, Former President of AHNA, Principal, Health Education Network, Inc., GlenArm, Maryland.
— 3rd ed.
p. ; cm.
Includes bibliographical references and index.
ISBN 978-1-4496-9421-0 (pbk.)
1. Holistic medicine. 2. Medicine, Chinese. I. Title.
R733.E425 2014
615.5—dc23
                              2013001118

6048

Printed in the United States of America
17 16 15 14 13   10 9 8 7 6 5 4 3 2 1

Contents

Appendix   491

Index   505

Preface

Nearly a decade has passed since the first edition of *Invitation to Holistic Health* was published, and during this time, health care has continued to evolve. Rather than being in the realm of a few offbeat practitioners and consumers, holistic and integrative care has become a common practice. Consumers understand the advantages of whole-person care in which they actively participate. Armed with a growing body of sound evidence, providers are utilizing therapies beyond traditional medical approaches and respecting consumers' right to make decisions related to their care. The current edition of this book recognizes these strides and provides resources for practitioners and consumers alike to engage in evidence-based holistic health practices.

From a Medical to a Holistic Model

The holistic health movement has been steadily gaining momentum over the past several decades. The term *holism* refers to a whole that is greater than the sum of its parts. In other words, 1 + 1 + 1 = 3 or 5 or more. When applied to health, holism implies that the health and harmony of the body, mind, and spirit create a higher, richer state of health than would be achieved with attention to just one part, such as physical functioning. Although some people equate it with the use of complementary and alternative therapies, holistic health is a philosophy of care in which a wide range of approaches are used to establish and maintain balance within an individual. Complementary and alternative therapies may be part of the approach to holistic health promotion but so can healthy lifestyle choices, counseling, prayer, conventional (Western) medical treatments, and other interventions.

A challenge that has influenced the acceptance and use of holistic health practices relates to the U.S. healthcare system that has functioned within the

Integrative

medical model. Actually, *health care* is a misnomer, as Western medicine has functioned under a sick care model that we have come to know as the biomedical model. The biomedical model was built upon certain tenets, highly valued by scientific minds, that include:

- *Mechanism.* This belief advanced the concept that the human body is much like a machine, explainable in terms of physics and chemistry. Health is determined by physical structure and function, and disease is a malfunction of the physical part. Malfunctions and malformations are undesirable. Disease is treated by repairing the malformed or malfunctioning organ or system with physical or chemical interventions (e.g., drugs, surgery). Nonphysical influences on health status are not considered, and healing, dysfunction, and deformity serve no purpose.
- *Materialism.* This thinking considers the human body and its state of health as being influenced only by what is seen and measurable. Physical malfunction is the cause of illness; therefore illness is addressed by concrete treatments. Emotional and spiritual states have no impact on health and healing.
- *Reductionism.* This thinking reduces the human body to isolated parts rather than a unified whole. Treatment of a health condition focuses on the individual organ or system rather than the whole being. Good health is judged as having body systems that function well, despite one's feelings or spiritual state.

[handwritten margin note: Newton, Descartes]

The first major challenge to the biomedical model occurred in the 1960s, when the relationship of body and mind began to be discussed. In retrospect, it is difficult to believe that the medical community was skeptical that the mind could cause illness, yet the resistance to accepting the body–mind connection was real. Similarly, recognition of the role of the spirit in the cause and treatment of illness has met similar skepticism. As the dust settles, however, healthcare practitioners are understanding the profound, dynamic relationship of body, mind, and spirit to health and healing and moving toward a holistic model of health care.

About This Book

This book offers guidance for the journey to holistic health. There are no special formulas provided that guarantee eternal youth and freedom from illness. There is no revolutionary diet or plan that will change your life in 30 days or less. No exotic substances that you can use to develop a new you will be found among the pages of this book. Instead, solid principles for building a strong foundation for optimal health are presented with practical advice that you can easily adopt and integrate into your life.

The book is divided into four parts. In the first part, "Strengthening Your Inner Resources," practices that can build your body's reserves and help it to function optimally are discussed. You will be guided through a self-assessment of your health habits so that you can determine areas that may need special attention. The realities of good nutrition are examined along with an in-depth look at dietary supplements. Exercise is approached from a body, mind, and spirit perspective. Likewise, the important activity of enhancing your immune system is considered from a mind–body framework. Methods to flow with the inescapable reality of stress are discussed.

"Developing Healthy Lifestyle Practices," is the second part of this book. The many complex factors that influence your health status as you interact with the world beyond your body are addressed in chapters on topics such as growing healthy relationships, family survival skills, spirituality, humor, touch, and the environment. In recognition of the significant impact of work and money on one's total being, chapters are dedicated to each of these topics.

The third part of this book, "Taking Charge of Challenges to the Body, Mind, and Spirit," offers information to equip you to be proactive in keeping yourself in balance. The hidden meaning of symptoms is explored to help you learn about the factors behind your health conditions. Practical advice is offered on how you can work in partnership with your healthcare provider to ensure you get the best care possible. Transitions associated with menopause are examined, along with a wide range of approaches to manage the symptoms that may be experienced. Interesting insights into gambling, drugs, overeating, and other addictions are shared. Skills for being an effective caregiver are presented.

With the growing use of complementary therapies, the fourth part of this book, "Safe Use of Complementary and Alternative Therapies," addresses these therapies based on current evidence. The purpose, benefits, and related precautions of the major types of therapies are presented to promote the sensible use of these products and practices. Chapters discuss the use of alternative medical systems, nutritional supplements, herbal medicine, aromatherapy, mind–body therapies, manipulative and body-based therapies, and energy therapies to offer practical insights into the safe, effective use of these popular therapies. The appendix offers extensive resources that can aid you in being an informed healthcare consumer.

A useful approach to using this book is to give it an initial fast read from cover to cover. This can be followed by a focus on chapters that address specific interests and needs. Although some of the chapters may not pertain to you directly (for instance, if you're not a caregiver, you may not have a keen interest in the "Surviving Caregiving" chapter), you may find that a quick review of the chapter could acquaint you with its content so that you'll recall it in the future if you or people you know are faced with this issue. You'll probably find that the rich facts and resources provided make this book a great reference for your personal library.

Acknowledgments

Appreciation is given to the following professionals who contributed content to the first edition of this book and who continue to be wonderful holistic health leaders:

Cynthia Aspromonte, RNC, NP, HTP-I, HNC
Julia Balzer Riley, RN, MN, HNC
Genevieve Bartol, RD, EdD, HNC
Irene Wade Belcher, RN, MSN, CNS, CNMT, HNC
Jane Buckle, RN, PhD
Charlene Christiano, RN, MSN, CS, ARNP, CHTP
Linda Coulston, RN, BSN
Myra Darwish, RN, MSN, CS, HNC
Joyce Deane, RNC, MS, CCM, CMC, CALA
Carole Ann Drick, RN, DNS, TNS, CP
Joan Efinger, RN, CS, MA, MSN, DNSc, HNC
Sue Fisher-Mustalish, RN, HNC
Eve Karpinski, MA, RNC, HNP, HNC
Marsha McGovern, RN, MSN, FNP, CS
Ann McKay, RNC, MA, DI Hom, HNC
Joyce Murphy, RN, BS, MSN, HNC
Natalie Pavlovich, RN, PhD, DiHt, CNHP
Marie Fasano Ramos, RN, MN, MA, CMT
Barbara Ann Stark, MSN, FNP, HNC
Marilee Tolen, RN, CHTP/I, HNC
Linda S. Weaver, RN, MSN, CCRN
Anneke Young, RN, BSN, CNAT
Katherine Young, RN, MSN

PART I

Strengthening Your Inner Resources

Introduction to Holistic Health

1

[handwritten:] 6 Pillars - Physical ⟩ Environmental
Emotional⟨ Spiritual
Mental ⟩ Social

OBJECTIVES

This chapter should enable you to

- Identify basic human needs and factors that enable them to be satisfactorily met
- List at least six features of an ideal health profile
- Perform a comprehensive self-evaluation of holistic health status

---

Welcome to your journey to optimum holistic health!

Often, health is thought of as the absence of disease. While that certainly is an important aspect of health, it is hardly complete. A comprehensive consideration of health includes all facets of an individual: physical, mental, emotional, social, and spiritual. This *whole person* view of the individual is what holistic health is all about.

As you most likely have experienced yourself, when one facet of your life is not functioning satisfactorily, your total well-being is affected. For example, you may eat nutritious foods, exercise regularly, and ace a physical exam with no abnormal findings. However, if you just lost a relationship with someone significant to you, you may feel fatigued, have a poor appetite, experience insomnia, be unable to focus, and, generally, feel poorly. Such situations remind us that a disruption to any one aspect of ourselves impacts general health and well-being.

Self-care is a term that is used to describe the active role people take in maintaining or improving their health. It is an aspect of health that is often overlooked when health care is discussed. Even in the arena of preventive medicine, which aligns close to the idea of self-care in modern medicine, the emphasis is more on the early detection of disease than the active promotion

> ### KEY POINT
>
> Americans have come to accept the World Health Organization's definition of health as a state of physical and mental well-being and not just the absence of disease.
>
> There is a broader view in traditional Chinese medicine (TCM). TCM includes the belief that the human being is composed of and surrounded by an energy system or field. This energy system is understood to resemble an electromagnetic field, expressed on the minute level as the behavior of electrons and neurons and on the gross level as the experience of vitality. The energy system is made up of energy pathways, often referred to as meridians. The pathways are believed to carry energy and information throughout the human organism to unite body, mind, and spirit. Health is seen as having a sufficient amount of energy circulating freely in the organism.

TCM = Energy

of health. Although there is a focus on health screening, less attention is given to educating people about healthy living habits, such as exercise, stress management, and nutrition and on the factors affecting the ability to achieve them.

Making minor adjustments in health practices to prevent diseases is easier than caring for diseases after they have developed. Prevention starts with taking stock of health habits and comparing them with those consistent with optimum health. By identifying the behaviors that lead to poor health, individuals can address those unhealthy practices and sources of imbalance and begin taking steps to change.

Basic Human Needs

To maintain a healthy state, people need to assure they are meeting basic human needs, which include the following:

Respiration
Circulation
Nutrition
Hydration
Elimination
Rest
Movement
Comfort
Safety

Connection with significant others, culture, the environment, and a
   higher power

Purpose

Although these needs appear straightforward and simple, their fulfillment
depends on some complex factors, such as:

- *Physical, mental, and socioeconomic factors.* A person who is para-
  lyzed and unable to lift a utensil to her mouth or someone who has
  Alzheimer's disease and cannot remember what to do when food is
  placed before him may be able to chew, swallow, and digest food,
  but lack the ability to get food into his or her mouth due to physical
  or mental impairments. Likewise, a senior citizen on a fixed income
  may omit the medications that her body needs to function normally,
  because she lacks adequate funds to pay for the prescription.

- *Knowledge, skills, and experience.* A pregnant woman who is unaware
  that alcohol can be dangerous to her baby may continue drinking and
  threaten the safety of her child. A person who lacks an understanding
  of the significance of a relationship with Christ may experience hopelessness and depression in an existence without spiritual meaning.

- *Desire and decision to act.* An individual could describe the recommended dietary intake and list foods that are harmful, yet continue to
  consume junk foods. A person may know that an adulterous relationship is loaded with problems and risks that could destroy health, job,
  and family, yet be unwilling to terminate the affair.

Exploring the factors that impact the basic need for nutrition demonstrates the complexities at play. To maintain a healthy nutritional state, an
individual needs to do the following:

Know what constitutes a healthy diet
Have the cognitive ability to plan, prepare, and consume meals
Have the money to purchase food
Be physically able to shop for, handle, prepare, and consume food
Know how to cook or have access to someone who can
Be motivated to eat properly
Have an emotional state that is conducive to proper food intake
Make sound dietary choices
Organize activities to have the time to eat

When deviations from health are identified, it is useful to consider what
factors could be contributing to the problem so that appropriate plans of cor-
rection can be developed. For example, someone with an obesity problem who

eats too much of the wrong foods may do so because he or she is depressed. Although classes that review healthy foods could be beneficial, behavioral changes may be more likely to occur if the person receives counseling and other treatment for depression.

Self-Assessment

An overall evaluation of health begins with a review of the current health status and health practices. An ideal health profile is one in which an individual:

- Consumes an appropriate amount of high-quality food.
- Exercises regularly.
- Maintains weight within an ideal range.
- Has effective stress coping mechanisms.
- Balances work and play.
- Looks forward to activities with energy and enthusiasm.
- Falls asleep easily and sleeps well.
- Eliminates waste with ease and regularity.
- Has meaningful relationships.
- Enjoys a satisfying sex life.
- Feels a sense of purpose.
- Feels safe.
- Is free from pain and other symptoms.

When the ideal is not being met, there needs to be an exploration into the reasons so that strategies to improve health habits can be identified and implemented.

Self-Assessment

An important first step to your journey to optimum holistic health is to take stock of your current status. This process takes time, effort, and serious evaluation of your current status and function. The following pages offer a comprehensive assessment tool to help you gain insight into your health status. Try to answer the questions as thoroughly as possible as they will help you later when you consider habits that you can acquire to improve your health in a holistic manner.

SELF-ASSESSMENT OF HEALTH

Age_____ Marital status_____ Children_____ Occupation_____
Height_____ Current weight_____ Weight range_____

Diet
Describe your food intake in a typical day:

Check all items present and describe:
_____Indigestion, heartburn
_____Regurgitation
_____Use of antacids
_____Poor appetite
_____Nausea, vomiting
_____Chronic halitosis

Condition of teeth:

Do you fast? If so, describe:

Nutritional supplements (vitamins, minerals, herbs, enzymes) used:

     Give amount and type:

Please check the frequency of intake of the following foods:

	Daily (amount)	*Sometimes*	*Rarely*	*Comments/Related Factors*
Fruit				
Fruit juices				
Vegetables				
Vegetable juices				
Red meat				
Poultry				
Fish				
Milk				
Cheese				
Pasta				
Bread, rolls				

Cereal	_____	_____	_____	_____
Beans, peas	_____	_____	_____	_____
Coffee	_____	_____	_____	_____
Tea (caffeinated)	_____	_____	_____	_____
Soda	_____	_____	_____	_____
Candy	_____	_____	_____	_____
Cakes, pies	_____	_____	_____	_____
Ice cream	_____	_____	_____	_____
Chocolate	_____	_____	_____	_____
Salty snacks	_____	_____	_____	_____
Table salts	_____	_____	_____	_____
Sugar	_____	_____	_____	_____
Sugar substitute	_____	_____	_____	_____
Beer	_____	_____	_____	_____
Wine	_____	_____	_____	_____
Hard liquor	_____	_____	_____	_____
Water	_____	_____	_____	_____

Comments:

Activity
Describe all checked:
_____Difficulty walking or moving
_____Joint pain or stiffness
_____Muscle cramps, pain
_____Muscles too loose, too tight
_____Frequent fractures, sprains
_____Brittle bones, osteoporosis
_____History of falling
Type and frequency of exercise:

Breathing and Circulation
Describe all checked:
_____Allergies
_____Nasal stuffiness
_____Chronic runny nose
_____Shortness of breath
_____Cough
_____Wheezing, asthma
_____Frequent colds

_____Chest pain
_____Palpitations
_____Numbness
_____Dizziness, light-headedness
_____Leg cramps
_____Varicose veins
_____History of smoking

Sleep Pattern
Usual bedtime_____ Usual wake-up time_____
Napping pattern:
Do you awaken refreshed?
Insomnia? Describe:
Fatigue? Describe
Sleep aids:
Quality of sleep:
Factors interrupting sleep:

Elimination Pattern
Describe all checked:
_____Urination difficulty, dribbling
_____Pain or burning with urination
_____Voiding during night
_____Inability to pass urine, hesitancy
_____Incontinence
_____Blood in urine
_____Constipation
_____Diarrhea
_____Gas (flatus)
_____Irritable bowel syndrome
_____Blood in stool
_____Hemorrhoids
_____Laxative use
_____Enema use, colonic irrigations
_____Regular   Frequency of bowel movements:

Skin and Hair
Describe all checked:
_____Rashes
_____Itching
_____Unusual sensations

_____Foul body odor
_____Dry skin
_____Oily skin
_____Unusual marks or moles
_____History of shingles
_____Hair loss, breakage
_____Dry scalp
_____Unhealthy-looking hair
_____Brittle nails
_____Soft nails
_____Other problems:

Reproductive
Female
Describe all checked:
_____Vaginal discharge
_____Vaginal dryness
_____Hysterectomy
_____Problems with sexual function
_____Change in sex drive, interest
_____Pain during intercourse
_____Breast abnormalities
Perform monthly self-exam of breasts?_____
Date of last mammogram:
Date of last gynecological exam:
If menopausal:
Year began:
_____Symptoms:
_____Hormonal replacement therapy
If menstruating:
_____Regular menstruation
_____Painful menstruation
_____PMS
Male
_____Prostate exam
_____PSA

Sensory

Describe all checked:

_____Wear eyeglasses

_____Poor vision

_____Cataracts

_____Glaucoma

_____See halos around lights

_____Cloudy vision

_____Pain in eyes

_____Dry eyes

_____Watery eyes

_____Poor hearing

_____Excess ear wax

_____Unusual sensations, tingling

_____Numbness

_____Paralysis

_____Decreased taste

_____Unusual taste in mouth

_____Inability to smell

_____Smell unusual odors

_____Sensitive to scents/odors

Date of last eye exam:

Date of last hearing exam:

General Symptoms

Describe all checked:

_____Frequent colds, infections

_____Headaches

_____Pain

_____Unusual fatigue

_____Swelling

_____Other

Emotional and Spiritual

Describe all checked:

_____Depressed

_____Anxious

_____Moody

_____Mood swings

_____Hyperactive
_____Suicidal
_____Episodes of confusion
_____Inability to focus
_____Easily cry
_____Never cry
_____Feel hopeless
_____Paranoid, suspicious
_____Argumentative
_____Passive
_____Difficulty maintaining relationships
_____Marital conflict, problems
_____Difficulty coping
_____High level of stress in life
       Measures to manage stress:
_____Belief in God, higher power
_____Connection with faith community
_____Feel spiritually empty, distressed
_____Feel worthless
_____Feel life has no meaning
Changes I would like to make in my life:

Known Health Conditions/Diagnoses   Treatment/Management

_____

_____

_____

_____

_____

_____

Prescription and Nonprescription Medications Used

Medication	Dosage	Frequency Taken	Reason Used

Complaints

List major complaints you have about your health in order of importance:

Landmarks in Your Life History

Often, significant events, positive and negative, can provide an understanding of your current health status and needs. Divide your life into decades and remember the significant occurrences during each decade. These can include the loss of a significant person, change in school or job, relationship started or terminated, illness of self or significant others, period of spiritual growth or distress, etc.

List the occurrences in the appropriate decade.

Age	Description of Significant Occurrence
1–9	
10–19	
20–29	
30–39	
40–49	
50–59	
60–69	
70–79	
80+	

©Charlotte Eliopoulos. Reprinted with permission.

You may feel that completing an assessment such as this one is a tedious process. Perhaps you've never had to participate in such a comprehensive assessment of your health status. Unfortunately, the realities of our healthcare system are that many practitioners are too busy to spend time getting to know the minds, bodies, and spirits of their clients, and insurance reimbursement favors the treatment of symptoms and diseases rather than the nurture and care of the whole person. This presents a challenge for you to be an informed, proactive healthcare consumer so that you will be able to:

- Understand the many influences on your health.
- Identify problems and relationships among your mind, body, and spirit that may not be readily apparent to your healthcare provider.
- Be able to seek the assistance you need from the source best able to help you (e.g., physician, clergy, nutritionist, counselor, etc.).

Go through your self-assessment and highlight or circle signs, symptoms, and unusual or abnormal habits. Now, think about the specific need that is affected by the signs and symptoms and write them under the appropriate heading in column A on the form that follows. Some signs and symptoms can affect several needs. For example, "Use of antacids" can be listed across from *Food and water* and *Safety*; "Unusual fatigue" can be listed across from *Movement and activity, Sleep and rest, Connection, Safety,* and *Normality.*

Now, examine the signs, symptoms, and habits and try to consider the underlying reason(s) that could be responsible, such as eating a lot of fried foods for use of antacids and eating poorly and having stressful job for unusual fatigue. Jot down what you believe the underlying reason to be in column B. In some circumstances, you may not know the underlying reason; it is fine to put a question mark in the column.

Lastly, in column C, write an action you can take to change or reduce the sign, symptom, or habit, such as "reduce meals at fast food restaurants to once a week" or "discuss excessive workload with supervisor." For some signs, symptoms, and habits, your action may need to be to obtain a medical evaluation, seek the counsel of a professional, or pray for insight and guidance into the situation.

Following the blank action plan for your use is one that shows some options to consider under each category. As you progress in this book, you will find additional suggestions to assist you in developing your actions.

YOUR ACTION PLAN TO IMPROVE YOUR HEALTH

Need	A Sign/symptom/ habit	B Underlying reason	C Action
Respiration/ circulation			
Food and water			
Elimination of wastes			
Movement and activity			
Sleep and rest			
Comfort			
Safety			
Connection with significant others, culture, environment, higher power			
Safety			
Purpose			

SAMPLE ITEMS TO INCLUDE IN YOUR ACTION PLAN

Need	*A* *Sign/symptom/* *habit*	*B* *Underlying* *reason*	*C* *Action*
Respiration/ circulation	Chronic cough Shortness of breath when climbing > 15 stairs	Smoking Poor physical condition	Enroll in smoking cessation plan Begin exercise program Do deep breathing exercises several times throughout the day
Food and water	Frequent heartburn High intake of snack food	High intake of fried food Eat while working stressed mealtime Don't have time to go to cafeteria at lunch time; rely on vending machine items	Eliminate fried foods Increase fresh foods, broiled and baked items Schedule time to eat in cafeteria Keep healthy snack foods in office
Elimination of wastes	Frequent constipation	Low fiber and fluid intake Low activity level	Include bran cereal at breakfast Eat at least five fresh fruits daily Eat a salad at lunch Adhere to exercise program

SAMPLE ITEMS TO INCLUDE IN YOUR ACTION PLAN (CONTINUED)

Movement and activity	Stiff joints in morning Difficult to walk and engage in physical activity	Lack of exercise	Get physical exam to determine safety of exercise program Begin exercise program Park car in farthest space from building Perform yoga stretches several times each day
Sleep and rest	Poor quality of sleep Awake tired, difficult to get out of bed Nod off after meals	High consumption of caffeine Spouse snores loudly Consume high amount of sweets	Eliminate caffeine after 4 p.m. Suggest spouse get evaluated for snoring Sleep in separate room every other night Change diet
Comfort	Stiff joints	Insufficient exercise	Adhere to exercise program Stretching exercises Heat application
Safety	Overmedicate with pain medications	Try to find quick and easy means to control joint pain	Engage in exercises to keep joints flexible Use heat, massage

SAMPLE ITEMS TO INCLUDE IN YOUR ACTION PLAN (CONTINUED)

Connection with significant others, culture, environment, higher power	Often neglect prayer life	Allow worldly demands to take priority	Discuss with friend and ask friend to hold accountable
Safety	Take higher than recommended doses of medications for headaches Overuse antacids	Fail to manage stress and eat well	Eat healthier diet Practice stress management techniques daily Eliminate foods that trigger heartburn
Purpose	Don't feel inspired or excited by anything	Working at job that is not challenging and that I do not feel good about	Begin to explore other jobs that use more of my skills and that offer meaningful work

Summary

In Western medicine, self-care primarily implies preventing illness and recognizing symptoms early; however, from a holistic perspective self-care refers to the active role individuals assume in maintaining and improving their physical, mental, emotional, social, and spiritual health. From that perspective, an ideal health profile is one in which a person consumes an appropriate quality and quantity of food, exercises regularly, maintains weight within an ideal range, has good stress-coping skills, balances work and play, looks forward to activities with energy and enthusiasm, falls asleep easily and sleeps well, eliminates waste with ease, enjoys a satisfying sex life, feels a sense of purpose, and is free of pain and other symptoms. When deviations from the norm are experienced, factors that may be responsible must be explored, and, again, the approach must be holistic. For example, a physical symptom such as chest pain could be due to a medical condition of the heart or lungs, but it could also be related to stress, fatigue, guilt, a dysfunctional relationship, or other nonphysical causes. An understanding of underlying factors affecting the health state is essential to developing individualized, effective health plans. Other chapters in this book will offer guidance in this process.

Suggested Reading

Balch, J., & Stengler, M. (2011). *Prescription for natural cures: A self-care guide for treating health problems with natural remedies including diet, nutrition, supplements, and other holistic methods.* Hoboken, NJ: John Wiley & Sons.

Brvada, D. M., Smith-Spangler, C., Sundaram, V., Gienger, A. L., Lin, N., Lewis, R., Stave, C. D., Olkin, I., and Sirard, J. R.(2007). Using pedometers to increase physical activity and improve health: A systematic review. *Journal of the American Medical Association, 298*(19), 2296–2304.

Callaghan, D. M. (2003). Health-promoting self-care behaviors, self-care self-efficacy, and self-care. *Nursing Science Quarterly, 16*(3), 247–254.

Clark, C. C. (2003). *American Holistic Nurses' Association guide to common chronic conditions. Self-care options to complement your doctor's advice.* Hoboken, NJ: John Wiley and Sons.

Forkner-Dunn, J. (2003). Internet-based patient self-care: The next generation of health care delivery. *Journal of Medical Internet Research, 5*(2), e8.

Funnell, M. M., & Anderson, R. M. (2003). Patient empowerment: A look back, a look ahead. *Diabetes Education, 29*(3), 454–458, 460, 462.

Halcon, L. L., Robertson, C. L., Monson, K. A., & Claypatch, C. C. (2007). A theoretical framework for using health realization to reduce stress and improve coping in refugee communities. *Journal of Holistic Nursing, 25*(3), 186–194.

Hertz, J. E., & Anschutz, C. A. (2002). Relationships among perceived enactment of autonomy, self-care, and holistic health in community-dwelling older adults. *Journal of Holistic Nursing, 20*(2), 166–185.

Hunt, A. (2010). *Holistic lifestyle: A layman's guide to eating and living your way to better health and happiness.* Victoria, BC, Canada: Friesen Press.

Kim, H. (2007). *Handbook of Oriental medicine* (3rd ed.). San Francisco, CA: Harmony and Balance Press.

Lu, H. C. (2006). *Traditional Chinese medicine.* Laguna Beach, CA: Basic Health Publications.

Murray, R. B., & Zentner, J. P. (2000). *Health promotion strategies through the lifespan* (7th ed.). New York, NY: Prentice Hall.

Oyserman, D., Fryberg, S. A., & Yoder, N. (2007). Identity-based motivation and health. *Journal of Personality and Social Psychology, 93*(6), 1011–1027.

Son, J. S., Kerstetter, D. L., Yarnal, C., & Baker, B. L. (2007). Promoting older women's health and well-being through social leisure environments: What we have learned from the Red Hat Society. *Journal of Women and Aging, 19*(3–4), 89–104.

Spero, D. (2002). *The art of getting well: A five step plan for maximizing health when you have a chronic illness.* Alameda, CA: Hunter House.

Sutherland, J. A. (2000). Getting to the point. *American Journal of Nursing, 100*(9), 40–45.

Waller, P. (2010). *Holistic anatomy: An integrative approach to the human body.* Berkeley, CA: North Atlantic Books.

Healthful Nutrition

OBJECTIVES

This chapter should enable you to

- Define nutrition
- Discuss factors related to the emotional, psychological, cultural, and traditional aspects of food
- List the components of a food journal
- Describe a healthy style for meal intake
- Outline recommended dietary intake according to MyPlate
- Describe the information that can be found on nutrition facts labels
- Define macronutrients and micronutrients
- List at least six tips for good nutrition

Sensible nutrition is a primary factor in leading a life that will allow for ample energy, productivity, and overall health. The old common sense adage, "You are what you eat" is now regarded as definitive, scientific-based fact. Each year brings new information and scientific research in the areas of nutrition and nutritional supplementation. The results continue to demonstrate that the foods we eat determine health, well-being, and longevity and that some foods offer medicinal qualities—something our foremothers and forefathers knew hundreds, even thousands of years ago!

Consumers are increasingly aware of the importance of nutrition to their health and growing numbers are working to improve this area of their lives; nevertheless, many people believe eating nutritiously will mean sacrifices, such as having to invest more time and money and forfeit good taste in order to eat healthfully. The overload of nutrition information from the media can be so overwhelming that many people simply give up thinking about their

> **KEY POINT**
>
> The idea that balanced nutrition is directly related to health, wellness, longevity, and the ability to heal the body is ageless. During the Stone Age, plants were used for medicinal purposes; the Chinese have used food for prevention and cures for centuries. Hippocrates was at the forefront of holistic health and wellness by suggesting that diet and nature should be taken into account when treating illness. Florence Nightingale also believed that "selecting and preparing healing foods, in addition to fresh air, quiet, and 'punctuality and care in administration of diet'" were necessary to keep the body working properly and for healing in times of injury or illness (Nightingale, 1860, p. 2). Samuel Hahnemann (1982), the father of homeopathy, believed that including foods that were most medicinal was integral to health and wellness. These great minds had an innate understanding that nutrition was directly related to health, wellness, and healing.

nutrition and diet altogether. Moreover, food companies run 60-second commercials that promote 60-second meals. The focus keeps us eating certain foods for their taste, texture, and quick preparation time rather than their nutritive value. Americans have lowered their fat and salt intakes yet continue to have a considerable gap between recommended dietary patterns and what they actually consume.

What Is Nutrition?

Nutrition refers to the ingestion of foods and the relationship of food to human health. Sensible nutrition requires the intake of nutrients from good quality, wholesome foods that support and maintain health throughout the life span. The need for sensible nutrition is essential throughout life because all humans require the same basic nutrients regardless of their stage of life. What varies is the amount of nutrients needed at each growth stage. There are also special needs because of growth and development, pregnancy and lactation, age, and disease or injury.

Proper nutrition works primarily through sound food choices. In order for proper digestion, absorption, metabolism, and elimination to take place, you must have high-quality food that contains optimum nutrients.

Nutrients are not immediately available as your food is eaten but must be broken down by the digestive process and transported by the blood and lymph to be used as needed; finally, their waste must be eliminated. The food you put on your plate must be transformed into proper condition and shape

> **KEY POINT**
>
> Nutrients are substances the body needs to provide you with energy, allow you to maintain your health, and repair and regenerate your tissues and cells.

for use by the body. In other words, it must be digested. This begins the process; then it must be further assimilated, metabolized, and finally eliminated.

Eating slowly and with few distractions, masticating (chewing) the food thoroughly, and drinking (any beverage) minimally during meals will allow the gastric juices to accomplish their proper function, and healthy digestion can occur. If food is swallowed nearly whole, a longer time will be required for its digestion and assimilation.

> **KEY POINT**
>
> Imagine the body as a group of workers who are building a house. Each substance (food), like pieces of wood, must be cut to just the right size (chewing) and prepared for use (digestion). Next, these pieces (nutrients), after due preparation in the workshop, must be taken by the different groups of workmen (blood and lymph) to their appropriate localities in the house (muscles, organs, tissues, and cells) and there be fitted into their proper places—this is assimilation.

Refuel, Reload, Rejuvenate

Energy is the most important reason to keep the body nutritionally sound. Each day, you must refuel the body so that it can move and function. The body is continually undergoing changes; worn-out tissues and cells are constantly being repaired and renewed. The elimination of digestive waste continually requires new supplies of energy, vitamins, minerals, and other nutrients that are derived from food. Other reasons for proper diet and nutrition include fighting infection, balancing hormones, assisting in better quality sleep, and keeping the body running smoothly in times of stress. Sensible and proper nutrition is important in fulfilling these demands.

Paramount to the value of nutrition in your life are the emotional, psychological, spiritual, cultural, and traditional aspects of food. How food is presented, its smell, taste, and the emotional climate as a meal is eaten all have a connection to how food is digested. Many people use food to help ward off anxiety, tension, depression, or boredom. Certain negative feelings can cause a physiological (bodily) response in which the hypothalamus (the brain's appetite control center) sets up a chain reaction in the autonomic (self-controlling) nervous system. Additionally, the meaning food has for each individual, from early childhood experiences to the present, and how that impacts nutritional and digestive habits must also be considered.

REFLECTION

How do your food preferences and eating patterns relate to your childhood experiences?

All of the senses are stimulated when you eat through the
- Visual presentation of the food.
- Surroundings in which the food is consumed.
- Aroma or odor associated with the food.
- Texture and taste of the food.
- Conversation and environmental sounds that are present while the food is being consumed.

The chemical reactions that take place in the body differ according to the combination of experiences. A delicious meal with friends that is filled with beautiful sights and sounds will elicit calmness and ease of the digestive process. Relaxing, enjoyable meals are a goal to work toward for health and wellness.

Consistency in your life, whether positive or negative, usually reigns. If food is consistently eaten quickly, poorly, and under stressful conditions, then the body, mind, and spirit will be quickly depleted, lack energy, and eventually become stressed, unhappy, uncomfortable, and diseased.

The gathering, preparing, eating, and sharing of food offer more than just nutritional value. Food and diet have social and cultural aspects. Traditions, family gatherings, and religious ceremonies that include food probably have

KEY POINT

From June Cleaver to Mickey D's

As late as the 1950s, many people were consuming food that was grown locally, bought fresh, and eaten in a mindful, respectful, and unstressed atmosphere. Today, you may find yourself eating in your car, on a bench at a sporting event, standing at the kitchen counter or refrigerator, or as you are walking from one meeting to the next. The demand for convenience in eating has skyrocketed to the point that more than 40% of our food expenditure is for food outside the home (USDA, 2012).

been part of your life on a regular basis. These activities help you to carry on tradition, culture, and values. Food is considered an expression of your individuality, history, values, and beliefs. Nutrition then is much more than what food group is eaten on any particular day at any particular meal.

A Healthy Diet

Recognizing that each individual is unique, some general recommendations can be followed to promote health and well-being. First, it is important to consider the times of day that food is eaten. It is generally not a good idea to skip breakfast, as it provides the foundation for the day. Eating a hot breakfast such as oatmeal or any other warm cereal or toast in the morning creates a warming and nourishing effect and supplies necessary energy to sustain daily activities. It is best to eat the heaviest meal in the middle of the day when the digestive energies are the strongest. Eating while in a rush or while conducting business is not a good practice; taking the time to eat in a relaxed manner without being engaged in any other activity is more healthful. In general, fresh organic foods (foods not treated with pesticides, hormones, and antibiotics) are good choices. A basic food guide is provided in **Table 2-1**. Natural vitamins and food supplements can provide additional benefits.

A Nutritional Evaluation

In order to assess how nutritional habits impact health, eating habits need to be evaluated for a period of time. Factors to consider include not just the type of foods consumed, but also the following:

- *When food is consumed.* Late-night eating puts extra stress on the digestive system. According to traditional Chinese medicine (TCM), at night, the yin (which is associated with rest, darkness, and stillness) predominates, and consequently, the digestive system slows down. The circulation slows down as well, conserving the amount of blood circulating to all of the digestive organs. Clearly, nighttime—when the body is supposed to be resting and preparing for restoring and repairing tissues—is not a good time to offer the body the challenge of digesting a big meal. Viewed from within the context of TCM, this behavior can lead to stomach yin deficiency, which is a condition in which the fluids of the stomach diminish, causing a sensation of heat in the stomach, manifesting as heartburn and indigestion. When this condition is allowed to persist, more serious stomach problems, such as a hiatal hernia or ulcer, may develop.

TABLE 2-1   BASIC DAILY FOOD PLAN

BREAKFAST
- Warm cereals, whole-grain muffin, or toast
- Fresh fruit
- Herbal tea or a grain beverage such as Postum, Cafix, or Bambu (1 cup of coffee a day)

LUNCH
- Fresh organic salads
- Homemade soup or a sandwich made with organic turkey, chicken, hard-boiled organic eggs, or nut butter on whole-grain bread
- Fresh fruit

DINNER
- Protein, such as organic poultry, fish, beef (not more than once a month), or beans
- A complex carbohydrate, such as a vegetable, whole grains, peas, and beans
- Organic vegetables (meaning they are free of contaminants, synthetic pesticides and herbicides, hormones, preservatives, and artificial coloring)
- Fresh fruit

- *What one does while eating.* When there is emotional tension while eating, energy is diverted and less available for digestion. When stressed by heated conversations or upsetting news on the television, a person's energy is drawn away from the digestive system, causing indigestion. For optimum digestion, adequate supplies of enzymes, coenzymes, and hormones are needed, and in order for these substances to be available, adequate amounts of blood must be circulating. Free-flowing energy, or qi, promotes blood flow. (In TCM, qi is considered the vital life force or energy that circulates throughout the body.) This emphasizes the importance of relaxing during mealtime so that there will be sufficient resources for digestion.

KEY POINT

Late night eating stresses the digestive system.

- *The amount that is consumed.* When too much food is consumed during a meal, the stomach gets stressed. This kind of behavior not only leads to indigestion, but also creates an energy deficit as energy is pulled from other areas to meet the demands of digestion. Between-meal nibbling creates the same kind of energy deficit, as energy is constantly required by the digestive organs to digest and not enough energy is available for other activities. Each time food is being consumed, blood is routed to the digestive organs, and less is available for other physiological activities, creating an imbalance in the body. In addition, the many muscle layers of the stomach need to rest for a certain amount of time.

Table 2-2 provides some questions for you to consider in assessing your nutritional status.

Identifying Patterns

Daily food journals that record food consumed, when it is consumed, and how one feels during food consumption can be beneficial in identifying patterns. Everything that enters the mouth—whether it is a piece of candy or a few sips of juice—should be recorded. Keeping this type of food diary increases self-awareness, which in turn can become the catalyst for positive changes in nutritional habits. After keeping a food journal for a couple of weeks, sufficient data will be available to determine the pattern and content of nutritional habits. Are meals being eaten on a regular basis? How much snacking is taking place? Are the foods chosen basically nutritious? Is one kind of food eaten in excess? Are some nutrients missing from the diet?

A review of beverage consumption is useful. What kinds of beverages are consumed? Is caffeine consumed from coffee and carbonated beverages, and if so, how much? Carbonated beverages are high in sugar or artificial sweetener, neither of which has any nutritional value. Caffeine is addictive, is a mild stimulant to the central nervous system, and is a diuretic (causes fluid loss through urine). Mixed scientific reports exist about the role caffeine plays in health and wellness. Because caffeine affects many body systems and can interfere with certain drugs, it is best taken in moderation—no more than 1 or 2 cups of caffeinated beverages should be consumed per day.

Alcoholic beverage intake needs to be captured as well. What is the alcohol intake on an average day? Alcohol supplies no nutrients, but it does supply calories. High levels of alcohol intake increase the risk of stroke, heart disease, certain cancers, high blood pressure, birth defects, and accidents. Women should drink no more than one alcoholic beverage per day, and men should drink no more than two alcoholic beverages per day.

TABLE 2-2   NUTRITIONAL SELF-ASSESSMENT

- How would you rate your general health status?
- What are your height and weight? Are they within normal limits? Have there been any recent changes in height or weight?
- What health conditions do you presently have?
- Do you have any problems with your blood sugar? Elevated cholesterol or triglyceride levels? High blood pressure? Osteoporosis? Irritable bowel syndrome?
- Are you aware of any food intolerances or allergies?
- Do you consume adequate amounts of protein, fruits, and vegetables?
- Do you limit your intake of saturated fats and simple carbohydrates?
- Are you taking nutritional supplements, and if so, which ones?
- What is your pattern of eating? How many meals do you eat per day? What is the size of those meals?
- Do you snack regularly throughout the day or evening? During the night?
- How many snacks do you have per day? Of what do they consist?
- What is your energy pattern? Do you have any slumps during the day?
- How much caffeine through coffee, tea, and carbonated drinks do you drink per day?
- What is your alcohol intake?
- What is your coping style when under stress? Do you consume unhealthy foods or large quantities of food when you are feeling stressed, depressed, anxious, or unhappy?
- Do you tend to make healthy food choices?

An evaluation of body weight is important, with attention to possible gradual increases in weights that may have been experienced with age. Maintaining weight within an ideal range is important to general health. Obesity has been on the rise for the last 2 decades, and approximately 30% of the adult population is obese. Obesity increases the risk of hypertension; heart disease; diabetes; arthritis; and uterine, breast, colon, and gallbladder cancers. People should be taught to use a weight chart to learn where their weight is in respect to their height (**Figure 2-1**). The higher the weight for a specific height, the greater is the risk of health problems. Obese individuals need to be encouraged and assisted in implementing a reduction program; at minimum, they should commit to not gain any additional weight. On the other hand,

BMI	19	20	21	22	23	24	25	26	27	28	29	30	31	32	33	34	35
Height	Weight (in pounds)																
4´10˝ (58˝)	91	96	100	105	110	115	119	124	129	134	138	143	148	153	158	162	167
4´11˝ (59˝)	94	99	104	109	114	119	124	128	133	138	143	148	153	158	163	168	173
5´ (60˝)	97	102	107	112	118	123	128	133	138	143	148	153	158	163	168	174	179
5´1˝ (61˝)	100	106	111	116	122	127	132	137	143	148	153	158	164	169	174	180	185
5´2˝ (62˝)	104	109	115	120	126	131	136	142	147	153	158	164	169	175	180	186	191
5´3˝ (63˝)	107	113	118	124	130	135	141	146	152	158	163	169	175	180	186	191	197
5´4˝ (64˝)	110	116	122	128	134	140	145	151	157	163	169	174	180	186	192	197	204
5´5˝ (65˝)	114	120	126	132	138	144	150	156	162	168	174	180	186	192	198	204	210
5´6˝ (66˝)	118	124	130	136	142	148	155	161	167	173	179	186	192	198	204	210	216
5´7˝ (67˝)	121	127	134	140	146	153	159	166	172	178	185	191	198	204	211	217	223
5´8˝ (68˝)	125	131	138	144	151	158	164	171	177	184	190	197	203	210	216	223	230
5´9˝ (69˝)	128	135	142	149	155	162	169	176	182	189	196	203	209	216	223	230	236
5´10˝ (70˝)	132	139	146	153	160	167	174	181	188	195	202	209	216	222	229	236	243
5´11˝ (71˝)	136	143	150	157	165	172	179	186	193	200	208	215	222	229	236	243	250
6´ (72˝)	140	147	154	162	169	177	184	191	199	206	213	221	228	235	242	250	258
6´1˝ (73˝)	144	151	159	166	174	182	189	197	204	212	219	227	235	242	250	257	265
6´2˝ (74˝)	148	155	163	171	179	186	194	202	210	218	225	233	241	249	256	264	272
6´3˝ (75˝)	152	160	168	176	184	192	200	208	216	224	232	240	248	256	264	272	279

Directions: Find your height in the left column. Go across the row to find your weight. Go up the column to the top row to find your BMI. Healthy BMI = 18-24.99. Overweight BMI = 25-29.99. Unhealthy BMI ("obese") = 30+.

Source: NIH/National Heart, Lung, and Blood Institute (NHLBI).

Figure 2-1   Body Mass Index (BMI) Table
Source: NIH/National Heart, Lung, and Blood Institute (NHLBI).

being underweight for height or having a recent unexplained weight loss may be a sign of other health problems and warrants further diagnostic evaluation.

A deeper understanding of nutritional patterns and habits can be done by examining a person's childhood relationship with food. What were his or her habits and patterns of nutrition and diet during childhood? Does the individual continue to carry these patterns and habits? How does the childhood experience with food and eating affect current nutritional choices and lifestyle?

REFLECTION

Do you believe that if you make sensible nutrition choices you will live a healthier, more energetic life and reduce your risk of disease? Do you think that you can deviate from sound nutritional habits without consequences? What has influenced your beliefs? If your beliefs do not support good nutritional habits, what can you do to change them?

Often, small actions are all that are needed to realize significant change. This can be explored by evaluating the way lifestyle impacts eating habits, food preferences, and food choices, similar to the nutritional self-assessment (Table 2-2) mentioned earlier. Writing the answers is important so that there will be a baseline to use for comparison after changes have been made. The answers will help in determining whether there are nutritional imbalances (excesses or deficiencies). Decisions can then be made as to which foods need to be added to and/or removed from the daily diet to improve nutrition.

The next step is to begin the process of eliminating or replacing foods that contribute to poor health. The intake of foods that are laden with artificial colorings, sweeteners, and preservatives should be reduced by half for the first few weeks and eventually eliminated altogether. Other foods that lead to ill health when consumed in excess are foods high in fat, refined sugar, salt, and dairy products.

KEY POINT

Paying attention to diet and nutritional habits does not mean that people need to become obsessed with eating to the point that they become stressed when they have an occasional slip from healthy eating. This stress could create an emotional imbalance, which has a negative effect on general health.

Fat

Eating high-fat foods such as ice cream, sour cream, cream cheese, hard cheese, heavy-butter sauces, red meat, pork, duck, oil, and whole milk is a contributing factor in conditions such as atherosclerosis, heart disease, and cancer. In addition to the studies that have been done from within a reductionist framework (ones in which answers are sought by breaking substances down into their smallest particles), there also is an understanding in TCM that fatty foods create heat in the system, which exceeds humans' needs given their ecological situation. Eskimos eat large amounts of fat because their bodies need to produce high amounts of heat. When average Americans eat the same amount of fat as Eskimos, they most likely will develop severe cholesterol problems. Furthermore, too much heat in the system creates an imbalance between hot and cold, or yin and yang, according to TCM. When this balance is disrupted, problems in the bioenergy system begin to manifest.

Sugars

Like saturated fats, refined sugars cause imbalances in the body when eaten in excess. Refined sugars are the simple sugars such as white, raw, brown, or turbinado sugar, as well as honey, corn syrup, corn sweeteners, dextrose, and fructose. Although a sweet flavor has a strengthening effect on the digestive system, according to TCM, it must come from complex carbohydrates, such as grains, fruits, and beans. Unlike the refined carbohydrates, these foods provide a more lasting energy, facilitating a more balanced physical, emotional, and intellectual experience every day. Foods filled with refined sugars create an excess in the system, as they overstimulate the endocrine system in the production of enzymes and hormones to deal with the sudden onslaught of glucose into the cells. When cookies, cakes, candies, ice cream, and other foods laden with refined sugars are consumed, an initial burst of energy occurs, and shortly thereafter a feeling fatigue and lethargy occurs. This kind of eating pattern, when continued for a period of time, can negatively affect health because it stresses the endocrine system unnecessarily.

KEY POINT

Fresh fruit and malt barley, rice syrup, or blackstrap molasses are good to use as sweeteners because as complex sugars they stress the body less.

Dairy

Dairy is another important group to consider. Although milk is touted as the complete food by the dairy industry, it is not without its problems. Dairy products can negatively affect the mucous membranes and contribute to digestive difficulties. Humans are the only species that drinks milk as adults, and it is milk of a different species. According to TCM, excess consumption of dairy products produces a condition called dampness, displayed as abdominal distention, edema, cysts, and allergies.

The Tune-Up

Good general health requires that there be balance in all areas of life—mind, body, spirit, family, and community. Physically it means that people eat healthful foods that provide energy and balance, without the problems of overeating, indigestion, or food intolerances. Psychologically and spiritually it means that people enjoy meals in peace, cherish and respect their food and those they share it with, and practice intention (belief and faith that all is right and as it should be) to a higher power. Family and community are areas where traditions and events usually occur with food and eating. Family get-togethers and social events are ways for us to feel connected to others; however, if established patterns of eating are not beneficial to general health, some changes may be necessary (**Exhibit 2-1**).

EXHIBIT 2-1   SMART SNACKING

INSTEAD OF	TRY
Potato chips or pretzels	Mini bagels or breadsticks
A candy bar	A piece of fruit or a glass of juice
Cookies	Graham crackers or raisins
Fried meats	Baked or grilled meats
Whole milk	Skim or 1% or 2% milk
Butter or syrup on pancakes	Fresh fruit
Ice cream	Frozen low-fat or light yogurt
French fries or home fries	Baked potato with herbs
Sour cream on baked potato	Nonfat yogurt and chives
Butter or cheese on vegetables	Lemon juice or herbs

A Nutritional Lifestyle for the Ages

The holistic approach to nutrition and diet considers self-care, healthful food selection, moderate intake, and balance. It suggests listening to one's own inner wisdom, being present (paying attention to what is happening at the present), and following a healthful lifestyle and a diet that includes foods that work in synergy (together) with other aspects of life.

Nutritional Intake

A varied, balanced diet from wholesome, high-quality foods will provide much of what is needed to live a healthy, productive, long life. Varying colors, tastes, textures, and temperatures from good quality, organic food will yield the best results.

Time should be taken at meals to enjoy food, masticate (chew) it, and allow the action of the digestive powers to be fully utilized. Relaxation, enjoyable company, tranquility of mind, and pleasant conversation while eating help to fulfill the psychological, social, and cultural needs associated with food intake. The spiritual aspects of eating can be addressed through rituals and family traditions that are incorporated into each day's meals.

Regular Meals

Consistency of food intake is important. The body must have intervals of rest from eating or its energies are soon exhausted, resulting in impaired function, dyspepsia (stomach upset), and other problems. Constant munching, whether on pastries and candy or apples and carrots, will lead to a digestive tract that is almost constantly at work; poor and weak digestion will follow, causing nutritional imbalances. Eating six small daily meals, beginning with a good breakfast, is best. Meals should be approached with an attitude of self-caring that will allow the mind, body, and spirit a much needed respite from the regular schedule.

REFLECTION

Do you have rituals—such as prayer, candle lighting, or sharing time—that you incorporate into your main meals with significant others in your life? If not, how could you incorporate at least one?

Amount and Timing

People generally eat too much rather than too little. It is an excellent plan to rise from the table before the desire for food is quite satisfied. The body's nutrition does not depend on the amount eaten but on the quality of food consumed. Eating too much is nearly as bad as swallowing food before it is properly chewed. Those who dine late should wait 2 or 3 hours before retiring. Late-evening meals usually lead to a poor night's rest, with organs such as the liver being unable to detoxify properly.

MyPlate

The U.S. Department of Agriculture and the U.S. Department of Health and Human Services have developed recommendations and tools to help consumers understand and apply sound nutritional principles (**Table 2-3**). One such tool that illustrated recommended dietary intake is MyPlate (**Figure 2-2**), which replaced the Food Guide Pyramid. This guide shows preferable amounts of consumption from each food group. Although MyPlate has replaced the Food Guide Pyramid, food pyramids continue to be used by several healthy diets, such as the Vegetarian, Mediterranean, and Asian Food Guide pyramids. Whether you are guided by the pyramid or plate, the key is to choose the foods from each group that will provide an individualized, tasty, nutrient balance and at the same time fit your lifestyle, food preferences, and cultural needs. Balance, variety, and moderation are the key to these guidelines. The number of servings from each group varies, depending on your age, sex, and level of physical activity. The website at www.ChooseMyPlate.gov provides recommendations for specific populations.

TABLE 2-3   DIETARY GUIDELINES: KEY RECOMMENDATIONS

BALANCING CALORIES TO MANAGE WEIGHT

- Prevent and/or reduce overweight and obesity through improved eating and physical activity behaviors.
- Control total calorie intake to manage body weight. For people who are overweight or obese, this will mean consuming fewer calories from foods and beverages.
- Increase physical activity and reduce time spent in sedentary activities.
- Maintain appropriate calorie balance during each stage of life—childhood, adolescence, adulthood, pregnancy and breastfeeding, and older age.

FOODS AND FOOD COMPONENTS TO REDUCE

- Reduce daily sodium intake to less than 2,300 milligrams (mg) and further reduce intake to 1,500 mg for persons who are 51 and older and those of any age who are African American or have hypertension, diabetes, or chronic kidney disease. The 1,500-mg recommendation applies to about half of the U.S. population, including children and the majority of adults.
- Consume fewer than 10 percent of calories from saturated fatty acids by replacing them with monounsaturated and polyunsaturated fatty acids.
- Consume fewer than 300 mg per day of dietary cholesterol.
- Keep trans fatty acid consumption as low as possible by limiting foods that contain synthetic sources of trans fats, such as partially hydrogenated oils, and by limiting other solid fats.
- Reduce the intake of calories from solid fats and added sugars.
- Limit the consumption of foods that contain refined grains, especially refined-grain foods that contain solid fats, added sugars, and sodium.
- If alcohol is consumed, it should be consumed in moderation—up to one drink per day for women and two drinks per day for men—and only by adults of legal drinking age.

FOODS AND NUTRIENTS TO INCREASE

Individuals should meet the following recommendations as part of a healthy eating pattern while staying within their calorie needs.

- Increase vegetable and fruit intake.

TABLE 2-3   DIETARY GUIDELINES: KEY RECOMMENDATIONS (CON'T.)

- Eat a variety of vegetables, especially dark green and red and orange vegetables and beans and peas.
- Consume at least half of all grains as whole grains. Increase whole-grain intake by replacing refined grains with whole grains.
- Increase intake of fat-free or low-fat milk and milk products, such as milk, yogurt, cheese, or fortified soy beverages.
- Choose a variety of protein foods, which include seafood, lean meat and poultry, eggs, beans and peas, soy products, and unsalted nuts and seeds.
- Increase the amount and variety of seafood consumed by choosing seafood in place of some meat and poultry.
- Replace protein foods that are higher in solid fats with choices that are lower in solid fats and calories and/or are sources of oils.
- Use oils to replace solid fats where possible.
- Choose foods that provide sufficient amounts of potassium, dietary fiber, calcium, and vitamin D, which are nutrients of concern in American diets. These foods include vegetables, fruits, whole grains, and milk and milk products.

RECOMMENDATIONS FOR SPECIFIC POPULATION GROUPS

Women capable of becoming pregnant

- Choose foods that supply heme iron, which is readily absorbed by the body, additional iron sources, and enhancers of iron absorption such as vitamin C–rich foods.
- Consume 400 micrograms (mcg) per day of synthetic folic acid (from fortified foods and/or supplements) in addition to food forms of folate from a varied diet.

Women who are pregnant or breastfeeding

- Consume 8–12 ounces of seafood per week from a variety of seafood types.
- Due to its high methyl mercury content, limit white (albacore) tuna to 6 ounces per week and do not eat the following four types of fish: tilefish, shark, swordfish, and king mackerel.
- If pregnant, take an iron supplement, as recommended by an obstetrician or other healthcare provider.

(continues)

TABLE 2-3   DIETARY GUIDELINES: KEY RECOMMENDATIONS (CON'T.)

Individuals ages 50 years and older
- Consume foods fortified with vitamin B$_{12}$, such as fortified cereals or dietary supplements.

BUILDING HEALTHY EATING PATTERNS
- Select an eating pattern that meets nutrient needs over time at an appropriate calorie level.
- Account for all foods and beverages consumed and assess how they fit within a total healthy eating pattern.
- Follow food safety recommendations when preparing and eating foods to reduce the risk of food-borne illnesses.

Source: U.S. Department of Agriculture, U.S. Department of Health and Human Services. (n.d.). Dietary guidelines for Americans. Retrieved from www.dietaryguidelines.gov

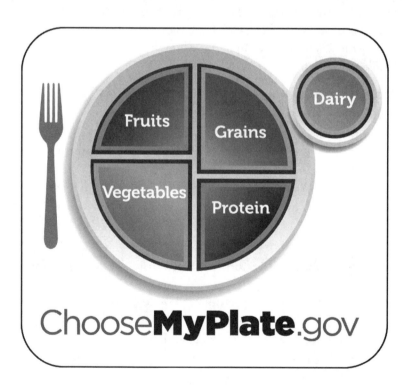

Figure 2-2   MyPlate

Source: United States Department of Agriculture, choosemyplate.gov

Nutrition Facts Label

The regulation of food dates back to the beginning of the last century. Under regulations from the Food and Drug Administration (FDA) of the U.S. Department of Health and Human Services and the Food Safety and Inspection Service of the U.S. Department of Agriculture (USDA), the food label was designed to give more information about nutrition. By law, nearly all food labels must contain a nutrition facts panel that contains information as to how the food can fit into an overall daily diet. The nutrition facts label (**Figure 2-3**) on the side or back of a package states the amount of saturated fat, cholesterol, fiber, and other nutrients each serving contains. By checking the serving size on several products, you can compare the nutritional qualities of similar foods.

The % daily values on the panel is based on a 2,000-calorie diet and shows the percentage of a nutrient provided in one portion. The aim is to meet 100% of the daily value for total carbohydrate, fiber, vitamin, and minerals listed. The percentage for fat, sodium (salt), and cholesterol can add up to less than 100. The amount of fat, cholesterol, sodium, carbohydrates, and protein (in grams or milligrams) are listed to the immediate right of these nutrients.

Nutrition at Different Life Stages

The Dietary Guidelines for Americans and nutrition facts labels are meant to be used by the average adult who is healthy, active, and within the guidelines for height and weight. There are other factors that affect the amount and type of food needed daily.

Children from birth to 5 years require more vitamins, calcium, and other nutrients for growth and development. The Dietary Guidelines generally are not meant to be used for children under 2 years old. In childhood, these

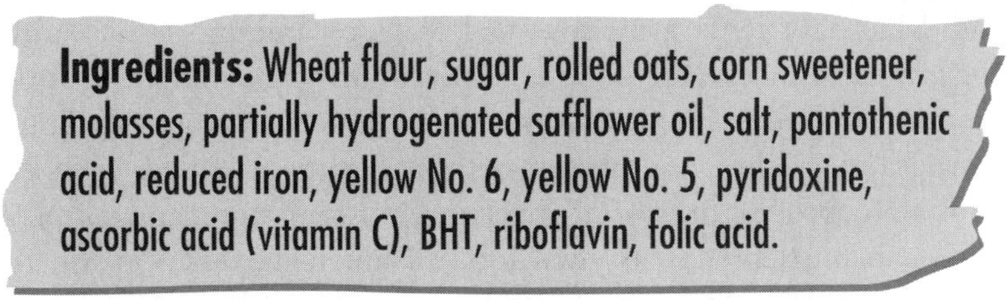

Ingredients: Wheat flour, sugar, rolled oats, corn sweetener, molasses, partially hydrogenated safflower oil, salt, pantothenic acid, reduced iron, yellow No. 6, yellow No. 5, pyridoxine, ascorbic acid (vitamin C), BHT, riboflavin, folic acid.

Figure 2-3   The Nutrition Facts Label

guidelines can be applied, using smaller portions for younger children and increasing portions as they get older. Adolescents, especially pregnant adolescents, and pregnant and lactating women need more breads, fruit, vegetables, milk, and meat because of higher nutrient needs.

Good nutrition for those 50 years of age and older can decrease the effects of health problems that are prevalent among older Americans. It also can improve the quality of life for people who have chronic disease. Older people are likely to have poor appetites, experience health problems, be taking medications, and have sedentary lifestyles. They may also limit beneficial foods because of chewing, digestive, or intestinal health problems. This may lead to poor dietary intake and nutritional imbalances.

Diversity in Nutritional Lifestyles

Ethnic cuisine is becoming more and more popular in the United States, with the Mediterranean diet perhaps being the best known and most used alternative diet. It includes bread and pasta, vegetables, legumes, fresh fruit, breads, and some unsaturated oils (olive), a few red meats and fish, and more eggs and poultry than MyPlate indicates. Garlic is usually a staple, and the food is cooked fresh with the use of other herbs for flavor. Diet pyramids have emerged for Asian, Latin American, and vegan diets, representing what epidemiologic (development of a particular issue within a certain population) studies have associated with optimum health.

Interest in vegetarianism is increasing and is leading the way for Asian—specifically Japanese, Chinese, and Native American (hunter–gatherer)—diets. The Japanese diet is one of the healthiest in the world, whereas other Asian cuisines offer low-fat diets with various selections from plant and some

---

KEY POINT

Nutrients are elements of our food that make up macronutrients and micronutrients. Macronutrients are consumed in large amounts—tens and hundreds of grams (ounces and pounds). Micronutrients, minerals, and trace elements are ingested in thousandths and millionths of a gram (milligrams and micrograms). Fiber and water, although not classified as nutrients, are essential to proper nutrition, digestion, assimilation, and elimination. Although each nutrient has its own job, all nutrients function together to help the body function at optimal levels. The combined action is far greater than if taken separately (synergy).

animal (fish) foods. Some studies have suggested that these diets lower risk factors for heart disease and some cancers.

The Macronutrients

Macronutrients include water, fiber, carbohydrates, protein, and fats. These are essential for the body and mind to work efficiently, function properly, and maintain and repair itself.

Water

Water, not truly a macronutrient, is fundamental for life. It has a role in every major bodily function and is the substance that allows chemical reactions to occur from the nutrients we ingest. This makes it essential to proper diet and nutrition. Water maintains the body's proper temperature, transports nutrients to and toxins from the cells, lubricates our joints, transports oxygen through the blood and lymphatic systems, and comprises from one half to two thirds of our body. Without water, death is imminent within days. The body requires between 9 and 12 cups of pure drinking water daily, in addition to water obtained through soups, fruits, and other foods.

Tap water quality varies by locale in the United States. Various processes are used to purify water and remove major contaminants from water supplies. These processes are filtration, distillation, and reverse osmosis. The results of all water testing done by your local water supply board are part of the public record, and you can request copies of recent monitoring reports as part of the Freedom of Information Act. If you have a private well, you can have the water tested by a certified laboratory.

Fiber

Like water, fiber is not a true macronutrient, although it is another important element in the nutrition equation. A diet containing a minimum of 20 to 30 grams of fiber per day is recommended. The typical American is said to eat only about 11 grams of fiber per day.

Vegetables, fruits, peas, beans, and whole grains contain fiber. Proper bowel elimination can be attained by eating fiber because it adds bulk and hastens the course of food and waste through the intestinal system. Fiber, in conjunction with a low-saturated-fat, low-cholesterol diet, has been found to reduce the risk for heart disease, digestive disorders, certain cancers, and diabetes. The sustained absence of fiber in the diet can lead to problems of the gastrointestinal tract, including diverticular disease and constipation.

> **KEY POINT**
>
> Following the suggestions of the USDA's MyPlate, eating two to four servings of fruit (fresh, unpeeled) and five servings of vegetables (some raw and unpeeled) will allow for an average of 20 to 30 grams of fiber in the diet—the recommended daily intake.

Carbohydrates

Carbohydrates provide the body with energy and are composed of two classes: complex and simple carbohydrates. Foods that contain starches and fiber are called complex carbohydrates; sugars are simple carbohydrates. Fruits, vegetables, and grain products (breads, cereals, and pasta) are complex carbohydrates. Avoiding simple carbohydrates—foods containing added sugar such as candy, soft drinks, cookies, cakes, ice cream, and pies—will increase energy and lower risks for health problems. The body cannot differentiate between complex and simple carbohydrates; therefore, the body will metabolize candy, which has a high added sugar content (and is high in saturated fat), and a piece of fruit, which has natural sugar, the same way. The candy, however, contains few nutrients, and the excess sugar will be immediately stored as fat in the body. The fruit contains many required nutrients as well as fiber and will be used for energy more quickly than the candy.

Carbohydrates are necessary for protein to be digested. They provide the blood with glucose, which is formed during the digestive process and is needed by the body and brain to give energy to muscles, tissues, and organs. When carbohydrates are present in the system, the body is able to use protein for regeneration and repair.

Protein

Protein is the body's secondary energy source and has many functions. Protein must be present for the body to grow, repair damaged or injured tissue, create new tissue, and regulate water balance. It is the component that lays the foundation for major organs, blood and blood clotting, muscles, skin, hair, nails, hormones, enzymes, and antibodies. It maintains a proper balance of acid and alkaline in the blood.

The essential elements in protein are amino acids. Foods that consist of all of the essential amino acids are considered complete proteins. These usually come from meat, fish, fowl, eggs, and dairy. Those that come from vegetable sources such as beans, grain, and peas are considered incomplete proteins because they supply the body with only some of the essential amino acids.

KEY POINT

There are 22 amino acids, 8 of which are called essential amino acids and must be consumed through the diet because the body is unable to manufacture them.

Fats

Lipids (fats) are another source of energy for the body. The lipid group of macronutrients is vital for health but is needed in small amounts. Fats supply essential fatty acids, such as linoleic (omega-6) and linolenic acid (omega-3). They also help maintain healthy skin; regulate cholesterol metabolism; are precursors to prostaglandin (a hormone-like substance that regulates some body processes); carry fat-soluble vitamins A, D, E, and K, and aid in their absorption from the intestine; act as a cushion and stabilizer for our internal organs; and supply a protective layer that helps regulate body temperature and maintain heat. Fats supply greater energy than carbohydrates or protein. Fat has nine calories per gram, whereas carbohydrates and protein contain only four calories per gram. One molecule of fat can be broken down into three molecules of fatty acids and one molecule of glycerol. This structure is known chemically as triglycerides, which make up approximately 90% of all dietary fat.

Fatty acids are generally classified as saturated, monounsaturated, and polyunsaturated. In general, fats that contain a majority of saturated fatty acids are solid at room temperature. Fats containing mostly unsaturated fatty acids are usually liquid at room temperature and are called oils. Saturated fatty acids are found in meats, cream, whole milk, butter, cheese, coconut oil, palm kernel oil, and vegetable shortening. Monounsaturated fatty acids are found in olive and canola oils. Polyunsaturated fatty acids are found in other vegetable oils, nuts, and some fish. Both types of fatty acids, when used to replace saturated fats, can help to reduce the level of bad cholesterol in blood.

Cholesterol is not fat, but is rather a fat-like substance classified as a lipid. Cholesterol is vital to life and is found in all cell membranes. It is necessary for the production of bile acids and steroid hormones. Dietary cholesterol is found in only animal foods and is abundant in organ meats, egg yolks, meats, and poultry. Low-density lipoprotein (bad cholesterol) and high-density lipoprotein (good cholesterol) are the two types. Low-density lipoproteins are those that increase the risk of cardiovascular disease.

Because all fats contain different amounts of the three fatty acids, the ratio of polyunsaturated to saturated fats in what is eaten is important. The

---

KEY POINT

When buying foods that contain fat, be sure to read the labels. Some foods say that they are low cholesterol but are still high in fat.

---

recommendation from the National Institutes of Health is that overall intake should be in the highest ratio of monounsaturated and polyunsaturated fats with no more than 30% of daily calories from fat sources in the diet and reducing saturated fat to fewer than 10% of calories.

The Micronutrients

Vitamins and minerals are the micronutrients (elements) in food that help the body to function properly. They are needed in trace or small amounts and occur naturally in food.

Vitamins and minerals work in synergy. This word comes from Greek *synergos*, meaning work. Synergism is the working together of all the nutrients that are in food, in the needed ratio, for the optimal functioning of the body/mind.

Vitamins and minerals should be taken in whole foods as much as possible because foods contain hundreds of compounds that assist the body to use these nutrients.

Vitamins

Vitamins do not have any calories and cannot be used as a food source, but they are necessary for life. They are needed for the body to grow, develop, maintain its metabolic processes, and assist in digestion and assimilation. Vitamins are assisted in the body by enzymes. **Table 2-4** lists signs of deficiency, functions, signs of toxicity, and food sources for each vitamin.

There are two types of vitamins: water and fat soluble. Water-soluble vitamins are those that dissolve in water and are excreted through our skin (perspiration), lungs (breathing), intestines (bowel), and mainly through the kidneys (urine) if not needed or used by the body. They include the B-complex group, which includes B_1 (thiamin), B_2 (riboflavin), B_3 (niacin), B_6 (pyridoxine), folic acid (folacin, folate, or pteroylglutamic acid [PGA]), B_{12} (cobalamin or cyanocobalamin), biotin, pantothenic acid, and vitamin C.

Each vitamin has its own function. The B-complex group's functions are many. They range from normal neurologic (nervous) system functioning; synthesizing (linking) nonessential amino acids; helping oxidize (burn) glucose;

TABLE 2-4   VITAMINS

VITAMIN	DEFICIENCY MAY CAUSE	HOW IT WORKS	EXCESS MAY CAUSE	BEST FOODS TO EAT
A	Night blindness, skin problems, loss of appetite, emotional upset, nerve damage to legs (late sign)	Aids bone and teeth growth, vision, keeps cells, skin, and tissues working properly	Birth defects, bone fragility, vision and liver problems	Eggs, dark green and deep orange fruits and vegetables, liver, whole milk
B-1	Tiredness, weakness, loss of appetite, emotional upset, nerve damage to legs (late sign)	Helps release food nutrients and energy, appetite control, helps nervous system, and digestive tract	Headache, rapid pulse, irritability, trembling, insomnia, interference with B_2, B_6	Whole grains and enriched breads, cereals, dried beans, pork, most vegetables, nuts, peas
B-2 Riboflavin	Cracks at corners of mouth, sensitivity to light, eye problems, inflamed mouth	Helps enzymes in releasing energy from cells, promotes growth, cell oxidation	No known toxic effect. Some antibiotics can interfere with B_2 being absorbed	whole grains, enriched bread and cereals, leafy green vegetables, diary, eggs, yogurt
Niacin	General fatigue, digestive disorders, irritability, loss of appetite, skin disorders	Fat, carbohydrate and protein metabolism, good skin, tongue, and digestive system, circulation	Flushing, stomach pain, nausea, eye damage, can lead to heart and liver damage	Whole wheat, poultry, milk, cheese, nuts, potatoes, tuna, eggs
B-6 Pyridoxine	Dermatitis, weakness, convulsions in infants, insomnia, poor immune response, sore tongue, confusion, irritability	Necessary protein metabolism, nervous system functions, formation of red blood cells, immune system function	Reversible nerve injury, difficulty walking, numbness, impaired senses	Wheat and rice bran, fish, lean meats, whole grains, sunflower seeds, corn, spinach, bananas
Biotin	Rarely seen since it can be made in body if not consumed. Flaky skin, loss of appetite, nausea.	Cofactor with enzymes for metabolism of macronutrients, formation of fatty acids, helps other B vitamins be utilized	Symptoms similar to vitamin B_1 overdose	Egg yolks, organ meats, vegetables, fish, nuts, seeds, also made in intestines by normal bacteria there
Folic Acid	Anemia, diarrhea, digestive upset, bleeding gums,	Aids red blood cell formation, healthy pregnancy, metabolism of proteins	Excessive intake can mask B_{12} deficiency and interfere with zinc absorption	Green leafy vegetables, organ meats, dried beans

(continues)

TABLE 2-4   VITAMINS (CONTINUED)

B₁₂ Cobalamin	Elderly, vegetarians, or those with malabsorption disorder are at risk of deficiency–pernicious anemia, nerve damage	Necessary to form blood cells, proper nerve function, metabolism of carbohydrates and fats, builds genetic material	None known except those born with defect to absorb	Liver, salmon, fish, lean meats, milk, all animal products
Pantothenic Acid	Not usually seen; vomiting, cramps, diarrhea, fatigue, tingling hands and feet, difficult coordination	Needed for many processes in the body, converts nutrients into energy, formation of some fats, vitamin utilization, making hormones	Rare	Lean meats, whole grains, legumes
C Ascorbic Acid	Bleeding gums, slow healing, poor immune response, aching joints, nose bleeds, anemia	Helps heal wounds, collagen maintenance, resistance to infection, formation of brain chemicals	Diarrhea, kidney stones, blood problems, urinary problems	Most fruits, especially citrus fruits, melon, berries, and vegetables
D	Poor bone growth, rickets, osteoporosis, bone softening, muscle twitches	Calcium and phosphorus, metabolism and absorption, bone and teeth formation	Headache, fragile bones, high blood pressure, increased cholesterol, calcium deposits	Egg yolks, organ meats, fortified milk, also made in skin when exposed to sun
E	Not usually seen, after prolonged impairment of fat absorption, neurological abnormalities	Maintains cell membranes, assists as antioxidant, red blood cell formation		Vegetable oils and margarine, wheat germ, nuts, dark green vegetables, whole grains
K	Tendency to hemorrhage, liver damage	Needed for prothrombin, blood clotting, works with Vitamin D in bone growth	Jaundice (yellow skin) with synthetic form, flushing & sweating	Green vegetables, oats, rye, dairy

> ### KEY POINT
>
> The opportunity for toxicity is greater with fat-soluble vitamins because they are stored in the body.

assisting in the digestion of carbohydrates; promoting protein metabolism; and assisting in the production of various hormones, red blood cells, and genetic materials (such as RNA and DNA).

Vitamin C (ascorbic acid) has its best known source in citrus fruits but can also be found in most fresh fruits and vegetables, especially tomatoes, broccoli, and potatoes. It is essential for the formation of collagen, the protein that helps form skin, bone, and ligaments. It is also needed for iron to be absorbed, to prevent hemorrhaging, to help wounds to heal, and to help with allergic reactions.

Fat-soluble vitamins attach to protein and are carried throughout the body by the blood. Unlike water-soluble vitamins, fat-soluble vitamins can be stored in the liver and adipose (fatty) tissues of the body if taken in excess. Fat-soluble vitamins include vitamins A, D, E, and K.

Small amounts of vitamin D and K can be made by the body. The sources for these vitamins are green/deep yellow/orange vegetables, deep yellow/orange fruits, whole grains, low-fat dairy, vegetable oil, seeds, nuts, eggs, and liver. Their functions include preventing night blindness; helping vision; promoting healthy skin, hair, teeth, and nails; boosting the immune system; assisting the absorption of calcium; maintaining mucous membranes; and blood clotting.

Minerals

Minerals (**Table 2-5**) are further broken down into the following groups: macrominerals, those required in milligrams (larger amounts) by the body; trace minerals, required in micrograms (smaller amounts) by the body; and other trace minerals or elements. The body is unable to synthesize minerals, and thus, they must be taken in through the diet on a regular basis.

The macrominerals are calcium, magnesium, phosphorus, sodium, and potassium. The trace minerals include iron, iodine, manganese, chromium, selenium, copper, fluoride, molybdenum, boron, and zinc. These are only needed in the body in minute amounts, but if a deficiency or imbalance exists, it can lead to serious health problems and, if left unchecked, sometimes even death.

TABLE 2-5   MINERALS

MINERALS	DEFICIENCY MAY CAUSE	HOW IT WORKS	EXCESS MAY CAUSE	BEST FOODS TO EAT
Calcium	Rickets, soft bones, osteoporosis, cramps, numbness and tingling arms and legs	Strong bones, teeth, muscle and nerve function, blood clotting	Confusion, lethargy, blocks iron absorption, deposits in body	Dairy, salmon and small bony fish, tofu.
Phosphorus	Weakness and bone pain, otherwise rare	Works with calcium, helps with nerve, muscle, and heart function	Proper balance needed with calcium	Meat, poultry, fish, eggs, dairy, dried beans, whole grains
Magnesium	Muscle weakness, twitching, cardiac problems, tremors, confusion, formation of blood clots	Needed for other minerals and enzymes to work, helps bone growth, muscle contraction	Proper balance needed with calcium, phosphorus, & vitamin D	Nuts, soybeans, dried beans, green vegetables
Potassium	Lethargy, weakness, abnormal heart rhythm, nervous disorders	Fluid balance, controls heart, muscle, nerve, and digestive function	Vomiting, muscle weakness	Vegetables, fruits, dried beans, milk
Iron	Anemia, weakness, fatigue, pallor, poor immune response	Forms of hemoglobin and myoglobinsupplies oxygen to cells, muscles	Increased need for antioxidants, heart disorders	Red meats, fish, poultry, dried beans, eggs, leafy vegetables
Iodine	Goiter, weight gain, increased risk of breast cancer	Helps metabolize fat, thyroid function	Thyroid-decreased activity, enlargement	Seafood, iodized salt, kelp, lima beans
Zinc	Poor growth, poor wound healing, loss of taste, poor sexual development	Works with many enzymes for metabolism and digestion, immune support, wound healing, reproductive development	Digestive problems, fever, dizziness, anemia, kidney problems	Lean meats, fish, poultry, yogurt
Manganese	Nerve damage, dizziness, hearing problems	Enzyme cofactor for metabolism, control blood sugar, nervous, and immune functions	High doses affect iron absorption	Nuts, whole grains, avocados
Copper	Rare, but can cause anemia and growth problems in children	Enzyme activation, skin pigment, needed to form nerve and muscle fibers, red blood cells	Usually by taking supplements. Liver problems, diarrhea.	Nuts, organ meats, seafood
Chromium	Impaired glucose tolerance in low blood sugar and diabetes	Glucose metabolism	Most Americans have low intakes	Brewer's yeast, whole grains, peanuts, clams
Selenium	Heart muscle abnormalities, infections, digestive disturbances	Antioxidant with vitamin E, protects against cancer, helps maintain healthy heart	Nail, hair, and digestive problems, fatigue, garlic odor of breath	Meat and grains, dependent on soil in which they were raised
Molybdenium	Unknown	Element of enzymes needed for metabolism, helps store iron	Gout, joint pains, copper deficiency	Beans, grains, peas, dark green vegetables

Calcium and phosphorus are the two most abundant minerals, and they work together in the body. Calcium is found predominantly in the bones, where it is needed for structure. The bones store calcium for its release into the blood as needed. It assists in blood clotting, transmitting of nerve conduction, and helping with muscle contractions.

Phosphorus is found in nearly all cells of the body. It is used for energy production, metabolizing some vitamins and minerals, and building and renewing tissue and cells. The best source of phosphorus is animal protein, although milk and legumes also contain this mineral. Deficiencies of phosphorus are not usually seen because of its abundance in foods consumed by most Americans.

Magnesium is stored in bone, where it can be used by the body as needed. It is important for calcium, potassium, and vitamin D assimilation and is necessary for the relaxation phase of muscle contraction. Magnesium also assists in the proper functioning of the heart, liver, and other soft tissue.

Potassium, sodium, and chloride are the three minerals that are sometimes called electrolytes. Potassium is essential for life and necessary for heart, nerve, muscle, and digestive functioning. Sodium is necessary for the balance of fluid outside of the cells and is important in maintaining nerve and muscle conduction. Because sodium is abundant in food, there rarely is a chance of deficiency. Chloride mainly occurs with salt in foods and is needed by the body with sodium for fluid balance of the cells and also to digest proteins. Sodium and chloride are found together in common table salt.

The microminerals (trace elements) include iron, iodine, zinc, copper, manganese, molybdenum, selenium, fluoride, chromium, and silicon. Iron is a most essential mineral because it is needed to carry oxygen from the lungs to all the cells. Vitamin C helps with iron absorption from food, whereas calcium and phosphorus have been found to inhibit it. When there is too much iron in the body, it can cause liver and oxidative damage and lead to iron toxicity. Low iron levels can cause iron-deficiency anemia.

Iodine is needed in miniscule amounts in the body to maintain the thyroid gland. Deficiency is rare, but if it occurs, it leads to goiter (enlargement of the thyroid gland). Seafood and salt are high in iodine.

Every organ in the body uses zinc. Zinc has an effect on immune function, healing wounds, digestion, and converting vitamin A to a usable form. A zinc deficiency can lead to many problems, such as poor immune function, digestive problems, and poor growth and development. Although rare, zinc toxicity can interfere with iron absorption and alter cholesterol metabolism.

Selenium has been found to work together with vitamin E as an antioxidant. Because it is found in the soil, deficiencies are rare because any plant that has been grown in selenium-rich soil provides selenium. Toxicity may occur when too much selenium is added to the diet through supplementation.

Chromium has recently been linked to insulin in controlling blood glucose levels. The American diet often is deficient in chromium, attributed to diets high in sugar and refined foods. Deficiencies in this vital mineral can lead to severe health problems.

The idea that what you eat will impact your health and longevity has been gaining momentum for decades. Scientific research seems to have finally caught up with it. The best nutritional strategy for reducing the risk of chronic disease and living a healthy productive life is to follow a basic, sensible nutritional plan, as outlined in Table 2-1.

Some basic tips for good nutrition are also listed in **Exhibit 2-2** and **Exhibit 2-3**. Even with good intake, however, many factors suggest that diet alone cannot meet the total nutritional needs for some individuals.

Summary

Nutrition refers to the ingestion of foods and their relationship to human health. All humans require the same basic nutrients, although the amount and type can vary based on age, diseases, and special needs.

Changes to nutritional habits can begin with just a few new actions that are achievable. A food journal can be a useful tool for assessing nutritional status and could include what is eaten, time consumed, with whom, and feelings associated with food consumption.

EXHIBIT 2-2   SOUND NUTRITION STRATEGIES

- Consume different foods from each group to improve your chances of receiving all of the nutrients that your body needs in the proper balance.
- As you head to the checkout at the supermarket, look over your choices. Determine if you have a good representation of all the foods groups in your cart—cereal, bread, rice, and pasta in good number; a variety of types and colors of vegetables and fruit, enough for five servings per day per person; some dairy (low fat), fish, poultry, or meat/meat alternatives; and a limited number of candy, cookies, cakes, pies, or other rich desserts, chips, or salty snacks.
- Use the nutrition facts label on food packages to guide you in meeting 100% of the recommended daily allowance nutrients.

EXHIBIT 2-3   TEN TIPS FOR GOOD NUTRITION

1. Use the MyPlate as a guideline for what to eat every day.
2. Read the labels on everything you consume that has one.
3. Choose plenty of whole-grain products (bread, pasta, and cereals).
4. Eat at least five fruits and vegetables per day.
5. Choose foods low in fat, added sugars, and salt.
6. Enjoy meals (this helps them digest better).
7. Have dinner in a quiet atmosphere (eating mostly complex carbohydrates will help ease you into the evening and promote better sleep).
8. If you take dietary supplements, take them with meals.
9. Keep coffee, tea, alcohol, and carbonated beverages to a minimum.
10. Drink at least eight glasses of water per day.

Six small meals, beginning with a good breakfast, provide a regularity of intake that helps maintain good energy. MyPlate, which promotes a diet high in plant foods and low in animal foods and fat, is a guide for the types of foods to include in the diet. When planning dietary intake, consideration should be given to the macronutrients, which include water, carbohydrates, protein, and fat, and the micronutrients, which consist of vitamins and minerals.

References

Hahnemann, S. (1982). Organon of medicine. In J. Kunzli, N. Alain, and P. Pendleton (Eds.), *The first integral English translation of the definitive sixth edition of the original work on homoeopathic medicine.* Blaine, WA: Cooper Publishing.

Nightingale, F. (1860). *Notes on nursing.* New York, NY: D. Appleton and Co.

U.S. Department of Agriculture (USDA). (2012). *Food CPI and expenditures.* Retrieved from http://www.ers.usda.gov/Briefing/CPIFoodAndExpenditures/

Suggested Readings

American Dietetic Association & Duyff, R. L. (2006). *American Dietetic Association complete food and nutrition guide.* Hoboken, NJ: Wiley.

Atkins, R. C. (2003). *Atkins for life: The complete controlled carb program for weight loss and good health.* New York, NY: St. Martin's Press.

Balch, P. A. (2002). *Prescription for nutritional healing: The A-to-Z guide to supplements.* New York, NY: Avery Press.

Balch, P. A., & Balch, J. F. (2000). *Prescription for nutritional healing* (3rd ed.). New York, NY: Avery Press.

Balentine, R. (2007). *Diet and nutrition: A holistic approach*. Honesdale, PA: Himalyan Press.

Macwilliam, L. (2007). *NutriSearch comparative guide to nutritional supplements*. Vernon, BC, Canada: Northern Dimensions Publishing.

Mahan, K., & Escott-Stump, S. (2000). *Krause's food, nutrition, and diet therapy* (10th ed.). Philadelphia, PA: W.B. Saunders.

PDR staff. (2011). *Physicians' desk reference (PDR) for nonprescription drugs and dietary supplements and herbs*. Montvale, NJ: PDR Network.

U.S. Department of Agriculture and U.S. Department of Health and Human Services. (2005). *Nutrition and your health: Dietary guidelines for Americans* (6th ed.). Washington, DC: U.S. Government Printing Office.

Wang, Y. C., Colditz, G. A., & Kuntz, K. M. (2007). Forecasting the obesity epidemic in the aging U.S. population. *Obesity, 15*(11), 2855–2865.

Whitney, E. N., & Rolfes, S. R. (2007). *Understanding nutrition*. New York, NY: Wadsworth Press.

Willett, W. C. (2005). *Eat, drink, and be healthy: The Harvard Medical School guide to healthy eating*. New York, NY: Fireside, Simon and Schuster.

Resources

Academy of Nutrition and Dietetics
216 West Jackson Blvd.
Chicago, IL 60606-6995
800-366-1655
Chicago Area 312-899-0040 ×4653
www.eatright.org
Formerly known as the American Dietetic Association, this organization provides information about a variety of nutrition resources and programs and includes a tip of the day, as well as numerous areas for reading.

American Diabetes Association
505 8th Ave.
New York, NY 10018
212-947-9707
www.diabetes.org
This website offers education, information, and referrals regarding diabetes.

American Heart Association
7320 Greenville Ave.
Dallas, TX 75231
214-750-5300
www.americanheart.org
This is an excellent site with areas called Healthy Tools and Healthy Lifestyle, particularly geared to nutrition.

Food and Nutrition Information Center, National Agricultural Library, Agricultural Research Service, USDA
10301 Baltimore Ave., Room 304
Beltsville, MD 20705-2351
www.nal.usda.gov/fnic
This resource offers a way to search any food-related topic on their website. Information on nutrition is given, along with links to other sites; it is updated daily.

National Institute of Diabetes and Digestive and Kidney Diseases
Weight Control Information Network
1 WIN Way
Bethesda, MD 20892-3665
800-WIN-8098
www.niddk.nih.gov/health/nutrit/win.htm
This is a nicely laid out website that has information about nutrition and obesity.

Nutrient Data Laboratory USDA Agricultural Research Service
Beltsville Human Nutrition Research Center
4700 River Rd., Unit 89
Riverdale, MD 20737
301-734-8491
www.ars.usda.gov/nutrientdata
This USDA nutrient database gives facts about the composition of food, a glossary of terms, and links to other agencies.

U.S. Department of Agriculture, Center for Nutrition Policy and Promotion
1120 20th Street, NW
Suite 200, North Lobby
Washington, DC 20036
202-418-2312
www.choosemyplate.gov
A variety of resources related to MyPlate can be downloaded.

U.S. Department of Health and Human Services Consumer Information Center, Department WWW
P.O. Box 100
Pueblo, CO 81009
www.pueblo.gsa.gov/food.htm
Booklets on food and nutrition can be downloaded (free) or ordered through this site.

U.S. Food and Drug Administration, Center for Food Safety and Applied Nutrition Food Labeling

5600 Fishers Ln. (HFE-88)

Room 1685

Rockville, MD 20847

301-443-9767

www.fda.gov/Food/default.htm

The *FDA Consumer Magazine* is accessible through this website, as is a plethora of information about health, nutrition, and dietary choices.

Wheat Foods Council

1100 Connecticut Ave., NW, Suite 430

Washington, DC 20036

www.wheatfoods.org

This organization is devoted to help increase awareness of dietary grains as an essential component to a healthy diet.

Exercise: Mindfulness in Movement

OBJECTIVES

This chapter should enable you to

- List at least 10 benefits of regular physical activity
- Describe rib cage breathing
- Define aerobic exercise
- Calculate target heart rate
- List at least six activities that provide an aerobic workout
- State an example of a muscle-strengthening exercise
- Describe the benefits of Hatha yoga and T'ai Chi Ch'uan

The body is designed to move. As you read this page, your heart is pumping and blood is coursing through hundreds of miles in your cardiovascular network. Your lungs are expanding and contracting. Your eyes are moving, eardrums are vibrating, and neurons are firing. Thousands of processes are transpiring to promote the essence of your being . . . and you have not even lifted a finger to turn a page.

The body is not intended to be stagnant. Every movement, each effort of the muscles to pump blood brings the life force to every cell and flushes or removes from the body all that is no longer needed.

What does it mean to exercise? How does it change us? Why does the body require exercise? What does it look and feel like? These are questions to explore when using a model for holistic movement that is life sustaining and promotes wellness.

REFLECTION

What attitude do you have toward exercise? What about your family background, experiences, and education contributed to that attitude?

What Does It Mean to Exercise?

For the last several decades, the public has been bombarded with the importance of exercise. Much of the effort to promote fitness was launched in the 1960s when President John F. Kennedy implemented the President's Council on Physical Fitness. Children were tested and rewarded for their ability to climb a rope, do sit-ups and push-ups, throw a softball, and speedily run the 100-yard dash. This was the country's attempt to address the health hazards of a sedentary lifestyle for children. Today, physical fitness of youth remains an unmet goal. The United States has an increasingly overweight population with evidence of hypertension and atherosclerosis beginning in the childhood years. Lifestyle-related diseases are noted earlier as people become increasingly physically inactive and technologically dependent. Labor-saving devices are not life saving, and significant physical and spiritual deterioration is evident among all ages.

Improved physical fitness also has been a priority for those interested in positively impacting the aging process. Congress, in 1975, broadened the definition of the Older Americans Act to include "services designed to enable older persons to attain and maintain physical and mental well being through programs of regular physical activity and exercise." In the current millennium, it is now more evident than ever that the country must reconsider its commitment to a healthy population for all citizens—young, old, rich, poor, and all ethnic populations that make up the diverse tapestry of the United States.

Fitness is not limited to youth; aging is not synonymous with a loss of good physical condition and function. Furthermore, it is a misconception that persons with physical limitations cannot engage in a health-promoting movement activity. Advanced age or disease does not mean weakening or sacrificing the abundance life has to offer. Hypertension control, bone and muscle strength, recovery from illness and injury, weight management, functional ability, and a general sense of well-being are all enhanced with regular physical activity.

KEY POINT

Research shows benefits of regular physical activity to include improved cardiopulmonary function, reduced risk of coronary artery disease, lowered risk of colon cancer, heightened immune function, decreased susceptibility to depression, reduced risk for obesity, increased self-esteem, and improved quality of life.

Gentle movement programs can produce an enhanced immune response that is vital for resilience and physical/emotional integrity. Muscles respond to movement—the demand to contract and relax. Muscles are made to work, and just as the human spirit thrives when given a task or job to do, so does the body when put into motion.

> **KEY POINT**
>
> It is a myth that a person must experience physical decline, dysfunction, disability, and dependency as he or she ages. These outcomes are related to inactivity, disease processes, and a sedentary lifestyle—not the person's chronological age.

Commonly held, although limited, definitions of exercise bring images of calisthenics, sweaty workouts to loud music, strenuous weight lifting, and sessions of breathlessness to near exhaustion. It is time to evolve into a new view of exercise that is gentler and deeper and lasts a lifetime. This revised view emphasizes the significance of movement that engages the muscles and enhances the flow of body fluids and energy. Insufficient movement can show its effects in all areas of the self. Immobility causes muscles to shorten and weaken and joints to become stiff and less able to move or rotate smoothly. Sluggish digestion, slower elimination, and removal of toxins and waste products are outgrowths of inactivity. Inadequate physical movement affects the mind as well—evidenced by mild depression or a lack of enthusiasm and zest for living.

> **KEY POINT**
>
> To keep the body in good health is a duty. . . . Otherwise we shall not be able to keep our minds strong and clear. —Buddha

Benefits of Exercise

There are many benefits to exercise. Regular physical activity improves muscle strength and tone. Regular deliberate movement tones muscles, which has a positive impact on appearance, posture, body image, and the ability to engage in self-care activities. Exercise improves the efficiency of the body's metabolism and helps use fat for fuel. When you eat nutritious meals, energy becomes available for activity.

Regular aerobic exercise helps to regulate blood sugar or glucose levels. The increased demand for oxygen during exercise (which is what makes movement aerobic) improves the regulation and use of insulin, necessities for bringing glucose into the cells for metabolism or energy production. Exercise helps stabilize and keep the body's blood sugar levels balanced. Developing a regular movement program is a major preventative measure for adult-onset diabetes, heart disease, and atherosclerosis, as well as disorders of mood and thinking.

Endurance is enhanced by regular aerobic exercise. Exercise helps increase the efficiency of the heart and lungs, aiding in creating a greater capacity for coping with life's challenges.

KEY POINT

When regularly challenged through aerobic activity (jumping, walking, jogging, biking, swimming, etc.), the heart and lungs become better able to accommodate the body's increased demand for oxygen.

Oxygen is the essence of the life force; every cell of the body requires it for proper functioning. The gentle exertion of exercise challenges the organs to work more efficiently to draw in the oxygen more rapidly. Heart rate, breathing rate, and circulation increase, thereby allowing each cell to receive nourishment and eliminate wastes.

There is also a dilation or widening effect to the blood vessels because the heightened demand for oxygen causes the vessels to open wider to accommodate the increased fluid flow. Openness promotes flow. Again, let metaphor bring wisdom and empowerment to a holistic lifestyle. When the mind/body is open, it can receive new information and new possibilities, more of what is necessary and good. Strength and endurance become characteristics for the whole person, not just of the physical being, but also of emotional and spiritual facets.

Exercise increases regularity and flow of the gastrointestinal tract, which assists with digestion and bowel elimination. An increase in activity is among the first nonlaxative approaches to preventing and managing constipation.

Blood carries the oxygen and nutrients necessary for the feeding of every cell. In addition, it serves as the vehicle for removal of unnecessary and unwanted waste products from the body. The increased demand for blood flow stimulates the bone marrow to produce more blood cells to carry the oxygen. In a very short period of time, usually less than 1 month, improvements in cardiopulmonary function can be experienced as exercise challenges the heart and lungs, making them stronger and more efficient. At rest, their rate goes down, meaning that they need to work less to get the same amount of work done, and resting blood pressure decreases.

Weight-bearing/resistance activities, such as walking, swimming, weight training, jumping, running, yoga, and T'ai Chi, enhance the ability of the bones to keep a healthy level of calcium in the skeleton. A regular movement program is a major factor in the prevention of osteoporosis.

Exercise stimulates the circulation of endorphins—the neurohormonal transmitters responsible for feelings of well-being and psychospiritual hardiness, which can relieve stress.

Regular exercise helps reduce chronic physical pain. Premenstrual cramping, headache, and joint stiffness can be relieved by movement. It can also help to prevent falls or injuries as the fit body is more resilient to the effects of gravity. Having the strength and balance to recover from falls is an important advantage to being in good physical condition.

Social, psychological, and emotional benefits are derived from exercise. The release of hormones and various neuropeptides from regular activity help decrease pain, alleviate anxiety, promote feelings of well-being, and suppress fatigue. Exercise improves circulation to the brain, which enhances alertness, clarity of thought, and memory; it sharpens the mind.

Exercise promotes the flow of lymph fluid. Lymph fluid is not driven by a pump like the heart; rather it undulates or moves in wavelike fashion, in response to muscle contraction. The lymph pathways are laced within the body much like a fishnet stocking from the top of the head to the bottom of the toes. Lymph fluid flows from deep within to the superficial layers of the skin and returns back to the thoracic duct in the neck. From the thoracic duct, it joins with the general circulation. Circulation of fluid lubricates the joints, moistens the body—keeps us flowing within.

Exercise normalizes hormonal balance of the body. Not only is insulin better used for blood sugar stability, but cortisol, from the adrenal glands, is better modulated through aerobic activity. The modulation of cortisol is extremely important for reduction of the detrimental effects of stress. (Modulating cortisol is one of the important ways of keeping calcium in the bones and out of the bloodstream where it tends to make the vascular system hard [atherosclerosis].)

What Does Exercise Look and Feel Like?

You do not need to sweat, pant, or hurt to benefit from action that is taken deliberately and in a context of wellness promotion. You simply need to develop a movement program that circulates fluid, contracts muscles, resists gravity, and symbolizes fun and value for you. This idea transcends the notion that exercise is done to produce the outside appearance of a perfect body. The commercial world promotes an ideal body that is an unrealistic image for most people.

Movement is one of the nonnegotiable laws of life. The body, like all life forms, thrives on movement. It is a metaphor of life itself. There is an automatic rhythm within and outside us that does not cease until death—the cyclic motion of breathing, the expanding and contracting of lungs. It is only through the breath that we have life, and the more breath we have, the more vital and alive we feel.

REFLECTION

Take a moment right now, put down the book, get comfortable in your seat, and inhale deeply through your nose down to your lungs. Let your lungs be so full that you feel the entire rib cage rise and fill to capacity. Slowly let the air leave escape through your mouth and let your body sink in the exhalation. What do you observe about the effects?

The Mechanics of Breathing

The mechanics of breathing primarily involve the diaphragm—a large, dome-shaped muscle that separates the abdominal and chest cavities—and the intercostal muscles, which are the muscles between the ribs. The process of inhaling and exhaling depends on the surface tension between the alveoli (air sacs of the lungs), the elasticity of the lungs within the chest wall, and the integrity of the large airways or bronchial tree to support the transport of air into the body.

When the diaphragm contracts or shortens, it flattens downward, increasing the chest cavity and creating a negative pressure that draws air into the lungs and produces the inspiration phase of the breathing cycle. Every breath inhaled can be considered an opportunity to inspire the life force or to bring in spirit (inspiration) for the soul to be fed or nurtured through the energizing of the body (see **Exhibit 3-1**).

A conscious effort to deep breathe can be incorporated into activities, such as walking, yoga, T'ai Chi, dancing, swimming, jogging, or cycling. Diaphragmatic breathing is a potent health-promoting exercise and has benefits beyond helping to bring greater amounts of oxygen into the body.

EXHIBIT 3-1   RIB CAGE BREATHING
Inhale deeply while raising your extended arms from your side to straight above your head. Exhale as your arms are returned to your side. Enjoy the sensation of a fully expanded rib cage. Keep the movement of the arms smooth and slow. Count 1 and 2 inhaling and moving the arms up; count 3 and 4 exhaling and moving the arms down. This is an excellent exercise for pulmonary hygiene—to expand the lungs fully, bringing oxygen to the very deepest aspects of the lungs. It is a wonderful way to cleanse lung tissue, bringing in life-giving oxygen and flushing out waste products that no longer serve the body. Do rib cage breathing four to six times to a session. It is a powerful way to renew and refresh as well as to put the lungs and ribs through their "full range of function."

There is a large collection of lymph nodes in the belly region. Each full-bellied breath massages this lymphatic center. The movement of the diaphragm milks the lymph fluid back up to the heart.

Aerobic Exercise

Aerobic exercise is one of the most important and efficient methods of attaining muscular and cardiovascular fitness. Aerobic exercise is accomplished when enough demand is put on the muscles to increase their need for oxygen, causing the heart to beat faster and the lungs to work harder. This not only increases cardiovascular endurance, but also helps to prevent heart attacks by strengthening the heart and increasing the flow of blood through the vascular system, thereby keeping the arteries open and elastic. Fresh oxygen in the blood improves the functioning of all cells in the body. It helps to burn away fat from the muscles and to build new, lean muscle tissue. This increases the metabolic rate of the entire body even during sleep and is valuable for losing weight and keeping it off.

In order to gain aerobic benefits, the heart rate must be elevated and maintained for the duration of the workout within the target heart rate. The target heart rate range is the range between the maximum and minimum calculated heart rate based on age. When an aerobic exercise is started, a person should keep the heart rate down toward the lower end of the range, gradually moving toward the higher.

To get aerobic exercise, it is easiest to use the legs; because the quadriceps are the largest muscles of the body, they require the most oxygen and thus

KEY POINT

To find your target heart rate, subtract your age from 220. This gives you your maximum heart rate. Multiply your maximum heart rate by 0.50 and 0.75 to get your target heart rate range.

will burn the greatest amount of energy in the shortest amount of time. Examples of moderate activities that use the legs and provide an aerobic workout are listed in **Exhibit 3-2**.

Rebounding or jumping on a trampoline is a particularly beneficial form of exercise. Most forms of aerobic exercise demand the body move forward, or horizontally, along the earth. Running, cycling, swimming, or walking allows the body to move in a measurable distance. Jumping on a rebounder or trampoline, on the other hand, gives the body the unique experience of moving vertically, allowing gravity to act on the cells, tissues, organs, and muscles in a way that literally squeezes out toxins and waste products. This activity also challenges every cell in the body (approximately 60 trillion) to improve integrity, strength, and function. The action of jumping up and down causes the body to adapt to and resist the force of gravity. This act of resistance promotes stronger bones, firmer muscles, and improvement in the circulation of all body fluids. Jumping or rebounding puts exceptional challenge on the venous and lymphatic systems to return the blood and lymph back to the heart. The main restriction to someone developing a rebounding program

---

EXHIBIT 3-2   ACTIVITIES FOR AEROBIC EXERCISE

- Fast dancing for 30 minutes
- Swimming laps for 20 minutes; water aerobics for 30 minutes
- Walking—this includes on a treadmill—for 2 miles in 30 minutes
- Bicycling—either outside or on a stationary bike—for 5 miles in 30 minutes
- Stair walking for 15 minutes
- Jumping rope or jumping on a trampoline for 15 minutes
- Ball sports, such as tennis, handball, racquetball, soccer, or basketball for 15 minutes
- Jogging or running 1.5 miles in 15 minutes

> **KEY POINT**
>
> Any aerobic activity you choose will serve you well if you enjoy doing it. There is no right or wrong movement, if you pay attention to your body. Pain, injury, and discontent are the symptoms of an inappropriate activity. Joy, enthusiasm, and commitment to the habit are indications that an aerobic activity is a well-suited one.

is lower back pain or injury. Lumbar back injury or strain would prohibit safe and enjoyable jumping.

By engaging in a moderate aerobic activity for at least 20 minutes a day for 3 to 5 days a week, improvements in all areas of mind, body, and spirit will be noticed. Changes can usually be seen after 3 to 5 weeks of starting the exercise program.

Muscle-Strengthening Exercises

Strength is an essential, functional component of much of our activities. The lack of strength is responsible for many injuries. Weight training is useful in developing muscle strength. Lifting weights is good for increasing the size and strength of specific muscles; some points to keep in mind in relation to this are:

- Muscle bulk is attained by using weights, doing 12 to 20 repetitions.
- Muscle strength is developed by using the heaviest weights manageable by the person, doing 2 to 6 repetitions.
- Muscle endurance and definition is increased by lighter weights, while doing 40 to 50 repetitions.

This kind of training can be useful for people who wish to develop particular muscles or muscle groups for a specific sport in which they are involved. It also is used for bodybuilding.

> **KEY POINT**
>
> When working out, the principle of overload must be employed. That is, the number of repetitions, the amount of weight, and the speed and intensity of the effort must continually increase if there is to be any real benefit from the practice.

Energy-Building Exercises

Exercise systems such as Hatha yoga and T'ai Chi Ch'uan are aimed at the development of energy that flows through the external body structure. The process of qi (or chi) development produces an effect in the physical body often seen as increased strength, attention, endurance, and vitality. As you work on your energy, you will enhance your physical body.

Hatha yoga develops poise, balance, strength, and amazing agility and limberness. The postures (asanas) massage and revitalize the internal organs and harmonize the qi of the body, imparting internal strength and youthfulness. There is no strain as one assumes the postures and lets the muscles relax and stretch into place. Special attention is paid to alignment of the spine and development of spinal flexibility because the spine is the center of the energetic and nervous systems, and energy blocked here affects the entire system.

Through T'ai Chi Ch'uan, some of the world's most advanced techniques for the training of the mind and body in harmony are available. In Taoist philosophy, the T'ai Chi, or the source and terminus of the universe manifest as a unity, is composed of two interacting and complementary forces called yin and yang. In T'ai Chi Ch'uan, this idea is expressed through a beautiful, coordinated series of postures that through regular practice develop and coordinate the body, under the control of qi, to a level of perfection not otherwise attainable. Proper alignment of the spine is maintained, and the mind is stilled through the slowness of movement with the focus of attention placed on the lower abdomen.

Any Activity Can Benefit Your Health

Regular structured activity, such as walking, yoga, swimming, trampoline jumping, T'ai Chi, or any number of other movement programs enhances health by causing:

- Muscles to grow
- Metabolism to increase, causing more efficient use of the energy of food
- Fat stores to be reduced
- Blood vessels to multiply
- Bones to stay harder
- Thinking processes to be sharper
- Mood to be positive and elevated

An exercise program also promotes vitality and enthusiasm in your relationships. Just as movement of fluids flushes toxins out and allows energy to

---

KEY POINT

The yin/yang concept of balance is promoted in many Eastern philosophies. Applying some of these concepts to movement is helpful when developing a lifetime fitness program.

Yang energy is assertive, outward, masculine in nature, and supportive. It is strength and protection—the ability to stand up for oneself. Developing an aerobic exercise program that uses the largest muscles of the body (thighs or quadriceps) most efficiently circulates blood, strengthens the cardiovascular system, releases toxins through sweat, and in general promotes strength, endurance, and resilience. It is yang activity.

Yin activity, the softer type, is just as necessary for a balanced life. Examples include T'ai Chi and other forms of moving meditation and yoga. Yin activity brings oneself inward and promotes quiet. It develops focus, balance, and a sense of center.

Lifetime fitness means to be active throughout the life span. All that is needed for a lifetime fitness program is finding the activity that resonates with the heart's desire to move. It is best to do the chosen activity at least three times a week, month to month, season to season. It does not take long for the activity to become a habit. Maintaining a regular, realistic, and pleasurable movement program is a key to radiant health. Each individual is the best care provider and custodian of his or her own body, heart, and soul.

---

flow more freely into you, it also opens your social experiences by increasing your confidence and sense of competence.

Summary

Physical fitness is important for people of all ages. Exercise strengthens and tones muscles, improves efficiency of the body, enhances cardiovascular function, prevents bone loss, relieves stress, helps reduce chronic pain, speeds healing, elevates mood, sharpens the mind, increases regularity and flow of the gastrointestinal system, promotes lymphatic fluid flow, and normalizes the body's hormonal balance.

Aerobic exercise involves putting ample demand on the muscles to increase their oxygen requirement, thereby causing the heart and lungs to work harder. To gain benefit from aerobic exercise, the heart rate should be maintained within the target heart rate range during exercise.

Target heart rate range is the range between maximum and minimum heart rate, calculated by using the individual's age. Other forms of exercise include weight training, which is useful in developing muscle strength, and Hatha yoga and T'ai Chi Ch'uan, to enhance energy flow.

Exercise plans that are sustainable and most effective are those that are regular, realistic, pleasurable, and individualized.

Suggested Reading

Anderson, B., & Anderson, J. (2010). *Stretching*. Bolinas, CA: Shelter Publications.

Brooks, L. (1999). *Rebounding to better health*. Albuquerque, NM: Ke Publishers.

Campbell, A. (2009). *The men's health big book of exercises: Four weeks to a leaner, stronger, more muscular you*. New York, NY: Rodale Press.

Campbell, A. (2009). *The women's health big book of exercises: Four weeks to a sexier, healthier more muscular you*. New York, NY: Rodale Press.

Chia, M. (2005). *The inner structure of Tai Chi: Mastering the classic forms of Tai Chi Chi Kung*. Huntington, NY: Healing Tao Books.

Hahn, F., Eades, M. R., & Eades, M. D. (2002). *The slow burn fitness revolution: The slow motion exercise that will change your body in 30 minutes a week*. New York, NY: Broadway Books.

Hutchinson, A. (2011). *Which comes first, cardio or weights? Fitness myths, training truths, and other surprising discoveries from the science of exercise*. New York, NY: HarperCollins.

Isacowitz, R. (2006). *Pilates*. Champaign, IL: Human Kinetics.

Jahnke, R. (2002). *The healing promise of qi: Creating extraordinary wellness through Tai Chi and Qigong*. New York, NY: McGraw-Hill Contemporary Books.

Khalsa, S. K. (2000). *Kundalini yoga*. Darya Ganj, New Delhi, India: DK Publishers.

Kirsh, D. (2004). *Sound mind, sound body: David Kirsh's ultimate 6-week fitness transformation for men and women*. New York, NY: Rodale Press.

Kisner, C., & Colby, L. A. (2007). *Therapeutic exercise: Foundations and techniques*. Philadelphia, PA: F.A. Davis.

Siler, B. (2000). *The Pilates body: The ultimate at-home guide to strengthening, lengthening, and toning your body without machines*. New York, NY: Doubleday.

Simon, H. (2007). *The no sweat exercise plan: Harvard Medical School guides*. New York, NY: McGraw-Hill.

Resources

American Council on Fitness
www.acefitness.org/

Medline Plus: Exercise and Physical Fitness Resources
www.nlm.nih.gov/medlineplus/exerciseandphysicalfitness.html

National Center for Complementary and Alternative Medicine
http://nccam.nih.gov/health/yoga/introduction.htm

Sleep and Rest

OBJECTIVES

This chapter should enable you to:

- Describe challenges the average person has in resting
- Describe the stages of sleep
- Identify signs of deficits in personal sleep requirements
- List four causes of sleep disorders
- Discuss measures to promote an adequate quality and quantity of sleep

Overcommitted and connected are common states for many individuals. We not only have something scheduled for every day, but we may microschedule activities in 15-minute blocks. Kids barely complete their school day when they are scurried off to club meetings or sporting events. Instead of listening to relaxing music during our commute, we now use the time to catch up on phone calls. Not only do we multitask, but we try to do so as fast as we can. Whereas in the past, there was some degree of patience displayed as people allowed reasonable times for telephone and mail responses, now texting and emailing have resulted in expectations that responses will be immediate. Technology also has enabled us to be accessible 24 hours a day, wherever we may be. Ironically, at the same time that there has been growing interest and engagement in exercise, the attention and time dedicated to rest has been declining.

Rest

The pace of our lives certainly has become accelerated. The average American worker today is working more and sleeping and resting less than those

> **KEY POINT**
>
> The term hurry sickness was introduced more than a half century ago by Dr. Meyer Friedman and Dr. Ray Rosenman, who identified the type A personality. In their popular 1974 book, *Type A Behavior and Your Heart*, they described the cardiovascular risk associated with this behavior.

of previous generations. Although this may be viewed as the new norm, we need to ask ourselves, at what price? We have become more focused on *doing* rather than *being*.

Research in the area of neuroscience is compiling evidence that the stimulation of the brain by the increasing exposure to technology is accelerating the brain's inner clock and increasing the need for fast-paced activities (Naughton, 2010). We become victims of what has been called hurry sickness, in which we are living in a constant state of overdrive, feeling like we're not getting things accomplished fast enough and focused on getting to the next task, with no time to experience events in a meaningful way. An additional unfortunate outcome is that we begin to see people around us as interferences with our getting tasks accomplished; the potential for nurturing rich relationships is diminished as we see interactions merely as unwelcomed interruptions.

There are some things you can do to reduce the pressures that interfere with adequate rest and integrate rest into your life. These things include:

- *Be realistic in your expectations.* Evaluate your schedule to determine if you are trying to accomplish more than can fit into a given period of time. If you find that you are regularly overscheduling and overcommitting, do some honest soul searching to determine the cause.
- *Focus on one thing at a time.* It may seem that you are accomplishing more when you multitask, but evidence shows quality suffers.

> **KEY POINT**
>
> In addition to its benefits to the body and mind, rest fosters spiritual well-being. In fact, rest is of such importance that it is included among the Ten Commandments: Remember the Sabbath day and keep it holy. Six days shall you labor and do all your work, but the seventh day is a Sabbath of the Lord thy God on which you shall do no work (Exodus 20:8–11).
>
> During this time, people of faith are to put their routine work and activities aside to reflect, pray, and worship. The quiet, restful time thereby strengthens their relationship with God.

- *Designate work and play times.* Give yourself permission to relax and engage in playful activities. This includes not checking emails and answering phones at home, avoiding performing work tasks during vacations, and respecting the time you and your significant others designate for recreational activities.
- *Give undivided attention.* When speaking with people face to face or on the phone, focus on the conversation and avoid multitasking.
- *Commit to having a Sabbath.* Particularly if you are a person of faith, schedule a day to put work aside and spend time on the care of your spirit.
- *Build stress-reducing activities into your life.*

Sleep

> *The death of each day's life, sore labor's bath,*
> *Balm of hurt minds, great nature's second course,*
> *Chief nourisher in life's feast.*
> —WILLIAM SHAKESPEARE

In these words from Macbeth, Shakespeare is eloquently describing the healing and nourishing importance of sleep. Although most people understand that the body must achieve sleep, often this aspect of a healthy lifestyle receives less attention than exercise and other requisites for health.

Rather than being a time when nothing much happens, sleep is a time when crucial activities take place that are vital to the proper functioning of the body and mind; for example, during sleep (National Sleep Foundation, 2011):

- Muscles relax and blood supply to them increases.
- Tissue growth and repair occurs.
- Energy is restored.
- Important hormones are released.
- The immune system's function is enhanced.

KEY POINT

Sleep can affect appetite by helping to regulate levels of the hormones ghrelin and leptin, which play a role in our feelings of hunger and fullness. Without adequate sleep, our sense of hunger increases, which can lead to weight gain.

Normal Sleep

Normal sleep consists of two major states: REM (rapid eye movement) sleep and NREM (nonrapid eye movement) sleep. NREM sleep accounts for approximately 75% of total sleep time and is divided into four stages (**Table 4-1**). Sleep progresses and deepens with each stage. The fifth stage is REM sleep. Sleep is regulated by the body's circadian rhythm.

Most adults sleep 7–8 hours in a single period. With this said, there is variation in sleep pattern from person to person.

TABLE 4-1   STAGES OF SLEEP

NREM

Stage 1

Nodding off begins. Light sleep occurs and the person is between being awake and entering sleep.

Stage 2

Deeper relaxation is achieved and the person falls into sleep. Breathing and heart rate are regular, body temperature goes down, and some eye movement is noted through closed lids. The person can be easily awakened.

Stage 3

This is an early phase of deep sleep. Blood pressure drops, breathing slower, muscles are relaxed, hormones are released for growth and development, and temperature and heart rate decrease. The person is more difficult to arouse than in stage 2.

Stage 4

Deep sleep and relaxation occurs. All bodily functions are reduced. Considerable stimulation is required to awaken the person.

REM

REM first occurs about 90 minutes after falling asleep and increases over the later part of the night. Rapid eye movements occur. Breathing and heart rate increase and may become irregular. The brain is active and dreams occur as the person's eyes dart back and forth. This stage is important for providing energy to the body and brain. Alcohol, barbiturates, and other drugs can decrease REM sleep.

REFLECTION

Do you know what your personal sleep needs are? One means to determine this is the following: Over a 1- or 2-week period of time, go to bed at least 8 hours before your scheduled time to awaken. Record the time that you went to bed and the time that you naturally awoke without an alarm. If you find you need the alarm, establish an earlier bedtime. Also note how alert and rested you feel during the day. This should help you to identify your general sleep requirement and the time you should plan to go to sleep daily.

There also is variation in sleep requirements with age, with newborns having the greatest need for sleep and the requirements decreasing with age until adulthood (**Table 4-2**). Lifestyle and health impact sleep needs, as does the aging process (**Box 4-1**).

Rather than base the assessment of rest and sleep just on the hours sleeping, people need to look for signs that indicate how effectively their personal sleep requirements are being met. Signs that a sufficient quantity or quality of sleep is not being achieved could include the following:

- Difficulty awakening in the morning or feeling groggy on awakening rather than refreshed
- Frequent yawning
- A tendency to nod off easily when sitting inactively for awhile
- Sleepiness throughout the day
- Becoming easily stressed and irritable
- Difficulty concentrating, solving problems, and being creative
- Fatigue, drowsiness

TABLE 4-2   SLEEP REQUIREMENTS WITH AGE

AGE	AVERAGE AMOUNT OF SLEEP PER DAY
Newborn	Up to 18 hours
1–12 months	14–18 hours
1–3 years	12–15 hours
3–5 years	11–13 hours
5–12 years	9–11 hours
Adolescents	9–10 hours
Adults, including elderly	7–8 hours

Source: National Sleep Foundation.

BOX 4-1   SLEEP DIFFERENCES IN OLDER ADULTHOOD

Older adults:

- Are more likely to fall asleep earlier in the evening and awaken earlier in the morning; this is referred to as phase advance (Ancoli-Israel & Martin, 2006)
- Have a reduction in NREM and REM stage sleep
- Sleep less soundly
- Are more easily awakened by noises, lights, and changes in room temperature
- Shift in and out of stage 1 sleep more than younger adults
- Have a decline in the proportion of time spent in the deeper stages 3 and 4 sleep

---

Common Causes of Sleep Disorders

There is a variety of reasons for people having difficulty falling or staying asleep; these can include restless legs syndrome, sleep apnea, health conditions, and medications.

Restless Legs Syndrome

Restless legs syndrome is a distressful sensation in the legs that is described as electrical, itching, pins and needles, or like insects crawling that causes a strong urge to move the legs. It is believed to be associated with changes in dopamine and iron metabolism. Moving the legs brings relief but also interferes with sleep.

Low blood sugar, caffeine, alcohol, antihistamines, antidepressants, and antipsychotics can contribute to this syndrome.

People with restless legs syndrome may experience sensations and discomfort that are relieved with leg movement. In addition, they may have anxiety, depression, excess daytime sleepiness, and reduced quality of life as a result of the sleep interference. Medications can be prescribed to help but are not an ideal solution because in some people the drugs cease to work after a period of time or present uncomfortable side effects.

Sleep Apnea

This sleep-related breathing disorder is characterized by snoring and breathing stopping at least five times for a period lasting at least 10 seconds during an hour of sleep. Snoring and sudden awakening and gasping for air also are classic features. Symptoms result from a partial or complete collapse of a narrowed pharynx (throat) that reduces oxygen intake. Daytime sleepiness and fatigue are typically present.

Sleep apnea is three times more common in men. In addition to gender, other risk factors include family history and obesity.

Alcohol and drugs with a depressant effect can aggravate the problem. Sleeping in a flat position should be avoided because it allows the tongue to fall back and block the airway. Sleep disorder clinics can be helpful in evaluating sleep apnea and determining the best treatment plan, which could include weight loss, surgery to remove obstructions or realign the bite, or nasal continuous positive airway pressure (CPAP), which involves a device worn during sleep that fits over the nose and mouth.

Health Conditions

The symptoms caused by health conditions can interfere with sleep. For example:

- Cardiovascular conditions can cause difficulty breathing and pain.
- People with diabetes can have blood sugar levels fluctuate, which interferes with sleep.
- Gastrointestinal problems, such as gastroesophageal reflux disease (GERD), can cause stomach pain that awakens people from sleep.
- Respiratory diseases can cause disruptive coughing and difficulty breathing.
- Arthritis can cause joint pain.
- Depression and other emotional conditions can alter sleep.
- Infections and age-related changes of the urinary system can cause sleep to be interrupted by frequent trips to the bathroom to void.
- Pain associated with a variety of conditions can awaken a person from sleep.

Medications

Medications can be useful in controlling some symptoms but may have the undesired effect of interfering with sleep as well; these can include drugs from

KEY POINT

It could be that the symptoms that interrupt sleep are caused by a condition that has not yet been diagnosed. For this reason it is important that sleep problems be recognized and discussed with one's medical provider.

the following groups: analgesics, anticholinergic agents, antidepressants, antihypertensive agents, benzodiazepines, beta-blockers, diuretics, drugs used to treat dementia, Parkinson's disease drugs, statins, sleep aids, steroids, and thyroid preparations.

Nightmares are another way that drugs can interfere with sleep. Sometimes this occurs when the drug is being used and at other times after the drug is discontinued. Although the exact reason for this isn't fully understood, some sleep experts believe it is due to interference with REM sleep (Foral, Knezevich, Dewan, & Malesker, 2011).

> **KEY POINT**
>
> Although short naps can offer an energy boost, naps longer than 30 minutes can cause you to feel sluggish.

Promoting a Good Quantity and Quality of Sleep

Obtaining a satisfactory quantity and quality of sleep is important to overall health. It is better to take steps to encourage good sleep patterns than to suffer the consequences of poor sleep or resort to sedatives as the primary means to obtain adequate sleep. Some actions that can be taken to promote good sleep include:

- *Try to establish a regular bedtime.* Ideally, you will go to bed approximately 8 hours (or the time you established as your personal sleep time) prior to the time you need to awaken.
- *Limit the use of your bed to sleep and sex.* Avoid making phone calls, watching television, or checking emails.
- *Use measures to relax you prior to going to bed.* Light stretches, relaxation exercises, a hot bath, aromatherapy with essential oil of lavender, and a high-carbohydrate snack can put you in a state that fosters sleep.
- *Watch the caffeine intake.* The effects of caffeine can last as long as 10 hours in some people. Be aware that not all caffeinated sodas contain the same amount of caffeine and that some drugs and even some decaffeinated products contain caffeine (See **Table 4-3**).
- *Exercise during the day.* Physical exercise, preferably outdoors, approximately 6 hours prior to bedtime can foster sleep.
- *Sleep on the right mattress.* Make sure your mattress offers comfort and good support.

TABLE 4-3   CAFFEINE CONTENT OF POPULAR PRODUCTS

	MILLIGRAMS OF CAFFEINE IN AN 8-OUNCE SERVING
BEVERAGES	
Cappuccino	120
Coffee	85–110
Tea, brewed	40–60
Shasta cola	45
Pepsi ONE	37
Mountain Dew	34
Diet Coke	31
Sunkist orange soda	28
Iced tea	25
Diet Pepsi	24
Coca-Cola	23
Snapple iced tea	21
Decaffeinated espresso	10
SINGLE DOSES OF MEDICATIONS	
NoDoz maximum strength	200
Excedrin maximum strength	65
Anacin	13

Source: National Sleep Foundation. (n.d.). Caffeine calculator. Retrieved from www.sleepfoundation.org

- *Control the environment.* Assure that the room is dark; if bright lights shine in the windows from the street, install room-darkening shades or drapes. Control noise; if noise cannot be controlled, consider buying a white-noise generator that produces soothing sounds to mask noise. Keep the temperature comfortable.
- *Manage stress.* Stress is part of normal living, but if it is poorly managed, it can interfere with proper rest and sleep.
- *Consider taking melatonin supplements.* There is some thinking that melatonin (**Box 4-2**) can aid in encouraging sleep, but research results are inconclusive at this time.

BOX 4-2   MELATONIN

Melatonin is a natural substance that is made in the pineal gland, a tiny, pine-cone-shaped organ just above the middle of the brain. The secretion of melatonin is controlled by the light–dark cycle. During the day, the pineal gland is inactive. When the sun goes down and darkness occurs, melatonin begins to be secreted. As the level of melatonin rises, you feel less alert. Blood levels of melatonin stay elevated for about 12 hours, after which they return to normal levels during the daytime. The production of this hormone decreases with age.

Melatonin, promoted as a sleep aid, can be purchased without a prescription in most stores that sell nutritional supplements. Although it has been effective for reducing jet lag and helping people adjust to night shift work, research has not supported its ability to help people fall and stay asleep.

Source: National Sleep Foundation. (n.d.). Melatonin and sleep. Retrieved from http://www.sleepfoundation.org/article/sleep-topics/melatonin-and-sleep

---

- *Consult a medical professional if insomnia is present for more than a few months.* If you are practicing all of the measures to support good sleep and continue having ongoing problems falling or staying asleep, it may be beneficial to have the problem evaluated. There could be undiagnosed health issues or other problems that need attention.

REFLECTION

Do you have difficulty falling or staying asleep? If so, explore factors that could be responsible, such as:

Time allocated for sleep

Activity level throughout the day

Caffeine and alcohol consumption

Eating pattern before bedtime

Stress-producing activities before bedtime

Snoring and other disruptions from person sharing one's bed

Environmental factors (light, noise, and room temperature)

What plans to correct sleep problems, based on the identified factors, can you develop and realistically commit to?

Summary

The average American faces more challenges to obtaining adequate rest and sleep than previous generations. This is compounded by increased exposure to technology that heightens the brain's desire for fast-paced activities. More than ever, individuals need to be proactive in establishing measures that promote rest.

Sleep disorders are common and can include restless legs syndrome, sleep apnea, health conditions, and medications. It is important to identify underlying causes of sleep disorders so that they can be addressed and to establish habits that facilitate good sleep, such as going to bed at a regular time, not using the bed for activities beyond sleep and sex, engaging in relaxation measures prior to bedtime, limiting caffeine intake, building exercise into the day, assuring the mattress is comfortable and supportive, controlling stress, and assuring the environment is conducive to sleep. Insomnia present for more than a few months despite following these measures warrants consultation with a medical professional.

References

Ancoli-Israel, S., & Martin, J. I. (2006). Insomnia and daytime napping in older adults. *Journal of Clinical Sleep Medicine, 15*(2), 333–342.

Foral, P., Knezevich, J., Dewan, N., & Malesker, M. (2011). Medication-induced sleep disturbances. *The Consultant Pharmacist: Journal of the American Society of Consultant Pharmacists, 26*(6), 414–425.

National Sleep Foundation. (2011). *What happens when you sleep.* Retrieved from http://www.sleep foundation.org/article/how-sleep-works/what-happens-when-you-sleep

Naughton, J. (2010, August 14). The Internet: Is it changing the way we think? *The Observer.* Accessed online 1/5/13 http://www.guardian.co.uk/technology/2010/aug/15/internet-brain-neuroscience-debate

Suggested Reading

Avlund, K., Rantanen, T. & Schroll, M. (2007). Factors underlying tiredness in older adults. *Aging Clinical Experimental Research, 19*(1), 16–25.

Bain, K. T. (2006). Management of chronic insomnia in elderly persons. *American Journal of Geriatric Pharmacotherapy, 4*(2), 168–192.

Cole, C., & Richards, K. (2007). Sleep disruption in older adults. Harmful and by no means inevitable, it should be assessed for and treated. *American Journal of Nursing, 107*(5), 40–49.

Cuellar, N. G., Rogers, A. E., & Hisghman, V. (2007). Evidenced based research of complementary and alternative medicine (CAM) for sleep in the community dwelling older adult. *Geriatric Nursing, 28*(1), 46–52.

Gleick, J. (2011). *Faster: The acceleration of just about everything.* New York, NY: Random House.

Johnson, E. O., Roth, T., & Breslau, N. (2006). The association of insomnia with anxiety disorders and depression: Exploration of the direction of risk. *Journal of Psychiatric Research, 40,* 700–708.

Kornblatt, S. (2010). *Restful insomnia: How to get the benefits of sleep even when you can't.* San Francisco, CA: Red Wheel/Weiser.

Liberman, J. (2011). *The gift of rest: Rediscovering the beauty of the Sabbath*. New York, NY: Simon and Schuster.

Merrill, R. M., Aldana, S. G., Greenlaw, R. L., Diehl, H. A., & Salberg, A. (2007). The effects of an intensive lifestyle modification program on sleep and stress disorders. *Journal of Nutrition, Health, and Aging, 11*(3), 242–248.

National Institute of Neurological Disorders and Stroke. (2011). *Brain basics: Understanding sleep*. Retrieved from http://www.ninds.nih.gov/disorders/brain_basics/understanding_sleep.htm

Olds, T. S. (2011). Sleep duration or bedtime? Exploring the relationship between sleep habits and weight status and activity patterns. *Sleep, 34*(10), 1299–1307.

Reed, S. D., Newton, K. M., LaCroix, A. Z., Grothaus, L. C., & Ehrlich, K. (2007). Night sweats, sleep disturbance, and depression associated with diminished libido in late menopausal transition and early postmenopause: Baseline data from the Herbal Alternatives for Menopause Trial (HALT). *American Journal of Obstetrics and Gynecology, 196*(6), 593.

Roth, T. (2007): Insomnia: Definition, prevalence, etiology, and consequences. *Journal of Clinical Sleep Medicine, 3*, S7–S10.

Tuya, A. C. (2007). The management of insomnia in the older adult. *Medical Health, Rhode Island, 90*(6), 195–196.

Watson, R. (2010). *Future minds: How the digital age is changing our minds, why this matters and what we can do about it*. Boston, MA: Nicholas Brealey Publishing.

Resources

American Sleep Apnea Association
1424 K Street NW
Suite 302
Washington, DC 20005
(202) 293-3650
www.sleepapnea.org

National Sleep Foundation
1522 K Street NW
Suite 500
Washington, DC 20005
(202) 347-3471
www.sleepfoundation.org

Immunity Enhancement: Mind/Body Considerations

OBJECTIVES

This chapter should enable you to
- Describe the purpose and components of the immune system
- Describe the peripheral lymphatic system
- Describe diaphragmatic breathing
- List at least three signs of an imbalanced immune system

The germ is nothing. . . . The terrain is everything.
—LOUIS PASTEUR

It is your immune system's resilience that protects you from overwhelming infections that can knock you out cold in the ring of life's daily matches. Your susceptibility to infections and diseases, from the simple cold to catastrophic cancers, is deeply influenced by the health and integrity of the immune system. Likewise, chronic diseases and imbalances can threaten the immune system.

The reflective saying "as within—so without" represents a way of understanding immune function. Immune integrity can be viewed as a metaphor of your ability to defend yourself. The immune system represents an understanding of boundaries and harmonious living in community with others. Your integrity, resilience, and support lie within the immune system's ability to mobilize, defend, communicate, and hold peace and balance within. A primary role of the immune system is to serve and protect. The capacity and success of this system to function optimally and be ever vigilant are important aspects of radiant health.

Components of the Immune System

The immune system consists of a lacy network of pathways capable of transporting immune cells throughout the body. It also has a collection of organs, tissues, and cells dispersed strategically throughout the body (**Table 5-1**). This system is intricately connected to the nervous system (brain) and the endocrine system (hormonal).

The lymph fluid of the body is constantly oozing toward the heart from the farthest reaches of the body and is then reintroduced into the general lymphatic circulation. Two layers of lacy lymph networks are just under the skin, and these return the lymph fluid to the heart. The purpose of these redundant lymphatic pathways is to provide a passage for the return of lymph fluid to the heart.

Keep in mind that lymph fluid is the consistency of an egg white—it is quite thick and moves very slowly. Lymphatic fluid does not have a pump to force it through the body like the heart forces blood with every beat. Lymph fluid movement is dependent on the muscles that provide movement.

Function of the Lymphatic System

All lymph fluid passes through lymph nodes. The lymph nodes are depots where special white blood cells, called T cells, wait on alert for foreign material, such as bacteria or viruses, to be brought into the nodes for identification and security check.

Lymph nodes are located strategically throughout the body. Seventy percent of the body's immune system surrounds the abdominal area. A large number of lymph nodes and vessels are in the gut to make sure that all the foreign, nonself material that is ingested becomes user friendly and beneficial. Imagine the amount of infection and disease that you could suffer if you did not have strong, vigilant immunity to counter all of the bacteria and other foreign material that is carried on food or produced by the process of digestion.

The remainder of the body's lymph nodes are located where major bones articulate, or meet, and where the body has openings to the outside world. Lymph nodes are at the ankles, knees, around the groin area, elbows, armpits, and chest, and chains of lymph nodes are along the neck and collarbone.

There are two reasons lymph must pass through nodes on its return trip to the heart. One is to carry protein molecules to the general circulation because proteins are too big to be circulated back through the venous circulation. This helps keep the fluid levels of the body balanced. The second reason that all

TABLE 5-1   THE IMMUNE SYSTEM

COMPONENT	FUNCTION
Spleen	Bloody organ in the upper left quadrant of the abdomen that produces antibodies, maintains cellular immunity, recirculates white blood cells, and receives B cells, T cells, antigens, macrophages, and antigen-reactive cells from the blood.
Bone marrow	Located in the hollow interior of the long bones, produces red blood cells and macrophages; B and T cells undergo development here.
Lymph nodes	Pea-shaped organs throughout the body that are connected by a network of vessels that receive drainage and filter antigens from this lymphatic fluid.
Thymus gland	Located beneath the breastbone, this gland reaches its full size in early childhood and then progressively shrinks. It produces and stores T cells.
Other organs	Tonsils are groups of lymphoid tissues located in the throat that contain B and T cells. The appendix, Peyer's patches (accumulations of lymphoid cells under mucous membranes that produce nodules), and intestinal nodes are sites of B-cell maturation and antibody production for the intestinal region.
CELLS	
Macrophages	Large white blood cells produced in bone marrow, responsible for phagocytosis.
B cells	Bone marrow–derived cells that produce antibodies that neutralize or destroy antigens.
T cells	Thymus-derived cells consist of T-helper cells that induce B cells to respond to an antigen and T-suppressor cells that halt specific activity of immunologic response. T-helper and T-suppressor cells are in a delicate balance that must be maintained for adequate immune response.
NK (natural killer) cells	NK cells kill foreign invaders on direct contact without B cell involvement by producing cytotoxin, a cell poison.

lymph fluid passes through lymph nodes is to identify pathogens (bacteria, fungi, viruses) that are foreign to the body.

The purpose of immune cells is to recognize what is part of your body's normal composition and what is not. Just as the eyes sense or recognize what is outside of your normal self and retain a memory of that image for life, so, too, the immune cells recognize and remember for a lifetime an encounter with a particular organism—whether be it a bacteria, virus, fungi, food, or an environmental allergen. Your immunity normally provides you with the surveillance mechanism to defend and protect from the day you are born until your last breath.

The immune system is intimately connected to the nervous and endocrine systems. Not only does it respond to physical factors, such as invading pathogens or germs, but also it is very sensitive to your thoughts and emotions. How you think and decide to interpret the world around you influences the kind of activity that either enhances immune resilience or promotes immune disorders.

The immune cells are on patrol and in action every day of your life. They perceive and remember the biochemical interactions between the body and substances foreign to it. The cells are mobile and, when optimal conditions exist, able to transport unwanted materials from the body, keeping the host victorious against infection or compromise.

KEY POINT

It is the job of the immune system to protect the body from disease.

Enhancing Lymphatic Flow

The most important muscle for the movement of your immune system is the diaphragm—the thin, dome-shaped muscle that separates the lungs from the abdominal cavity. Every deep breath and every step that you take have the effect of massaging or pressing lymph fluid along its way. Vigorous deep breathing, as occurs during brisk walking or any aerobic activity, or conscious breathing, such as that done in yoga and other meditative practices, enhances the flow of lymph fluid through a type of breathing called diaphragmatic or belly breathing. Babies come into the world belly breathing, and it is something that needs to be relearned to promote optimal immune function. Deep breathing with the diaphragm is an activity that is extremely useful in improving immune integrity.

> **KEY POINT**
>
> Diaphragmatic breathing is the process of contracting the diaphragm, the thin, dome-shaped muscle covering the stomach and liver, to create a deep inhalation. During the in breath, an effort is made to push the stomach out as the diaphragm flattens down onto the abdomen. This enables the lungs to expand more fully. The rhythm produced by this breathing enhances lymphatic fluid movement and helps remove toxins and waste products from lymph fluid.

Making a habit of practicing belly breathing is a powerful yet subtle means of stress management. Deep breathing helps the heart beat more regularly and perform more competently. Carbon dioxide is more efficiently removed with diaphragmatic breathing. One will be more alert and fit when the breath is attuned with other rhythms of the body. Changing the breathing style to belly breathing rather than chest breathing will bring more oxygen into the cells, increase the energy available for activity and performance, and enhance the innate harmony between breath, heart rate, sense of well-being, and enthusiasm for life. Belly breathing promotes relaxation and maintains calmness in situations of perceived stress through the action of the diaphragm synchronizing its rhythm with the heart's rhythm and other processes of the body. A state of peace and harmony helps to conserve the immune system.

Physical exercise has the ability to increase the vessels that carry blood and lymph throughout the body. The more vessels available to carry blood and lymph, the more efficiently the heart functions and fluids flow. Just as the Dan Ryan Expressway in Chicago opens its collaterals or extra lanes to accommodate the increased number of vehicles, the body also has the ability to develop collateral circulation to relieve congestion and keep the flow moving easily and effortlessly.

Signs of Imbalanced Immune Function

You may have noticed a relationship between immune integrity, exercise, nutrition, positive attitude, and the other aspects of healthful living. The body and mind make up an interconnecting network of systems affecting each another and comprising the total being. It is the immune system's lacy network, covering the body from the top of the head to the tips of the toes, that links the nervous and endocrine systems with thoughts and perceptions. This is why consideration of how well people are in rhythm with themselves and life around them is of equal importance as the quality of the air they breathe

> **KEY POINT**
>
> The job of the peripheral lymphatic system is to clear germs and cancer cells from the body. This lacy network accomplishes this through the massaging movement of the muscles of motion and breathing. Regular physical activity and deep breathing help the efforts of the immune system.

or the amount of exercise they get. Social alienation is compromising to the immune system. For example, it has been shown that spouses have a much greater chance of becoming gravely ill during the first year after the loss of their mate than do other persons in the same age group. Loneliness and the lack of feeling that you belong to others is as depleting of the immune system as any other essential nutrient deficiency.

REFLECTION

In addition to the quality of your air and water, the quality of your relationships impacts your immune system. Are you as concerned about your psychosocial environment as you are with your physical environment?

In addition to physical signs of disease, there are additional symptoms affecting the mind and spirit that can help people to realize that they are not in balance. Some questions that can aid in exploring the presence of imbalance are described in **Exhibit 5-1**.

Boosting Immunologic Health

Diet

In addition to a good basic diet, some foods can positively affect immunity. These include milk, yogurt, nonfat cottage cheese, eggs, fresh fruits and vegetables, nuts, garlic, onions, sprouts, pure honey, and unsulfured molasses. A daily multivitamin and mineral supplement is also helpful; specific nutrients that have immune-boosting effects are listed in **Exhibit 5-2**. Because of their negative effect on the immune system, the intake of refined carbohydrates, saturated and polyunsaturated fats, caffeine, and alcohol should be limited.

Fasting

Fasting, abstaining from solid foods for 1 to 2 days, is becoming increasingly popular as a means to promote health and healing. The effects of fasting on the immune system include the following:

- Increased macrophage activity and neutrophil antibacterial activity
- Raised immunoglobulin levels
- Improvement of cell-mediated immunity, ability of monocytes to kill bacteria, and natural killer cell activity
- Reductions in free radicals and antioxidant damage

For most persons, a day or two without food is safe; however, an assessment of health status is essential before beginning a fast because some health Note

---

EXHIBIT 5-1   QUESTIONS TO REVEAL IMBALANCES ASSOCIATED WITH POOR IMMUNE SYSTEM FUNCTION

- Do you catch colds easily?
- Do you experience frequent or chronic infections (e.g., urinary tract, pneumonia, thrush, shingles, abscesses, ear infections, sore throats)?
- Do you experience frequent allergic reactions?
- Are you frequently fatigued or drained?
- Do you often have fevers?
- Do you lack enthusiasm for social activities and hobbies?
- Have your work responsibilities become more difficult for you to fulfill?
- Are you feeling like you lack the energy or motivation to engage in activities that express your interests, passions, and life calling?
- Is your home disorganized or uncared for?
- Do you lack the energy or interest to engage in activities to nourish your spirit and soul?

EXHIBIT 5-2   IMMUNE-ENHANCING NUTRIENTS

Protein
Vitamins A, E, B_1, B_2, B_6, B_{12}, C
Folic acid
Pantothenic acid
Iron
Magnesium
Manganese
Selenium
Zinc

conditions and medication needs can be affected. Also, it is essential that good fluid intake be maintained during a fast.

Exercise

Any form of exercise, done regularly, can be of benefit to the immune system. Exercise need not be strenuous; low-impact exercise, such as yoga and T'ai Chi, has a positive effect on immunity.

Stress Management

The thymus, spleen, and lymph nodes are involved in the stress response; therefore, stress can affect the function of the immune system. Some stress-related diseases, including arthritis, depression, hypertension, and diabetes mellitus, cause a rise in serum cortisol, a powerful immunosuppressant. Elevated cortisol levels can lead to a breakdown in lymphoid tissue, inhibition of the production of natural killer cells, increases in T suppressor cells, and reductions in the levels of helper T cells and virus-fighting interferon.

Individuals need to identify stress reduction measures with which they are comfortable so that they will practice them on a regular basis. It makes no sense for a person to attempt to engage in meditation if he or she is uncomfortable with that activity because it will be more stress producing than stress reducing. Some stress-reduction measures that could be used are progressive relaxation, meditation, prayer, yoga, imagery, exercise, diversional activity, spending time with nature and pets, and substitution of caffeine and junk foods with juices and nutritious snacks.

> **KEY POINT**
>
> The ability of our psychological state to affect physical health is recognized; in fact, the specialty of psychoneuroimmunology has emerged in recognition of the fact that thoughts and emotions affect the immune system.

Psychological Traits and Predispositions

Studies have identified traits consistent with strong immune systems to include the following (Cohen & Miller, 2001; Friedman, 2007):

- Assertiveness
- Faith in God or a higher power
- Ability to trust and offer unconditional love

- Willingness to be open and confide in others
- Purposeful activity
- Control over one's life
- Acceptance of stress as a challenge rather than a threat
- Altruism
- Development and exercise of multiple facets of personality

Individuals could improve their immune health by developing and nurturing some of these characteristics.

REFLECTION

How many traits consistent with a strong immune system do you possess? What can you do to nurture those traits and to develop additional ones?

Caring for the Immune System by Caring for Self

As people support and nurture themselves, their immune systems will respond by helping them to feel

- *Reoriented.* A renewed sense of belonging and purpose. Perceptions of belonging and connectedness help people feel grounded and secure. This is in contrast to having perceptions of alienation and aloneness, which cause the immune system to stay vigilant and on the defense, which can be an exhausting stance over time.
- *Reorganized.* Being able to discern what is and is not truly needed for one's highest good. This can include letting go of what no longer serves one whether it is old clothes, appliances, or relationships that tear down instead of building up and making revisions on priorities in life. These are ways that people can empower themselves without the stress of trying to take charge over those things that cannot be controlled.
- *Reidentified.* A renewed definition of one's identity and purpose. This can be achieved through developing ways of nourishing your body, mind, and spirit. Meditation, solitude, and prayer are among the practices that can assist with this.
- *Reintegrated.* A renewed belief and confidence in self. As the mind/body is supplied with what it needs for optimal function, there is renewal of hope and zest for life. Reducing the fear and anxiety that people feel about daily life eases the burden on the immune system by reducing its need to protect and defend.

It is the ability to adapt and endure that gives people healing powers to recover from disease and move from darkness into light. Beliefs can lay the foundation for the body to restructure or reform its physical self. How people think and feel connected has an enormous influence on the strength and vitality of immune function. The health and well-being of the mind and spirit can be just as important to the immune response as nutrition or immune-boosting herbs such as *Echinacea*.

A variety of additional measures can assist in enhancing the function of the immune system. Some of these are discussed in the chapters about healthful nutrition, exercise, flowing with the reality of stress, herbal remedies, and the environment.

Summary

The immune system is a lacy network consisting of organs, tissues, and cells. It monitors the body for disease-producing organisms and initiates defenses to eliminate them. It is helpful for people to be concerned about the health of their immune systems and to engage in practices to promote immune health. Physical exercise and diaphragmatic breathing promote the movement of lymphatic fluid throughout the body.

Individuals can take action to enhance their immune function, such as eating specific foods, fasting, exercising, managing stress, and developing psychological traits consistent with strong immunity.

An imbalanced immune system can create a variety of physical signs, such as increased ease and frequency of infection. In addition to physical signs, an imbalanced immune system can affect mood, intellectual activities, spiritual state, and general function.

References

Cohen, S., & Miller, G. E. (2001). Stress, immunity, and susceptibility to upper respiratory infections. In R. Ader, D. Felten, D., & N. Cohen (Eds.), *Psychoneuroimmunology*, (3rd ed., pp. 499–509). New York, NY: Academic Press.

Friedman, H. S. (2007). *The multiple linkages of personality and disease. Brain, behavior and immunity.* Retrieved from www.sciencedirect.com

Suggested Reading

Ader, R. (Ed.). (2006). *Psychoneuroimmunology* (4th ed.). New York, NY: Academic Press.

Alford, L. (2007). Findings of interest from immunology and psychoneuroimmunology. *Manual Therapy, 12*(2), 176–180.

Avitsur, R., Padgett, D. A., & Sheridan, J. F. (2006). Social interactions, stress, and immunity. *Neurologic Clinics, 24*(3), 483–491.

Cohen, N. (2006). Norman Cousins lecture. The uses and abuses of psychoneuroimmunology: A global overview. *Brain, Behavior, and Immunity. 20*(2), 99–112.

Cott, A. (2007). *Fasting: the ultimate diet.* Winter Park, FL: Hastings House.

Daruna, J. (2004). *Introduction to psychoneurology.* Burlington, MA: Elsevier Academic Press.

Fuhrman, J. (2011). *Super immunity: The essential nutrition guide for boosting your body's defenses to live longer, stronger, and disease free.* New York, NY: HarperCollins.

Godbout, J. B., & Johnson, R. W. (2006). Age and neuroinflammation: A lifetime of psychoneuroimmune consequences. *Neurologic Clinics, 24*(3), 521–538.

Gold, S. M., & Irwin, M. R. (2006). Depression and immunity: Inflammation and depressive symptoms in multiple sclerosis. *Neurologic Clinics, 24*(3), 507–519.

Goldsby, R. A., Marcus, D. A., Kindt, T. J., & Kuby, J. (2003). *Immunology.* New York, NY: W.H. Freeman and Company.

Goulart, F.S. (2009). *Super immunity foods: A complete program to boost wellness, speed recovery, and keep your body strong.* New York, NY: McGraw-Hill.

Jason, E., & Ketcham, K. (1999). *Chinese medicine for maximum immunity.* Three Rivers, MI: Three Rivers Press.

Langley, P., Fonseca, J., & Iphofen, R (2006). Psychoneuroimmunology and health from a nursing perspective. *British Journal of Nursing, 15*(20), 1126–1129.

Moldawer, N., & Carr, E. (2000). The promise of recombinant interleukin-2. *American Journal of Nursing, 100*(5), 35–40.

Novack, D. H., Cameron, O., Epel, E., Ader, R., Waldstein, S. R., Levenstein, S., Antoni, M. H., & Wainer, A.R. (2007). Psychosomatic medicine: The scientific foundation of the biopsychosocial model. *Academic Psychiatry, 31*(5), 388–401.

Opp, M. R. (2006). Sleep and psychoneuroimmunology. *Neurologic Clinics, 24*(3), 493–506.

Schniederman, N., Ironman, G., & Seigel, S. G. (2005). Stress and health: Psychological, behavioral, and biological determinants. *Annual Review of Clinical Psychology, 1*, 607–628.

Flowing with the Reality of Stress

OBJECTIVES

This chapter should enable you to

- List the three stages of response to stress that Selye identified
- Define psychoneuroimmunology
- Describe different types of stress
- Outline the response of the sympathetic nervous system to stress
- Describe factors to consider in the self-assessment of stress
- List four common elements of stress-reduction measures
- Describe a progressive muscular relaxation exercise
- List at least three measures that can aid in stress reduction

Stress is an inescapable reality of the average life. On a daily basis, people are exposed to numerous events, issues, and circumstances that challenge them. When faced with stress, some people seem to rise to the occasion and thrive, whereas others experience a myriad of negative physical and psychological effects. Why is this so? The answer, despite much research on the subject, is not clearly understood but is strongly connected to how an individual *manages* stress.

The Concept of Stress

In the 1950s, Dr. Hans Selye, recognized as the father of stress research, laid the foundation for much of the work that has since unfolded in the field of stress (Selye, 1984). His premise was that all organisms have a similar response when confronted with a challenge to their well-being, regardless of whether

> **KEY POINT**
>
> The three stages of response to stress that Selye identified are the *alarm reaction*, the *stage of resistance*, and the *stage of exhaustion*.

that challenge was seen as positive or negative. He called that response the general adaptation syndrome, and identified three stages of it.

The first stage is the alarm reaction, more commonly known as the fight-or-flight response, a physiologic process first described by psychologist Dr. Walter Cannon in the early 1900s. In this stage, the body gears up physically and mentally for battle or energizes to escape the threat. Often referred to as an adrenaline rush, it can be recognized as the pounding heart, dry mouth, cold hands, and knot in the stomach felt when you perceive yourself to be threatened.

In the stage of resistance, the body maintains a state of readiness, but not to the extent of the initial alarm reaction. Selye believed that if the threat is not eliminated and this heightened state of readiness persists, the stage of exhaustion would be reached. At this point, the body, having spent its existing energy reserves, is no longer able to sustain the workload of constant readiness. It is here that it may begin to fail, resulting in the onset of illness and possibly death.

Decades of continuing research into the mechanisms and effects of stress have yielded much information; however, interpretations of that information vary greatly and are sometimes considered controversial. A major development in the area is the field of psychoneuroimmunology—the study of the interaction between psychologic processes and the body's nervous and immune systems (Irwin & Vedhara, 2005). This has brought new definitions of stress that address the mind–body connection, such as one of the early ones offered by Seward that stress is "the inability to cope with a perceived or real (or imagined to be real) threat to one's mental, physical, emotional, or spiritual well-being, which results in a series of physiologic responses and adaptations" (Seward, 2005, p. 5).

Most authors and researchers now agree that there is a difference in the body's response to good stress and bad stress. Good stress (termed *eustress* by Selye) motivates and has pleasant or enjoyable effects, such as that resulting from a job promotion or a surprise birthday party. Although it causes an alarm response, the strength and duration of that response is usually short lived. Conversely, bad stress (termed *distress* by Selye), such as that experienced when involved in a confrontation with a spouse or being involved in a car accident, most often fully initiates the fight-or-flight response and may

> **KEY POINT**
>
> Psychoneuroimmunology is the in-depth study of the interaction of the mind, the central nervous system, and the immune system, and their impact on health and well-being.

also have a prolonged impact on your well-being. This distress is what people usually are speaking of when they use the word *stress*.

Stress can also be viewed as acute or chronic. Acute stress has a sudden onset and is usually very intense but ends relatively quickly. The body quickly recovers, and the symptoms subside. Chronic stress lasts over a prolonged period of time but may not be as severe or intense as the acute type. Chronic stress is believed to be a major culprit in the development of stress-related diseases (Girdano, Everly, & Dusek, 2008).

> **KEY POINT**
>
> An example of acute stress could be losing your wallet containing your paycheck. Initially, when you discover you have lost your wallet, your stress level is very high. After you find your missing wallet under the front seat of your car, the crisis is over.
>
> A prolonged illness of a loved one or lengthy unemployment can cause chronic stress, exhausting considerable coping resources over time.

The Body's Physical Response to Stress

When your brain perceives a threat to your well-being, a series of events made up of chemical reactions and physical responses occur rapidly. The first of these is the activation of your sympathetic nervous system, which stimulates the release of epinephrine from the outer layer of the adrenal gland (medulla) located on top of the kidney and norepinephrine, also from the adrenal glands and from the ends of nerves located throughout the body. When these hormones are released, the fight-or-flight response is triggered. Your heart, blood vessels, and lungs are strongly impacted by these hormones. The force and rate of the heart's contractions increase, as does the rate and depth of breathing. The arteries, vessels carrying oxygen and nutrient-rich blood to your vital organs, widen or dilate to ensure extra blood flow to the heart, lungs, and major muscles. At the same time, the arteries to areas that are not essential (the skin and digestive tract) narrow or constrict. This provides extra blood for the vital organs. One of the other major outcomes of sympathetic nervous system stimulation is a large increase in the production of glucose, your body's primary energy source. The overall net result is an increase in the available amount of glucose and oxygen for the organs and tissues that need it.

Additionally, the pituitary gland, located in the brain, is actively involved in the stress response. The anterior pituitary gland releases a hormone called adrenocorticotrophic hormone. This hormone stimulates the outer layer of the adrenal gland (cortex) to release aldosterone and cortisol. Aldosterone along with vasopressin or antidiuretic hormone, a hormone produced by the posterior pituitary gland, work to preserve blood volume by limiting the amount of salt and water the kidney is allowed to excrete. Cortisol increases the production of glucose and assists in the breakdown of fat and proteins to provide the additional energy needed to protect the body from the perceived threat. The hormones released during the stress response have many effects on the body (**Exhibit 6-1**).

Sources of Stress

The sources of stress in daily life are different for each individual. One person may find a 20-mile drive home through a mountain pass after work tedious and frustrating, whereas another may view it as a source of pleasure and relaxation. Sources of stress can be associated with the physical environment, job, interpersonal relationships, past experiences, work, finances, and psychological makeup. Identifying what stresses them and how they react to that stress is the first step for people to take in developing effective personal stress-management strategies. A variety of tools can help people to identify the stresses in their lives, one of which is offered in **Exhibit 6-2**.

REFLECTION

Take a few minutes to complete the self-assessment in Exhibit 6-2. What are the three major stresses in your life that you have identified?

Stress and Disease

As mentioned, the recognition of the link between the mind and the body is not new; however, the specific mechanism to explain the link between stress and disease is still unclear despite years of scientific research. It is widely believed that the impact of stress—especially chronic stress—on the human body greatly increases the risk of developing a variety of diseases such as

---

EXHIBIT 6-1   EFFECTS OF STRESS

PHYSIOLOGIC

Increased heart rate, grinding of teeth, rise in blood pressure, insomnia, dryness of mouth and throat, anorexia, sweating, fatigue, tightness of chest, slumped posture, headache, pain, tightness in neck and back, nausea, vomiting, urinary frequency, indigestion, missed menstrual cycle, diarrhea, reduced interest in sex, trembling, twitching, and accident proneness

EMOTIONAL

Irritability, tendency to cry easily, depression, nightmares, angry outbursts, suspiciousness, emotional instability, jealousy, poor concentration, decreased social involvement, disinterest in activities, bickering, withdrawal, complaining, criticizing, restlessness, tendency to be easily startled, anxiety, increased smoking, increased use of sarcasm, and use of drugs or alcohol

INTELLECTUAL

Forgetfulness, errors in arithmetic and grammar, poor judgment, preoccupation, poor concentration, inattention to detail, reduced creativity, blocking, less fantasizing, reduced productivity

WORK HABITS

Increased lateness, absenteeism, low morale, depersonalization, avoidance of contact with coworkers, excess breaks, resistance to change, impatience, negative attitude, reluctance to assist others, carelessness, verbal or physical abuse, poor quality and quantity of work, threats to resign, resignation

Source: Eliopoulos, C. (2011). *Nursing administration manual for long-term care facilities* (7th ed.). Health Education Network. Reprinted with permission.

---

asthma, arthritis, cancer, hypertension, heart disease, migraine headaches, strokes, and ulcers. Statistics from a variety of sources state that 50% to 90% of health-related problems are linked to or aggravated by stress (Seward, 2005; Slade, 2007). Nearly every consumer-oriented publication from hospitals, public health departments, health maintenance organizations, and physicians' offices recommends or offers some type of stress-management program.

Most people probably are able to recognize the major physical symptoms of stress in their lives. They also need to be aware of other important

EXHIBIT 6-2   HOLISTIC SELF-ASSESSMENT OF STRESS

LIST MAJOR FORMS OF STRESSES IN YOUR LIFE:	IDENTIFY THE SOURCE OF EACH FORM OF STRESS[a]	LIST ACTIONS THAT CAN BE TAKEN TO ADDRESS EACH SOURCE OF STRESS[b]

[a]Sources of stress can be

    Mental (e.g., feeling bored, overloaded)

    Physical (e.g., disease, injury)

    Emotional (e.g., anger, grief, fear)

    Relational (e.g., altered feelings between self and significant others, violation of trust)

    Spiritual (e.g., values conflict, feeling God is not listening)

[b]Examples of actions that can be taken to address stress

    Obtaining medical attention

    Asking for help with a project

    Being realistic in what can be achieved

    Seeking counseling

    Developing a hobby

    Exercising

    Meditating

    Praying

behavioral, emotional, or mental symptoms that may be stress related, such as compulsive eating, drinking or smoking, restlessness, irritability or aggressiveness, boredom, inability to focus on the task at hand, trouble thinking clearly, memory loss, or inability to make decisions. The self-assessment tool shown in **Exhibit 6-3** contains common physical symptoms often related to stress to help people assess how they are affected by stress in their lives. It is helpful for healthcare professionals to encourage people to engage in a self-assessment as a means to gain insight into the impact of stress in their lives. This is not only important to maintaining a state of wellness, but also as part of living with chronic conditions.

Stress-Reduction Measures

After people recognize the symptoms of stress in their lives, they can minimize the impact of stress on their physical, mental, emotional, and spiritual well-being by using one or a combination of measures designed for stress reduction. The ultimate goal of stress reduction or stress management is the relaxation response, or a state of profound rest and peace. The term *relaxation response* was first used by Dr. Herbert Benson in his book of the same name. He cites four elements that are essential to and found in most stress-reduction measures (Benson, 2000, pp. 110–111):

1. A quiet environment
2. A mental device, such as a word or a phrase that should be repeated over and over again
3. The adoption of a passive attitude (which is perhaps the most important of the elements). A *passive attitude* is one in which a person is open to a free flow of thoughts without analysis or judgment.
4. A comfortable position

KEY POINT

Numerous studies of the results of stress reduction have demonstrated positive findings to include the reduction of blood pressure in individuals with hypertension, improved sleeping patterns in individuals suffering from insomnia, decreased nausea and vomiting in chemotherapy patients, and reduction in the multiple symptoms of women diagnosed with premenstrual syndrome or who are experiencing menopausal symptoms (Kwekkeboom & Gretarsdottir, 2006).

EXHIBIT 6-3   STRESS AND DISEASE: PHYSICAL SYMPTOMS QUESTIONNAIRE

Look over this list of stress-related symptoms and circle how often they have occurred in the past week, how severe they seemed to you, and how long they lasted. Then reflect on the past week's workload and see if you notice any connection.

		How often? (number of days)	How severe? (1 = mild; 5 = severe)	How long? (1 = 1 hour; 5 = all day)
1.	Tension headache	0 1 2 3 4 5 6 7	1 2 3 4 5	1 2 3 4 5
2.	Migraine headache	0 1 2 3 4 5 6 7	1 2 3 4 5	1 2 3 4 5
3.	Muscle tension (neck and/or shoulders)	0 1 2 3 4 5 6 7	1 2 3 4 5	1 2 3 4 5
4.	Muscle tension (lower back)	0 1 2 3 4 5 6 7	1 2 3 4 5	1 2 3 4 5
5.	Joint pain	0 1 2 3 4 5 6 7	1 2 3 4 5	1 2 3 4 5
6.	Cold	0 1 2 3 4 5 6 7	1 2 3 4 5	1 2 3 4 5
7.	Flu	0 1 2 3 4 5 6 7	1 2 3 4 5	1 2 3 4 5
8.	Stomachache	0 1 2 3 4 5 6 7	1 2 3 4 5	1 2 3 4 5
9.	Stomach/abdominal bloating/distention/gas	0 1 2 3 4 5 6 7	1 2 3 4 5	1 2 3 4 5
10.	Diarrhea	0 1 2 3 4 5 6 7	1 2 3 4 5	1 2 3 4 5
11.	Constipation	0 1 2 3 4 5 6 7	1 2 3 4 5	1 2 3 4 5
12.	Ulcer flare-up	0 1 2 3 4 5 6 7	1 2 3 4 5	1 2 3 4 5
13.	Asthma attack	0 1 2 3 4 5 6 7	1 2 3 4 5	1 2 3 4 5
14.	Allergies	0 1 2 3 4 5 6 7	1 2 3 4 5	1 2 3 4 5
15.	Canker/cold sores	0 1 2 3 4 5 6 7	1 2 3 4 5	1 2 3 4 5
16.	Dizzy spells	0 1 2 3 4 5 6 7	1 2 3 4 5	1 2 3 4 5
17.	Heart palpitations (racing heart)	0 1 2 3 4 5 6 7	1 2 3 4 5	1 2 3 4 5
18.	TMJ	0 1 2 3 4 5 6 7	1 2 3 4 5	1 2 3 4 5
19.	Insomnia	0 1 2 3 4 5 6 7	1 2 3 4 5	1 2 3 4 5
20.	Nightmares	0 1 2 3 4 5 6 7	1 2 3 4 5	1 2 3 4 5
21.	Fatigue	0 1 2 3 4 5 6 7	1 2 3 4 5	1 2 3 4 5
22.	Hemorrhoids	0 1 2 3 4 5 6 7	1 2 3 4 5	1 2 3 4 5
23.	Pimples/acne	0 1 2 3 4 5 6 7	1 2 3 4 5	1 2 3 4 5
24.	Cramps	0 1 2 3 4 5 6 7	1 2 3 4 5	1 2 3 4 5
25.	Frequent accidents	0 1 2 3 4 5 6 7	1 2 3 4 5	1 2 3 4 5
26.	Other (please specify)	0 1 2 3 4 5 6 7	1 2 3 4 5	1 2 3 4 5

Score: Look over the entire list. Do you observe any patterns or relationships between your stress levels and your physical health? A value over 30 points may indicate a stress-related health problem. If it seems to you that these symptoms are related to undue stress, they probably are. Although medical treatment is advocated when necessary, the regular use of relaxation techniques may lessen the intensity, frequency, and duration of these episodes.

Source: Seward, B. L. (2011). *Managing stress: Principles and strategies for health and well-being.* Sudbury, MA: Jones and Bartlett Learning. Reprinted with permission.

Individual preferences and circumstances will influence the selection of stress-reduction measures. Some stress-management techniques are simple, whereas others require some initial instruction. Regardless of the choice of method or combination of methods, *all* require practice and need to be used on a regular basis to be effective. Nurses and other healthcare professionals provide a valuable service by instructing, assisting, and coaching people in the use of stress-reduction techniques. Some of the specific measures that can be employed are described in the remainder of this chapter.

Exercise

In the early days of human existence, most threats were physical and demanded an immediate, intense physical response to ensure survival. The response was literally fight or flight. All of the stress hormones released were quickly consumed, and their physical effects were diminished in that burst of activity. In today's environment, the majority of the sources of stress are much less physical and more complex. They usually result from cumulative factors, such as multiple, often simultaneous demands at home and at work.

Physical, emotional, and mental well-being depends on finding a way to dissipate the negative effects of those stressors. Aerobic exercise burns off existing catecholamines and stress hormones by directing them toward their intended metabolic functions, rather than allowing them to linger in the body to undermine the integrity of vital organs. A consistent exercise program has also been demonstrated to help decrease the level of reaction to future stressors. The key to using exercise for stress reduction is to develop an individualized program that is tailored to one's physical abilities, time constraints, and finances. Additionally, to reduce stress, it is beneficial to select an activity that is relaxing and enjoyable rather than competitive.

Progressive Muscular Relaxation

When you are stressed, anxious, angry, or frightened, your body automatically responds by increasing muscle tension. You may have experienced the effects of that response resulting in muscular aches and pains in various parts of your body after an unpleasant encounter or a hectic day. Progressive muscular relaxation (PMR) was developed in the 1930s by Dr. Edmund Jacobson, a physician–researcher at the University of Chicago, as a method to reverse this tension and elicit the relaxation response. Moving sequentially from one major muscle group or area of the body to another, for example, from head to toes or vice versa, muscles will be consciously tensed and then relaxed. This conscious muscular activity interrupts the stress response by interfering with the transmission of stress-related tension via the sympathetic nervous system to the muscle fibers.

Among the benefits of PMR are decreases in the body's oxygen use, metabolic rate, respiratory rate, and blood pressure. These effects can have benefits for people diagnosed with hypertension and chronic obstructive lung disease. Additionally, PMR has been shown to be a useful pain management tool in some patients with cancer and chronic pain (Kwekkeboom & Gretarsdottir, 2006).

PMR is relatively easy to learn. The method cited **Exhibit 6-4** is only one of the variations on the original technique designed by Dr. Jacobson.

EXHIBIT 6-4   PROGRESSIVE MUSCULAR RELAXATION EXERCISE

PMR can be done from a sitting or lying position and usually takes 20–30 minutes to complete. As you take in a deep breath, tighten or tense individual groups of muscles to the count of five. A common sequence to follow is this:

- Forehead
- Eyes
- Jaw
- Neck
- Back
- Shoulders
- Upper arms
- Lower arms
- Hands
- Chest
- Abdomen
- Pelvis/buttocks
- Upper legs
- Lower legs
- Feet

As you focus on each area inhale to the count of five and then exhale slowly, allowing the muscles to relax. Repeat this process twice for each major muscle group or body area.

Meditation

Meditation is one of the oldest known techniques for relaxation, dating back to the 6th century BC. There are as many definitions of meditation as there are types of meditation, but basically, it can be understood to be a practice that quiets and relaxes the mind. During a meditative state, the individual strives to let go of all connections to physical senses and conscious thoughts. All focus and awareness is turned inward, with the ultimate goal of calmness and harmony of mind and body. Studies have shown that anxiety, a major symptom of acute stress, is significantly reduced with the regular use of meditation. As stated earlier, there are many techniques for meditation, some with rituals or rules for that specific technique. In *Minding the Body, Mending the Mind*, Joan Borysenko, PhD, presents a simple, eight-step process for meditation that is easily understood. The process is (Borysenko, 2007):

1. Choose a quiet spot where you will not be disturbed by other people or by the telephone.
2. Sit in a comfortable position.
3. Close your eyes.
4. Relax your muscles sequentially from head to feet.
5. Become aware of your breathing, noticing how the breath goes in and out, without trying to control it in any way.
6. Repeat a focus word silently in time to your breathing.
7. Do not worry about how you are doing.
8. Practice at least once a day for between 10 and 20 minutes.

REFLECTION

Have you built a time for meditation into your daily routine? If not, why? Consider developing a plan to incorporate this practice into your days for the next week. Designate a quiet space. Schedule the time period, and leave yourself a written affirmation (I honor my body, mind, and spirit by taking time to meditate daily). Evaluate your efforts and responses after 1 week.

Imagery

Imagery is a mental representation of an object, place, event, or situation (Post-White, 2010, p. 63). In guided imagery, a person is led with specific words, symbols, and ideas to elicit a positive response. People use images regularly when they describe feelings or concepts in their conversations. For example, when stressed, they might say that they feel tied up in knots, or if they receive recognition from an employer for closing a major deal, they may say they feel like they are on top of the world. Those images can convey powerful messages and can be used as a stress-reduction method.

Other Stress Reducers

Other measures can be used to help reduce stress, such as music therapy, biofeedback, and diversional activities. Other chapters in this book (titled "Herbal Medicine," "Aromatherapy: Common Scents," and "The Therapeutic Benefits of Humor") offer additional insights into measures that assist with stress management. It is beneficial for the nurse or other healthcare professional to help people understand the dynamics of stress, identify individual sources and responses to stress, develop effective stress-management strategies, and reinforce the significance of efforts to manage stress.

Summary

Hans Selye identified the three stages of stress as the alarm reaction, the stage of resistance, and the stage of exhaustion. Some stress is positive (eustress) in that it motivates or has pleasant effects. Distress is the bad stress and has negative effects.

The sympathetic nervous system is activated when a person is stressed, which produces many effects in the body. *Psychoneuroimmunology* is the study of the interaction of the mind, the central nervous system, and the immune system and their impact on health.

Identifying and understanding stressors is essential in changing the way stress is managed. Elements of most stress-reduction exercises include a quiet environment, a mental word that can be repeated, adoption of a passive attitude, and a comfortable position. Exercise, meditation, and imagery are among the measures beneficial in stress reduction.

Not of —
Music,
E.O.'s &
Journals

a couple only

References

Benson, H. (2000). *The relaxation response*. New York, NY: HarperTorch.

Borysenko, J. (2007). *Minding the body, mending the mind*. New York, NY: Bantam Books, pp. 42–46.

Girdano, D., Everly, G. S., & Dusek, D. E. (2008). *Controlling stress and tension*. (8th ed.). Upper Saddle River, NJ: Benjamin Cummings.

Irwin, M., & Vedhara, K. (2005). *Human psychoneuroimmunology*. New York, NY: Oxford University Press.

Kwekkeboom, K. L., & Gretarsdottir, E. (2006). Systematic review of relaxation interventions for pain. *Journal of Nursing Scholarship, 38*(3), 269–277.

Post-White, J. (2010). Imagery. In M. Snyder & R. Lindquist (Eds.), *Complementary/alternative therapies in nursing* (pp. 63–90). New York, NY: Springer Publishing Company.

Selye, H. (1984). *The stress of life*. New York, NY: McGraw-Hill, pp. 29–40.

Seward, B. L. (2005). *Managing stress: Principles and strategies for health and well-being* (5th ed.). Sudbury, MA: Jones and Bartlett.

Slade, T. (2007). The descriptive epidemiology of internalizing and externalizing psychiatric dimensions. *Social Psychiatry and Psychiatric Epidemiology. 42*(7), 54–560.

Suggested Reading

Appel, L. J. (2003). Lifestyle modification as a means to prevent and treat high blood pressure. *Journal of the American Society of Nephrology, 14*(7, Suppl 2), S99–S102.

Banga, K. (2000). Stress management: A step-by-step process. *Nurse Educator, 25*(3), 130, 135.

Bartol, G. M., & Courts, N. F. (2009). The psychophysiology of bodymind healing. In B. M. Dossey & L. Keegan (Eds.), *Holistic nursing: A handbook for practice* (5th ed., pp. 601–614). Sudbury, MA: Jones & Bartlett.

Clark, A. M. (2003). "It's like an explosion in your life": Lay perspectives on stress and myocardial infarction. *Journal of Clinical Nutrition, 12*(4), 544–553.

Forester, A. (2003). Healing broken hearts. *Journal of Psychosocial Nursing & Mental Health Services, 41*(6), 44–49.

Greenberg, J. S. (2006). *Comprehensive stress management*. New York, NY: McGraw-Hill.

Haight, B. K., Barba, B. E., Tesh, A. S., & Courts, N. F. (2002). Thriving: A life span theory. *Journal of Gerontological Nursing, 28*(3), 14–22.

Jefferson, D. K. (2012). *Stress relief today: Causes, effects, and management techniques that can improve your life*. Charleston, SC: Amazon.com CreateSpace.

Lambert V. A., Lambert, C. E., & Yamase, H. (2003). Psychological hardiness, workplace stress and related stress reduction strategies. *Nursing and Health Sciences, 5*(2), 181–184.

Mate, G. (2011). *When the body says no: Exploring the stress–disease connection*. Hoboken, NJ: John Wiley & Sons.

Mimura, C., & Griffiths, P. (2003). The effectiveness of current approaches to workplace stress management in the nursing profession: An evidence-based literature review. *Occupational & Environmental Medicine, 60*(1), 10–15.

Posen, D. (2009). *The little book of stress relief*. Buffalo, NY: Firefly Books.

Richardson, S. (2003). Effects of relaxation and imagery on the sleep of critically ill adults. *Dimensions of Critical Care Nursing, 22*(4), 182–190.

Rofe, R. J. (2012). *Meditation: How to reduce stress, get healthy, and find your happiness in just 15 minutes a day*. Charleston, SC: CreateSpace.

Roizen, M. F., & Oz, M. C. (2011). YOU: Stress less; The owner's manual for regaining balance in your life. New York, NY: Free Press.

Seaward, B. L. (2006). *Stressed is desserts spelled backwards*. New York, NY: Barnes & Noble Books.

Seaward, B. L. (2011). *Managing stress: Principles and strategies for health and well-being* (2nd ed.). Sudbury, MA: Jones & Bartlett Learning.

Shealy, C. N. (2002). *90 days to stress-free living*. Boston, MA: Element Books.

Sloman, R. (2002). Relaxation and imagery for anxiety and depression control in community patients with advanced cancer. *Cancer Nursing, 25*(6), 432–435.

Wheeler, C. M. (2007). *10 simple solutions to stress: How to tame tension and start enjoying your life*. Oakland, CA: New Harbinger Publications.

Yonge, O., Myrick, F., & Hanse, M. (2002). Student nurses' stress in the preceptorship experience. *Nurse Educator, 27*(2), 84–88.

P A R T I I

Developing Healthy Lifestyle Practices

Growing Healthy Relationships

OBJECTIVES

This chapter should enable you to

- Identify the characteristics of a healthy relationship
- Discuss the difference between one's little ego and higher ego
- List four defense mechanisms that people use to protect themselves
- Describe what is meant by a body memory
- List at least five personal characteristics that aid in developing healthy relationships

Throughout life, we encounter and form innumerable relationships. Each relationship, no matter how loving or distasteful, how short or long term it may be, contains within it a storehouse of information about our own and others' unique personalities. Should we choose to view it with a new openness of mind, each relationship also offers a rich opportunity to reexamine our innermost selves and to reform, if necessary, the ways in which we perceive and behave in our relationships with other people and in given situations.

But how do we determine healthy versus unhealthy relationships—constructive versus destructive ones? How do relationships become so out of control and out of balance, perhaps to the point of ultimately becoming toxic and disease producing to a person's core being? Precisely what can be done to change these uncomfortable or intolerable relationships in which people tend to find themselves repeatedly?

Identifying Healthy Versus Unhealthy Relationships

To begin to answer these questions, we first need to know how to identify a healthy relationship.

> ### KEY POINT
>
> A healthy relationship is one in which there is ongoing mutual trust, respect, caring, honesty, sharing, and acceptance and that supports physical, mental, emotional, and spiritual growth for all persons involved.

A *healthy relationship* can be defined as "a healthy sense of connection in which two or more persons agree to share hurts, failures, learning, (and) successes in a nonjudgmental fashion (in order) to enhance each other's life potentials" (Hover-Kramer, 2004, p. 670). A healthy relationship requires a balance of dependence and healthy independence (see **Table 7-1**).

In order to form healthier relationships, it is necessary for people to learn about

- Their own unique personalities and idiosyncrasies
- The defensive coping mechanisms they use to react or respond to unconscious or conscious emotions, such as anxiety, anger, fear, conflict, loneliness, envy, and jealousy

This learning may best take place in the context of a professional yet trusting relationship with a mental health professional and involve the gradual conscious revealing of previously unrecognized feelings. This process or journey, although at times painful and difficult, can lead to new insights and ways of feeling and being that, perceived through a positive lens, can fill one's life with an expanded consciousness, creating richer meaning and healthier patterns or ways of living and relating to others.

> ### KEY POINT
>
> To fully understand relationships, you must also explore those you have with yourself, expressed through your work, your hobbies, behaviors (such as eating, smoking, and alcohol consumption), and your response to bodily illnesses and pain.

Your Relationship to Self

The most important relationship that you will ever form is the one you create with your true self. This is also the only relationship over which you will ever have total control. Because the way you relate to your own self significantly impacts how you choose to relate to others, it is imperative, before delving

TABLE 7-1   TASKS IN HUMAN DEVELOPMENT AND FORMATION OF HEALTHY AND UNHEALTHY BOUNDARIES

APPROXIMATE AGE OCCURS	TASKS	HEALTHY BOUNDARIES AND RELATIONSHIPS	UNHEALTHY BOUNDARIES AND RELATIONSHIPS
Birth to 1 year	Be connected	Infant believes he/she is part of and extension of parents	Same as healthy
1 year	Trust	Helps release food nutrients and energy; appetite control, helps nervous system, and digestive tract	Headache, rapid pulse, irritability, trembling, insomnia, interference with B_2, B_6
	Cracks at corners of mouth, sensitivity to light, eye problems, inflamed mouth	Helps enzymes in releasing energy from cells, promotes growth, cell oxidation	No known toxic effect Some antibiotics can interfere with B_2 being absorbed
Feel	Mother–infant symbiosis helps organize perceptions and feelings through healthy giving and receiving	Narcissistic or otherwise distracted parents mistreat and mold infant to be an extension of their wants and needs. Parents also may neglect children creating an insecure environment	Flushing, stomach pain, nausea, eye damage, can lead to heart and liver damage
Love	Separate	Necessary protein metabolism, nervous system functions, formation of red blood cells, immune system function	Reversible nerve injury, difficulty walking, numbness, impaired senses
2 years	Rarely seen since it can be made in body if not consumed. Flaky skin, loss of appetite, nausea.	Cofactor with enzymes for metabolism of macronutrients, formation of fatty acids, helps other B vitamins be utilized	Symptoms similar to vitamin B_1 overdose
Initiate	Anemia, diarrhea, digestive upset, bleeding gums,	red blood cell formation, healthy pregnancy, metabolism of proteins	Excessive intake can mask B_{12} deficiency and interfere with zinc absorption
Explore	Begins to recognize that he or she is separate from parents; begins to explore world	Parents disallow exploration of world; parents set boundaries that are too rigid or too loose	
Think			
3 years	Cooperate	Models behaviors and thinking after parents and others who are close	Lack of positive models

(continues)

TABLE 7-1   TASKS IN HUMAN DEVELOPMENT AND FORMATION OF HEALTHY AND UNHEALTHY BOUNDARIES (CONTINUED)

4 years	Master	Continues learning how he or she is similar to and different from others	Distortion of sameness into co-dependence and differentness into low self-esteem
6 years	Evaluate Create Develop morals, skills, values	Continues exploring with growing sense of self	Parents and others stifle healthy exploration and self-esteem
13 years	Evolve Grow	Begins to separate self from parents and family; struggles for further self-identity	Unhealthy separation; distorted boundaries and sense of self
19 years	Develop intimacy	Explores and engages in intimate relationships	Dysfunctional attempts at relationships
Mid- and late-life	Create Extend love Self-actualize Transcend ego	Continued search for God and self; recycles many earlier issues	Spiritual distress; unfulfilled search for God and self

into external or other relationships, to take a look at the meaning of *self* and the inner relationships you have formed with aspects of your many selves rooted in your earliest experiences, learning, perceptions, and understanding.

You might think of yourself as consisting of two major selves: your lower, false self or little ego, and your authentic, divine self or higher ego. The little ego is the part of you that is dominated and controlled by the emotion of fear. It is the part that creates such feelings as not good enough, not deserving enough, not smart enough; all of those not enoughs prevent you from moving forth to pursue and manifest your dreams. In contrast, your higher ego is defined and directed by the emotion of unconditional love (Bradshaw, 2005). With your higher ego as your guiding force, you find that you genuinely love yourself and others. This kind of love allows you to believe in yourself and have the confidence that whatever you seek to do or be in life can be envisioned and created in abundance.

REFLECTION
Are there any not enoughs that limit you? How do they affect your life?

Love and fear are basic emotions from which all other emotions evolve, and it is difficult for these two emotions to coexist within one simultaneously. Unfortunately, many people operate from the perspective of their little ego. They choose to view possibilities as impossibilities. Relationships often have an underlying theme of "sure, they like me now, but when they find out who I *really* am, they won't want anything to do with me." Their little ego is deathly afraid of being found out, and thus, they may go about presenting a false face and, consequently, a false self to the world. This classically represents the wisdom of the words, "as within, so without." In the fear of confronting and coming to know the true self, they hide it both in their relationship with their inner selves and in their outer relationships with others.

The Search for the Authentic Self

We begin to develop our relationships as infants and young children as we watch and interact with significant people in our lives—parents, siblings, grandparents, extended family, friends, and authority figures—and to our environment in general. As we grow, we listen, observe, and absorb what is being said and acted out around us. We see how the authority figures in our lives respond to different situations and people, and with them as our role models, we tend to embrace the similar kinds of patterns in our own lives. If there is a lot of fear and anxiety within the family unit, we may tend

> **KEY POINT**
>
> Carl Jung, one of the most renowned and respected psychiatrists of our time, contended that each of our personalities is structurally made up of four archetypes or models and that these archetypes significantly influence our interpersonal behaviors and (therefore) our relationships. His four models include our:
>
> - *Persona*: The social mask or face that we reveal to the public
> - *Shadow*: The parts of ourselves that we disown or deny
> - *Anima-animus*: The part of us that contains both our male and female characteristics
> - *Self*: The part of the personality Jung considered the most significant because it embodies each person's longing for unity and wholeness
>
> Jung also believed that the goal of personality development is self-realization, as can be interpreted as a remembering, realizing, and reclaiming of our true, authentic self. In the process of seeking and finding our selfhood, Jung proclaimed that we as human beings are transformed from biological creatures into spiritual individuals (Jung, 2006).

to react with that same kind of fear. If there is trust and contentment, our responses, rather than spontaneous defensive reactions, tend to be on a more trusting, calm, and balanced plane. In most families, there will be a mixture of these types of feelings, behaviors, reactions, and responses. Depending on their individual makeups, people perceive, think, and feel about situations and relationships in their own unique ways.

Numerous theories are available as to why individual differences exist. One, not surprisingly, has to do with genetic makeup. One child may take after one side of a family or individual in that family more than the other side or other individuals. Another contributing factor is the manner in which one chooses to perceive given events, situations, or people. What one person chooses to take personally and get extremely upset or fearful about, another may take in stride with an attitude of calmness and trust and a certain *knowing* that all will work out as it needs to. If the stimulus in a given

> **KEY POINT**
>
> Children from the same households with the same parents and circumstances and with very little, if any, notable differences in upbringing often can be extremely different personality-wise.

situation is overwhelmingly negative, people may develop misperceptions about the world as a whole. They may categorize *all* people as good or bad and *all* situations as either black or white with no gray in between. When young, people are essentially powerless over the circumstances of where and with whom they live. If people are abused or perceive themselves as being wounded in some way, it is natural to look for ways to deny, escape, manipulate, react, or do whatever it takes to ultimately survive in settings that may otherwise be intolerable or literally destructive to their very bodies, minds, souls, and spirits.

Through ages 7 or 8, children are much like sponges, soaking up whatever is taking place within the world. They have not yet formed any type of protective psychological barrier or screen through which they can filter or sort out the tremendous amount of information and stimuli that is being received. They are left to deal with overwhelmingly negative people, circumstances, and stimuli, with both the good and the bad in life, in the best ways they can manage.

> **KEY POINT**
>
> When adults fail to heal from the wounds of their childhood, there is a strong probability that their wounds will impact their children.

Few people enter adulthood with a totally positive self-esteem. Within some people, there dwells a significant amount of unhealed toxic shame. This shame is the result of low self-esteem born out of painful childhood situations in which people had no ability or permission to choose and during which times their innermost selves felt vulnerable and inappropriately exposed (Bradshaw, 2005). The psyche was stripped naked. Over and over they may have been told negative things about themselves and even called names that were demeaning. This type of relating from role models not only cultivated fear and distrust, but because of the childhood inability to see themselves apart from their role models, it also created fear and distrust within themselves. An individual can grow to hate aspects of himself or herself that were identified as the bad, negative, or ugly child. Throughout their lives, until they begin to recognize and heal those self-misperceptions, they are doomed to repeat the same words, feel the same feelings, and, in essence, repeat the same comments and situations over in their minds that prove to them again and again that they are not good or competent human beings and that no matter what they do it is never enough.

KEY POINT

Some of the more common defense mechanisms used by people in efforts to protect themselves include denial, displacement, rationalization, and regression.

Denial allows people to refuse to see anything that may cause pain or that they don't want or know how to deal with.

Displacement causes people to place blame on other people or situations rather than taking on personal responsibility for whatever is happening in their lives.

Rationalization is simply mental justification for the inappropriate actions, feelings, and thoughts people have.

Regression is a reverting back to more primitive or childish behaviors, such as whining, pouting, yelling, cursing, and even physical hitting or beatings. It occurs when one is feeling out of control and is connected to feelings of loss of control or power.

These unhealthy coping mechanisms, along with a score of others, are fear based and create barriers between people that block the ability to form loving, intimate relationships.

Even if people did not grow up in the worst of circumstances, they face problems and difficulties that require the development of some type of coping strategies. It must be emphasized that in given circumstances, whether what has happened in the past or what is happening in the present is perceived rightly or wrongly, people develop their own unique coping mechanisms that actually work or at least work well at certain times and in given situations.

In the physical realm, people *appear* to be totally separate from others, and given that perception, they naturally form certain defense and coping patterns that serve to act as barriers to anyone or anything that may seem to be a threat to their well-being; however, from the holistic perspective, there is no such thing as true separation. The body is an energy field that interconnects energetically to others in the vast universe in which we live.

Intimacy Versus Isolation, Love Versus Fear

The coping mechanisms that people develop, as noted previously, may have helped them to survive the otherwise intolerable times in their lives. It is only later, when they are older and free from a toxic family and/or environment, that they may undergo an ah-ha experience. At this time, they realize that

> ### KEY POINT
>
> Old patterns of thinking and behaving can become so entrenched, so much a part of who a person is, that even the suggestion of change can feel too overwhelming, frightening, and threatening.

these particular ways of relating or these old patterns of behavior are simply not working for them any longer. They may feel a desperate need to change, to relate differently, but not know how.

The process of change can feel as though some sort of death is occurring. In a very real sense, this is true. Even though people are not really dying—they actually are healing and transforming old parts of themselves—but the very act of moving through that incredible process can feel extremely scary. Any time people go from what is familiar and "fits like an old shoe" and metaphorically step into a new pair of shoes representing a new, unknown, and unfamiliar world, their natural human instinct is to resist. Intellectually, they may realize that the changes they are choosing to make are all in their own best interest. Still, their emotional tendency is to hold tight to what has felt right for such a long time.

It is at times such as these that the soul is searching to find what has been missing from life all along. People begin to develop an awareness that there are others who appear to have something missing from them. That something, although it may not be easily identified or defined right away, is the ability to love and form intimate relationships without the anxiety and fear of rejection or abandonment.

REFLECTION

Intimacy is the ability to form close or intense (as in marriage) relationships in which you share your life easily and openly with others who have also developed this capacity. With whom do you share intimate relationships? Are they satisfying? Nurturing?

Intimate relationships do not have to be limited to other humans. People can form them with pets, certain causes that they support, or their creative efforts. When we think about marriage or any relationship akin to a lifelong partnership between two people, intimacy grows out of the capacity these two people have for mutual love and their ability to pledge a total commitment to each other. This intimacy goes far beyond any sexual relationship they may have. Rather than a person, this type of commitment could be formed with a particular career, cause, or endeavor that one chooses to devote his or her life.

No matter with whom or what one decides to become intimately involved, personal sacrifices will be made in the giving of self. The ability for this kind of intimacy is learned when, as children, people have been the recipients of unconditional love, nurturing, and sharing within their family unit.

The inability to form intimate relationships results in withdrawal, social isolation, and loneliness. People may seek what they think is intimacy through numerous superficial friendships or sexual contacts. No career is established; instead, they may have a history of frequent job changes, or they may so fear change that they remain most or all of their adult lives in undesirable job situations.

Body Memories as Blocked Manifestations of Relating

In the holistic framework, your body is not confined to that which is within your skin. Surrounding you are mental, emotional, and spiritual subtle bodies that are invisible to the average human eye. These bodies are sometimes referred to as auras and are reported to be seen by certain people who are sensitive to that particular energy. When all of these bodies are in harmony, both with your internal and external worlds, there is a constant circulating free flow of energy; however, when that balance is disturbed and the energy is blocked on one or more levels, you experience imbalance and disharmony. Left unattended and untreated, the imbalance eventually leads to discomfort and/or disease on all levels. These energetic holding patterns, or body memories (events stored within the physical body at the cellular level), can be and often are stuck emotions within your physical, mental, emotional, and/or spiritual bodies. These memories may be of a pleasant or unpleasant quality, or of a conscious or unconscious nature. Illness can result from disharmony among these various aspects of the self, reinforcing the importance of addressing body, mind, emotions, and spirit to affect healing.

Every relationship encountered, be it with a friend or perceived foe, brings back a part of self that was missing. When that part resonates within, it is transformative, drawing a person closer to wholeness. Healing seldom, if ever, occurs in a vacuum. Healing does not involve or affect just one person. Rather, people are continually, whether knowingly or unknowingly, consciously or unconsciously, participating together as copartners in healing—in their relationships with others and their relationship with the planet and a higher power.

The Role of Forgiveness in Relationships

He drew a circle that shut me out—Heretic, rebel, a thing to flout.
But love and I had a way to win: We drew a circle that took him in.
—EDWIN MARKHAM

Perhaps more than any other quality in your life, the ability to forgive is the key to inner peace. Mentally, spiritually, and emotionally, it transforms fear into love. Many times the perceptions of other people and situations become a battleground between the ego or the lesser self's desire to judge and find fault, whereas the higher authentic self desires to accept people as they are. The lesser ego is a relentless fault finder in both self and others; however, the places in which a person estranges from love are not faults, but wounds. The authentic self seeks out its innocence and never seeks to punish, but rather to heal self and others.

Why, a person might ask, would I not judge or find fault with something I know to be wrong? The emphasis here needs to be placed on the behavior or the action as opposed to the person. When one perceives the action and the person as the same, it is impossible to see the person as innocent and his or her actions as unhealed wounds.

KEY POINT

Forgiveness is selective remembering—a conscious decision to focus on love and let the rest go (Williamson, 2005). It is easy to forgive those who have never done anything to make you seriously upset or angry; however, it is the people who trigger you, who push your buttons, who create fear and self-doubt, who are your best teachers. They measure your capacity to love unconditionally, and it is this very love that brings healing to all the lives involved, including your own.

Personal Characteristics in the Development of Healthy Relationships

Dorothea Hover-Kramer has identified eight major personal characteristics that can be tremendously helpful in developing healthy relationships. The following is a modification of her list (2004, p. 670):

1. Being willing to take a genuine look at one's own faulty personal defenses and blind spots and begin the process of identifying and letting go of these defense patterns
2. Creating a correct sense of self-worth, confidence, and self-esteem; having no grandiosity, yet not putting oneself down
3. Developing flexibility; looking at people and situations from different perspectives; a willingness to walk in another's shoes
4. Developing a willingness to take personal responsibility for all of one's feelings or actions
5. Setting conscious awareness of one's spiritual essence and wholeness experienced as a sacred space of inner calm and setting boundaries that allow for a clear sense of purpose, goal orientation, and direction
6. Making the effort to be understood, persevering to find common ground; seeking and integrating (genuinely taking in) feedback
7. Developing sincere empathy and mutual respect for others without appeasing, complying, or attempting to be overly pleasing
8. Committing oneself to a willingness to revisit, rethink, and redefine previous decisions; accepting the possibility of being wrong and allowing others the space to acknowledge their mistakes

These are characteristics that can be easily identified in effective communicators and negotiators. They bring integrity and balance to the art and skill of successful communication, which is necessary to the building, maintenance, and enhancement of healthy relationships.

The suggestions for having, maintaining, and growing healthy relationships that have been discussed in this chapter correlate to many of the principles that Deepak Chopra (2007) has promoted in his writing and teaching. Chopra strongly emphasizes that success is a journey, not a destination, and you should never expect to arrive, and that you will continue throughout your life to learn and hopefully grow from your experiences. He suggests that a basic need that people have to build healthy relationships is for daily *stillness*—meditation or prayer time during which they can go within and listen to the silence from which comes the wisdom of their spiritual and innermost beings. It is through this silence that they ultimately get in touch with their ability to heal old faulty ways of thinking and behaving, to manifest and create their bliss, and it is also through this practice that, over time, they come to realize that low self-esteem (relationship to self) combined with a large dose of fear and negative thinking are the only barriers between themselves and what they desire most in life.

Giving is another crucial ingredient for successful relationships. What is given does not have to be extraordinary, but it does have to be genuine. It may take time, energy, and monies that could be spent elsewhere and experiencing life as more pleasant and less demanding in that moment. This type of giving does not demand nor expect anything in return and is given in the spirit of unconditional love.

Receiving can be the hardest for many people. If people grew up never feeling worthy or deserving, then accepting anything given freely to them by others can be most difficult. In order to grow, they need to have the ability to receive. Chopra notes that it is helpful to affirm oneself with such thoughts as:

> Today I will gratefully receive all the gifts that life has to offer me, including the gifts of sunlight, birds singing, spring showers, or the first snow of winter. I will be open to receiving from others and will make a commitment to keep (positive relationships) circulating in my life by giving and receiving life's most precious gifts: the gifts of caring, affection, appreciation, and love. (Chopra, 2007, p. 36)

Next comes *acute awareness* of the choices made in each moment and how, in the mere witnessing of these choices, they are brought into conscious realization. One begins to understand that the best way to prepare for any future moment is to be fully conscious in the present.

Acceptance of people, situations, circumstances, and events as they occur is another key part of fully entering into a state of peace with self, others, and the world. One can know that the moment is as it should be and consciously choose not to struggle against what cannot be changed.

KEY POINT

When people accept, they *take responsibility* for problems, choosing not to blame anyone or anything (including themselves) for what is happening. Rather, they recognize all events and situations as opportunities in disguise . . . opportunities that they can take and transform into a greater benefit for all involved.

Defenselessness is another important component, according to Chopra (2007). People must let go of the need to defend their point of view. The need to persuade or convince others to accept a point of view is a use of energy that could be much better spent elsewhere. Remaining open to all points of view and not rigidly attaching to any one of them allow for a plethora of choices and opportunities that people may have been blind to within their defensiveness.

Detachment is the last ingredient. Detachment, he says, allows an individual and others the freedom to be who they are and to travel their life paths as they must while remaining unattached to the outcome. This detachment does not mean that people do not care. In contrast, they deeply care and are concerned about those with whom they relate, and they are there for others in loving, caring ways. Once again, however, they must not interfere or try to do others' processes for them. People must learn in their own way and at their own pace what feels right for them. They may have later come to the conclusion that a person who gave them advice was right all along, but what is significant is that they chose to act on it in their own time and at their own pace.

REFLECTION

Do you tend to force your opinions and advice on others when you are confident that you know what is in their best interest? If you think for just a moment about how resentful and angry you felt when others have attempted to take over or force your choices in a particular direction, it does not take long to see how your attempts to do the same with others can so easily backfire.

As people begin to understand and absorb the principles Chopra outlines, they can gradually unblock painful and disease-producing energy that has been kept in and, in doing so, allow that energy to become more balanced and free flowing. This action frees them to discover their unique talents and to create and manifest long-held dreams. At the same time, they also are freed up to serve and respond to others with an unconditional and unbounded love.

In this global, multicultural world, people are continually encountering a wide variety of people from different cultures, beliefs, backgrounds, skills, and education. Access to advanced technology through the Internet and other avenues has given people the means to bring different parts of the world right into their own homes. In doing so, they are forced to a greater extent than ever before to deal with other people, cultures, ideas, information, and beliefs that may be profoundly different from their own; however, if individuals are willing to crack the door to the possibility of experiencing a new consciousness, they can begin the step forward toward the creation of richer, healthier relationships and lives filled with an abundance of experiences, learning, and growth that previously would have been unimaginable.

Summary

A healthy relationship is one in which there is ongoing mutual trust, respect, caring, honesty, and sharing that is exchanged in an environment of nonjudgmental and unconditional love. Learning about oneself fosters the building of healthy relationships.

Healthy relationships are not built on neediness, but on love and caring. Major personal characteristics are helpful in developing healthy relationships, such as a willingness to look at oneself honestly, creating a realistic sense of self-worth, having flexibility, being aware of spiritual essence, making an effort to be understood, having empathy and respect for others, and acknowledging one's own mistakes.

Deepak Chopra describes the requirements for successful relationships, which include the need for stillness, giving, receiving, awareness of choices, acceptance of people and circumstances, defenselessness, and detachment. With an understanding of these principles, people can gradually unblock painful and disease-producing energy that has been long held within and be freed to serve and respond to others with an unconditional and unbounded love.

References

Bradshaw, J. (2005). *Healing the shame that binds you*. Deerfield Beach, FL: Health Communications, Inc.

Chopra, D. (2007). *The seven spiritual laws of success: A practical guide to the fulfillment of your dreams*. San Rafael, CA: Amber-Allen Publishing and New World Library.

Hover-Kramer, D. (2004). Relationships. In B. M. Dossey, L., Keegan, & C. E. Guzzetta (Eds.), *Holistic nursing: A handbook for practice* (4th ed., p. 670). Sudbury, MA: Jones and Bartlett.

Jung, C. G. (2006). *The undiscovered self*. New York, NY: New American Press.

Williamson, M. (2005). *A return to love: Reflections on the principles of a course in miracles*. New York, NY: HarperCollins.

Suggested Reading

Autry, J. A. (2008). *The spirit of retirement. Creating a life of meaning and personal growth*. New York, NY: Prima Press.

Barnum, B. S. (2010). *Spirituality in nursing* (3rd ed.). New York, NY: Springer Publishing Company.

Bender, M., Bauchham, P., & Norris, A. (1999). *The therapeutic purpose of reminiscence*. Thousand Oaks, CA: Sage.

Borysenko, J. (1998). *A woman's book of life: The biology, psychology, and spirituality of the feminine life cycle*. New York, NY: Riverhead Books.

Carson, V. B., & Arnold, E. N. (Eds.). (2000). *Mental health nursing: The nurse–patient journey* (2nd ed.). Philadelphia: W. B. Saunders & Co.

Carter-Scott, C. (2000). *If life is a game, these are the rules*. New York, NY: Broadway Books.

Cox, A. M., & Albert, D. H. (2003). *The healing heart: Communities*. Gabriola Island, Canada: New Society Publishers.

Felton, B. S., & Hall, J. M. (2001). Conceptualizing resilience in women older than 85: Overcoming adversity from illness or loss. *Journal of Gerontological Nursing, 27*(11), 46–53.

Gibson, F. (2011). *Reminiscence and life story work: A practice guide.* London, UK: Jessica Kingsley Publishers.

Hertz, J. E., & Anschutz, C. A. (2002). Relationships among perceived enactment of autonomy, self-care, and holistic health in community dwelling older adults. *Journal of Holistic Nursing, 20*(2), 166–186.

Hillman, J. (2000). *The force of character and the lasting life.* New York, NY: Random House.

Lemme, B. H. (2005). *Development in adulthood* (4th ed.). Boston, MA: Allyn & Bacon.

McLemore, C., & Parrott, L. (2008). *Toxic relationships and how to change them: Health and holiness in everyday life.* San Francisco, CA: Jossey Bass.

Moore, S. L., Metcalf, B., & Schow, E. (2000). Aging and meaning in life: Examining the concept. *Geriatric Nursing, 21*(1), 27–29.

Puentes, W. J. (2000). Using social reminiscence to teach therapeutic communication skills. *Geriatric Nursing, 21*(3), 318–320.

Quadagno, J. S. (2007). *Aging and the life course: An introduction to social gerontology.* Boston, MA: McGraw-Hill College.

Richo, D. (2008). *When the past is present: Healing the emotional wounds that sabotage our relationships.* Boston, MA: Shambhala Publications.

Shaffer, G. (2010). *Taking out your emotional trash: Face your feelings and build healthy relationship.* Eugene, OR: Harvest House.

Semmelroth, C. (2002). The anger habit workbook. Proven principles to calm the stormy mind. Carlsbad, CA: Writers Club Press.

Snowden, D. (2002). *Aging with grace: What the nun study teaches us about leading longer, healthier, and more meaningful lives.* New York, NY: Random House.

Thomas, E. L., & Eisenhandler, S. A. (1999). *Religion, belief, and spirituality in late life.* New York, NY: Springer.

Wallace, S. (2000). Rx RN: A spiritual approach to aging. *Alternative and Complementary Therapies, 6*(1), 47–48.

Wilt, D. L., & Smucker, C. J. (2001). *Nursing the spirit.* Washington, DC: American Nurses Publishing.

Wolf, T. P. (2003). Building a caring client relationship and creating a quilt: A parallel and metaphorical process. *Journal of Holistic Nursing, 21*(1), 81–87.

Survival Skills for Families

OBJECTIVES

This chapter should enable you to

- List at least five types of families
- Describe at least three assumptions about families
- Discuss the body, mind, and spirit of a family
- Outline at least six questions that can be used to explore the balance of work and family needs
- Describe seven major components of a healthy options assessment
- List the six steps of goal setting
- Describe what is meant by holistic parenting

Family is a word that represents something very personal, yet common, to each of us. There was a time when the word family conjured up an image of a husband, wife, and dependent children; however, recent decades have revised this image as a result of relaxed sexual mores, increased acceptance of alternate lifestyles, the women's movement, and other factors. Families take a variety of forms, including the following:

- Blended families
- Single-parent families
- Families with one child
- Families with many children
- Same-gender partner families
- Grandparents raising grandchildren
- Young-parent families
- Older parent families
- Couples without children

> **KEY POINT**
>
> Each family
>
> - Is unique in its beliefs, identity, composition, gifts, and challenges.
> - Influences and is influenced by its community and its culture.
> - Shares many things in common with other families.
> - Deserves respect and opportunities to be successful.
> - Has the responsibility to respect and care for each of its members while accepting each person as an individual and encouraging their personal health and growth.

These families can be youthful or seasoned, multigenerational, legally married, or partnered without the benefit of marriage. With these changes have come less clear rules and boundaries in the way families behave.

REFLECTION

What meaning does family hold for you? Examining your personal beliefs about the meaning of family and the cultural and social influences that surround family is one step toward developing successful survival skills as a family.

Just as there is a cultural evolution redefining the description of family, there is a personal one as well. Each family changes with time. Even with the same people and the same composition, individual family members age, develop new interests, gain new friends, move to other communities, or change jobs. Each change that happens to one person within a family has an effect on the whole. Accepting and acting on this principle of holism is a step toward developing successful survival skills for families.

Family Identity

What's in a name? That which we call a rose,
By any other name would smell as sweet.
—WILLIAM SHAKESPEARE, *ROMEO AND JULIET*

When was the last time you looked at your birth certificate? Among the spaces that include the date, time, and place of your birth are the spaces for your mother's maiden name and your father's name. This record of birth, completed or not, forms part of your identity. You are far more than a piece of paper, yet this document follows you through life and represents your heritage.

What meaning do those names hold for you? Were you named after someone, living or deceased? Do you have a middle name? Are you called by the name inscribed there, or over the years have you gained and lost nicknames? Did any of those nicknames stick and become the name to which you now respond?

The names on a birth certificate may give a clear picture of the origins of a family going back to the "old country." Perhaps names have been altered to fit the late 19th-century view of what it meant to be an American. Somewhere in the dusty stacks of a family's history are the storytellers who know. They carry the tales and the details of where a family originated and how long a family has been in this country. There are the tales of courage of those who traveled, learned a new language, received an education, and raised a large, successful family. The challenges faced by ancestors are amazing; by preserving and recalling them, younger generations gain an appreciation for their roots and connection to a larger story.

Even when the stories are shadows or memories created from a need to know more about family background, people gain an appreciation for the courage and life of their ancestors who brought them here and gave them life. Those names and stories, whether clear or hazy, contribute to their identity and that of their family.

REFLECTION

Sunday Dinner: A Guided Reflection

How far back in your childhood can you remember? Were you 2, 3, or 4 years old? If someone asked you to talk about Sunday dinner when you were 10, how would you respond? You may have clear memories of special foods, aromas, the time of day, your place at the table, who was present, and who was not. In your family, who prepared the meal, set the table, cleared the dishes, and cleaned up? Compare that with your Sunday dinner last week. Answer the same questions. Are there similarities between Sunday dinner then and now? What were the differences? Ask each adult in your family to reflect on those similarities and the differences, then and now.

What does identity have to do with balancing family, work, and leisure? Balance—be it on ice skates, a bicycle, or juggling the many needs and desires of a family—happens when people know where they are in space; they have focus and understand their purpose. They know who they are, where they have come from, and where they are headed.

Family identity is based on many things and may change over time as individuals mature, children leave the nest, and life presents new opportunities,

> **KEY POINT**
>
> One interesting dynamic about families is that each represents a merging of the identity, needs, and desires of two or more people. People may be different as individuals, but when joined together as a family, their success depends in part on their ability to develop a unique definition of family that all members can accept and value. Finding and maintaining a balance between the needs of the family and those of each individual member contribute to a family's survival skills.

challenges, and surprises. This identity is reflected in where families live, what foods they prefer, religious or spiritual beliefs, political values, lifestyle preferences, educational interests, and vocational choices. It can include their rest/activity patterns, favored forms of entertainment and recreation, and the nature of their personal relationships.

Articulating family identity requires people to sit in peace and equality, each as individuals contributing to the family unit. Each person brings together what he or she desires personally and what each is able to contribute to the whole.

Exhibit 8-1 offers a family identity exercise. It could be beneficial for you to complete this yourself before using it with others. As you complete this exercise, focus on your family and what is needed for its health and success as a unit while acknowledging each person's need for individuality, respect, and opportunities for growth. Maintaining such a focus contributes to the holistic awareness of your family's unique composition. As you proceed with this exercise, think in terms of body, mind, and spirit.

The body of the family describes the physical connection—how they are joined genetically, legally, and emotionally. Body acknowledges how things get done within the system—the responsibilities of child care, transportation, home maintenance, meal preparation, financial management, and the myriad of daily things a family has to do.

Mind includes the common belief systems and how the family thinks, reviews the past, plans for the future, and adjusts to life through learning. Mind touches on the many realms of human life; it influences communication and relationships and the ability to adapt to change.

Spirit acknowledges the awareness that there is something beyond the here and now. Some families manifest their spirituality by adopting a specific lifestyle; others do so through their religions or by contributing to their communities. Still, others define themselves as perpetual seekers, seeking an intense something that links them to others beyond the family.

EXHIBIT 8-1   FAMILY IDENTITY EXERCISE

The following categories of body, mind, and spirit include topics to think about and discuss in relation to your family's identity. These lists can be lengthened or shortened. The purpose is to stimulate conversation about who you are and who you want to be as a family. Try reviewing them as a family unit.

BODY

Type of area where you live (country, city, suburbs; region of the nation):

Your dwelling place is a(n) (apartment, condominium, house, ranch):

Your food preferences are:

Your physical recreation is:

Your overall health is:

You receive your health care from:

You describe your financial situation as being:

MIND

Your educational achievements include:

Your educational interests and goals for your children (if applicable) include:

Your reading interests, individually, and as a family are:

The forms of arts and entertainment you enjoy are:

You use these types of media and technology:

Your political preference is:

Voting habits of adults in the family are:

Your favorite topics of conversation are:

Favorite activities that nurture your creativity include:

You participate in the following groups and community activities:

SPIRIT

Traditions, rituals, and spiritual practices that are honored by your family include:

You realize meaning and purpose in your life through:

Your religious preference, if any, is:

You intentionally engage with nature in these ways:

You consider the following volunteer activities to be part of your spiritual experience:

People who are very dear to you include:

As families explore their unique connections with and contributions to the whole, words and practices that help to clarify a family's unique expression of their spirituality may become apparent. Defining that which is meaningful from the heart and uplifting in daily life supports a family's sense of spirit.

As you go through this exercise, think of key words and phrases to describe your family in selected categories. For *food preferences*, a family who has some vegetarian members and others who eat meat may describe themselves as being mixed or varied. This description acknowledges that variety and choice are welcome. Its food choice statement might mention that "variety is the spice of life" and then go on to list favorite menus.

A family who limits use of technology by using television only for special broadcasts and the Internet for just fact searches could include the word *selective* in their media use description. Those who read newspapers, magazines, and journals can find words or phrases that reflect their reading style. Its identity may be varied, reflective of popular culture, or aligned with their recreational interests.

You will notice that some areas overlap. This is holism at work; those topics that thread between body, mind, and spirit are supporting wholeness within your family's life. A family who lives in the country, reads magazines focused on country living, and spends leisure time outdoors feels very much a part of the natural world. Its identity with nature is strong and influences the choices the members make throughout their lives.

As you complete and reflect on the family identity exercise, think about your current situation and what you want for the future. The family in the country may be longing for a move to the city. The members want more social diversity and a greater variety of arts and entertainment right at their doorstep. Those who live in the city may wish for the peace and open space of the country. Such interests have the potential to determine choices they will make in their work, dwelling place, and community activities.

The descriptions chosen in the family identity exercise will offer guidance in determining how to achieve balance between family, work, and leisure.

Family and Work

Work for most families is a necessity. Livelihood is our bread and butter, the means for the roofs over our heads. For some, work or career becomes an expression of who they are. A person's job can affect the family, and in turn, family needs may affect the job.

Creating balance between work and family occurs in an atmosphere of mutual respect. There are workplaces that are family friendly, offering flexible hours, health insurance, vacation hours, child care benefits, and employee-assistance programs. The best employers are understanding of families while having reasonable work expectations.

When the family is respectful of the work of its members, people are able to be punctual, have good attendance, and maintain focus while at work. Family members are respectful of each other's work responsibilities and limit unnecessary interruptions (e.g., texts, phone calls) during the workday. In turn, when a valid family need calls an employee away, it should be expected that the employer demonstrates sensitivity and understanding. Salary or hourly wages are important but may not account for the most desirable aspects of a job. Questions that explore how work life aligns with your family life are offered in **Exhibit 8-2**.

Creating a balance between work and family responsibilities can be achieved, even partially, by having a clear understanding of the needs and

---

EXHIBIT 8-2   EXPLORING WORK AND FAMILY NEEDS

1. Are work hours a reasonable match for family responsibilities and desires?
2. What benefits, written and unwritten, are provided to employees?
3. Is there flexibility in where I do my work? Am I able to work at home when my child has a day off from school?
4. Is my employer one that meets or exceeds the conditions of the federal government's Family and Medical Leave Act?
5. Is my workplace a reasonable commute from home?
6. Is my workplace free of hazards, and does it actively promote staff health and safety?
7. Is there support for continuing education with both allocation of time and financial reimbursement?
8. Is there a wage and benefits package that allows me to provide for my family as I wish?
9. Is this a workplace and job that supports my personal and family values and beliefs?
10. Does this workplace embrace employee suggestions and involvement?
11. Are customers of this business viewed with respect and dignity?
12. Am I empowered to make decisions and take actions related to my work activities?

limitations of each. After these are established, it is important to convey your limits to both family and work. Sample statements are these:

- "I'll be able to go on your field trip if I can get someone to fill in for me that day."
- "I'm available to attend one evening meeting each week. I'll need to check with my family if I'm requested to work extra evenings."
- "I prefer flexing my hours to working overtime."
- "I'll be working a little late on Thursday and need you to fix supper that evening."

While patterning your comments after these, it is advisable for people to know and understand their employee rights and obligations. An employer's attitude toward family and flexibility can vary over time or with a change in supervisors. Ultimately, it is the responsibility of the individual employee to know the employer's policies and to advocate for change if warranted.

Another point to consider in balancing family and work is that each family/work scenario is unique. What works for one family may not work for another. Part-time employment may not have exciting financial benefits but may provide the time to become more involved in a child's activities and education. The family as a whole can determine the best mix for its current situation. When circumstances change, family/work balance can be reevaluated, and new combinations developed as desired. **Exhibit 8-3** lists questions that can guide families in exploring these issues.

Families and Leisure

Leisure is defined as time free from work or duties (Merriam Webster, n.d.). Families today are very much on the go, continually on the move, around town and across the country. Everyone has much to do—responsibilities to home and family, a job or career, and health and fitness. When and how can anyone plan for fun? With all that needs to be accomplished, how can leisure happen?

People have busy lives, and thus, creating leisure, whether personal or for the family, often requires planning. Spontaneity would be wonderful fun, but it often does not fit in with the structure of the average life.

Making a commitment to leisure is making a commitment to oneself. Relaxation is a practice, as are playing the piano, meditating, and woodworking. Daily practice encourages people to become skilled and proficient in the art of leisure and relaxation. By pausing, even for a brief moment, during the busiest days to focus on breath, an individual takes the time to renew.

EXHIBIT 8-3   QUESTIONS TO REFLECT ON REGARDING YOURSELF, YOUR FAMILY, AND FAMILY ROLES IN RELATION TO WORK

- What days and hours do I want or need to be home with my family? Are there certain times of day that I want to be available for them?
- Does my family understand my profession/career/job and its responsibilities?
- What messages do I convey to my family about my workplace, my career, and my work-related goals?
- Do I/we have backup plans for child care, transportation, and illness?
- Who assumes the responsibilities for chores, meal preparation, and family coordination? Are household and family tasks shared cooperatively?
- Do I foresee that I will need to plan for family changes over the coming years? These may include the needs of dependent adults or a change in the number and ages of children.
- Are there established ground rules related to my work (e.g., accepting phone calls and text messages, leaving during work hours for appointments)?

REFLECTION

Say the words <u>unhurried ease</u> aloud three times, and say them slowly. What sorts of images arise for you—time to smell the roses, ladies twirling parasols in a green-grassed Monet painting, the sweet schuss of powdered snow beneath skis? As your imagination creates more pictures of what you would do with free time, what you would do with unhurried ease, shift to your body. Has your breathing relaxed? Have your muscles softened? Hopefully the answer to these questions is yes. If so, you are experiencing the importance of leisure. Daydreaming, spacing out, doing nothing, taking a nap, and taking your time are simple yet important acts that are a forgotten art for many. Wildlife and animals are excellent models for how to enjoy each moment, time to live with unhurried ease. Observe the bird as it perches to preen its feathers, a cat lounging in the afternoon sun, a moose and her young browsing on rich swamp delicacies.

A family commitment to leisure acknowledges that needs related to free time vary from one person to another. Quiet contemplation is meaningful to one while creating garden space is the height of joy for another. Rock music and engaging in social media symbolize unhurried ease to some, but not to others.

Ideally, a plan to create balance through leisure includes individual as well as family relaxation periods. Scheduling unstructured family time works because it affirms commitment to self and the family unit. When a block of time is set aside for family, individual members can go with the flow or plan an activity that has group meaning. They may go somewhere for hours or days or stay at home and plan something out of the ordinary. The key is to remain focused on their together time and to avoid old patterns of distraction.

Some families set aside a family day, a time to be together, exceptions occurring only for special reasons or occasions. This form of structure opens the way for group leisure and time together that can lead to spontaneous moments such as a cookout, a walk in the park, or a drive to the shore. Families can be encouraged to make a wish list of activities to do with unhurried ease and tailor it to the moment.

REFLECTION

As you think about leisure and how to use it to create balance in your life, go back to your family identity descriptions. Who are you now as a family? Are there ways in which you want to change? Do you see where leisure will enhance your overall well-being, uplifting your spirit while moving your body and quieting your mind?

As people take steps toward creating balance between family, work, and leisure, they will begin to note shifts in their behaviors or in those of others around them. It becomes easier to say no to extra duties . . . anywhere. Sunsets seem to last longer. People do learn to focus on their breath. They do share a kiss and hug when leaving for work. The roses somehow do smell sweeter than they did in the past.

Healthy Habits for Families

Habits are patterns of behavior. They are established or discarded over time and with practice. After habits are firmly in place, they can be hard to change because people perform them almost without thinking. When seat belts first came into use, many people would forget to wear them, buckling up only after a friendly reminder from a fellow passenger; later, a reminder light on the dashboard may have facilitated the fastening of a seat belt. Fastening a seat belt is now as automatic as starting the car.

Decision making and choices enter into this area as well. How do people decide when it is time to adopt one behavior over another? Is it a behavior that they feel free to choose, or does it tend to result from the suggestion of another person? Are they choosing this habit because they were told it would be good for them or because it is something they believe in and embrace with a full heart? Is it something they are trying for someone else's benefit in hopes that it will improve their relationship? Is the habit they are adopting healthy for their entire family, and have they been respectful of others' needs while seeking to meet their own?

The discussion of healthy habits for families will align with and build on the family identity exercise. Comments and thoughts summarized in this activity will help to assess if current habits or those being sought contribute to the family's individuality, thereby promoting overall health.

In the family identity exercise, *body* includes where people live, their type of dwelling, food choices, physical activity, and finances. These are the most basic needs—to have food and shelter and the ability to pay for them. People need to be able to physically move from one place to another and may depend on others or physical means to do so.

The city or town in which they live and the dwelling in which they reside can help to promote family identity and lifestyle. They may find that their neighborhood has some limitations or influences that are unhealthy for what they want in their lives. **Exhibit 8-4** depicts the numerous elements of country, city, or village life to consider when assessing the current or future situation. As each item is reviewed, evaluate how the community contributes to or detracts from the family's identity, goals, and well-being.

When considering a future move, a family should consider its dream town. The qualities that are most important for the family's health and well-being need to be prioritized and used as a guide for their search.

The sense of place is then moved from a community focus to a personal family focus. This includes household members and home. Families need to think about the living space, both common areas and private areas. Does the home provide designated places for people to gather for meals and conversation? Does it assure that people have rooms or corners to call their own? Even in crowded situations, the surroundings can be adapted by consciously choosing behaviors that contribute to privacy. To do so is respectful because expectations are clarified about space, times for bathing, quiet hours, and meals.

EXHIBIT 8-4   HOW DOES YOUR COMMUNITY'S PROFILE CONTRIBUTE OR DETRACT FROM YOUR FAMILY'S IDENTITY, GOALS, AND WELL-BEING?

- Environment/climate/geography
- Population density
- Overall safety
- Infrastructure: roads, government, utilities, transportation, and fire/ police/rescue
- Service base: education, health care, and social services
- Spiritual life/civic organizations
- Recreation/arts/cultural events
- Employment options/economic base/sustainability
- Tax base
- Overall quality of life/acceptance of diversity

Healthy Options

Attending to their relationships while planning for privacy contributes to the manner in which family members nurture or nourish each other. The topic of nourishment includes providing for the mind and spirit as well as feeding the body. Each statement in the healthy options assessment (see **Exhibit 8-5**) represents the behaviors and practices of families who exhibit balanced living. When surveying the healthy options list, you will notice that following these guidelines requires commitment, communication, and coordination. Each family must decide which guidelines have meaning and which to disregard. Feel free to add any that will create an individualized profile of nourishing habits.

Patterns/Repatterning

After the healthy options assessment has been completed, it should be reviewed and put away for a few days. Resist the temptation to judge or analyze behaviors when you return to it. Survey the responses, looking for patterns that emerge. **Exhibit 8-6** offers issues that can be reviewed to help identify patterns within the family.

It is useful for people to identify those healthy options areas that reflect their family's optimal functioning. This can be followed by noting those areas in which they would like to create change over the coming weeks or months. They should be guided in staying focused on keeping their goals reasonable

EXHIBIT 8-5   HEALTHY OPTIONS ASSESSMENT

Consider the following nourishment statements, and note your level of agreement (most of the time, sometimes, or rarely) with each. This activity is not intended to be scored or to rate you on your abilities; the purpose is to determine your family's individual patterns and establish if there are changes you would like to make.

	Most of the time	Sometimes	Rarely
A. Food			
• We are satisfied with our diet and believe that it is healthy.	___	___	___
• We eat together as a family every day.	___	___	___
• We eat our meals at the table.	___	___	___
• We eat without distractions such as television, phone calls.	___	___	___
• We honor each other's food choices.	___	___	___
B. Communication and displays of affection			
• We communicate about our schedules (work, recreation, etc.).	___	___	___
• We routinely check in with each other by phone, notes, or e-mail.	___	___	___
• We support each other with positive statements.	___	___	___
• We each have opportunities to express our opinions.	___	___	___
• We use hugs, kisses, and healthy touch.	___	___	___
• We use and practice effective listening skills.	___	___	___
• We practice the arts of apology and forgiveness.	___	___	___
C. Physical activity			
• We exercise 3 to 5 times a week for at least 20 minutes a session.	___	___	___
• We incorporate activity in our family and work lives as much as possible.	___	___	___
• Our fitness plan includes stretching and rest periods.	___	___	___
• We choose exercise that is enjoyable.	___	___	___
• Our exercise plan includes activities that can be done individually or as a family.	___	___	___
• We use safety equipment, such as helmets and pads, when exercising.	___	___	___
• We limit use of television and other sedentary activities.	___	___	___

(continued)

EXHIBIT 8-5   HEALTHY OPTIONS ASSESSMENT (CONTINUED)

D. Rest and relaxation
- We honor each individual's sleep, rest, and relaxation patterns.
- Our children have regular bedtimes with calming bedtime rituals.
- Adults get 7 to 9 hours of sleep; children sleep 9 to 12 hours, depending on their age.
- We set aside time every week to relax and have fun as a family.
- Nap or quiet time is planned for those who choose or need it.
- We practice daily relaxation, such as breath work and meditation.

E. Financial security/household responsibilities
- Household chores and responsibilities are shared.
- Our income allows us to pay our bills on time.
- We pay our bills on time.
- We have a long-range financial plan.
- Children have chores in addition to care of their possessions.
- We continually upgrade our work-related skills.

F. Family/friends/community
- We have a nearby network of family and/or friends.
- We feel connected to our community.
- We participate in community groups and/or activities.
- Our community offers a positive quality of life for our family.
- We have regular get-togethers with family and/or friends.
- We are spiritually fulfilled through our religion and/or other spiritual activities.

G. Creative expression
- Family members are encouraged to use their individual talents.
- Family artwork/crafts are honored and displayed.
- Financial resources are allocated for lessons, equipment, or materials used in creative expression.
- All forms of creativity are valued, including culinary, financial, and athletic skills.
- We have rituals and celebrations that are unique to our family.

EXHIBIT 8-6   ISSUES FOR DISCUSSION TO AID IN IDENTIFYING PATTERNS WITHIN THE FAMILY

- Identify choices that could promote greater balance and simplicity.
- Do you have a healthy amount of time together as a family? Do these occasions contribute to your well-being?
- Note how your choices support individual and family wellness.
- Determine whether your responses align with your family identity. Observe where they vary and explore the meaning of that variation.
- Ask yourself if your choices contribute to the health of your community.
- Which areas elicit an intuitive response of satisfaction, consternation, or anything in between? An intuitive response arises when you have observed something that has significance for you. Such an observation can indicate a pattern to continue because of its benefits. Conversely, your response may be rooted in a pattern that does not support health and warrants change.

---

and in alignment with their family identities. They may want to begin their change process with small goals that can be readily accomplished. After they have celebrated success and developed an atmosphere that fosters growth, they can select more challenging goals.

Goal Setting

There are steps families can follow to make concrete action plans that will help them to create what they want for the future (see **Exhibit 8-7**). This process is enriched by establishing equality. Each person has a chance to be heard, to state opinions, to offer suggestions, and to describe their role in family goal setting and goal attainment.

Changing habits and adopting new life patterns can often be accomplished through small shifts in behavior. For example, if people want to become more physically active, they can increase activity at work by going for a 10-minute walk rather than taking the usual coffee break. If there is a desire to increase the quality of family time, periods in which family members will refrain from engaging with technology can be designated.

These ideas are not new. Most people have heard them before, time and again. If behavior changes have been attempted in the past without success,

---

EXHIBIT 8-7   GOAL-SETTING STEPS

1. Plan a time for the family to meet about a specific purpose, such as planning meals, choosing a pet, or selecting a vacation destination.

2. Set the goal. What is it you want to do? Does it fit within the picture of your family identity? Is it a reasonable goal given your resources of time, skills, and finances? Do you have a clear picture of what your goal will look and feel like once it has been reached?

3. Plan the process for reaching your objective. Design a long-term plan with steps that can be accomplished in 2 to 4 weeks. This will make your efforts seem more reasonable because your gains will be readily visible. Working on short-term steps allows room for course corrections as your project unfolds.

4. Set aside time to review and celebrate your successes. It makes it easier to focus on your next steps when you can look back and pat yourself on the back for a job well done.

   Reflection serves to remind the entire family that everyone is in this together, no matter what the individual roles and responsibilities are.

5. Review and revise goals as needed. Sometimes in the course of a project the goal changes.

   It may be altered enough to look like a whole new goal. It may no longer be relevant, being dropped entirely or reframed into a new goal. For instance, a family was in the process of planning to purchase a larger home. They explored their options and decided that, given low interest rates and love of their current home, it was better to build an addition than to purchase a new home.

6. Recognize goal attainment. This is an opportunity to review your change process and to identify alterations you will make in it when you establish goals for future projects. It also is a time to celebrate success in reaching desired outcomes.

---

perhaps there has been an attempt to alter too much at once or to effect the change during a time when major challenges were faced. Change takes time and a modest approach. Look for small, healthy changes that can be easily accomplished. Remember that habits are developed over time. Weeks or months of practice may be needed before positive effects of actions are noted.

> **KEY POINT**
>
> Adopting small actions that can be readily integrated into the daily routine increases the success of a plan.

REFLECTION

One action-oriented example of a healthy change involves your daily, routine travel. Whether you travel by car, train, or bus, take three breaths when you arrive at your destination. The 15 or 20 seconds you invest in this routine will clear your mind and cleanse your body. Do you think you are worthy of this modest investment?

Holistic Parenting

Holistic parenting is based on mindfulness, kindness, respect, and reflection. Through the parent–child relationship, people learn more about themselves while learning about their child. They come to honor their individuality as much as they cherish their common traits. They learn to listen to their hearts, the wisdom centers that will guide them to parent in a way that is respectful and prepares their children and themselves for the future. They learn that discipline is another way of caring.

Parenting, once started in a holistic and loving way, lasts a lifetime. Parents become the family elders. As people age and grow in their parenting experiences, they will reflect more on their own lessons; they will see that their children also have lessons to learn through their own experiences. They will stand as guides and supporters, pointing the way and offering a heart and hand when the going is rough.

Topics that are discussed in this section will build on or reflect back to the exercises completed in the section about balancing family, work, and leisure. Through holistic awareness, people will develop a realization that there are many paths to follow when parenting a child. Although there are those who will suggest, prod, coerce, even mandate that people parent in a specific manner, ultimately, all parents are responsible to their own children and to themselves for the form their parenting will assume.

Holistic parenting is based on a strength-oriented approach. It maintains a focus on personal competence while promoting positive traits and behaviors. The child is honored for who he or she is and the attributes radiated as a unique person while being guided with clear, reasonable, and healthy

> ## KEY POINT
>
> Mindful parenting involves being attentive, setting limits, having realistic expectations, and displaying fairness, consistency, and flexibility.

expectations. This practice does not disregard undesired behaviors, actions, and words; rather, it educates on how to keep them in perspective. It is centered on the art of offering love and respect to a child while discussing challenging, frustrating, or painful situations.

Holistic parenting can begin at any time; it is never too late. Humans are resilient and respond well to those who focus on their gifts and treat them with respect. When people parent from a heart center, they learn to apologize and to forgive with integrity and without the burden of guilt for either themselves or their children. This is a discipline; practice is necessary. Mindfully, they will do the best they can for today, and tomorrow they will arise and engage in their heart-centered practice again. At times, parents may feel that their parenting skills are developing more slowly than desired. Parents need to be encouraged to hold on to their intentions and to focus on their desires to have open, balanced relationships with their children.

Parenting Education from a Holistic Perspective

Parenting education, until recent years, has been primarily experiential. People learned about parenting as children from their parents' examples. They took those lessons and applied them to their children, making adjustments as each child grew or another came along. Anticipatory guidance, or what to expect in a child's development and how to prepare for and manage it, came primarily from well-meaning family and friends and information gathered during pediatrician visits. Points of view often varied, sometimes being contradictory.

Today there are parenting classes and support groups, books and videos, and community resources that provide specialized support and guidance for families. Many of these will ask parents to do the same thing: begin at the beginning. How do parents begin at the beginning when their child is 6 years old and they feel bogged down by patterns that are already established? Part of the beginning is going back to their own childhoods, to those earliest memories of their families. Parents can be guided to recall the faces and relationships, the good times and the bad. The following questions can be asked: When were you happy? What made you run and hide? What do you do or say now that seems like a flashback from those years?

Remembering those times and family patterns that brought them joy—or pain—can offer parents hints for what to do and avoid with their own children. Thinking back to those years can help people to become mindful parents. Moment by moment they will become aware of verbal habits, facial expressions, and mannerisms that have become embedded in their behavior patterns. The manner with which their children respond to their words and actions will give parents clues about which behaviors to continue and which to change. When parents release an unhealthy behavior from their past, they may be ending a family cycle that has been in place for generations. The legacy of their actions has the potential to extend past the lives of their children.

Another way to begin at the beginning is to study child development. This commitment to learn about development from the earliest moments of life in utero (in the uterus/womb) can help parents to understand children's behaviors. They will be less confused by the once-friendly 7-month-old who now melts into tears as a reserved 10-month-old. They will understand why their practical 4-year-old has suddenly developed an intriguing relationship with an imaginary friend.

KEY POINT

There are predictable patterns of growth and development at various ages that parents need to understand.

Learning about development helps parents appreciate their children in relation to others of the same stage while gaining a sense of each child's temperament and talents. They learn to respect and have compassion for those parents whose child is intense, very shy, or has special needs. Sharing their joys and frustrations with other parents can help them to learn new techniques while guiding others by the example of their lessons. They come to realize they are not alone and that being with someone who listens fully while acknowledging their feelings is a precious gift.

When parents gain an understanding of child development they are better able to:

- Understand their children's words and actions.
- Adjust their expectations to match their children's current level of ability and understanding.
- Use behavioral guidance and discipline that is appropriate for a child's developmental level.

- Anticipate what stages will come next and how to prepare for coming changes.
- Provide a safe environment for a child.
- Plan family activities that match a child's interests and abilities.
- Offer foods that are interesting, safe, and nutritionally sound.
- Help others to understand their children's individual talents, preferences, and challenges.
- Understand the variations in development from one child to another and typical milestones (a milestone is the average age at which most children will develop a skill such as sitting up, talking, and walking).
- Know when they, their children, or their family need professional guidance and support regarding health, development, behaviors, or relationships.
- Accept their children and others for the people they are and the individuals they are becoming.
- Identify, acknowledge, and balance their personal needs with those of their children.

REFLECTION (FOR PARENTS)

As you reflect on your own childhood experiences and what you have learned about typical (usual, average, expected) child development, pause to reflect on the thoughts that arise as you complete these statements:

- I became a parent because
- I knew I wanted to be a parent when I
- I hoped for a child who

Sit with your responses, breathing gently into your heart. As you continue to breathe, release emotions that have manifested with your statements. Release any fear, sadness, and anger and allow yourself to be filled with peace, suspended in this eternal moment.

Hold that sense of peacefulness and proceed by completing these statements:
- As a parent I wish I could
- The greatest gift I have given my child is
- The greatest gift my child has given me is
- The best advice I could give a new parent is
- The person who is most supportive of me as a parent is
- This person shows their support by

Put your responses aside and breathe quietly for a few minutes. Review your statements without judgment. As you reflect on your statements, consider the following:

- Those statements that are strength based, containing words that are positive, offering affirmation and a feeling of hope. Note where there are patterns of similarity or contradiction.
- Those statements that surprise you
- Those statements that bring up emotions—your joy, sadness, and anger

Write a page about what you have learned about how your original desires and expectations of parenting differ from your present reality. Define steps you can take to become a more effective parent over the coming year. Use your affirmative statements as the building blocks for what you want to accomplish. For instance, you could say, "I will use supportive listening and open dialogue to guide my child in seeking new solutions for problems at school."

Seeking Guidance

Disruption may signify that professional guidance is needed to help resolve a situation. That support could be available from a healthcare provider, a rabbi or minister, or a counselor. Today, there are numerous support groups and therapies for families who are seeking wisdom and solace.

Mutual trust and respect are two of the cornerstones on which a therapeutic relationship is based. People need to give themselves permission to search for the right match for their family. They should talk with the professional person about their beliefs and values and the hopes they have for their parent–child relationship. They can then invite that person to partner with them as their parenting develops and as their children grow in life awareness.

Parents need to take time to center and find renewal in personally satisfying ways. When they feel whole and refreshed they have greater resources to bring to their relationships with their children. The children, in turn, will learn from their parents' example that personal time is important for maintaining health and balance. Parents need to encourage their children to find their own methods for relaxing and attaining personal fulfillment.

Family Comes Full Cycle

Our days go from dawn to dusk; our years flow from spring through winter. Thus it is with our lives and those of our family members.

We have touched on survival, balance, and each family's unique qualities. A person's individuality is embraced in the family's inherent rhythms and cycles. By pausing to recall the past and plan for the future, a person becomes more keenly aware of those patterns.

Summary

Families come in a variety of forms, including blended, single parent, one child, multiple children, same-gender partner, grandparents raising grandchildren, younger parents, older parents, and couples without children. There are some basic assumptions when working with families that can guide a professional's actions. Each family is unique, influences and is influenced by its community and culture, shares many things in common with other families, deserves respect and opportunities for success, and has responsibility to respect and care for its members while accepting the individuality of its members. Each family has a unique body, mind, and spirit.

The healthy options assessment offered in this chapter can be useful in guiding families in the discussion of their patterns in regard to food, communication, displays of affection, physical activity, rest and relaxation, financial security, household responsibilities, family, friends, community, and creative expression. From this, plans can be made for repatterning.

Parenting should be viewed as a holistic process. It is based on mindfulness, kindness, and reflection. It is a strength-oriented practice. With reflection, knowledge, and practice, parenting patterns can improve.

References

Merriam Webster Online Dictionary. (n.d.). Retreived from http://www.merriam-webster.com/dictionary /leisure

Suggested Reading

Bolles, R. N. (2011). *What color is your parachute? 2012: A practical manual for job-hunters and career-changers.* Berkeley, CA: Ten Speed Press.
Covey, S. (1999). *The seven habits of highly effective families.* New York, NY: St. Martin's Griffin.
Covey, S., & Curtis, S. (2008). *The seven habits of happy kids.* New York, NY: Simon and Schuster.
Fosarelli, P. D. (2006). *Ages, stages, and phases: From infancy to adolescence; Integrating physical, social, emotional, intellectual, and spiritual development.* Liguori, MO: Liguori Press.
Gluck, B. R., & Rosenfeld, J. (2005). *How to survive your teenager: By hundreds of still-sane parents who did and some things to avoid, from a few whose kids drove them nuts.* Atlanta, GA: Hundreds of Heads Books.
Goldberg, H., & Goldberg, I. (2007). *Family therapy: An overview.* New York, NY: Brooks Cole.
Schor, E. L. (Ed.). (1999). *Caring for your school age child.* New York, NY: Bantam Books.
Shelov, S. (Ed.). (2009). *Caring for your baby and young child* (5th ed.). New York, NY: Bantam Books.
St. James, E. (2000). *Simplify your life with kids.* Kansas City: Andrews McMeel Publishing.
Wright, H. N. (2004). *How to talk so your kids will listen.* Ventura, CA: Regla Books.

The Spiritual Connection

9

OBJECTIVES

This chapter should enable you to

- Describe the differences between spirituality and religion
- Describe the process of letting go of preconceived notions
- List two elements of spiritual preparation
- Discuss the healing value of service to others

In recent years, there has been considerable emphasis on eating a nutritious diet, exercising regularly, and taking other actions to live a healthy lifestyle. Many people have responded by adopting positive practices and taking an active interest in their health. Even conventional medicine has slowly, yet progressively, incorporated some common sense lifestyle suggestions into patient care.

Health, however, is influenced by more than just the maintenance of the body. From a holistic perspective, the individual is comprised of body, mind, and spirit. Each of these components impacts the others and influences health.

Increasingly, research is shedding light on the relationship between spirituality and health. People who are spiritually fulfilled and engage in regular spiritual practices tend to enjoy higher health states than those who do not. On the other hand, persons with ideal physical bodies who are spiritually empty or distressed fall short of meeting an optimal health status.

> **KEY POINT**
>
> Spirituality is not synonymous with religion.

Spirituality and Religion

Looking past the worship practices and belief systems of organized religion, you can move into an integrated dimension of wholeness known as spirituality. Being essential to life, spirituality reflects the common threads that flow through all religions and belief systems. Spirituality embraces all aspects of every living thing—the mind, the body, and the spirit. From this larger perspective in consciousness, you embrace life and life's issues from a place of collective being and knowing. This is different from the choices involved in practicing a religion.

REFLECTION

How are spirituality and religion viewed and expressed differently in your life?

Spirituality is simply being. Religion is like following a map that was charted to keep people on a predetermined single path toward a destination. Spirituality is similar to sitting on top of a mountain where insight is gained from a large viewpoint that draws one to the best path. As perspective widens from religion to spirituality, people begin to realize the unity and connectedness of all things. This innate state of being is essential to life. It is that place that enables people to know their authentic selves. Spirituality does not negate religion; rather, it shows where the commonalities or universal truths among all beliefs are founded.

When comparing spirituality and religion, some interesting differences can be noted. Words defining religion relate to belief systems and practices of worship that can vary from culture to culture. Words used to define spirituality have a wider universal expression that goes beyond religion and opens one to a greater awareness of one's true being (see **Exhibit 9-1**).

Spirituality as a Broad Perspective

From a broad perspective of spirituality, commonalities among all religions and beliefs can be seen. There is a unifying force, a power, a presence, an essence to life that is greater than the individual yet encompasses all human beings and all living things.

Spirituality impacts everything people say, think, and do. This ornate tapestry of interconnectedness expands from the individual to other people, places, and things. Seeing a beautiful picture or experiencing great music can

---

EXHIBIT 9-1   COMMON WORDS ASSOCIATED WITH SPIRITUALITY AND RELIGION

SPIRITUALITY	RELIGION
Ethereal	Belief
Airy	Creed
Holy	Faith
Higher Power	Church
Sacred	Cult
Essence	Sect
Transcendence	Theology
Being	Doctrine

KEY POINT

That which represents the spiritual dimension is represented by many names and ideas, such as God, Jesus Christ, chi, Allah, Great Spirit, infinite intelligence, unconditional love, Goddess, universal life force, intuition, higher power, and, most simply, from India, the energy.

fill one with love and rapture of emotions. Still expanding in unity and connectedness, the connection with nature, the planet, and the universe is also realized. One can become lost in a beautiful sunset, enjoying and connecting with it so much that for a few moments the sunset is all that there is. This connectedness also means experiencing pain and hurt as well as beauty and inspiration. Pictures of a war-torn country can cause people to feel the pain, sorrow, and hardship of its victims.

From this place of seeing and experiencing, people begin to comprehend on a deeper level the spiritual level. Seeing that what happens to the one affects everything, and what happens to everything affects the one. This often results in a sense of awe or inspiration as people discover this creative force/ god in everything.

Planetary Connectedness

In this place of awareness and connectedness, many people are discovering a deep reverence for the planet. There is an awareness that the abuse of the planet not only affects the earth, but each of its inhabitants. The individual affects the planet and the planet affects the individual. The individual's health affects the community and the community's health affects an individual's health. The individual, the community, the nation, and the world are all connected.

Being, Knowing, and Doing

In considering spirituality, being, knowing, and doing take on new meanings. People become more aware of their inner selves and begin to intuitively sense what to do and how to do it. They begin to feel more comfortable acting on these feelings. Herein lies the essence of the spiritual connection and the beginning of healing.

Being is the art, the awareness of the here and now, or the present. There are no thoughts—only stillness—and with stillness comes awareness through all the senses. The mind becomes still and just *is*. This stillness heightens the awareness of each of the senses even more. This place of stillness with self, others, and the environment enables one to experience the moment on a deep level. It is as though the body is responding to observations from all of the senses. This state is often referred to as being centered or balanced; yet, it is more than that.

REFLECTION

Experience being:

Take a deep breath and gently release it.

Feel how the mind releases with the body.

Take another deep breath and another.

Notice how each breath permits the body and mind to relax even further.

Take a breath, and as you exhale, linger in the exhalation.

From this larger place of being, people begin to express with greater awareness, peace, and connectedness than from desires, self-centeredness, or envy. It is in this place of awareness and peace that healing takes place.

Knowing comes from being open to all things, not just the ones we have stored in our memory. There is a deep inner sense of awareness of things and their interactions and effects. This knowing is accompanied with a feeling of openness or lightness in the entire body.

REFLECTION

Experience knowing:

Close your eyes.

Take some slow, deep breaths with gentle exhalations.

Scan your body for any tight places.

Gently breathe into the tight areas.

Remain in this easy breathing.

Notice that your body is becoming lighter and lighter.

Feel the freedom and lightness with each breath.

Feel the gentle lightness—almost tingling.

Continue with this breathing.

Stay in this place of freedom and healing as long as you desire.

Often people come to recognize that they reach a place of clarity and right thought when the knowing and the light feeling come at the same time. It is like the tingling or lightness is validating or confirming the clarity of the knowing. This is a natural intelligence unrelated to physical memory. This natural intelligence is intuition—one's energy talking. It lets people know what they need to be doing. It tells them what is and is not supportive, what opens or constricts them, and what makes them feel heavy or light. Knowing comes with sensing and feeling this natural intelligence.

From this inner place of being and knowing emerges doing. This outward expression is spirituality made visible. Doing starts with individuals taking care of themselves in healthy, loving, life-giving ways. Honoring knowing and being is sometimes a challenge, especially when being and knowing conflict with expectations imposed by self or others.

Learning From Relationships

The ability for people to experience and express themselves from a wider consciousness is an evolving process. Relationships are the school for this growth in spirituality. People are in a relationship with everything in their lives . . . other people, places, things, themselves, and their higher powers. All desires for fame, fortune, and to get ahead are transient happiness as they are nonrelational. Ultimately, the nonrelational aspects of life increase the hunger for the real relationship, the inner connection, the spiritual connection with a power greater than self.

Some people spend considerable time trying to fix things on the outside when their need is to respond to their deepest calling. They may want to change jobs, improve their image, live in an upscale community, or lose weight. The changing and fixing of these outer things brings some satisfaction for a brief period, but then the novelty wears off and they must be attended to once again. Striving for satisfaction through outer means is indicative of a need for a deeper inner relationship with being.

The deepest calling, the deepest yearning, is to know and remember one's connectedness and oneness to something greater than oneself. As people begin to realize the power and peace of this unity, this oneness in their lives, they begin to desire and experience connectedness more and more. They begin to realize that this connectedness is almost magical and most certainly sacred. In seeking this inner light or higher power, individuals find that the unity that feeds the spirit is more important than their physical pleasures or dreams of wealth and prestige. They are able to let go of having to create images and control others and events, and, as they let go, their lives and health begin to change.

KEY POINT

Releasing all need for control, one can move with love, joy, and clarity in the moment.

Abundance is a simple yet profound truth of the universe. When people want and move into the eternal more than the ephemeral, they begin to experience a peace that surpasses all understanding. They discover the unconditional love that sees beyond all outward appearances and behaviors. They feel the inner joy that is always there, independent of what is happening in their lives.

How people deal with outer relationships reflects their state of consciousness and inner awareness of who they really are. The emotional personality, that constricted ego place in consciousness, is like blinders on a horse that keep it looking and moving only in one direction while ignoring many of the things nearby. Living without blinders allows people to experience the expansive spiritual place and see the joy and beauty of everything around them. Spirituality assists in examining emotions, behaviors, and attitudes. Without the blinders, individuals can see more clearly the limitations that constrain them in self-serving, destructive thinking. The individual can be his or her own best teacher. By observing how they respond to others and their deeper motivations, people can begin to understand the purposes behind their actions. This insight can guide them in discovering the blocks and hindrances to healthy, joyous relationships. They begin to identify the ideas that keep them locked in limitation, disease, and imbalance.

Individuals' relationships with others demonstrate the limitations that

> **KEY POINT**
>
> Native Americans instruct their youth to look beyond to discover truth. After that truth has been discovered, one looks beyond, again and again until one has reached the essence of universal truth.

they hold in their minds. Their judgments and pronouncements only lead to a frustrating cycle of trying to fix someone else. Rather than thinking and acting in a limiting and tightening manner, they need to loosen up and allow themselves to be open, accepting, and free. When they release the preconceived notions of how things should work or how others should act, life begins to flow easily. The mind and body respond by being in balance; this is health.

REFLECTION

Experience letting go:

Consider a recent situation that caused you to become unsettled.

Look at the reason for the emotion.

Now look beyond that reason to the underlying reason.

Keep looking beyond to the next reason and the next until you reach the core reason and can go no further.

Take slow deep breaths and gently release them.

Remain in this place of peace and ease as long as you like.

Being Honest

Being honest with oneself is an important part of healthy living. Honestly going within may reveal things that people may not feel comfortable discovering, but this process could help them to discover the blinders and falsehoods that have been accepted as being real. Take this example:

A college administrator was sitting at her desk perplexed at the response she was receiving from her faculty regarding the proposed curriculum changes. She kept attempting to figure out how to present the curriculum so that the changes would be accepted. In her silence and quiet breathing, she realized that she was manipulating the faculty to do things her way, rather than being open to their recommendations for changes. It felt awful to see herself in the role of a manipulator when she had thought of herself as an empowering leader. Releasing this, she stayed with her breathing and went deeper. She realized that she needed to present the facts as clearly as possible and allow the faculty to make decisions. The administrator let go of any need to have the curriculum changes fit her preconceived idea and opened herself to the faculty's suggestions. The faculty moved together and improved the curriculum beyond everyone's expectations.

When issues are seen for what they really are, people are in a better position to release them. These issues no longer have control and people can move deeper into a place of clarity, healing, and being.

There is a saying that insanity is doing the same thing and expecting different results. Stopping the insanity is paramount to being honest. If a person's joints are stiff each morning after drinking wine the previous evening, why does he or she keep drinking the wine and expecting different results? If life is not working, if a person is not healthy and not moving in a healthy direction, it is time to do something different. Often, insights and guidance provided by a healthcare professional can help individuals clarify unhealthy habits and initiate actions to change them.

Gratitude

Gratitude comes in continually larger and larger doses as one takes the time to go within and be quiet. The vicious cycles of anger, fear, and resentment are replaced with love, knowing, being, and increasing awareness. Spirituality soon permeates every part of life, giving guidance, comfort, healing, purpose, and direction. The body and mind are taken to new levels of connectedness, and a person is able to see the unity and beauty in everything.

Prayer is an expression of gratitude. People can express appreciation that they are being protected and blessed as they face that which is before them and also give thanks for the protection and blessings that have been experienced. Prayers of thanksgiving reflect a connection with a higher power and contribute to feelings of abundance.

Forgiveness

Virtually all human beings have had the experience of someone hurting or wronging them in some manner. Sometimes the offense was so severe and the

> **KEY POINT**
>
> As spiritual awareness is heightened, a person begins to feel thankful for the ordinary—a new day, the ability to get out of bed independently, a job that can provide the means to keep a roof over one's head and food in one's stomach, the support of loved ones. These aspects of everyday life that often are taken for granted can, instead, be appreciated as special gifts.

subsequent pain so strong that it is difficult to accept an apology and forgive the person. Nevertheless, there are other times when the offending individual fails to apologize or ask for forgiveness.

Forgiveness is not only the process of forgiving those who come to you asking for forgiveness but also offering forgiveness without being asked. It may seem difficult to consider forgiving a person who has done something bad to you or hurt you when that individual has not even apologized or sought forgiveness. Nevertheless, by not forgiving, you keep the wound of pain open and cause yourself continued suffering. For persons of the Christian faith, forgiveness is an act of obedience to their faith; they forgive others as they have been forgiven by God.

Forgiveness is an act that yields more benefit for you than for the one forgiven. It is liberating and a necessary step to healing from the incident. Also, the stress relieved through the act of forgiveness offers multiple health benefits, including reducing depression and anxiety and improving heart rate and blood pressure (Lansky, 2007; Mayo Clinic, 2008; Ufema, 2007).

Forgiveness does not imply that the act was condoned or excused but, rather, that you have made the decision to not allow the emotional cloud of unforgiveness to hang over you and continue to control you. You may not forget the incident, but you do not have to be imprisoned by the emotions it generates, such as anger, bitterness, and resentment.

Spirituality as Lived Experience

Spirituality is a daily experience highlighted with some peak experiences. Within the unity and connectedness, a sense of order and natural balance in life and in the world exists. One discovers that there is no single path to spirituality, no road map. Spirituality simply is. This is a lived experience and a moment-by-moment awakening and discovering of one's true being. Things that used to baffle a person become understood and manageable. Problems

> **KEY POINT**
>
> When moving toward the process of forgiveness, it is useful to ask yourself these questions:
>
> - What burden am I carrying by not releasing the feelings associated with the incident?
> - Is it worth occupying my mind and spirit with the negative feelings arising from continuing to be angry or resentful about the incident rather than freeing that space for positive feelings?
> - Have I ever wronged someone and been given the gift of forgiveness?
> - What example do I want to set for others?

are viewed as opportunities to learn and grow. Unhealthy habits begin to melt away. An individual begins to respond from a different place in consciousness.

REFLECTION

You have probably heard of situations in which people risked their lives to help others in emergency situations. They immediately responded and did what was necessary at the time without consideration of the personal risks and dangers. Can you think of a time when you or someone close to you took this type of risk?

When guided by spirit, there is no fear but rather, being, knowing, and doing. Fear enters when the mind leads. Being is a place of no fear—no peripheral thoughts or emotions—just total presence in the moment. Fear comes from one's memory of personal experiences or what others have led him or her to fear (e.g., warnings from persons in authority or situations described in the media). The memory collects everything that is happening in the environment whether a person is aware of it or not. This collection of memory is what is used as a standard against which current situations are measured. This memory, although helpful at times, can keep a person locked into certain beliefs, ideas, and concepts.

Resisting the temptation to be limited by the memories that have guided their actions in the past, people can face the present without fear. They will learn to trust that they will not be harmed or overwhelmed, which can help them to discover happiness, joy, and spontaneity unlike anything experienced before.

REFLECTION
During the day, observe your thoughts when something unexpected happens. Do you respond with anxiety and fear or with confidence that you will be able to handle the situation?

Moving to Solutions

Throughout life, people experience many challenges in their relationships, health, work, and family lives. When challenges continue beyond what people feel is reasonable they may begin to get concerned, worried, and fearful. Prayers do not seem to be answered, and they may begin to question, usually silently, whether there really is a God, an infinite intelligence. At these times, it feels as if they will never be free of the challenge. These experiences offer an opportunity to either be part of the solution or part of the problem. Moving to solutions expands and opens people to positive ideas, thoughts, and a way through the situation, whereas remaining in the problem keeps them locked in the role of victim. Helping people to move to solutions empowers them to act, change, and discover the path to managing the situation effectively.

Complaining is an example of an activity that causes people to remain in their problems. For example, a person may telephone several friends complaining about a situation, repeating the same story each time. She may be so consumed with complaining that she does not take the time to reflect and consider possibilities for changing the situation, somewhat like a person who complains about being hot and thirsty but who fails to take a drink of water! It is when the person is willing to let go of the problem that she can move into the solution.

REFLECTION
Moving to solutions:

Close your eyes and begin to take long, slow breaths.

Gently allow the exhalations to become longer than the inhalations.

Focus on your breathing. If necessary say, "Breathe in, breathe out."

Feel the tension, fear, judgments, and blame leaving.

Accept and know that peace and harmony are filling you.

Remain in this place of peace until you naturally feel like returning your consciousness to the present.

When people have a challenge or a health problem in their lives, they need to take the time to be quiet and still and go within. In this place of stillness, there is no talking, just quiet. Focusing on their breathing, they become peaceful and let go of the fear, the judgments, the blame, and all of the defenses that may contribute to this challenge.

Simply feeling the experience of peace and harmony is a very powerful type of prayer—a very powerful healing process. It is this feeling of peace that works the magic.

KEY POINT

Life is about expanding. People become small when they remain in the problem. As they become still and peaceful, inner knowing emerges, and they become whole again. Wholeness begins in thought and then is expressed in the physical.

Spiritual Preparation for Health and Healing

Increasing time in conscious connection and stillness brings people into a place of expanding peace, love, and healing. They begin to feel extremely positive and safe, as though nothing can go wrong; however, their comfort can cause them to be surprised and taken off guard when difficulties arise.

Pilots take time before each flight to do a preflight check of the craft and its instrumentation. They know that this preparation helps to ensure a smooth flight because problems are handled before the plane leaves the ground. Like the pilot, people can spare themselves some problems and function more smoothly by starting each day with a check of their minds, bodies, and spirits. A daily time of quiet reflection can clarify the challenges that are present and assist in resolving issues. This time of daily reflection—sometimes called prayer, personal private time, getting quiet, or meditation—is a time of being reflective and honest with the self. It gives one the opportunity to work on unresolved emotional issues. When people begin to see the areas of life that are being threatened, then they are able to determine how these threats repeatedly appear in different forms. After they identify the blocks to health, peace, and love, they can turn these blocks over to a higher power and continue with living and being. They are taking action to control their threats rather than be controlled by them.

REFLECTION

Identifying patterns:

With paper and pencil in hand, find a quiet place.

Review your day.

Select a current issue or problem.

In two or three words, state why it is a problem.

Ask what area or areas of your life are being threatened: personal relations, sexual relations, financial security, emotional security, and self-esteem.

Select another issue, and repeat the process.

Continue repeating the process until all of the day's issues have been identified.

Look for similarities in causes among the issues. These are the patterns that restrict your life.

Imagine that you have a very special friend whom you hold in great esteem. This wonderful and fantastic friend takes care of your requests. After you tell this friend about whatever areas of your life are being threatened, you immediately feel secure, light, and free. Wouldn't it be great to have a friend like this? People do. This friend goes by many names and titles—God, Jesus Christ, higher power, infinite intelligence, the absolute, or creator.

The second part of spiritual preparation is having faith that everything is being taken care of and that there is nothing more that one needs to do. When you instruct the computer with a series of commands, you often get a "please wait" comment on the screen while the computer is working on the instructions. Likewise, when things are turned over to God or a higher power, people often get a "please wait" message, although this response may not be communicated as obviously as the message on a computer screen. Individuals must go about their lives and trust that the higher power is in charge and taking care of everything. When worry is released to a higher power, it is no longer necessary to think or obsess over it. This also means that not only is the outcome out of one's control but that it also may not be what one envisioned. Sometimes what we want is not what we really need.

Service as Healing

In the process of healing, people can become quite absorbed in changing the effects or symptoms in their lives. It soon becomes clear that as long as they stay focused on themselves, they will not be fully connected to the whole. In truth, they are part of everything—the planet, the universe, and the situations happening around them. All of these affect reality. When people personalize illness, it keeps them separated from the whole; they can blame others, their situations, and themselves. When they move illness from a personal to a larger perspective, their primary consciousness is altered—a change from the blaming, self-serving ego into the expanded awareness of the unity and connectedness of all things.

From this larger perspective, people can begin to realize that a critical part of healing is being of service to others. Focusing on oneself can limit the full potential to heal. They need to discover that to receive they must give; to be loved, they must be loving, and to heal, they must be healing to others. There can be no receiving without giving.

KEY POINT

The universal law of giving and receiving is interactive and circular; one cannot occur without the other.

Being of service comes in all shapes and forms—offering a glass of water to a thirsty traveler, paying the difference at the cash register for someone who is short of cash, calling someone who is feeling isolated and down, washing a daughter's new dress by hand, volunteering to clean the gutters of a neighbor's home. Service without motive creates an atmosphere of love and joy. The only reason for doing is the pure act of love and connectedness. In order to be healed one must be in the place of service to others, to administer to others without any motive except to give.

Health Benefits of Faith

Through the ages, faith has been accepted and recognized as very powerful. This includes faith in oneself, in what one is doing, in friends, and in God. Faith is much more than just the religious experience of faith, although the highest expression of faith is in the spiritual attitude.

REFLECTION

How often do you ask for assistance and then take your troubles back and worry about them, rather than trusting that help will be provided?

Frequently, the main element lacking in prayer life and daily life is faith. People pray, turn things over to a higher power such as God, and then worry about the issue and proceed to plow ahead in solving a problem without trusting that the matter is in God's hands.

Faith establishes a rapport with the infinite. There is the realization that whatever is happening, a higher power is in complete charge. Everything is moving in divine concert for the highest and best. Faith opens people not only to healing, but to all the activities that support that healing process—diet, exercise, environment, and thinking patterns.

Faith means a conviction that one can trust that there is a purpose for everything and that occurrences will ultimately yield a greater good. People must nurture faith and release thoughts that would in any way weaken this conviction. The attitudes of belief, acceptance, and trust build and nourish faith. Any thoughts that threaten or weaken any of these attitudes, even in the slightest, also weaken faith. When thinking is in line, then the healing process is accelerated.

Attention, Attitudes, and Healing

Attention, attitudes, and healing are invisible, yet their effects are constantly producing visible forms in life. If this thought life is not aligned and in balance, there will be painful consequences. Where individuals place their attention is what they bring into their lives. When they focus on problems, they potentially are inviting more problems. By focusing on pain, they experience more pain; by focusing on the relief of the pain, they experience comfort.

REFLECTION: WHERE IS YOUR ATTENTION?

Wait until you are thirsty.

Fill a glass half full with water.

Observe your thoughts.

Are you seeing the glass half empty or half full?

Are you thinking about how good the water will taste or that it won't be enough?

Are you thinking about how your thirst will be quenched or that you will still be thirsty?

Where is your attention?

Simply observing breathing will reveal much about how individuals are responding to a situation. When people become constricted in their thinking or are becoming fearful about something, their breathing becomes shallow. People need to be taught that when this change in breathing is observed, they need to breathe deeply and remind themselves that their higher power is in charge, not the fear or the doubts of the life situation.

People are living an endless and ever-expanding experience. Only by expanding the mind and spirit can one evolve, grow, and heal. When bound by the past with all its constrictions and predetermined ways to do things, people are contracting. When they breathe deeply and allow themselves to move past this, they expand consciousness and enter into new understandings of unity and connectedness with a higher power.

People have the opportunity to begin life anew—to have the opportunity to set into motion the deeper positive attitudes that heal. They can either stay where they are and repeat their mistakes of the past in different forms, or they can change their attitudes and the attention of their thoughts. Through attention and the observation of their thoughts, they come into alignment with awareness and its health and healing properties. They become where they place their attention.

REFLECTION

Read this story, and consider whether you have witnessed similar experiences in your life. Think about what a health professional could do to lead more people to behave like Beth.

There were three women who all lived in the same town; they all had the same type of breast cancer in the same location at the same time. They all went to the same doctor. When Sally heard the diagnosis, she cried and felt that her life had ended. She could think of nothing else but the dreaded course of the disease and how she was going to die. She was dead within 6 months. June received the news with mixed emotions and vacillated between being dreadfully depressed and feeling rather well about the prognosis and treatment. She had treatments on and off for 3 years, gradually going downhill and finally dying. Beth was shocked to hear the news of her breast cancer and went home and shared it with her husband. She spent time getting quiet, reading about the symptoms, and finding the treatment that felt the best for her. She discovered tapes, books, and friends who addressed and worked in positive thinking. Her attention was on healing—herself, her family, and others. Twenty years later she is doing well, still focusing on healing—herself, her family and others.

When attention is coupled with an attitude of healing, movement can begin toward new solutions. People will begin to feel and know that healing is taking place in every aspect of their lives. Every time they realize they strayed

> **KEY POINT**
>
> Be here *now*.

from this place of high watch they can return. The art of realization is enough to take them out of the negative thoughts. It is as though they have become the watcher of their thoughts rather than being their thoughts. They can move back into the center of all life and allow peace to substitute for judgment, anger, and impatience. Being here in the present moment is the place of peace, healing, truth, and oneness with the infinite.

The Spiritual Connection

We came from a perfect creator. Our essence is whole, perfect, and complete. We are our creator in microcosm. Health, healing, and balance are already present.

The task of every person is to know wholeness, health, perfection, and completeness in his or her mind and heart right now; to recognize his or her perfection as a spiritual being. There should be no doubts, no disbeliefs, no maybes—simply assurance in knowing. Any beliefs that do not support this must be released. As the mind becomes cleared from doubt and limitation, one is able to discover radiant health and healing.

Summary

Although spirituality is a thread within religion, it is a broader concept than religion. It entails the relationship with a higher power that can be identified by various names by different people. It is important to health and healing for individuals to get in touch with their spirituality. This can be facilitated by establishing a daily quiet time for reflection, letting go of preconceived notions of how things should go and how people should act, turning things over to a higher power, and serving others.

References

Lansky, M. R. (2007). Unbearable shame, splitting, and forgiveness in the resolution of vengefulness. *Journal of the American Psychoanalytical Association, 55*(2), 571–593.

Mayo Clinic. (2008). Finding forgiveness: A path to better well-being. *Mayo Clinic Women's Healthsource, 12*(1), 7.

Ufema, J. (2007). The power of forgiveness. *Nursing, 37*(12), 28.

Suggested Reading

Anderson, M., & Burggraf, V. (2008). Passing the torch: Transcendence. *Geriatric Nursing, 28*(1), 37–38.

Autry, J. A. (2002). *The spirit of retirement: Creating a life of meaning and personal growth.* New York, NY: Prima Press.

Barnum, B. S. (2010). *Spirituality in nursing* (3rd ed.). New York, NY: Springer Publishing Company.

Craig, C., Weinert, C., Walton, J., & Derwinski-Robinson, B. (2006). Spirituality, chronic illness, and rural life. *Journal of Holistic Nursing, 24*(1), 27–35.

Craigie, F. C. (2010). *Positive spirituality in health care.* Minneapolis, MN: Mill City Press.

Dass, R. (2002). *One-liners: A mini manual for a spiritual life.* New York, NY: Bell Tower.

Dean, A. (1997). *Growing older, growing better: Daily meditations for celebrating aging.* Carlsbad, CA: Hay House.

Dossey, L. (1997). *Prayer is good medicine.* New York, NY: HarperCollins.

Dossey, L. (1999). *Reinventing medicine: Beyond mind–body to a new era of healing.* San Francisco, CA: Harper.

Dossey, L. (2003). *Healing the body: Medicine and the infinite reach of the mind.* London, England: Random House.

Hillman, J. (2000). *The force of character and the lasting life.* New York, NY: Random House.

Kellemen, R. W., & Edwards, K. A. (2007). *Beyond suffering: Embracing the legacy of African American soul care and spiritual direction.* Grand Rapids, MI: Baker Books.

Kimble, M. A., McFadden, S. H., Ellor, J. W., & Seeber, J. J. (Eds.). (2004). *Aging, spirituality, and religion.* Minneapolis, MN: Fortress Press.

Kirkland, K. H., & McIlveen, H. (2000). *Full circle: Spiritual therapy for the elderly.* Binghamton, NY: Haworth Press.

Koenig, H. G. (1994). *Aging and God: Spiritual pathways to mental health in midlife and later years.* New York, NY: Haworth Pastoral Press.

Koenig, H. G. (2011). *Spirituality and health research: Methods, measurements, statistics, and resources.* Las Vegas, NV: Templeton University Press.

Lindberg, D. A. (2005). Integrative review of research related to meditation, spirituality, and the elderly. *Geriatric Nursing, 26*(6), 372–327.

McSherry, W. (2006). *Making sense of spirituality in nursing and health care* (2nd ed.). Philadelphia, PA: Jessica Kingsley Publishers.

Myss, C. (1996). *Anatomy of the spirit.* New York, NY: Three Rivers Press.

O'Brien, M. E. (2010). *Spirituality in nursing: Standing on holy ground* (4th ed.). Sudbury, MA: Jones and Bartlett.

Pargament, K. I. (2001). *The psychology of religion and coping.* New York, NY: Guilford Press.

Rice, R. (1999). A little art in home care: Poetry and storytelling for the soul. *Geriatric Nursing, 20*(3), 165–166.

Rybarczyk, B., & Bellg, A. (1997). *Listening to life stories.* New York, NY: Springer.

Schweitzer, R., Norberg, M., & Larson, L. (2002). The parish nurse coordinator: a bridge to spiritual health care leadership for the future. *Journal of Holistic Nursing, 20,* 212–231.

Shelly, J. A., & Miller, A. B. (1999). *Called to care: A Christian theology of nursing.* Downers Grove, IL: InterVarsity Press.

Stanley, C. (1997). *The blessings of brokenness.* Grand Rapids, MI: Zondervan Publishing House.

VandeCreek, L. (1998). *Scientific and pastoral perspectives on intercessory prayer.* New York, NY: Harrington Park Press.

Wallace, S. (2000). Rx RN: A spiritual approach to aging. *Alternative and Complementary Therapies, 6*(1), 47–48.

Wilt, D. L., & Smucker, C. J. (2001). *Nursing the spirit.* Washington, DC: American Nurses Publishing.

Wimberly, A. S. (1997). *Honoring African American elders: A ministry in the soul community.* San Francisco, CA: Jossey-Bass.

Balancing Work and Life

10

OBJECTIVES

This chapter should enable you to

- Define *work*
- List three purposes of work
- Describe at least eight characteristics of a positive work attitude
- List three negative work experiences
- Describe at least three measures to convert negative work experiences into positive work experiences

Work, the productive activity that often consumes a considerable part of most adults' lives, is hardly something that is easily compartmentalized and separated from all other facets of our lives. It influences our ability to meet physical, emotional, social, and spiritual needs and to contribute to the order and harmony of the family, community, and the universe.

Various Meanings of Work

Work generally is defined as an effort directed to produce or accomplish something. We work to provide ourselves with goods and services needed for life. Work is simply part of life; however, work is far from a simple concept.

Work has different meanings for different people, often as a result of their experiences with work. Some people say they are lucky because work does not feel like work to them. They claim their work is enjoyable and satisfying regardless of the monetary reward. One wonders why these people are reluctant to identify work as work. What does the word *work* mean to these people that they must view satisfying work as being something unique? Why is there not an expectation that work normally is enjoyable and satisfying? Others associate work only with drudgery, toil, travail, and slavery. They consider work as something to be avoided as much as possible. They might say, "If I could only win the sweepstakes, I would never work again." One wonders if they would really be satisfied to live the rest of their lives in total leisure. Is it really work they abhor or the work in which they are currently engaged?

Many people associate work with earning money and consider any activity for which they receive pay as work. Nevertheless, not all work is paid work. Those who devote their time to rearing children and caring for sick or disabled loved ones in their homes, for example, are often considered unemployed, even though they are surely working. As caretakers, they use their talents to provide service to others and may even be aware of the purposes of their service in the universe. Their work is sometimes fascinating, delightful, rewarding, and joyful, but at times it is also demanding, distressing, frustrating, sad, mysterious, and even frightening. Although caretaking may not be acknowledged as work, most people do understand that it is important and contributes to the good of society.

Through research that began several decades ago, Michael Macoby identified the values and motivations of workers and offered categories of work motivations that have some relevancy in the consideration of the various meanings of work. The categories are (Macoby, 2007):

- *Experts*. These individuals derive satisfaction from demonstrating competence and being sought for their expertise.
- *Helpers*. Those who place emphasis on caring and assisting people and humanizing the workplace fall into this group.
- *Defenders*. Often known for being advocates, these individuals derive job satisfaction by fighting for justice.
- *Innovators*. An entrepreneurial or inventive nature drives some people in their work activities.
- *Self-developers*. People in this group work best in organizations in which teamwork and an emphasis on ongoing improvement are present.

People of religious faith may have a view of work that has a transcendent meaning. For example, they may feel that their chosen work is consistent with God's call or offers an opportunity for them to serve others. Some forms of work, such as farming or environmental protection, could be considered a form of stewardship of God's creation. Work that is viewed as unfulfilling may be accepted by persons of faith who believe it to be the curse resulting from Adam's disobedience that his descendants must bear.

REFLECTION

What does work mean to you? What has influenced your thoughts about the meaning of <u>work</u>?

Work contributes to self-development, thereby serving a purpose greater than the collection of a paycheck or production of goods and services. Work causes most people to feel productive, satisfied, and worthwhile; therefore, the need to work remains even when sufficient goods and services can be produced with ease, as is witnessed in technologically advanced societies. Work, moreover, is essential to our well-being as humans.

A survey of older Americans showed that 40% of men and women between the ages of 50 and 75 years are working or plan to work in retirement (Geary, 2007). Another 40% served as volunteers or planned to volunteer when they retired. They did not want a full-time job, but they wanted to work and make a contribution in the larger scheme of society. Apparently, for many, the need to work does not automatically disappear at the age of retirement. They may quit their jobs, but they do not stop working.

> ### KEY POINT
>
> Work serves many purposes, among them providing necessary and useful materials and services, an opportunity for service, and—in cooperation with others—an opportunity to build community and a means to use and develop individual gifts and talents.

Positive Work Experiences

You probably have experienced the joys of work at one time or another, whether in a job for which you receive wages or in doing a household chore, such as cleaning your house or raking the yard. You drew satisfaction not just from the results of the work (a job well done or a clean house or yard) but also from the process. You inherently know what work is good regardless of what others may think.

Perhaps your work is not separated from the rest of life but permeates your entire manner of living. You see work a useful endeavor that is satisfying in itself and brings significant pleasure. Such work is an important part of a harmonious life and is not viewed as a duty or an obsession.

A positive attitude toward work contributes to a harmonious life–work spirit state. Some of the qualities that contribute to a positive work attitude

> ### EXHIBIT 10-1   QUALITIES CONSISTENT WITH A POSITIVE WORK ATTITUDE
>
> A person with a positive work attitude
>
> - Is rejuvenated rather than depleted by work
> - Gives full attention to the effort
> - Experiences an altered sense of time
> - Displays a creative and nurturing outlook
> - Is optimistic
> - Demonstrates open-mindedness
> - Has abundant energy
> - Feels a sense of purpose
> - Sees self as an important contributor to a great order
> - Appreciates the contributions of others

are described in **Exhibit 10-1.** People who display these qualities consider whatever work they are engaged in to be part of a greater whole. Their sense of purpose is not obscured by a particular task so they can shift their thinking and actions as needed. In the interests of furthering a project, they will put aside what they are doing to help a coworker when necessary.

- They see the unifying patterns and common threads, not just the conflicts and contradictions.
- They perceive the possibilities and draw on varied resources instead of becoming mired in old mind-sets.
- They recognize and appreciate the contributions and talents of their coworkers.
- They measure their effectiveness by the quality of their relationships.
- They delight in the work process as well as the product.
- They possess an enthusiasm in the work setting that suggests harmony of spirit and optimum balance of life and work.

Negative Work Experiences

Not everyone in today's society experiences harmony between their work and other aspects of their life. Job stress seems to be a common finding in all industrialized societies. Unemployment, underemployment, and overwork are major problems.

Work is commonly seen in negative terms in our modern industrial society. In the interests of efficiency, jobs sometimes are limited to a small segment of a task that does not afford us a sense of the whole. Our language reflects this idea. People are called "cogs in the wheels of industry" or a "minor functionary in the machinery of an organization," and compared with computers, their effectiveness is measured by their ability to multitask. Work becomes misdirected and no longer serves our purposes. Instead, work creates undue stress and absorbs an undue portion of our attention.

Unemployment

For those who are unemployed, the lack of work can have profound emotional and spiritual effects. These people are deprived of an opportunity to realize their unique talents, to contribute to the welfare of their families, and to pay taxes that will serve the greater community. When they are compelled to depend on public assistance for extended periods of time, their pride is further undermined. They receive the message that they are not needed or valued and in some measure almost become invisible.

Unemployment certainly affects society in more than an economic sense. The unemployed are unable to fully participate in the community. Is it in any wonder that people who are not able to contribute through working turn to violence, drugs, and other modes of self-hatred as a way of announcing their presence? Everyone needs a sense of feeling needed and making a contribution to the welfare of others within the scope of their capability.

Underemployment

Underemployment, although perhaps not as clearly harmful as unemployment, also takes a toll on the persons affected. The message for the underemployed, although more subtle, is the same as for the unemployed. Some derive limited satisfaction from being gainfully occupied and perhaps being able to pay bills. They often work in part-time jobs or only part of the year when they want full-time employment year-round. They believe they have no options, and thus, they stay in their jobs. Still others remain in jobs that do not truly engage them simply because they earn high wages. They turn to activities other than work for fulfillment or seek pleasure in what their wages can buy them. They may work in jobs that require little skill and afford almost no chance to develop any of their special talents or realize personal achievements.

Overwork

Overwork is an increasing problem. Sometimes this is driven by the desire for consumer goods. They might feel that nothing is good enough; there is always a better appliance or gadget to lure them, and these new and improved objects cost money. It is easy to get caught in the wheel of earning and spending.

As we increase the time devoted to work, we often sacrifice health, relationships, and other important areas of our lives. We spend so much time earning money to buy things that we seldom get to enjoy what we have bought.

Some of us become workaholics whereby our fundamental identities are

equated with our work; we see ourselves only in terms of our jobs, as in, "I am an accountant," or "I am a lawyer." Success is solely measured by the number of clients or cases and/or the amount of money earned. Workaholics have difficulty relaxing. There is a persistent need to complete a few more tasks before they can feel good and allow themselves a break. At the same time, they often feel resentful about their need to continue their frantic, compulsive working. If anything happens to prevent them from continuing this pattern, their basic identity is threatened. Being is completely absorbed by doing.

Workaholism can result from filling the emptiness of one's life with work. The workplace offers interactions with other employees, which can fill the void of insufficient satisfying relationships in an individual's personal life. The mission of the organization can be seductive and compensate for the lack of purpose and passion in one's life. Work may provide social activities and fun not obtained in other parts of life. An individual can become so dependent on a job, beyond the paycheck, that he or she is at a loss when not working and can't imagine life without it. Like a drug, work becomes an addiction for the workaholic.

Even when performing acts of charity, workaholics press on to do more and more. Guilt drives them on, never allowing them to rest.

Disconnection

Many people have separated their lives from their work and endure work in order to satisfy materialistic needs and desires. Work is not a natural part of their lives. They dream of retirement when they can be free of the burden of work.

Work for many is experienced in jobs that mostly consist of mindless and repetitive tasks for which they have little interest that are designed to produce profits for which they have little benefit. Discontent, passiveness, boredom, lack of joy, a sense of hopelessness, and a vague feeling that life is meaningless permeates these situations.

Values Conflict

Sometimes the values of the organizations people work for conflict with their personal values. For example, a salesperson may believe that honesty is important and takes care to avoid being dishonest in his personal interactions. The company the salesperson represents may have the practice of convincing people they cannot obtain the product through any other source when, in fact, there are other distributors of the product. The salesperson may struggle between being dishonest to customers or telling the truth and not getting the sale—or even losing his job. This type of conflict also can arise when a nurse who is Christian and holds the belief that abortion is morally wrong works in a clinic that counsels and refers pregnant women for abortions. In some circumstances, people can avoid or leave jobs that place them in such conflicts, but sometimes, that option is not possible.

Organizational Changes

Fortunately, there is emerging awareness that job situations that are fragmented and disconnected from a sense of the whole are not only disadvantageous for the workers, but also not good for the bottom line either. Flexible hours, child care, parent leave, and job sharing are helpful in some cases but do not serve most workers. Current trends of increasing burnout, decreased employee satisfaction, and rising healthcare costs are spurring changes in work settings. Greater attention is being paid to how an organization can improve the quality of work for employees, not just how employees can benefit the organization. New models for organizations that emphasize balance are being tried.

What Is to Be Done?

The problems associated with work in today's society are very complex. Thus, what can be done? The achievement of harmony and balance in work and life occurs when life and livelihood are reunited. People need to search

> **KEY POINT**
>
> Organizations, like individuals, need to achieve a balance between input and output. Like fossil fuels, individuals and organizations need to be replenished.

for ways to convert negative work experiences into positive ones. This begins with them taking the first step by changing their image of work.

Some people may have given little thought to the full impact and image of work in their lives. They may think of work in terms of a paycheck and benefits for the employee and productivity for the employer. In order for them to achieve a healthier work life, they need to reflect on the quality of their work experience. **Exhibit 10-2** offers a spirituality of work questionnaire that can be used to guide people through this process. The response can be used as a springboard for discussion leading to a deeper exploration of the meaning of work.

What Our Language Tells Us

Our language reflects the time pressure many of us feel on a regular basis. Phrases such as time savers, fast food, express lane, rapid transit, overnight delivery, and instant message control our lives. Colleges and universities have created accelerated programs so that students can complete the requirements for a baccalaureate in 3 rather than 4 years and even begin working for a graduate degree while still an undergraduate. Young children are pressured to get a head start in our highly competitive world. Parents rush around shuttling their children to gymnastics, soccer games, computer classes, and other assorted activities to ensure their admission to the better colleges. Free play is almost a thing of the past. We even have an expression for this need to hurry; we say we are experiencing a "time crunch." It is as though things are out of control and we are all traveling at warp speed.

The popular press offers many articles on how to cope with our accelerated pace and be more efficient. We are inundated with suggestions for setting goals, improving organization, and increasing efficiency. We are told, for example, to handle each piece of mail only once, to make lists and prioritize tasks, and to reduce the time in meetings by allotting specific, limited times for each item on the agenda. We are encouraged to speed read and are inundated with periodicals that provide digests of the literature of journals in our fields.

The motto in many work settings is this: "Do more with less and do it faster." Workers sometimes feel that they are treated like parts of a machine,

---

EXHIBIT 10-2   SPIRITUALITY OF WORK QUESTIONNAIRE

---

1. Do you feel drawn to the work you do? If yes, why did you choose this work? Has this feeling increased or decreased over time? Why?
2. Do you have a sense of purpose in the work you do?
3. Do you experience your work as a part of a greater whole?
4. What are the unifying patterns in your work?
5. Do you experience enthusiasm for your work? If so, under what circumstances? How frequently? What increases your enthusiasm? What dampens it?
6. How does your work enhance or interfere with your creative expression?
7. Do you continue to learn from your work? Why or why not?
8. How does your work affect others?
9. Who benefits from your work?
10. In what ways does your work distress others?
11. How do you recognize the talents and contributions of others in your work setting?
12. What is the quality of your relationship with coworkers?
13. How does your work affect the environment?
14. Is your work in harmony with the environment?
15. How does your work contribute to caring for the environment?
16. How will your work affect future generations?
17. What are the long-term effects of the work you do?
18. What are the unintended consequences of the work you do?
19. Would you continue to work if you suddenly became independently wealthy? What would you do instead of the work you are now doing? How would you use the money?
20. How do you rest from work and renew yourself?
21. What do you do with your leisure time?
22. What would you do if you had a sabbatical year?
23. What is sacred about the work you do?
24. What is your philosophy of work?
25. What do you value about your work?

---

and they must maintain the expected pace or be discarded as obsolete. Although some of the ideas can be helpful, there is the danger of overplanning and structuring life at the expense of living life.

REFLECTION

Notice how your life is affected by the messages that link a fast pace and busyness with importance. What do they do to you? Uncover, accept, and understand the patterns that influence your life. As you increase your awareness of these patterns, you will be in a better position to correct the imbalances you find.

The material and psychological rewards people receive from their increased efficiency may simply prompt them to invest more time and energy into work at the expense of their health, family, and community. They may end up working more efficiently, but also more hours. Workers in the United States already work more hours than workers in other industrialized countries. It is easier to become psychologically trapped by success as success begets greater expectations for performance.

REFLECTION

Reflect on your own work situation and consider what you can change to promote balance. Reestablish control of your life and schedule. You may need to review your spending habits and eliminate expenses that are not essential in order to reduce the amount of time you spend working. Avoid turning to shopping as therapy. Consider decommercializing holiday festivities and creating simpler but meaningful holiday rituals. Find ways of sharing with your neighbors through borrowing and lending. Practice saying "no." Restore the practice of providing Sabbath time for yourself, if you don't already do so. You may discover increased life satisfaction!

Balancing Work and Life

People are encouraged to live a balanced life, but what is a balanced life? Balance means the harmonious ratio between all parts, not necessarily equally divided parts. Just as people need to balance the time that they devote to sleep with the time they are awake, they need to balance the time spent in work with the time spent in other activities. Individuals determine this balance for themselves. The ratio of time spent in each activity creates a harmonious whole for them and fosters their well-being.

People need to be challenged with this question: "Why do you work?" All activities—sleeping, eating, playing, and working—are part of living. Work is not something done merely to acquire material goods. Work should be connected to one's being and be its expression; it is a spiritual activity. Work that is externally driven interferes with that connection.

> **KEY POINT**
>
> Work should serve a higher purpose than merely as the means to acquiring material goods.

Converting Negative Into Positive Work Experiences

There is much talk about stress and its effects on health. Some believe stress causes illness and should be decreased, if not totally eliminated. This is an oversimplification of the notion of stress. Stress is inevitable and can be a positive force. Stress, for example, can prompt you to look for solutions and solve problems.

Any number of events that people experience as normal—changes in living conditions, work, relationships, even ordinarily joyous occasions such as marriage or the birth of a child—can be stressful. These stressors, however, do not inevitably cause illness. People can learn to manage them by finding balance with nutrition, exercise, play, and relaxation.

Life-Balancing Skills

Although major problems associated with work can only be addressed on a societal level, some things can be done to manage many of the problems on an individual basis. Individuals have inner resources that can be drawn upon to cope with daily stresses. Researchers have found that certain personality

> **KEY POINT**
>
> Humans have varied natural, biologic rhythms like those found in nature. One of the important rhythms humans experience during one 24-hour period is known as the ultradian rhythm. The general pattern of this rhythm is 90 to 120 minutes of activity, followed by a 20-minute recovery period. Periods of higher energy regularly alternate with signals suggesting a need for rest. Ignoring signals calling for rest while continuing to work will upset the ultradian rhythm and lead to stress. Responding appropriately to the call for rest supports the normal pattern and allows for recovery and renewal. Even a small change in activity may be beneficial. Planning your work so that after approximately 90 minutes of concentration on a project you take a stretch break or engage in a different activity will help. A play break during work may help to restore us.

traits help people cope effectively with emotional wear and tear. These traits prevent people from breaking down emotionally and physically in the face of life crises.

The capacity to *attend to, connect,* and *express* (ACE factor) feelings can help people to maintain physical and mental health. Feelings (emotions) are gifts that provide people with information about the effects of what is happening at that moment. When people attend to those feelings, connect them to consciousness, and express them appropriately, they are able to cope successfully with stress and restore balance to their systems. For instance, when you notice a dull, aching pain in your right shoulder, you draw your attention to the pain (attend) rather than simply try to mask the pain with medication. You ask yourself questions to connect the pain with what is happening to you. When did the pain start? Can the pain be associated with a particular activity or movement? Is the pain worse at the end of the work day or work month? What about when the weather is cold or damp? What is the character of the pain? Does it come on suddenly or does it build up gradually over time? You may immediately recognize that the pain is related to the increased time you spend at the computer doing billing before the end of the month. A closer look at how you arrange your work area may suggest that you are stressing your right shoulder. You notice that you are repeatedly twisting your right shoulder as you do this monthly task. You are now in the position to look for creative solutions to your problem. You may realize that you can rearrange your work station to avoid the stress on your right shoulder. By doing so, you eliminate the source of your pain and still fulfill your job responsibilities.

Other troubles in the workplace can occur. Work can be a place of misunderstanding or undue competition that threatens well-being. Perhaps your work situation causes irritation or even anxiety. Skill in using the ACE factor can be helpful in such situations. For example, suppose you notice that you tend to make mistakes when you are assigned to work with a particular coworker. The same thing does not happen when you work with others. You feel that this one person does not like you; he or she acts unfriendly and seems to be impatient when things do not go smoothly. You notice that as soon as you receive your assignment with this coworker, your stomach seems to get tied up in knots, and you already sense that things will not go well. You may get stuck in blaming this coworker for your difficulties, complain to your supervisor or union, and/or do all in your power to discredit this person. You may be successful in changing the situation; you or the targeted coworker may be reassigned. You may find you are forced to work with someone who is even more difficult, and the situation may grow worse. Now you are faced with an even bigger problem. You are reluctant to call attention to the

The ACE factor can be good for your health. Research shows that people who lack the ACE factor have weaker immune systems, which results in less ability to defend against infectious diseases and cancer. Persons who have greater ACE abilities have stronger immune systems. Quite simply, neither the presence or absence of anxiety nor the degrees of anxiety are the deciding factors. Rather, it is how the person deals with the anxiety that makes the difference (Dreher, 1995; MacEoin, 2008).

problem for fear of being labeled "difficult." Your anxiety increases, and you find that your health deteriorates as a result. The ACE factor can be applied in this circumstance as well. You start to attend to the signs that your anxiety level is increasing. You connect the increased anxiety to your tendency to make errors. You acknowledge (express) the link between your anxiety and errors. Your awareness allows you to take steps to reduce your anxiety (e.g., take a few deep breaths). You may further consider the link and discover the underlying source of the anxiety. This coworker works at a faster speed than you do, and your anxiety increases when you cannot match his speed. You ask your coworker for help and you work together to find a solution to the problem.

You cannot ignore feelings, disregard their meanings, and repress expression without adverse consequences to your health. Anxiety, like pain, is a signal that tells us something is wrong and needs our attention.

People who are psychologically hardy have been found to have certain characteristics referred to as the three Cs—commitment, control, and challenge (Dreher, 1995, p. 52). People who are strong in commitment find meaning and purpose in their work as long as the work is useful and not harmful. They are wholeheartedly involved in whatever work they do. A person who works as a garbage collector, by knowing the meaning of his or her work and appreciating its contribution to society, has commitment. Drudgery may be part of the work, but drudgery does not negate its purpose or value. When we are committed, the meaning of our work will be present to us even when it is obscured by routine and only breaks through to our awareness from time to time.

REFLECTION

Do you feel you have control over your work activities, are committed to them, and are able to frame work problems as challenges to overcome rather than threats?

People who believe and behave as though they have influence over life circumstances demonstrate control. These people have a sense of mastery and confidence in their ability to deal with the challenges of life even when faced with limitations in external freedom. Their sense of control is healthy and should be distinguished from unhealthy attempts to control others' behavior and reactions. People who demonstrate control respond to reports that there will be major layoffs in their plant with plans for dealing with that possibility. They begin to make immediate changes in their lifestyles to set aside extra money in the event that they are among those laid off. They do not spend their energy railing against events over which they have no control.

People who are are motivated by challenges view problems resulting from change as trials to overcome and not as threats. These people recognize that change also represents opportunities for growth and not just loss of comfort and security. They creatively adapt to the fact that they may lose their job by exploring other employment possibilities. These are the people who usually have an exit plan in place before they receive their pink slip.

Commitment, control, and challenge, often referred to as the hardiness factors, are the building blocks of healthy coping. Healthy copers respond to loss, instability, and change by drawing on their own inner resources. Those who have the hardiness factor are not necessarily younger, wealthier, better educated, or in possession of greater social support.

People ought not to postpone living because of their work or because of their plans to buy something with the money they make. It is easy to become absorbed in acquiring more and more money and lose sight of the original goal. **Exhibit 10-3** lists questions that people can use to gain insight into their work–life balance. It can prove useful for people to discuss their responses with significant others in their lives or a professional who can assist them in making changes.

Downshifting is one response to these daily realities. People find that by working less they are able to do something more meaningful with their lives, including spending more time with their children. This requires a change in their basic thinking. A common tendency is to acquire things that allow us to live more luxuriously but that do not necessarily enhance our lives. These possessions can become anchors that hold us down rather than free us. A person may buy a boat with the dream of spending sunny days relaxing on the water; however, the reality of the attention a boat needs may soon hit. The person will either have to invest time and energy to maintain the boat or spend money for someone else to do so; either option could impose a burden sufficient to dilute the joy of boat ownership. People need to ask themselves whether their purchases will enhance the quality of their lives or be added

---

EXHIBIT 10-3   QUESTIONS TO USE IN GUIDING ASSESSMENT OF LIFE/ WORK BALANCE

- Are you doing that which energizes and renews you?
- What gives you the greatest joy?
- Do you continue to expand your business in response to the opportunity to make more money to the detriment of your health and personal life?
- Do you devote more time to work in order to escape problems at home?
- Do you spend wastefully and then work overtime to pay your bills?
- How would you really like to spend your time?
- What are the things you always wanted to do but do not because of lack of time? How can you find the time to do these things?

---

burdens. Perhaps trips to the local public beach with family and friends could provide equal or greater pleasure than owning a boat.

It is difficult, however, for people to resist the predominant culture on their own. Adjusting to a reduced income requires adjusting to decreased spending. One need not drop out of society to achieve the goal of simplicity. Refusing to buy name brand clothes for their children can initially create conflict for parents. Deciding to reduce spending on holiday presents and celebrations takes courage. Joining a group with similar goals can provide much-needed support in efforts to downshift. Newsletters, websites, and study circles are springing up around the country to help those interested in transforming their work lives, spending habits, and values to create life–work balance. Work is part of life, but people should not live to work. They need to be careful not to postpone living in order to work. Work life needs to be in balance with personal life in a way that causes people to experience a unity that celebrates their personhood. Current lifestyle habits need not be permanent. People need to ask the critical questions that will help them to create a harmonious life–work balance in their lives.

Summary

Work serves many purposes, including providing materials and services, offering services to others for the purpose of building community, and enabling people to use and develop their gifts and talents. A positive work attitude contributes to a harmonious life–work spirit state and is reflected by being rejuvenated rather than depleted by work, giving full attention to the effort, experiencing an altered sense of time, displaying a creative and nurturing outlook, showing optimism, having abundant energy, demonstrating open-mindedness, feeling a sense of purpose, seeing self as an important contributor to a greater order, and appreciating the contributions of others.

Negative work experiences include unemployment, underemployment, and overwork. Some ways to convert negative work experiences into positive ones include rearranging work schedules to coincide with ultradian rhythm, being attentive to feelings, developing psychological hardiness, and downshifting. People can achieve a higher quality of life by making work a meaningful, but not all-consuming experience.

References

Dreher, H. (1995). *The immune power personality.* New York, NY: Penguin Books.

Geary, L. H. (2007). *Work until you die: A new retirement goal. Financial literacy 2007.* Retrieved from www.bankrate.com/brm/news/Financial_Literacy/April07_retirement_poll_national_a1.asp

MacEoin, B. (2008). *Boost your immune system naturally: A lifestyle action plan for strengthening your natural defenses.* London, England: Carlton Books.

Macoby, M. (2007). *The leaders we need: And what makes us follow.* Boston, MA: Harvard Business School Publishing.

Suggested Reading

Aron, C. S. (2001). *Working at play.* New York, NY: Oxford University Press.

Bolles, R. N. (2007). *What color is your parachute?* New York, NY: Barnes & Noble.

Briggs, J., & Peat, D. (2000). *Seven life lessons of chaos.* New York, NY: Harper-Perennial.

Caproni, P. J. (2000). *The practical coach: Management skills for everyday life.* New York, NY: Pearson Education.

Damasio, A. (2000). *The feeling of what happens.* New York, NY: Harcourt, Brace.

Davidson, J., & Dreher, H. (2004). *The anxiety book.* New York, NY: Penguin Putnam.

Dossey, B. M., Keegan, L., & Guzzetta, C. E. (2005). *Holistic nursing* (4th ed.). Sudbury, MA: Jones and Bartlett.

Fassel, D. (2000). *Working ourselves to death. The high cost of workaholism. The rewards of recovery.* San Francisco, CA: iUniverse.

Moore, T. (2008). *A life at work: The joy of discovering what you were born to do.* New York, NY: Random House.

O'Neill, F. (2010). *Never enough: Lessons from a recovering workaholic.* San Francisco, CA: iUniverse.

Paddison, D. (2011). *Work, love, pray: Practical wisdom for young professional Christian women.* Grand Rapids, MI: Zondervan.

Robinson, J. (2003). *Work to live: The guide to getting a life.* New York, NY: Perigee.

Rutledge, T. (2002). *Embracing fear.* New York, NY: HarperCollins.

Swenson, R., & Margin, A. (2002). *The overload syndrome.* Colorado Springs, CO: NAV Press.

Wheatley, M. J. (2006). *Leadership and the new science.* San Francisco, CA: Berrett-Koehler.

Creative Financial Health

OBJECTIVES

This chapter should enable you to

- Define money
- Discuss the energetic principles of money
- Describe the universal laws that apply to money
- Discuss the difference between a mechanistic and holistic perspective on money
- List at least six measures that can promote harmony of body, mind, and spirit in achieving financial goals

Except for the payment of services, the area of finances is seldom regarded as a matter of concern in health care, yet it can significantly influence health status. Money affects all areas of life, including where you live, your education, the types of food you eat, and the amount of stress in your life.

It seems that everywhere we turn we are faced with television programs, articles, and seminars that correlate money with health and happiness. In actuality, there is some truth to this. The relatively new medical specialty of psychoneuroimmunology is confirming that positive emotions and happiness result in a higher functioning immune system and, consequently, better health. If the amount of money one has influences one's level of happiness and contentment in life, then we could surmise that money is an important factor in promoting health.

What Is Money?

Money is defined as standard pieces of gold, silver, copper, and nickel used as a medium of exchange. The word *money* is derived from the Latin word

> **KEY POINT**
>
> Minted coins first appeared about 650 B.C. in Asia Minor. The neighboring Greeks began minting coins soon thereafter and within a few centuries developed a system of banking that included borrowing and lending money.

moneta, which means *mint*. One of the definitions of *mint* is to create, and thus, you could reason that money serves to create. Money often is referred to as *currency*, in which it shares a definition with electricity.

You may think of money in terms of paper and coins that carry value, but it is more than that. Money is an energy that is neutral. Your consciousness gives money the charge, either negative or positive, resulting in how it serves your life. Just as your mind can affect your physical health, it also can affect your financial health.

> **KEY POINT**
>
> There is a common view among spiritual and conscious prosperity coaches that suggests that if all of the wealth in the world were distributed evenly to every man and woman, within a short time period, the rich would be rich and the poor would be poor again.

Beliefs About Money

What is wealth? Who is rich? How do you determine financial security? The answers are based on perceptions, and your belief system creates your perceptual field. Beliefs are derived from a variety of sources, such as societal and cultural influences, religious background, ancestral heritage, educational systems, and the mind-set of the adults in the household in which you were raised. Beliefs about money influence how you manage and use it and how much you have. For example, many people who carry negative beliefs about money find themselves in states of financial deficit and struggle.

Many cultural influences shape beliefs about money. For example, your work ethic may convey the message that you have to work very hard and struggle to obtain money. Or you could have been influenced by a message that it is wrong for some people to have great wealth in a society in which some are poor. Certain religious views imply that to be spiritual one must live in poverty. These views are very old and embedded into masses of minds and consciousness. What messages have influenced your views about money? An awakening moment can come when you gain an understanding of the factors that have contributed to the beliefs you hold about money.

REFLECTION

What does financial security mean to you? How did you develop your beliefs about this?

It may be difficult to identify your true beliefs about money. A good way to discover your beliefs is to observe your thoughts, feelings, emotions, and desires when in the act of a financial exchange. For example, when writing a check to pay a bill or giving a donation, what are your thoughts and feelings? People who see the glass as half empty often find themselves with ongoing thoughts, such as "I can't really afford this," or "I wish I had more." Emotions that often accompany these thoughts can be anger, fear, or guilt; they come from a negative perspective. The good news is that it is possible to change beliefs to heal the wounds of *scarcity consciousness* and move into a more positive place by acquiring *abundance consciousness*.

Energetic Principles of Money

As mentioned, money often is identified as currency, representing an avenue in which electricity or energy flows. Currency also carries a charge (another electrical/energetic term that has been adapted in our monetary transactions) that is given to it by the substance through which it is flowing. The way this applies to the way you deal with money is that the charge comes from your beliefs and attitudes (negative or positive). This charge magnetizes your energy field (through which you view the world), and consequently, through the use of universal law (definition and description to follow), you can put an energetic charge on money with your intentions and thoughts.

Become aware of the thoughts and intentions you have while you are handling money. If you note that they are negative, plan to create a change in your mind-set. One suggestion is to have a positive thought or consider your blessings each time that money passes through your hands. Gratitude is always a good energy to instill into money.

Money as Spiritual Energy

Many people who are seeking ways to develop an abundance consciousness use the affirmations, prayers, and other wisdom offered in ancient religious texts. For example, an ancient Hebrew prosperity affirmation is *Jehovah-Jireh*,

> **KEY POINT**
>
> Money's spiritual roots are represented in its physical form of the dollar bill. On the dollar bill is an eagle that holds in its beak a scroll that reads *E pluribus unum*, which means *From many, one*. Also pictured on the bill is a pyramid, which is a noted spiritual geometric structure that has a truncated tip that holds the all-seeing eye of the divine providence, the guiding power of the universe, and the motto, *Annuit coeptis*, which means *He has smiled on our undertakings*. Below the pyramid is a scroll that bears the inscription, *Novus ordo seclorum*, which is translated as *new order of the ages*. On every bill we read *IN GOD WE TRUST*.

meaning, *the Lord now richly provides*. In Deuteronomy 8:18, the insight that God is the source of prosperity is conveyed in these words: "You shall remember the Lord your God, for it is he who gives you your power to get wealth." The biblical admonition that "the love of money is the root of all evil" does not imply that money in itself is bad and harmful, but that the *love or worship of money* is. It is how people use money that matters. A healthy spiritual perspective realizes that money has its place and that it should not be idolized or used to harm others.

REFLECTION

Money can represent many things in a person's life. What does it mean to you?

Money as Metaphor

As money is a formless substance, it can take on various meanings in your life and represent different things. To many people, the degree to which they are loved and cared for is demonstrated with money. Money can represent something that has authority over you or that can be used as a means to control others. Also, the way in which you deal with your money is a reflection of how you take care of yourself. Money is an extension of who you are in all areas of your life; therefore, it is important when taking a holistic view of self and others that money be an integral consideration. The more ways in which this aspect is integrated and seen as a metaphor for different parts of yourself, the more you can come into your wholeness.

Universal Laws

Universal laws are the framework in which you can operate as you change the perspective of money in your life, especially if there is an interest in working on the cocreative process of manifesting more money. In actuality, universal laws exist whether you know about them or not—or whether you choose to work with them or not. They are similar to many physical laws that operate in your daily life of which you may not be aware—such as the law of gravity. Universal laws are nonphysical in nature, but they are fundamental laws of mind and spirit. Everyday life is based on these laws, much like physical laws and laws of nature (changing of seasons, the sun shining, turning on a light switch, taking a shower, and listening to the radio). As you rely on these laws in the physical world, so can you rely on universal laws of mind and spirit. People who consciously work in harmony with these laws find a high degree of fulfillment because they begin to align their actions with their purpose in life.

KEY POINT

Universal laws are unbreakable, unchangeable principles that operate in all phases of our lives and existence, for all people everywhere all the time; they are impartial, irrefutable, and unavoidable.

KEY POINT

Florence Nightingale wrote extensively about universal law in an attempt to unify science and religion in a way that would bring order, meaning, and purpose to human life. She was motivated to use this information to help the working class people of England who wanted and needed an alternative to atheism. Nightingale said that universal laws are expressions of God's thoughts and that law is a continuous manifestation of God's presence.

How Universal Laws Work

Universal laws work on an energetic premise with simple, subtle energy principles. There are two energetic fields in operation, coexisting and intermingling—the universal energy field and the human energy field. The universal energy field is that for which we have no control, often referred to as God or the divine. The human energy field, resulting from beliefs, thoughts, and attitudes, is a mutable force that can be created and formed (the free-will aspect). That which we carry in our human energy field in the form of beliefs, attitudes, and thoughts will create an energetic charge that will incur a response according to universal laws. We can charge our field with new beliefs, attitudes, and thoughts simply by bringing in new input in the form of thoughts, much like one would program information into a computer. The following examples of universal laws will help to further explain how they work.

The Law of Expectancy

The law of expectancy suggests that whatever you expect you will receive. This is not the same as wishful thinking, but rather something that is expected with certainty, much like a pregnant woman is expecting a baby. The degree of expectancy that is maintained in your consciousness is vital. What is it that you are expecting? Do you expect to go to work tomorrow? Do you expect to make a certain amount of money? Expectations can limit you, even when you learn to temper your expectations. It is suggested that when working with the law of expectancy an open-ended thought in which expectations exceed the minimum should always be maintained; for example, "this or something better" or "this or something more."

Faith is an important element in the law of expectancy. This wisdom is reflected in Matthew 9:29, where it is said, "according to your faith be it done unto you." In working with this law, it is important not to limit expectations.

KEY POINT

Working with the law of expectancy is similar to working with principles of the reemerging ancient art of Chinese placement called feng shui. Feng shui is an energetic approach to improving all aspects of one's life by working with the physical space in which people spend most of their time—either at home or in their offices. In feng shui, a *bagua* is a map used of physical space. The *bagua* is a standard template that has in it all areas of one's life and the location in their physical space that is connected to these areas of life. The section of the *bagua* that is connected to money is in the fortunate blessings location. If there is junk, clutter, or physical discord in this area of someone's home or office, then that may impede the flow of financial energy in one's life. A way to enhance financial abundance in one's life is to clear the physical space in this area of the home according to the *bagua* so as to prepare for the flow of blessings and prosperity.

Imagination helps with that. That is why Albert Einstein said that imagination is more powerful than knowledge.

The Law of Preparation

When you prepare for something, you often get it, and that includes negative things as well positive ones. Do you know people who have put money away for a rainy day or an emergency? In a sense, they are preparing for these negative experiences. The energy can be shifted, and instead, they could consider preparing for abundance and success. This can be as relatively minor as framing the saving of money to "I'll have the resources to be free from pressures in unexpected events," rather than "I need to save because something is always breaking down around this old house." It is important to prepare for something in your consciousness and then take action to achieve it.

The Law of Forgiveness

Energetically, forgiveness helps to remove blocks to receiving. It releases negative experiences in your consciousness and creates space that allows the entrance of divine consciousness in the form of abundance. One of the country's leading prosperity coaches, Edwene Gaines (2006), has identified four spiritual laws of prosperity as tithing, goal setting, forgiveness, and divine purpose. She suggests that all financial debt is about unforgiveness of past issues and, mostly, not forgiving ourselves. She teaches that one of the fastest ways to prosperity and abundance is forgiving on a daily basis. Everyone and everything in the past should be forgiven so as to move forward in one's life abundantly.

REFLECTION
Who and what do you need to forgive?

The Law of Tithing

The word *tithe* means *one tenth* and usually is related to offering a portion of income to spiritual or sacred purposes. A tithe is usually given to a person, place, or institution that contributes to one's spiritual nourishment and growth. Tithing is one of the fundamental laws of life and is a powerful prospering exercise. It is an ancient, spiritual practice mentioned throughout the Bible. In Genesis 28:20, Jacob said of "all that thou shalt give me, I will surely give one tenth unto thee." Tithing can be seen as simple as the inflow/outflow principle. As with breathing, where it is necessary to regularly rid yourself of air in order to receive fresh air into the lungs, so it is necessary to

> **KEY POINT**
>
> The intent with tithing is to plant a seed. A farmer tithes in order to reap harvest, returning one tenth of the grain to the soil.

give regularly if you wish to receive regularly. Biblical wisdom related to the belief that what one gives one also receives in return is reflected in Galatians 6:7, which states "Whatsoever a man soweth, that shall he also reap." If you do not sow, you do not reap.

Your Mission Becomes Your Money

Every person on this planet has a mission and a purpose in this lifetime. In order to fulfill your destiny, this work (work in the sense of contribution to the world not the source of a paycheck, although it may be delivered in that form) needs to be carried out. Too often, people are not connected to their purpose in life, and consequently, many imbalances may appear, including physical ailments, relationship disputes, and financial difficulties. Financial fulfillment can be derived through the process of following your own heart to perform the work of your dreams. More important, by tapping into and executing the right livelihood, spiritual aspects of life can be fulfilled.

The process of discovering one's life mission can be a very challenging one. Often, people find themselves going through the motions of a job that not only is ill suited for them, but may go against many of their natural energetic rhythms; however, the fear of not having enough and not being able to survive compels many to go against their natural flow and tolerate ill-suited work.

REFLECTION

In what ways is your attitude about work different from and similar to that of your parents and grandparents?

Discovering your mission in life is a spiritual and creative process. The approach to this process is very simple and begins by exploring your desires. It is that which is your greatest longing that takes you to your mission. Some people spend years not knowing what their purpose in life is, whereas others seem to know their purpose the moment they are born. Many job opportunities in our society are focused on the needs of the employer, rather than being designed to fill people's needs of expressing their creative essence and offering their God-given gifts to the world. The system is created in such a way that

> **KEY POINT**
>
> A midlife crisis can occur when people feel they are limited by their jobs and not fulfilling their missions. In this awakening, there is a yearning to follow one's longing. The crisis can result in some unfortunate consequences, such as divorce and unsound business ventures, or in new opportunities that realign with values and life purpose.

people end up redesigning themselves to fit the system's needs. This, combined with fears about survival, results in stifled missions. Fear drives people to be stuck in containers (e.g., jobs) that are not made for them.

Passion Creates Magnetism

There is a vital force connected with fulfilling life purpose and mission. Think of the vitality that people have when they are excited about what they are doing, even if it is interpreted as a mundane act to others. When one is excited, passionate, and satisfied with fulfilling an act of life, there is a certain charge that is being created in the vital force field (energy; electromagnetic field) of the individual. This energy creates magnetism, and this magnetism draws in universal energy, which manifests abundance.

Fulfilling a Mission Versus Collecting a Paycheck

Identifying your mission is integral to the process of attaining financial balance that leads to wealth and abundance, but equally important are the actions for implementation. There is a popular confusion with *work* being defined as paid employment. By expanding your thinking to defining *work* as any activity from which you derive purpose, fulfillment, and meaning—rather than limiting the view of *work* to paid employment—you can begin to value your activities and yourself differently.

Being limited in your ability to express your mission because of needing to meet the employer's agenda potentially can be disempowering to you. It is important to recognize this and allow your own mission to be expressed—even if it is through a hobby or volunteer activity rather than formal employment.

REFLECTION

It is possible that one's mission may not be expressed regularly in the daily place of employment. What are your avenues to expressing your purpose and mission?

From a Mechanistic to a Holistic Perspective on Money

Dealing with money from a physical perspective keeps it in the mechanistic model, which implies that money is a separate entity in and of itself. The making and utilization of money are confined to the physical realm. This suggests that there are no other influences on money other than cause and effect from the physical perspective—work, investments, savings, and budgeting increase available money.

There is another way to view money, however, involving a more holistic view. A crucial principle involved in the holistic paradigm is that the non-physical (thoughts, intentions, requests) influences the physical (concrete manifestation of the request). In other words, belief that a goal can be fulfilled can bring about desired results. Spiritual masters have promoted similar beliefs, including Jesus, who said, "All things whatsoever ye shall ask in prayer, believing, ye shall receive" (Matthew 21:22).

The nonphysical substance used in this holistic paradigm is a creative force that allows individuals to produce in the physical world what they are thinking. The creative force within people is not connected to personality or ego; however, it is important to identify the belief systems and perceptions under which personality operates to be aware of barriers that could impede the creative force.

After aligning with purpose, the next step is to focus on the request, desire, or goal that one wants to manifest. It is important to remember to leave results open ended by adding the words *or something better* to requests. At this point, it can be helpful to visualize what the end result may look like. For example, if you desire a specific amount of money, visualize a bank statement that shows that particular amount as your savings account balance or the exact item you wish to purchase with that amount of money.

KEY POINT

The scientific arena of quantum physics aids in understanding how the mind's creative force can manifest itself in the physical world by explaining that rapidly spinning atomic particles are invisible, but slower ones exist in the form of physical matter. Faster spinning particles can slow down and create physical forms. Consistent with this theory, thoughts can be manifested as physical matter.

EXHIBIT 11-1   MIND, BODY, AND SPIRIT HARMONY IN ACHIEVING FINANCIAL GOALS

- Recognize that there is a power greater than yourself that can affect your outcomes.
- Quiet your mind, and connect to the creative force of the universe.
- Clarify and align with your purpose and mission in life.
- State and visualize your desire and specific goal.
- Be aware of emotions connected to financial goals and actions.
- Release preconceived notions of the ways in which resources will be delivered and the forms in which they will come.
- Trust and have faith that your results will be achieved.
- Consider the ways in which you can share and use your money to benefit other people and purposes.
- Express gratitude for what you receive.

As it is important to visualize and see in the mind's eye what is desired (**Exhibit 11-1**), and at the same time, it is important to let go of the process and outcome. Releasing is the last and possibly the most difficult step in the creative manifestation process. The act of releasing the entire process and letting go of the outcome calls one to progress in faith and trust that a greater power will be in control.

We live in a time in which the ways we perceive the world and our lives are changing. We can access information from higher places and pay more attention to the higher senses of our being, such as intuition and belief in the nonphysical world. With this heightening in perspective comes more options and approaches for accessing creative solutions.

Summary

From a mechanistic view, money is viewed in terms of its physical characteristics; however, from a holistic view, money is more than a medium of exchange to purchase goods and services, representing an avenue through which energy flows. People's attitudes and beliefs about money and their financial well-being can influence the amount of money they receive and its use. To obtain financial goals successfully, people must have harmony of body, mind, and spirit.

Reference

Gaines, E. (2006). *The four spiritual laws of prosperity* [Audiocassette]. Valley Head, AL: Prosperity Products. Available from www.prosperityproducts.com

Suggested Reading

Allen, J. (2011). *Eight pillars of prosperity*. Seattle, WA: CreateSpace Publishers.

Babson, R. W. (2010). *Fundamentals of prosperity: What they are and whence they come*. Baltimore, MD: Wildside Press.

Barnett, E. W., Gordon, L. S., & Hendrix, M. A. (2001). *The big book of Presbyterian stewardship*. Louisville, KY: Geneva Press.

Blomberg, C. (2001). *Neither poverty or riches: A biblical theology of possessions*. Grand Rapids, MI: Eerdmans.

Chopra, D. (1998). *Creating affluence: The A to Z steps to a richer life*. Novato, CA: New World Library.

Chopra, D. (2007). *The seven spiritual laws of success. A pocketbook guide to fulfilling your dreams*. San Rafael, CA: Amber-Allen Publishing.

Dunham, L. (2002). *Graceful living: Your faith, values, and money in changing times*. New York, NY: Reformed Church in America Press.

Fillmore, C. (2009). *Prosperity*. New York, NY: Classic Books America.

Gawain, S. (2000). *Creating true prosperity*. Novato, CA: New World Library.

Gwartner, J. D., Stroup, R. L., Less, D. R., & Ferrarini, T. H. (2010). *Common sense economics: What everyone should know about wealth and prosperity*. New York, NY: St. Martin's Press.

Hill, N., Wattles, W. D., Allen, J., Frankllin, B. Holmes, E., et al. (2007). *The prosperity bible: The greatest writings of all time on the secrets to wealth and prosperity*. New York, NY: Tarcher.

Horton, C. (2002). *Consciously creating wealth*. Golden, CO: Higher Self Workshops.

Militz, A. R. (2011). *Both riches and honor* (Kindle ed.). Indianapolis, IN: Day Dream Company.

Murphy, L., & Nagel, T. (2002). *The myth of ownership: Taxes and justice*. New York, NY: Oxford University Press.

Ponder, C. (2003). *A prosperity love story: Rags to enrichment*. Camarillo, CA: DeVorss & Co.

Ponder, C. (2011). *The dynamic laws of prosperity*. BN Publishing. Available from www.bnpublishing.net

Ridley, W. (2011). *Rational optimism: How prosperity evolves*. New York, NY: Harper.

Walters, D. J. (2000). *Money magnetism: How to attract what you need when you need it*. Nevada City, CA: Crystal Clarity Publishers.

Wattles, W. (2009). *The science of getting rich*. Blacksburg, VA: Thrifty Publishers.

Promoting a Healthy, Healing Environment

12

OBJECTIVES

This chapter should enable you to

- Explain electromagnetic field
- Describe chakras
- Describe energy associated with different colors
- List five factors that can influence the vibrational fields
- List the names and locations of the seven main chakras of the body
- List at least six symptoms of toxicity
- List six sources of toxic substances
- Describe at least four ways that people can reduce environmental risks to the immune system
- Outline guidelines for detoxification

Everything in the environment—the food that you eat, the air you breathe, the sensations you experience—affects your body, positively or negatively. Maximizing the benefits of the environment, while reducing the negative impact, is important to promoting good health.

Internal and External Environments

You are affected by two types of environments—external and internal. *External environment* implies anything that is outside your body, such as the weather, elements in the air, food, sounds, and interactions with other people. Major environmental crises such as catastrophic floods, fires, and hurricanes affect you, as do the minor daily crises, such as waiting in long lines at the supermarket or getting stuck in traffic.

Internal environment pertains to that which is inside you. All that you encounter in your daily living is experienced within you and influences your thinking, feelings, and behavior. Often these events produce some degree of stress that you deal with uniquely based on your sensitivity, perception, awareness, flexibility, and adaptability. You use many skills and resources to balance the stressors from your external and internal environments and stay healthy.

REFLECTION

Think about times in your life when your stress levels were unusually high. Did you find that you had a higher incidence of illness or injury during or after those times?

Electromagnetic Fields

Life is a perpetual cycle as revealed in the constant changes that occur in nature—the changing tides, the seasons, the weather patterns, and the circadian rhythm of day and night followed by the final cycle of life and death. Being part of nature, all human beings are influenced by energy cycles within the physical body as well as the vibrational or universal body that corresponds to the electromagnetic field. The electromagnetic field is multidimensional and ranges from radio waves to gamma waves. The visible spectrum of the electromagnetic field ranges from infrared to ultraviolet.

The electromagnetic field is referred to as the vibrational field and an individual's energy body as the vibrational body. The field that surrounds the dense physical body is known by the scientific term *electromagnetic* or *bio-energy field*. This field is depicted by ancient Christian artisans who showed the presence of a light or halo emanating from the head or crown of religious men and women. The vibrational field is composed of at least seven layers that correspond to the physical, intellectual, and spiritual components of the visible body. Thus, both positive and negative external stimuli can have a profound impact on the body, mind, and spirit of an individual.

KEY POINT

During the 20th century, theorists such as Albert Einstein and Martha Rogers emerged with scientific theories that revealed man as an energy being. These theories challenged man's relationship with nature and the universe.

> **KEY POINT**
>
> People are more than skin and bones, thoughts and feelings. They are spectacular vibrational beings who resonate with the full-spectrum rays of the sun.

The vibrational field is hypothesized to be made of seven major chakras. A chakra (coming from the Sanskrit for *wheel of light*) is a spinning wheel or energy vortex that acts as a vibrational transformer within the field that extends outside the body. Although the chakra energy system is just now gaining recognition in the United States, ancient Indian Yogi literature described chakras approximately 5,000 years ago.

The vibrational body is, in part, also believed to be composed of seven major chakras, each of which connects with a nerve plexus, endocrine gland, and specific physiology and anatomy within a specific area of the body. The first five chakras extend from the base of the spinal column to the cervical spine, the sixth chakra to the pituitary gland, and the seventh chakra to the pineal gland. The seven major chakras resonate with the full-spectrum rays from red to violet. (**Exhibit 12-1** and **Figure 12-1**).

Each chakra resonates with a musical note, as well as the sequence of the seven colors in the rainbow from infrared to ultraviolet. The chakra energy system energetically connects with the associated endocrine glands, nerve plexus, and anatomy and physiology at the individual points of origin. For example, the fourth chakra—the heart chakra—is located mid-chest and corresponds to the thymus gland, cardiac nerve plexus, heart, and lungs.

> **KEY POINT**
>
> The environment consists of the objects, conditions, and situations that surround us.

Sunlight

Sunlight provides a complement of the full-spectrum rays that correlate with the vibration of the seven major chakras. Although artificial light does allow us to partake in life after dark, artificial light does not resonate with the colors of the chakras. A means to correct inadequate natural sunlight or poor artificial lighting is the use of a full-spectrum light bulb in primary rooms. These are not essential in every light source.

EXHIBIT 12-1   CHAKRAS

Chakra is the Sanskrit word for *spinning vortices* or *wheels of energy*. It is believed that chakras receive nourishing energy from the universe and send out energy from the body. They have distinct names, often described for the area of the body over which they reside.

 Root chakra comes downward from the base of the spine and relates to feeling secure in the world, feeling grounded to the earth, and having the will to survive. It deals with the sense of joy and vitality. The root chakra governs the adrenal glands, which are responsible for secreting hormones that, in frightening situations, help one to be more alert and escape danger if necessary. If the chakra is not balanced or unhealthy from prolonged stress, problems such as high blood pressure and anxiety can be seen. One often feels a sense of lacking presence, of not being here, or low on physical energy. Energy can be seen through color and heard through sound. The vibrational energy of the root chakra is associated with the color red (the slowest vibration on the color spectrum) and the sound of bumble bees. It can also be related to the first sound, *do*, in the scale, *do, re, mi, fa, sol, la, ti, do*.

 Sacral chakras are a pair of chakras located just below the umbilicus (belly button) on the front and lower back portion of the body and are related to the life force and vital body energy. The reproductive organs and the lower abdomen are affected by these chakras, especially the gonads (endocrine organs of sexuality and reproduction). Nutrients and the absorption of fluids through the intestine are also related to these chakras. The color vibration that relates to these chakras is orange and the sound is the one of a flute or the sound, *re*. Psychologically, these chakras deals with feelings and emotions. They also relate to the expression of sexuality and desires, pleasures, and feelings about reproduction.

 Solar plexus chakras are a pair of chakras located where the ribs form a V on the front of the body and the middle of the back of the body. They are associated with one's own power, strength, and ego identity. They also center around the issues of the will to think, control, exert authority, display aggression and warmth, as well as the physical components of digestion and metabolism. With these chakras intact, one has the ability to start things and see them through to completion. The color associated with these chakras is lemon yellow, and the sound is of the stringed instruments, such as the violin, or the sound, *mi*.

EXHIBIT 12-1   CHAKRAS (CONTINUED)

Heart chakras are a pair of chakras located between the breasts and between the shoulder blades and are the center of love. They have the qualities of loving self and others, harmony, compassion, openness, giving, peace, and grace. The heart chakras are called the transformation center because at the heart level one begins to deal with his or her spiritual side. Whereas the first three types of chakras deal with the physical, emotional, and mental components of the self, the heart center speaks to the soul. It is through these chakras that we experience the feeling of unconditional love or the deep love for others. This love is of a pure or spiritual kind of caring. The heart chakras are believed to be open when one cries tears of happiness or joy or in those instances where the heart strings are pulled. The color associated with the heart is emerald green, and the sound is of bells or fa.

Throat chakras are a pair of chakras located at the point of the Adam's apple in the neck and the curve of the neck on the back of the body. They deal with issues surrounding communication and expression. By using the voice, one is able to create sounds of speech or song to express oneself. The throat chakras relate to the neck, esophagus, throat, mouth, and ears. The sound of wind blowing through trees or the note *sol* is associated with these chakras. They resonate to the vibration color of turquoise or light blue. Some describe it as an electric blue. Physically, a healthy throat chakra is one revealing a clear voice that communicates thoughts, feelings, and ideas concisely. If the chakras are unbalanced, one may speak softly or with hoarseness. Having a "frog in the throat" or "choking on words" are descriptions of the free flow of speech being impeded. Other symptoms of imbalance are knots in the neck muscles, tight throats, and clenched jaws.

Brow chakras, also called the third eye, are a pair of chakras located in the middle of the forehead on the face and directly behind on the back of the head. They are associated with the pituitary gland, which controls most of the hormonal functions of the body. The brow chakras relate to memory, dreams, intuition, imagination, visualization, insight, and psychic perception. These chakras are all about seeing, within and without. The color associated with the brow chakras is deep indigo blue or deep purple. The sound of la or waves crashing on the beach help with this center.

(continues)

EXHIBIT 12-1   CHAKRAS (CONTINUED)

A *crown chakra*, a single chakra located on the top of the head, is the one with the highest energy vibration of the physical body. It is often called the spiritual center or center of connection to the oneness of the universe. The crown chakra relates to the divinity of self and our connection to a higher source however one defines it. The crown chakra also relates to intelligence, thought, consciousness, and information. An alignment to a higher power is seen through this center. It is associated with the colors white or lavender and the sound *om* or *ti*. The pineal gland is an endocrine gland that is not well understood. Its purpose is to deal with the body's rhythms and timing.

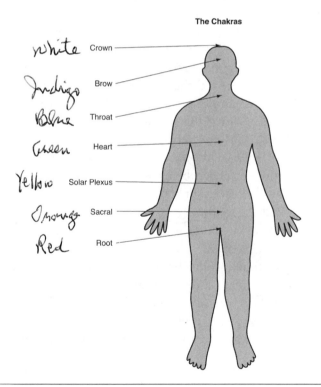

The Chakras

white — Crown

Indigo — Brow

Blue — Throat

Green — Heart

Yellow — Solar Plexus

Orange — Sacral

Red — Root

Figure 12-1   Chakras

> **KEY POINT**
>
> Seasonal affective disorder is a severe form of winter depression. It is characterized by lethargy, boredom, lack of joy, hopelessness, increased sleeping, and overeating. The primary factor responsible for this is the lack of natural sunlight. Natural sunlight affects the pineal gland (seventh chakra) to reduce its release of melatonin, which controls the circadian rhythm of the sleep–wake cycle. Although artificial light allows us to partake in life after dark, it does not resonate with the colors of the chakras.

Color

Color is a vibration that resonates with the chakra energy system, and it can influence the sense of well-being. The vibration flows in accordance with the full-spectrum of colors of the rainbow—red, orange, yellow, green, blue, indigo, and violet. Various colors are associated with different types of energy.

Violet: Spiritual energy connected to universal power and healing

Indigo: A blend of red and violet, stimulates intuition or the third eye

Blue: Invokes a sense of calm

Green: A healing quality, the color of nature that supports the heart chakra

Yellow: Associated with solar energy, power, and intellect

Orange: A nurturing energy related to creation, sunrise, and sunset

Red: Invigorating life force in small amounts and can be energetically overwhelming with large or continual exposure

It is useful for people to determine their favorite color. After the favorite color has been identified, people should determine how much this color appears in their personal environments. It is important that the favorite color be in the immediate environment either as an accent or as a primary color. With some planning, the color can be incorporated into the environment while respecting others who share the same space. For example, a bowl of fresh oranges could be sufficient to incorporate a favorite color of orange into an area rather than painting an entire room that color. Fresh flowers, plants, fruit bowls, pillows, jewelry, and clothing of the favorite color are useful vibrational remedies that can be incorporated into daily life. People also may want to pay attention to their life's circumstances to see whether their favorite color changes as result of their experiences.

KEY POINT

Color Therapy's Struggle for Acceptance
The history of color therapy dates from 1550 B.C. when Egyptian priests left papyrus manuscripts showing their use of color science in their healing temples. A renewed interest in color therapy developed over the past 2 centuries but was met with skepticism. For example, in 1878, Edwin Babbitt wrote a book that is referred to as the Materia Medica for color therapy or chromotherapy, yet during his lifetime, his thoughts and views were equally praised and criticized by his peers. In the early part of the 20th century, Dinshah Ghadiali developed Spectro-Chrome tonation, which was a system of attuned color waves that struggled for 40 years to gain acceptance from the American Medical Association. Roland Hunt promoted the meditative approach of color breathing and color visualization in his 1971 book, *The Seven Keys to Color Healing*; however, this had little impact in mainstream healthcare circles.

Clutter

Clutter is a major stress factor that potentially irritates the nervous system. It not only disturbs the visual field but the vibrational field as well. This is due to the fact that the vibrational field exists beyond the skin; therefore, clutter in a space can cause a restriction of that field.

Vibrational remedies for clutter begin with the simple act of examining the environment with a critical eye. People can be challenged to consider how they would like their homes to look if a special guest was visiting. They can be encouraged to begin the habit of not bringing a new item into their homes without discarding or giving away something in return. Reorganizing living space and donating unnecessary belongings can be beneficial, as can recycling (which not only reduces clutter, but helps to protect the earth). In addition to the reward of a more harmonious environment, people most likely will discover more floor space to use for relaxing or exercising.

Sound

The link between music and health is ancient and elemental. Music therapy influences the chakra energy system according to vibration and tone. The natural rhythm of the ocean, raindrops, wind-ruffled leaves, and birds chirping universally promote relaxation. Manmade music dates to the ancient cultures and mythology. Apollo, known as the god of medicine, also was the god of music.

Drumming correlates to the human heartbeat. New age music with synthesizers and string instruments connects with the higher vibrations of the chakra energy system. Today, classical music is incorporated into surgical suites, dentists' offices, and hospice facilities as a holistic, stress-reducing measure.

REFLECTION

Try closing your eyes and noticing the sounds in your immediate environment. List three pleasant sounds, for example, the voices of loved ones, sounds of nature, even the absence of sound (peace and quiet). Now, list all of the negative sounds in your home or office: unpleasant verbal interactions, loud music, machinery, persistent phone ringing, household appliances, and so forth. Be aware that noise pollution can disturb the nervous system, creating an increase in stress levels. Compare your lists and try to make adjustments to reduce auditory overload and enhance positive aspects of sound in your environment.

KEY POINT

The vibrational body can be soothed through the sound of a water fountain. People can create their own fountains with a fish tank, water pump, deep pottery bowl, and imagination.

Scents

The vibrational body adapts to most situations over time. The sense of smell is an ideal example of this. Normally, the sense of smell is keen when a scent or odor is first detected, but it then becomes less acute with continued exposure. This can create a problem in that toxic substances can permeate living areas without detection. Toxic chemicals found in the home or workplace can lead to a host of symptoms, such as headaches, nasal stuffiness, and shortness of breath related to chemical sensitivity.

Aromatherapy is a holistic approach that uses the essential oils extracted from plants, flowers, resins, and roots for therapeutic effects. It has an important role in promoting a healing environment.

To promote a healthy sense of smell, people should use all-natural cleaning agents and scents. It is important for people to learn to be wise consumers by reading labels looking for toxins in all products that contain artificial chemicals. Less is better when aromatics are used in the home.

KEY POINT

Radon is an odorless, inert radioactive gas that is the natural radioactive decay of radium and uranium found in the soil. Excessive levels of radon have been linked to an increased lung cancer risk. (There is a very high incidence of lung cancer in uranium miners.) Many factors determine the amount of radon that escapes from the soil to enter a house—weather, soil porosity, soil moisture, granite, and ledge. The incidence of radon is greater from water supplied from a private well versus a public water system. (Call the EPA hotline to learn more about this problem at 1-800-SOS-RADON.)

KEY POINT

The basic ingredients for natural cleaning agents are white vinegar and baking soda. Essential oils such as eucalyptus, lemon, lavender, and tea tree can be added to these agents to offer antibacterial and antiviral effects.

Feng Shui

Feng shui, which grew out of Chinese astrology, is the Chinese art of arranging the environment to affect a person's internal state positively. It offers ways to align pieces of furniture that are said to promote better health, prosperity, and overall well-being. The attributes of the five elements—water, wood, fire, earth, and metal—are considered in planning an environment that is best suited for an individual.

Flavors

Food preparation is one of the greatest expressions of love. Love and caring are especially evident when an effort is made to maintain the integrity, quality, and nutritional value of all the foods that are served. Good, old-fashioned home cooking can do wonders to restore the body, mind, and spirit.

A rainbow diet satisfies nutritional needs as well as the vibration of the seven chakras. To follow a rainbow diet, people need to eat as many natural foods (fruits and vegetables) in as many colors as there are in the rainbow. Examples of this could include red apples, orange squash, yellow lemons, green broccoli, blueberries, indigo eggplant, and purple potatoes.

The rainbow diet also provides a variety of flavors, which enhances food consumption.

Touch

Touch is imperative in the development of a healthy life; nevertheless, the advent of technology has contributed to an era of touch deprivation. The ease of computer communications reduces personal interactions, such as a friendly handshake or a hearty hug. Studies have revealed that babies in hospitals or orphanages who were not picked up or nurtured suffered from a life-threatening syndrome, failure to thrive.

REFLECTION

Consider the amount of touch in your own life and how you can increase opportunities for touch, such as classic hugs and handshakes.

A mere touch on the hand with the intention to send healing love to someone can make a world of difference. Pets provide opportunities to touch and be touched in return. The value of touch is vital to life enhancement.

Intuition or higher sense perception is accessible to everyone. It is that small inner voice—a knowing without knowing how or why—that nudges you along to pay attention to positive and negative aspects of daily life. Pay attention to caustic or harmful situations in your environment. Invite your intuitive thoughts to surface regularly as a barometer of your personal healing environment.

Immune System

The body's internal environment is safeguarded by the immune system whose primary purpose is to fight disease-producing microorganisms, such as bacteria, viruses, and fungi in order to protect the body from infections and illnesses. In other words, the immune system helps to maximize your health potential as well as to maintain your health. The weakening of the immune system frequently results in an increased susceptibility to unpleasant symptoms, illnesses, and diseases. If the immune system is severely weakened or suppressed, it cannot destroy or reprogram the cells in the body. Environmental conditions influence the normal function of the immune system and can threaten its optimum function.

Toxicity

Toxicity can result from environmental agents that produce poisonous substances that cause an unhealthy environment within the body, leading to physiological or psychological problems. The accumulation of toxins weakens the immune system. Initially, mild symptoms affecting the body organs, such as a headache, sneezing, swollen eyelids, or irritability may be experienced. Other physical symptoms often include fatigue, restlessness, insomnia, difficulty breathing, and confusion. At first, these symptoms are faint and may be missed or minimized; however, when such symptoms continue over an extended period of time, they are difficult to ignore. Toxicity that continues for a long time begins to impede the normal functions of the body's organs and systems. This results in an imbalance, leading to impairments in normal digestion, walking, and thinking.

KEY POINT

The word toxicity generally refers to a poisonous, disease-producing substance that is produced by a microorganism.

Heavy Metals

In your daily environment, you are constantly exposed to environmental factors that lead to the development of toxicity in your body. Of these, the major category of toxic substances is heavy metals.

Aluminum, arsenic, cadmium, lead, and mercury are examples of heavy metals. They are found in air, food, and water and have no function in the human body. Over a long period of time, these heavy metals accumulate in the body and reach toxic levels. Heavy metals tend to accumulate in the brain, kidneys, and immune system, where they can severely disrupt normal functioning. Heavy-metal toxicity causes unusual symptoms and ailments that tend to linger for an extraordinary length of time. People with symptoms related to this toxicity may visit several medical doctors to find out what is wrong, but often physicians are unable to identify and treat the problem. Frequently, medications are prescribed for symptomatic relief but may provide minimal to no relief.

Transportation companies, waste management companies, and manufacturers have been major sources of heavy-metal pollution. For example, many

KEY POINT

Toxic substances can be found in a variety of sources, such as heavy metals, chemicals, parasites, pesticides, and radiation.

KEY POINT

Overall, the heavy metals tend to (U.S. Department of Labor, 2008):

- Decrease the function of the immune system.
- Increase allergic reactions.
- Alter genetic mutation.
- Increase acidity of the blood.
- Increase inflammation of arteries and tissues.
- Increase hardening of artery walls.
- Increase progressive blockage of arteries.

manufacturing plants throughout the United States have polluted the air with poisonous toxins, particularly lead, from their industrial processes and productivity. **Exhibit 12-2** lists the common sources of the five heavy metals— aluminum, arsenic, cadmium, lead, and mercury.

Aluminum

The average person ingests between 3 and 10 milligrams of aluminum each day (Becaria, Campbell, & Bondy, 2002). Of the various sources of aluminum, the highest exposure comes from the chronic consumption of aluminum-containing antacid products. It is primarily absorbed through the digestive tract as well as through the other organs, such as the lungs and skin. Research has suggested that the heavy metal aluminum contributes to neurologic disorders such as Parkinson's disease, dementia, clumsiness of movements, staggering when walking, and the inability to pronounce words properly (Becaria et al., 2002). Aluminum tends to accumulate in the brain, bones, kidneys, and stomach tissues. Consequently, the common health problems caused by aluminum are colic, irritation of the esophagus, gastroenteritis, kidney damage, liver dysfunction, loss of appetite, loss of balance, muscle pain, dementia, psychosis, seizures, shortness of breath, and general weakness (Becaria et al., 2002).

EXHIBIT 12-2   COMMON SOURCES OF HEAVY METAL TOXINS

TYPE OF TOXINS

ALUMINUM
Source
Aluminum cookware
Aluminum foil
Antacids
Antiperspirants
Baking powder (Aluminum
containing)
Bleached flour
Buffered aspirin
Canned acidic foods
Cooking utensils
Cookware
Dental amalgams
Foil
Food additives
Medications—drugs
Antidiarrheal agents
Anti-inflammatory agents
Hemorrhoid medications
Vaginal douches
Processed cheese
"Softened" water
Table salt
Tap water
Toothpaste

TYPE OF TOXINS
ARSENIC
Source
Air pollution
Antibiotics (given to commercial livestock)
Bone meal
Certain marine plants

EXHIBIT 12-2   COMMON SOURCES OF HEAVY METAL TOXINS (CONTINUED)

Chemical processing
Coal-fired power plants
Defoliants
Dolomite
Drinking water
Drying agents for cotton
Fish
Herbicides
Insecticides
Kelp
Laundry aids
Meats (from commercially raised poultry and cattle)
Metal ore
Pesticides
Seafood (fish, mussels, oysters)
Smelting
Smog
Smoke
Specialty glass
Table salt
Tap water
Tobacco
Wood preservatives

TYPE OF TOXINS
CADMIUM
Source
Air pollution
Art supplies
Bone meal
Cigarette smoke
Fertilizers
Food (coffee, tea, fruits, and vegetables grown in cadmium-laden soil)
Freshwater fish

(continues)

EXHIBIT 12-2   COMMON SOURCES OF HEAVY METAL TOXINS (CONTINUED)

Fungicides
Highway dusts
Incinerators
Meats (kidneys, liver, poultry)
Mining
Nickel-cadmium batteries
Oxide dusts
Paints
Pesticides
Phosphate fertilizers
Plastics
Power plants
Refined foods
Refined grains
Seafood (crab, flounder, mussels, oysters, scallops)
Secondhand smoke
Sewage sludge
Soft drinks
Soil
"Softened" water
Smelting plants
Welding fumes

TYPE OF TOXINS
LEAD
Source
Air pollution
Ammunition (shot and bullets)
Bathtubs (cast iron, porcelain steel)
Batteries
Canned foods
Ceramics
Colored advertisements
Chemical fertilizers
Cosmetics

EXHIBIT 12-2   COMMON SOURCES OF HEAVY METAL TOXINS (CONTINUED)

Dolomite
Dust
Food grown near industrial areas
Gasoline
Hair dyes and rinses
Leaded glass
Newsprint
Paints
Pesticides
Pewter
Pottery
Rubber toys
"Softened" water
Solder in tin cans
Tap water
Tobacco smoke

TYPE OF TOXINS
MERCURY
Source
Contaminated fish
Cosmetics
Dental fillings

Sources: Adapted from Balch & Balch (1997), Marti (1995), Post-Gazette (10/14/99), Pouls (1999), Ronzio (1999).

Arsenic

Arsenic is another highly toxic substance. It has an affinity for most of the bodily organs, especially the gastrointestinal system, lungs, skin, hair, and nails. In acute arsenic poisoning, nausea and vomiting, bloody urine, muscle cramps, fatigue, hair loss, and convulsions are noted. Headaches, confusion, abdominal pain, burning of the mouth and throat, diarrhea, and drowsiness may occur in chronic arsenic poisoning. Arsenic toxicity can contribute to the development of cancer (lung and skin), coma, neuritis, peripheral vascular (vessels of the extremities) problems, and collapse of blood vessels (U.S. Department of Labor, 2008).

Cadmium

Cadmium is also considered an extremely poisonous heavy metal. Inhaling cadmium fumes causes pulmonary edema, followed by pneumonia and various degrees of lung damage. Another way that a person can acquire cadmium poisoning is by ingesting foods contaminated by cadmium-plated containers; these cause violent gastrointestinal symptoms. Cadmium levels also arise in individuals who have zinc deficiencies.

KEY POINT

The health problems that can arise from cadmium toxicity are anemia, cancer, depressed immune system response, dry and scaly skin, emphysema, eye damage, fatigue, hair loss, heart disease, hypertension, increased risk of cataract formation, joint pain, kidney stones or damage, liver dysfunction or damage, loss of appetite, loss of sense of smell, pain in the back and legs, and yellow discoloration of the teeth. These problems occur because cadmium tends to gravitate to tissues of specific body organs, including the brain and its pain centers, the heart and its blood vessels, the kidneys, and the lungs, as well as the tissues that influence the appetite (U.S. Department of Labor, 2008).

Lead

Of all of the common heavy metals, lead has attracted public attention since the early 1900s. At the turn of the 20th century, paint companies knew that lead was a serious risk to health; however, they continued to produce building paints that contained a lead base. These paints were used in homes and commercial buildings. Consequently, children were seriously affected by lead poisoning. In addition, the Environmental Protection Agency estimated that drinking water accounts for approximately 20% of young children's lead exposure (EPA, 2008). Health problems in children caused by lead poisoning include attention deficit disorder, hyperactivity syndromes, learning disabilities, loss of appetite, low achievement scores, low intelligence quotients, seizures, and, on occasion, death. Because of the public outcry against lead-based paint, it was banned in 1978. At the same time, the government issued regulations cutting out lead in gasoline.

Lead poisoning affects cognitive (intellectual) function. Studies have revealed that low-level lead exposure impairs children's IQ (Binns et al., 2007). In addition to the danger for young children, there is also a great risk of harm from lead exposure for infants and pregnant women. In pregnant women, lead toxicity may cause premature birth, miscarriage, and birth defects. Lead toxicity may produce a variety of additional symptoms, including abdominal pain, anemia, anorexia, anxiety, bone pain, brain damage, coma, confusion, constipation, convulsions, dizziness, drowsiness, fatigue, heart attack, headaches, high blood pressure, inability to concentrate, indigestion, irritability, kidney disease, loss of muscle coordination, memory difficulties, mental depression, mental damage, muscle pain, nervous system damage, neurological damage, pallor, and vomiting.

Mercury

As Exhibit 12-2 shows, mercury is found in many sources that we use in our daily life. This toxic metal accumulates in the bones, brain, heart, kidneys, liver, nervous system, and pancreas. Some studies have shown a relationship between unusual symptoms and mercury exposure from silver fillings, finding that a major source of mercury toxicity in the human body was the dental filling, and when mercury fillings were removed, the symptoms and related health problems disappeared; however, recent research is challenging the claim that mercury fillings are dangerous (Bellinger et al., 2006). Although mercury levels in blood and urine do increase somewhat with amalgam dental fillings, toxic levels apparently do not occur. Elevated mercury levels in urine and hair merely reflect the ongoing elimination of mercury from the body by these routes (Bellinger et al., 2006).

Reducing Environmental Risks to Health

Assess

To begin helping yourself, you need to take an objective look at yourself. One method of self-assessment is the Toxicity Self-Test, shown in **Exhibit 12-3**. It involves checking off symptoms frequently experienced within 15 common areas and the degree to which each symptom has been noticed; then the score is totaled. The final score may fall in the mild, moderate, or severe range, as indicated at the end of the Toxicity Self-Test checklist. After this assessment is completed, you can explore ways of reducing the toxicity that exists in your body. There are daily positive actions that you can take to reduce or prevent the accumulation of toxins in the body.

EXHIBIT 12-3   TOXICITY SELF-TEST

Rate each of the following symptoms based on your health profile for the past 30 days.

Point Scale:

0 = never or almost never have the symptom

1 = occasionally have the symptom; effect is not severe

2 = occasionally have the symptom; effect is severe

3 = frequently have the symptom; effect is not severe

4 = frequently have the symptom; effect is severe

1. Digestive Function

___ Nausea or vomiting

___ Diarrhea

___ Constipation

___ Bloated feeling

___ Belching, passing gas

___ Heartburn

___ TOTAL

2. Ears

___ Itchy ears

___ Earaches, ear infection

___ Drainage from ear

___ Ringing in ears, hearing loss

___ TOTAL

3. Emotions

___ Mood swings

___ Anxiety, fear, nervousness

___ Anger, irritability

___ Depression

___ TOTAL

EXHIBIT 12-3   TOXICITY SELF-TEST (CONTINUED)

4. Energy/Activity
___ Fatigue, sluggishness
___ Apathy, lethargy
___ Hyperactivity
___ Restlessness
___ TOTAL

5. Eyes
___ Watery, itchy eyes
___ Swollen, reddened, or sticky eyelids
___ Dark circles under eyes
___ Blurred vision/tunnel vision
___ TOTAL

6. Head
___ Headache
___ Faintness
___ Dizziness
___ Insomnia
___ TOTAL

7. Lungs
___ Chest congestion
___ Asthma, bronchitis
___ Shortness of breath
___ Difficulty breathing
___ TOTAL

8. Mind
___ Poor memory
___ Confusion
___ Poor concentration
___ Poor coordination
___ Difficulty making decisions

(continues)

EXHIBIT 12-3   TOXICITY SELF-TEST (CONTINUED)

___ Stuttering, stammering
___ Slurred speech
___ Learning disabilities

___ TOTAL

9. Mouth/Throat
___ Chronic coughing
___ Gagging, frequent need to clear throat
___ Sore throat, hoarse
___ Swollen or discolored tongue, gums, lips
___ Canker sores

___ TOTAL

10. Nose
___ Stuffy nose
___ Sinus problems
___ Hay fever
___ Sneezing attacks
___ Excessive mucus

___ TOTAL

11. Skin
___ Acne
___ Hives, rash, dry skin
___ Hair loss
___ Flushing or hot flashes
___ Excessive sweating

___ TOTAL

12. Heart
___ Skipped heartbeats
___ Rapid heartbeat
___ Chest pains

___ TOTAL

EXHIBIT 12-3   TOXICITY SELF-TEST (CONTINUED)

___

13. Joints/Muscles
___ Pain or aches in muscles
___ Arthritis
___ Stiffness, limited movement
___ Pain or aches in muscles
___ Feelings of weakness or tiredness
___ TOTAL

14. Weight
___ Binge eating/drinking
___ Craving certain foods
___ Excessive weight
___ Compulsive eating
___ Water retention
___ Underweight
___ TOTAL

15. Other
___ Frequent illness
___ Frequent or urgent urination
___ Genital itch and/or discharge
___ TOTAL

Add the numbers to arrive at a total for each section. Add the totals for each section to arrive at the grand total.

Mild toxicity = 0–96 grand total score

Moderate toxicity = 97–168 grand total score

Severe toxicity = 169–240 grand total score

Attend to the Basics

With the fast pace of life and all the demands of family and work, it is easy to neglect yourself. This negligence affects your sleep, rest, relaxation, and leisure. Reflect on your daily activities and ask, "Am I getting adequate rest? Do I awaken feeling refreshed? Am I taking time to relax and enjoy the simple pleasures of life?"

It is important to make sure you are getting sufficient water in order to flush toxins from the body. Be sure to drink at least eight glasses of chemical-free water daily.

REFLECTION

Are you drinking at least eight glasses of water daily to flush toxins from your body? How can you ensure the water you drink is free from contamination by lead and other chemicals?

Relax

Relaxation can be achieved by using a variety of techniques, such as those that follow.

- *Guided imagery.* Investigative findings have revealed a relationship between lowered immune function and health problems. Guided imagery is done by creating mental scenes in one's mind and has been shown to be helpful in strengthening the immune system. This imagery technique, also known as mental imagery or visual imagery, uses the body–mind connection to alleviate energy imbalances in the body.
- *Meditation.* Meditation is the practice and system of thought that incorporates exercises to attain bodily or mental control and well-being, as well as enlightenment. Studies have shown that a relationship exists between meditation and increased immunity. There are several ways to meditate. One person may prefer to meditate to a particular sound, whereas another may focus on breathing. You may need to try a few styles to determine your personal preference.
- *Deep breathing.* Deep breathing, or belly breathing, can promote relaxation.

Get a Massage

Deep-tissue massages or lymph massages are helpful for the release of toxicity and balancing energy. In deep-tissue massage, the therapist applies greater pressure and focuses on deeper muscles than in the Swedish massage, which

KEY POINT

Research has shown that declines in IQ are associated with higher blood concentrations of lead (Binns, Campbell, & Brown, 2007).

primarily aims to promote relaxation. The purpose of lymph massage is to increase the circulation throughout the lymphatic system. Through massage of deeper tissues and increased circulation, toxins are mobilized, released, and eventually eliminated from the body. Beware that deep-tissue massage is not recommended for individuals who have high blood pressure, a history of inflammation of the veins, or any other circulatory problem.

Sweat It Out

Engaging in regular saunas can help relieve the body of toxins through the skin. If you choose to use a sauna, it is best to avoid eating within 1 hour before the sauna to avoid nausea.

Brush Your Skin

Another valuable technique is brushing the skin with a dry brush that has firm natural bristles. The purpose of skin brushing is to rid the body of the poisonous substances that are excreted through perspiration.

Detoxify

Another simple method for the general elimination of toxins is cleansing the bowel. This is a form of detoxification therapy.

Detoxification, a noninvasive process, has been quite popular in many healthcare systems around the world, especially Europe—to a much greater extent than in the United States. Detoxification is essential as a first step in clearing the body of toxicity because a person cannot rebuild nor maintain health if the toxins remain stored in the body. The majority of toxins tend to accumulate in the bowel where waste matter in the intestines remains for lengthy periods of time. This allows the toxins to be absorbed into the bloodstream throughout the body, causing health problems. After the toxins are cleansed out of the bowel, the body can begin to heal itself. Typically, a fiber substance, such as ground flaxseeds or psyllium husks in powdered form is taken to aid the body in eliminating the toxins. **Exhibit 12-4** shows the effect of toxins on unhealthy and healthy functions within the body.

KEY POINT

Detoxification was strongly recommended by pioneers in nutritional and natural medicine, including Bernard Jensen, John H. Kellogg, Max Gerson, and John Tilden.

EXHIBIT 12-4   IMPACT OF TOXINS ON UNHEALTHY AND HEALTHY FUNCTIONS IN THE BODY

UNHEALTHY

- Toxins form internally, leak through the unhealthy intestine, and flow to the liver.
- Toxins are not completely detoxified in the unhealthy liver.
- Unchanged toxins leave the liver and are stored in tissues, such as fat, the brain, and the nervous system.

HEALTHY

- Few toxins are formed, and most of them are excreted, with only a small amount naturally transported to the liver.
- Toxins are transformed to an intermediate substance.
- The intermediate substance is transformed to a more water soluble substance and released to the kidneys.
- The water soluble substance is excreted via the urine.

Source: Detoxification, San Clemente, CA: Metagenics, Inc. August, 1994.

KEY POINT

General Guidelines for Detoxification

In using the detoxification process, be aware that it must be done slowly. The reason for this is that the body must adapt to the changes that are occurring. If the detoxification occurs rapidly, unusual symptoms can develop because of the bodily changes that are causing an imbalance. If the detoxification is performed at a slower pace, the body develops its equilibrium while it is eliminating the toxins. Following the detoxification, it is generally recommended that you:

- Drink at least 8 ounces of water each day. Water helps the body to get rid of the debris and also provides sufficient liquid to help with elimination.
- Begin eating raw foods, including vegetables. This provides fiber, a natural substance, that helps with rebuilding tissue in the body.

Continue to avoid eating the foods listed in **Exhibit 12-5**. The avoidance of the suggested foods will decrease the faulty bowel function that occurs with an imbalanced diet.

> ### EXHIBIT 12-5   RECOMMENDED NUTRITIONAL CHANGES TO REDUCE TOXICITY
>
> **AVOID THE FOLLOWING:**
> - Alcohol, drugs, cigarettes
> - Caffeinated drinks
> - Chlorinated (tap) water
> - Commercially prepared foods
> - Fats
> - Foods high in additives and preservatives
> - Hydrogenated and partially hydrogenated oils
> - Fried foods
> - Heated polyunsaturated fats (fast food oils, theater popcorn oil)
> - Monosodium glutamate
> - Refined sugars
> - Soft drinks
> - Softened tap water
> - Topical oils (cottonseed, palm)
> - White flour foods

REFLECTION

What can you do to reduce exposure to toxins?

Contact with environmental pollutants is a reality of daily life. You come in contact with them in the air you breathe, the foods you eat, and the water you drink. These toxic substances can threaten your health in several ways and lead to major health problems such as lung damage, mood changes, neurological dysfunction, tissue damage, and visual disturbances. By avoiding exposure as much as possible and minimizing or eliminating the buildup of toxins within your tissues and organs, you can protect your health and enjoy maximum function. **Exhibit 12-6** gives specific suggestions to help avoid exposure to toxicity in our work, homes, and community environments.

Summary

The body is more than physical substance. An electromagnetic field that offers energy to promote health surrounds the body. It is theorized that the body also has a vibrational field made of seven chakras, which are spinning wheels

---

EXHIBIT 12-6   BASIC SUGGESTIONS TO MINIMIZE TOXICITY IN THE WORKPLACE, HOME, AND COMMUNITY ENVIRONMENTS

- Safely store or remove, if possible, any potentially toxic materials, including the following:

  Acids

  Cleaning agents

  Dyes

  Glues

  Insecticides

  Paints

  Solvents
- Use an air purification system in the home, if materials cannot be removed.
- Wear protective clothing and/or breathing apparatus when using any toxic materials.
- Replace furnace and air conditioning filters in the home on a regular basis.
- Eat fresh, wholesome foods, including fruits, vegetables, and grains.
- Avoid using pesticides and herbicides.
- Avoid smoking cigarettes and being exposed to secondhand smoke.

---

of energy. Each chakra resonates with a musical tone and represents colors in the rainbow from infrared to ultraviolet. Various colors are associated with different types of energy. The vibrational fields are influenced by color, clutter, sound, scents, flavors, and touch. People can influence their health and well-being by selecting colors that promote the desired feeling, reducing clutter, controlling noise, selecting music that is soothing, using scents therapeutically, and increasing touch contact with other people and pets. People need to develop sensitivity to the reality that positive and negative external stimuli can have a profound impact on the body, mind, and spirit.

Internal and external environments affect the health of the immune system. The accumulation of toxins can weaken the immune system and can cause a wide range of dysfunctions of the body and mind. In normal daily life, exposure to heavy metals, parasites, pesticides, radiation, and geopathic stress (natural radiation that arises from the earth) can lead to toxicity and cause serious health consequences. It is important to identify symptoms associated with toxicity and implement measures to reduce it. The approaches to reduce or prevent the accumulation of toxins in the body include good basic health practices, relaxation techniques, deep tissue massage, sauna, skin brushing, consumption of nutritional supplements, and detoxification. Because exposure to toxins is a reality of life for the average person, active steps must be taken to prevent exposure and strengthen the body's ability to resist the ill effects of toxins.

References

Becaria, A., Campbell, A., & Bondy, S. C. (2002). Aluminum as a toxicant. *Toxicology and Industrial Health, 18*, 309–320.

Bellinger, D. C., Trachtenberg, F., Barregard, L., Tavares, M., Cernichiari, E., Daniel, D., & McKinlay, S. (2006). Neuropsychological and renal effects of dental amalgam in children: A randomized clinical trial. *JAMA, 295*(15), 1775–1783.

Binns, H. J., Campbell, C., & Brown, M. J., for the Advisory Committee on Childhood Lead Poisoning. (2007). Interpreting and managing blood lead levels of less than 10 µg/dl in children and reducing childhood exposure to lead: Recommendations of the Centers for Disease Control and Prevention Advisory Committee on Childhood Lead Poisoning Prevention. *Pediatrics, 120*, e1285–e1298. Retrieved from http://pediatrics.aappublications.org/cgi/content/abstract/120/5/e1285

Environmental Protection Agency (EPA). (2008). *Lead in drinking water.* Retrieved from http://water.epa .gov/drink/info/lead/index.cfm

U.S. Department of Labor, Occupational Safety and Health Administration. (2008). *Heavy metals.* Retrieved from http://www.osha.gov/SLTC/metalsheavy/index.html

Suggested Reading

Asante-Duah, K. (2002). *Public health risk assessment for human exposure to chemicals.* Norwell, MA: Kluwer Academic Press.

Ayres, J., Maynard, R. L., & Richards, R. (2006). *Air pollution and health.* Academic Press. Burlington, MA: Elsevier.

Babbitt, E. (1942). *The principles of light and color.* Whitefish, MT: Kessinger Publishing.

Bennett, P. (2001). *7-day detox miracle.* Rocklin, CA: Prima Publishing.

Blanc, P. D. (2009). *How everyday products make people sick: Toxins at home and in the workplace, updated and expanded.* Berkeley: University of California Press.

Fitzgerald, P. (2001). *The detox solution.* Santa Monica, CA: Illumination Press.

Friis, R. W. (2010). *Essentials of environmental health.* Sudbury, MA: Jones & Bartlett Learning.

Hunt, R. (1971). *The seven keys to color therapy.* San Francisco, CA: Harper & Row Publishers.

Krohn, J., & Taylor, F. (2002). *Natural detoxification* (2nd ed.). Point Roberts, WA: Hartley and Marks Publishers.

Landner, L., & Reuther, R. (2004). *Metals in society and in the environment: A critical review of current knowledge on fluxes, speciation, bioavailability and risk for adverse effects of nickel and zinc.* Norwell, MA: Kluwer Academic Press.

Lembo, M. A. (2011). *Chakra awakening: Transform your reality using crystals, color, aromatherapy & the power of positive thought.* Woodbury, MN: Llewellyn Worldwide.

Rana, S. V. S. (2006). *Environmental pollution: Health and toxicology.* Oxford, UK: Alpha Science.

Slaga, T. (2003). *The detox revolution.* New York, NY: Contemporary Books/McGraw Hill Co.

The Therapeutic Benefits of Humor

CHAPTER

13

OBJECTIVES

This chapter should enable you to

- Define *gelotology*
- List at least six physiological effects of laughter
- Identify positive from negative humor
- List at least three ways to develop a comic vision
- Describe at least four strategies for adding humor to work and home

Humor adds perspective to life, altering our perception of a potentially negative incident and providing preventive maintenance against the strain of hard times in personal and professional relationships.

A man was upset and taking it out on coworkers. A friend thought to query, "What's the matter, John, did you have nails for breakfast?" This quip got John's attention. He realized he was bringing other problems to work and was able to laugh at himself.

The nurse leader on a busy hospital unit would walk down the hall wearing a pair of huge sunglasses when staff was rushing and stressed. This was the sign to take a deep breath, have a good laugh, and regroup. One long-term patient would tell her, "Lois, get the glasses. We need them today."

As the previous situations demonstrate, humor can alter the perception of a situation by changing the expectation of a negative result. A humorous

> **KEY POINT**
>
> Laughter is the shortest distance between two people.

perspective helps you to deal with stressors in life, from minor irritations to life-threatening illness.

Humor and a playful attitude can build relationships at home and at work and prevent negative reactions in stressful times. In an atmosphere of general goodwill, where no one is expected to be perfect at all times, tough times become more manageable. The mother who can make a game out of power outages by roasting hot dogs in the fireplace models a positive coping strategy for unexpected inconveniences. The manager who organizes a baby picture contest for staff mobilizes positive energy, provides something to look forward to at work, and humanizes team members.

Health Benefits of Humor

Gelotology is the study of the physiological effects of humor and laughter. This body of knowledge tells us that humor and laughter benefits us in many ways (Bennet & Deane, 2006; Klein, 2008; Sullivan & Deane, 1988), including the following:

- Stimulation of the production of catecholamines and hormones, which enhance feelings of well-being and pain tolerance
- A decrease in anxiety
- An increase in endorphins (natural narcotic-like substances produced in the brain)
- An increase in cardiac and respiratory rates
- Enhancement of metabolism
- An improvement of muscle tone
- Perception of the relief of stress and tension with increased relaxation, which may last up to 45 minutes following laughter
- Increased numbers of natural killer cells, which fight viral infections of cells and some cancer cells
- Increased T cells (T lymphocytes) that fight infection
- Increased antibody immunoglobulin A, which fights upper respiratory infections
- Increased gamma interferon, which helps activate the immune system

Laughter exercises the breathing muscles, benefits the cardiovascular system by increasing oxygenation, and promotes relaxation. Laughter helps control pain by distracting one's attention, reducing tension, and changing or reframing one's perspective. During episodes of laughter, the blood pressure increases, but then lowers below the initial rate. Studies have shown that laughter can help in fighting the negative effects of stress. For example, in a study at Loma Linda University School of Medicine, blood samples were drawn after subjects were shown humor videos. Results were compared with those of a control group. The mirthful experience appeared to reduce serum cortisol, 3, 4-dihydroxyphenylacetic acid, epinephrine, and growth hormone. These changes are related to the reversal of the neuroendocrine and classical stress hormone response.

> **KEY POINT**
>
> In times of stress, the adrenal glands release corticosteroids, which are converted to cortisol in the blood. Increased levels of cortisol can suppress the immune system. If laughter reduces serum cortisol, it may diminish the chemical effects of stress on the immune system.

Hospitals, nursing homes, hospices, and rehabilitation centers have implemented humor programs with a humor cart or designated room equipped with humorous audiocassettes, videos, games, and toys.

Mention

> **KEY POINT**
>
> Attention to the relationship of laughter and healing was profoundly increased when author Norman Cousins shared his personal experience. Trying to cope with pain from the inflammatory disease ankylosing spondylitis, Cousins watched reruns of *Candid Camera* and old Marx Brothers films and had people read humorous material to him. He found that 10 minutes of belly laughter gave him 2 hours of pain-free sleep. Blood studies of his sedimentation rate (a measure of the inflammatory response), which were drawn after his laughter therapy, showed a cumulative positive effect. He had reasoned that if being in a bad mood makes people feel worse, perhaps being in a good mood would make him feel better (Cousins, 1979)!

Norm Cousins

Positive Versus Negative Humor

Not all humor is positive. Have you ever been on the receiving end of negative humor? Perhaps someone got a laugh at your expense. You may have felt hurt and angry and not seen anything funny about the situation. To make matters worse, your reaction may have been met with, "Can't you take a joke?" making it seem that you were the one with a problem.

REFLECTION

Negative humor ridicules, belittles, and distances people. It can be sexist, racist, or embarrassing. Have you been the victim of negative humor? If so, how did it make you feel?

Positive humor tends to bring people closer together. It is associated with hope, love, and closeness. It is a gentle banter, not a caustic, sarcastic barb. It is timely and adds perspective.

Consider three criteria for positive uses of humor. Ask yourself these questions:

1. *Is the timing right for this quip?* When someone is in the middle of a crisis or is in great pain, humor might not be appreciated even though you have often swapped one-liners.

2. *Is the other person receptive to humor?* Does this person already use humor to cope? Some people do not seem to have or value a sense of humor.

3. *Is the content acceptable? Is it in good taste? Does it make light of self, rather than others?* Jokes that could help relieve tension in a closely knit work group could be seen as insensitive to outsiders who do not understand the commitment of the group to the service it provides (Riley, 2003b).

Develop Your Comic Vision

Sometimes being able to laugh at life and at yourself relieves tension. Consider the mental self-talk that can alter the interpretation of minor irritants in life. Perhaps instead you could remember with a smile the one-liner from the children's story, *The Little Engine That Could* (Piper, 2012, p. 6): "I think I can, I think I can," instead of ruminating over such lines as, "I can't do this. It's too hard. It isn't fair." Begin to collect one-liners of your own to share. How about this one?

New clinical studies show there aren't any answers.
—Anonymous

How do you develop your comic vision (see **Exhibit 13-1**)? If you think you have no sense of humor or just want more joy in your life, try planning to have fun, looking for humor in your life, and exposing yourself to humorous stories and other resources. Taking time to read the cartoons along with the front page will help you reframe your day.

EXHIBIT 13-1   DEVELOPING YOUR COMIC VISION

1. Start with yourself. Laugh at yourself. Give yourself permission to be human. If you trip, laugh out loud.
2. Read the comics and political cartoons in newspapers as examples of comic vision. Look at local newspapers when you travel to get the community perspective and learn about regional humor.
3. Start an album with cartoons that track current work issues and encourage all team members to contribute.
4. Attend funny movies and comedy clubs. Rent classic comedy videos.
5. Listen to humorous audiobooks on the way to work to begin your day looking for humor.
6. Collect humorous one-liners that are inside jokes with your work team.
7. Experiment with building a humor kit at work. Start with a few items and encourage participation and a feeling of ownership.
8. Laugh with others for what they do, at the incongruities in life in which we all share.
9. Pay attention to your own self-talk. Replace negative thoughts with positive ones. Focus on being someone others find pleasant company.
10. Share your comic vision to make other people laugh. Laughter is contagious and adds much needed joy in all our lives.

Copyright © Julia Balzer Riley RN, MN, (2000). Humor and Health, home study, AKH Consultant, Inc., Orange Park, Florida. Used with permission.

Joel Goodman, founder of the Humor Project (n.d.) in New York, talks about aikido humor—humor that uses the momentum of the situation to roll with the punches. Aikido is a form of martial arts that uses the principle of nonresistance rather than pushing forward; it looks like a flowing dance. Goodman exemplified this with the story of an elementary school teacher. Precisely at 10 A.M., while the teacher had her back turned writing on the blackboard, all of the students knocked their books from their desks onto the floor. The teacher turned around, looked at the clock, said, "Oh, sorry. I'm late," and swept the books off her desk, too.

Think of things you can use to add a light touch (see **Exhibit 13-2**). Add colorful confetti, available for all occasions, to letters or memos. Give out

---

EXHIBIT 13-2   BUILDING A HUMOR KIT

Collect these items to use at work or school. Put them in a tote bag or box for easy access by all.

- Bubbles—small bottles, bubble bottles on necklaces, or wands that make giant bubbles. Take a bubble break to give yourself and others a laugh and to appreciate the small wonders in life.
- Whistles for stress relief
- Funny hats
- Clown noses and other wild noses
- Funny books, audiotapes, videotapes
- Cartoons collected in a photo album or for display on bulletin boards
- Children's games—wooden paddle and ball, pick-up sticks, coloring books and crayons—ahh, that smell! Splurge and get the box of 64 with the sharpener—they even have bigger boxes now!
- A laughter box
- A big teddy bear when a hug is needed
- Lapel buttons with funny one-liners
- A magic wand
- Be on the lookout for funny things. Sources include toy stores, clown supply stores, party stores, Halloween and other seasonal displays, souvenir shops, teacher/school supply stores, your children's discards, and garage sales.
- Bubble gum for a brief bubble-blowing contest for staff stress relief

Copyright © Julia Balzer Riley, RN, MN (1996). Used with permission.

---

stickers at home or at work when someone does something right or just needs a boost. Use a noisemaker to get people's attention when chairing a meeting. The element of surprise can be effective in changing the mood. This is positive humor—when the sound is not an irritating one and is used playfully. It is negative humor if it is used to interrupt someone who is speaking and embarrasses the person. Use toys and props to facilitate teams and to teach important content such as customer service (Riley, 2003a). Take a new look at garage sales and your children's cast-off toys. Take a field trip to a toy store or magic shop for props to add to your kit. Purchase inexpensive toys in quantity to use as incentives at work and at home.

> **KEY POINT**
>
> *He who laughs, lasts!*
> —ANONYMOUS

Plan to Laugh

If humor helps you stay healthy, how can you build it into your daily life? Be open to humor and actively pursue it. Remembering jokes and delivering a powerful punch line can be part of a rich sense of humor, but reading jokes or funny greeting cards can do wonders to stimulate laughter, too.

REFLECTION
How long has it been since you have laughed out loud?

The Internet can supply you with a steady source of laughter. Subscribe to a list of jokes. Check out funny videos on YouTube and share them with others. Find a humor buddy online or be one. Remember that not every sense of humor is alike. Try to open yourself to a variety of kinds of humor to see what tickles your funny bone (**Exhibit 13-3**). *PUNs . . . Please Forgive Me* offers examples of puns circulated via email. A pun is a playful use of words, using words with different meanings that sound alike or similar. It is the stretching that makes these examples funny. If you groan when you read this, consider the comic relief this can offer on a tough day (**Exhibit 13-4**). *OXYMORONS . . . Be on the Alert for More* gives examples of oxymora, also shared via email. An oxymoron is an amusing contradiction in language. Trying to construct puns or recognize oxymora is good practice for creative thinking.

Plan to Play

Evidence of play and toys is found as far back as relics of human existence in ruins of Egyptian, Babylonian, Chinese, and Aztec civilizations. Humor can shift our mental paradigms in problem solving to stimulate creative thinking. To be more open to creativity, plan to play. Put leisure activities on your calendar. Take time to just be and not do; lie on the ground and look at the clouds, fly a kite, take a break to play several games of solitaire when you have computer fatigue. See **Exhibit 13-5** for ideas to get you thinking playfully. This is quite a task for some grown-ups! Take some time now to

---

EXHIBIT 13-3   PUNS . . . PLEASE FORGIVE ME

Middle Age: When actions creak louder than words

Egotist: One who is me-deep in conversation

Income tax: Capital punishment

Archeologist: A man whose career lies in ruins

California smog test: Can UCLA?

Two Eskimos sitting in a kayak were chilly, but when they lit a fire in the craft, it sank—proving once and for all that you can't have your kayak and heat it, too.

Two boll weevils grew up in South Carolina. One went to Hollywood and became a famous actor. The other stayed behind in the cotton fields and never amounted to much. The second one, naturally, became known as the lesser of two weevils.

A mystic refused his dentist's Novocain during root canal work. He wanted to transcend dental medication.

A woman has twins and gives them up for adoption. One of them goes to a family in Egypt and is named "Amal." The other goes to a family in Spain; they name him "Juan." Years later, Juan sends a picture of himself to his mom. Upon receiving the picture, she tells her husband that she wishes she also had a picture of Amal. Her husband responds, "But they are twins—if you've seen Juan, you've seen Amal."

EXHIBIT 13-4   OXYMORONS . . . BE ON THE ALERT FOR MORE

Act naturally

Clearly misunderstood

Alone together

Airline food

Found missing

Resident alien

Genuine imitation

Almost exactly

Legally drunk

Small crowd

Taped live

Plastic glasses

Working vacation

Jumbo shrimp

And the number one listing . . . Microsoft Works (a little computer humor)

generate your own list. Ask friends and family members to add to the list. Contract to do one thing just for fun within the next 2 weeks and develop a plan for increasing the play in your life (see **Exhibit 13-6**).

EXHIBIT 13-5   PLAN TO PLAY

1. Find someone to take the children on an outing and have the house to yourself.
2. Go to a museum, a play, or a concert.
3. Look through a catalog and circle all of the things that you would buy if you could.
4. Buy a lottery ticket and fantasize how you would spend the winnings.
5. Just move. Turn on music you like and move with the rhythm. You do not have to know how to dance.
6. Go to a travel agency and get some brochures.
7. Go bowling or play miniature golf.
8. Go get an ice cream cone.
9. Take a novel. Go to a restaurant. Order a drink, and sit and read.
10. Bake bread or prepare your childhood comfort food.
11. Call a local recreation department and sign up for a class.
12. Find a new recipe. Buy the ingredients. Prepare it, and invite someone to join you for dinner.

EXHIBIT 13-6   A PLAN FOR FUN

List five things that you have enjoyed doing for fun in the past:

1.
2.
3.
4.
5.

Select one of the fun things from your list that you can realistically do within the next 2 weeks: _____

Set a target date for doing the fun activity: _____

Write this on your calendar and share your plan with someone who will hold you accountable.

EXHIBIT 13-7   HUMOR AT WORK . . . WE NEED IT

---

1. Remember to greet staff in person and on the telephone with a smile. Share a joke.
2. Post cartoons that make light of shared concerns at work.
3. Buy a bottle of bubbles. Start building a humor kit for the office.
4. Collect funny Post-it notes and use them with staff. Use stickers and gold stars.
5. Organize a fun activity at work, such as a staff baby picture contest, a pumpkin-carving contest at Halloween, a bowling night, or a funny movie night.

Humor and Play at Work

Do you know someone at work who always sees the funny side of life? Think of the popularity of Dilbert cartoons. How many Far Side cartoons have surfaced in the workplace? Do you have an employee picnic with games? Do you decorate for holidays? Do you celebrate birthdays? How do you feel reading this if your workplace includes such activities? What if it does not? Some settings do not seem to lend themselves to such formalized play; nevertheless, staff members appreciate comic relief in meetings. Consider activities that might actually help people to look forward to coming to work (see **Exhibit 13-7**).

KEY POINT

A light touch at work can help staff build team spirit—they who laugh together stay together, manage stress, and tap their creativity.

Humor at Home

We take ourselves and our lives so seriously. We feel we need to be perfect and expect others to have a positive attitude at all times and to appreciate our efforts. A helpful principle to remember is that whatever behavior a person offers at any given time is usually the best that he or she can do. We all have grumpy times. Give some thought to the last time you were able to laugh at yourself. Consider how much more peaceful life would be if we could lighten up (see **Exhibit 13-8**).

EXHIBIT 13-8   HUMOR AT WORK . . . WE NEED IT

HUMOR AT HOME

- Try sharing moments of laughter.
- Learn to laugh at your own seriousness.
- Forgive yourself and others.
- Remember that we do not always have to understand each other or even like each other all of the time.
- Lighten up!
- Agree to disagree.
- Do not try to be so perfect, and do not expect it of others.
- Add a little humor by intention. Go do something just for fun—get a life.

Work on keeping a positive attitude. Here are some tips to help keep a positive attitude:

- Focus on positive thoughts, such as "I'm full of energy" and "Today is a good day."
- Imagine good things happening to you. If you are trying to lose weight, picture yourself at your ideal weight.
- Stop looking for the negatives or the flaws in situations and people.
- Find the humor in a difficult situation.
- Smile. A study showed that using all your facial muscles in a smile can put you in a more cheerful mood.
- Exercise and keep active to keep up the flow of endorphins, a hormone that elevates mood.
- Just do it. . . . Eat right and take care of your body to feel better.

REFLECTION

What measures do you use to promote a positive attitude in yourself? List a few ways that you can improve this.

Be an Ambassador for Humor

As you search for ways to be healthy, to tap the power of the mind–body–spirit connection, some strategies intuitively feel right. If you are lucky enough to have a sense of humor, a comic vision, share it; the world needs you! If humor does not come easily, pursue it. Take life lightly. Humor is a great alternative.

Summary

Gelotology, the study of the physiology of laughter, reveals that laughter produces many physiological effects, including stimulation of the production of catecholamines, a decrease in anxiety, an increase in heart and respiratory rate, an increase in endorphins, an enhancement of metabolism, improvement of muscle tone, an increase in natural killer cells, and relief from stress. There are many practical ways that healthcare professionals can implement humor appropriately into healthcare settings; doing so is therapeutic and beneficial to health and healing. In addition, incorporating humor into daily life and work also is an important self-care measure for healthcare professionals.

References

Bennett, M. P., & Deane, D. M. (2006). Humor and laughter may influence health. *Evidence Based Complementary and Alternative Medicine, 3*(1), 61–63.

Cousins, N. (1979). *Anatomy of an illness as perceived by the patient.* New York, NY: WW Norton.

The Humor Project. (n.d.). Retrieved from http://www.humorproject.com

Klein, A. (2008). *Who says humor heals? Retrieved from* www.allenklein.com/articles/whosays.htm

Piper, W. (2012). *The little engine that could.* New York, NY: Grosset & Dunlap.

Riley, J. B. (2003a). *Customer service from A to Z . . . Making the connection.* Albuquerque, NM: Hartman Publishers.

Riley, J. B. (2003b). *Humor at work.* Ellenton, FL: Constant Source Seminars.

Sullivan, J. L., & Deane, D. M. (1988). Humor and health. *Journal of Gerontological Nursing, 14*(1), 20–24.

Suggested Reading

Korotkov, D., Perumovic, M., Claybourn, M., & Fraser, I. (2011). *Humor, stress, and health: Psychology of emotions, motivations and actions.* Huntington, NY: Novinka Books.

Lebowitz, K. R., Suh, S., Diaz, P.T., & Emery, C. F. (2011, July/August). Effects of humor and laughter on psychological functioning, quality of life, health status, and pulmonary functioning among patients with chronic obstructive pulmonary disease: A preliminary investigation. *Heart Lung, 40*(4), 310–319.

Marden, O. S. (2011). *Cheerfulness as a life power: A self-help book about the benefits of laughter and humor.* Seatlle, WA; CreateSpace.

Sparks, S. (2010). *Laugh your way to grace: Reclaiming the spiritual power of humor.* Woodstock, VT: SkyLight Paths Publishing.

Ulbrich, C. (2011). *How can you NOT laugh at a time like this? Reclaim your health with humor, creativity, and grit.* New Haven, CT: Tell Me Press.

Resource

Association for Applied and Therapeutic Humor
www.AATH.org

PART III

Taking Charge of Challenges to the Mind, Body, and Spirit

Understanding the Hidden Meaning of Symptoms

14

OBJECTIVES

This chapter should enable you to

- List at least six early warning signs of health problems
- Discuss lessons that can be learned by the choices people make
- Describe the meanings that symptoms can have

Recognizing warning signs is easy when driving a car. The yellow sign with the curved black arrow communicates clearly that the road ahead has a curve and that the driver needs to slow down and pay attention. If this warning sign is observed and the driver gives complete attention, the curve in the road makes the journey more interesting. If the warning sign is ignored, the driver could be forced to pay attention to a car accident, a much less pleasant diversion on the journey.

When people experience good health, they feel alive, trusting, enthusiastic, and full of expectation. When health is threatened, warning signs can develop that alert them to "watch the curve in the road." **Table 14-1** offers a list of warning signs of health problems that warrant attention.

These warning signs first appear as whispers, but if ignored, they may become shouts. A whisper may be a toothache that, if ignored, becomes the shout of an abscess, or a sore knee that unattended gives out and causes you to slip and break a bone. It may be the whisper of a spouse who is unhappy with your behavior and eventually disrupts your life with the shout of a divorce.

REFLECTION

Would you show the same attention to a whispered message from your own body as you would to an unusual sound from your car?

TABLE 14-1   WARNING SIGNS OF HEALTH CONDITIONS

- Fatigue
- Dizziness
- Significant rise or drop in blood pressure
- Unintentional weight loss of more than 5% of body weight within the past month
- Significant rapid weight gain
- Loss of appetite
- Painful urination, increased urination
- Blood in urine, stool, vomit, or sputum
- Chronic cough
- Wheezing
- Shortness of breath
- Pain
- Numbness
- Feeling of pressure on chest
- Difficulty breathing
- Chronic indigestion
- Confusion, memory loss, inability to concentrate
- Poor relationships with family, friends, coworkers
- Personality change
- Depression, feelings of hopelessness
- Anxiety, worry
- Swelling
- Rash
- Change in sleep pattern
- Change in vision, eye pain
- Frequent falls, accidents

Purposes of Warning Signs

Warning signs provide feedback for people to pay attention to the way they are living so that their energies will not be diverted to unnecessary complications, expense, and hardship. They remind that "a stitch in time saves nine."

Warning signs also teach people how to recognize the consequences of their choices. They help people learn how to be healthy and protect themselves. For instance, working out to stay physically fit is a wonderful way to stay healthy;

however, if people try to initiate the same exercise program at age 60 that was used at age 20, they may find themselves frustrated, dreading exercise time, and perhaps injured—warning signs that their exercise program needs to be adjusted. Warning signs can help individuals to recognize safe limits.

Good Health as a Process

When taking an automobile journey, people usually have some idea of where they are going and when they will arrive at their destination. What about health? How will they know when they have achieved a satisfactory level of health?

Some people have conceived of health as a continuum from 0% to 100%, in which 0% is severe sickness and debilitation and 100% is a perfect, disease-free state. This perspective does not account for the quality of life, nor does a continuum model address the rewards experienced through living. Individuals can have meaningful, satisfying lives even in the presence of physical illness. Many people living with chronic illnesses hold the view that they may have *xyz* disease, but *xyz* disease does not have them. A continuum does not account for this feeling of well-being or enthusiasm for living in the presence of an illness. A continuum does not account for the synergy of the physical, mental, emotional, and spiritual aspects of health. Furthermore, a continuum presupposes that death is bad, and, by implication, that old age is something to be avoided.

Good health is better conceived of as a process rather than a continuum. The American Heritage Dictionary (2011) defines *process* as "a series of actions, changes, or functions bringing about a result"—for example, the process of digestion and the process of obtaining a driver's license. Health can be conceived as the process by which people become more aware of themselves, recognize their level of well-being, and take actions that foster good health.

Sometimes, people learn to achieve good health by experiencing poor health. If one has always known good health, a person may not be aware of what elements make up this experience of well-being nor is he or she likely to be aware of how these elements are related. For example, if your computer is working, you may have little interest in what makes it work; however, as soon as you cannot access your email, you are calling tech support to find out how to fix the problem. After help is at hand, some people will want to learn more about what caused the problem and how they may be able to fix it in the future without having to call tech support, whereas other people may only be interested in having tech support tell them what to do to fix the current problem. Both approaches are valuable because they result in your email being accessible. Likewise, the process of health may lead to a problem being corrected, or it may lead to the problem being corrected *and also* a greater understanding of what constitutes well-being and the behaviors likely to sustain it.

If people want to take a vacation and tour the temples of ancient Egypt, they typically hope to enjoy good health on the trip. They may plan to enhance their physical endurance to have sufficient health as to walk moderate distances in the hot sun. To achieve their goal, they may engage in the behaviors of eating well, exercising, and getting enough rest. These behaviors dynamically interact to produce the experience of physical endurance and health sufficient to accomplish their goal. Each time they make choices about food, exercise, and rest, they have the potential to learn more about themselves, such as which attitudes and behaviors support their goal of physical endurance and which do not. They also may become more confident in their ability to recognize choices that result in a feeling of well-being. As individuals make decisions to support their goals, they feel at one with themselves. Health is the process of learning how to achieve this sense of wholeness. Health is recognizing, acknowledging, accepting, and continuing the behaviors that lead to a sense of well-being, peace, and wholeness.

People can also learn about themselves by making choices and engaging in behaviors that do not support their goals. When this occurs, people feel separated from themselves and conflicted. Health is then the process of recognizing, acknowledging, forgiving, and changing the behaviors that lead us to a sense of separation.

REFLECTION

Choices made about your health can give insight into values, commitment, perseverance, strengths, creativity, and patience. What do your health-related choices say about you?

In the process of making bad decisions, there are lessons as well. Here people can touch a part of self that feels cut off. It is a piece of self that is demanding expression and will most likely sabotage them until they either give it their attention or unhappily pay attention to it.

How people choose to give attention and expression to that which seems to be at odds with their goals of good health may, in fact, teach them more about health than their good choices. Forgiveness can be learned and practiced in the presence of bad decisions. Forgiveness has incredible power to help individuals feel whole and at peace. When people practice forgiving themselves, they find it is only a short step to feeling the forgiveness presented to them as a gift from their creator.

Warning signs are messages of conflicting choices that call for people to make single-minded decisions. When they shrink from the opportunity to examine their choices or see where they have made conflicting decisions, people pass up an opportunity to learn more about themselves. Likewise, when they expect someone else to fix them, they forfeit an opportunity to reflect upon their choices, make decisions, and learn. Reflection leads to greater self-awareness.

REFLECTION

In what ways does the healthcare system foster consumers' attitudes that someone else needs to fix their health problems?

Symptoms as Teachers

The presence or absence of physical symptoms may offer feedback about how alive people feel and how trusting, how enthusiastic, and how much expectation they hold. Symptoms can warn people of imbalances and problems that involve more than the body parts in which they are manifested.

Warning signs will have different meanings at different times along the life journey. This is not intended to make life more difficult. Life is not a textbook experience. Understanding the variety of possible meanings enables people to learn about themselves more fully and deeply. Furthermore, the many possible interpretations provide the opportunity to choose the specific meaning of a symptom. For example, suppose you have lost a front tooth, and while waiting to see the dentist, you reflect on this event. You begin to see a metaphor, and it strikes you that you must be experiencing difficulty "biting into something" or "taking a bite." You could think this: "This is another example of how I'm doomed to have one problem after another," or you could think "This dental problem is an example of how I do not pay ample

> **KEY POINT**
>
> When physical symptoms are addressed on more than the surface level and the other meanings they hold in life are explored, people have an opportunity to heal beyond the physical level.

attention to small matters that eventually accumulate into big problems for me. I need to seek help to do the tasks of daily living that need my attention and action." It is expected that a person would respond to a tooth problem by visiting the dentist; however, the tooth problem holds other meaning. The metaphor, if used, can lead to even greater self-knowledge and change. When you manage the physical symptom on more than the surface level (the dentist) and seek to work with the metaphor, you have an opportunity to heal on more than just the physical level. In fact, it may be that the physical symptom will not change, but that change will occur on a more profound level, bringing greater happiness, growth, and satisfaction than you could ever experience by simply treating the physical symptom.

Summary

Symptoms involving the mind, body, and spirit can indicate the presence of a health problem. They are warning signs that offer feedback as to the way one is living and lifestyle and health-related choices. Rather than treat or mask their symptoms, people can achieve higher levels of holistic health by exploring their deeper meaning and choosing attitudes and behaviors that will foster health and healing.

References

American Heritage Dictionary. (2011). *The American Heritage Dictionary of the English Language* (5th ed.). Online: Houghton Mifflin Harcourt Publishing Company. Accessed January 17, 2013.

Suggested Reading

Bauby, J. (2007). *The diving bell and the butterfly*. New York, NY: Vintage International.
Burkhardt, M. A., & Nagai-Jacobson, M. G. (2002). *Spirituality: Living our connectedness*. Albany, NY: Delmar, Thomson Learning.
Duff, K. (2000). *The alchemy of illness*. New York, NY: Random House.
Frank, A. W. (1997). *The wounded storyteller: Body, illness, and ethics*. Chicago, IL: University of Chicago Press.
McSherry, W. (2006). *Spirituality in nursing practice: An interactive approach*. New York, NY: Churchill Livingstone.
Morris, D. B. (2000). *Illness and culture in the postmodern age*. Berkeley; University of California Press.
Sacks, O. (1998). *A leg to stand on*. New York, NY: Touchstone Books.

Working in Partnership with Your Health Practitioner: Advocating for Yourself

OBJECTIVES

This chapter should enable you to

- List three responsibilities that consumers have when meeting with their health practitioner
- Describe a consumer's responsibilities when diagnosed with a medical problem, when hospitalized, and when facing surgery
- List at least five questions to ask of health insurance plans
- Describe at least six factors to evaluate in a primary healthcare practitioner

A partnership implies an active relationship of give and take with a mutual respect between the parties. Of the various partnerships that you may experience in your life, your partnership with your healthcare provider is among the more important ones that you can develop. Yesterday's model of blind obedience and dependency on one's physician or other healthcare provider is no longer appropriate or desirable. People have a right to be informed, to have a variety of healthcare options (conventional and complementary/alternative) made available to them, and to make decisions that are right for them—not only medically, but also emotionally, spiritually, and financially. They also have the responsibility to take an active role in their health care, educate themselves about health matters, and equip themselves with the necessary information to make sound healthcare decisions. They must be responsible advocates for themselves.

Responsibilities When Seeking Health Care

Choosing a Healthcare Practitioner

Because the relationship with a healthcare practitioner is an important one that should function with the same harmony as a good marriage, people need to choose a primary healthcare practitioner with whom they can comfortably communicate. They must feel free to ask questions and expect to get responses they understand.

When making an appointment, if people think they will need more time than usual to talk things over, they should let the office know in advance. The response to this request could offer insight into the type of relationship that can be expected with the healthcare provider. For example, if you ask that you be scheduled for a longer visit because there are some complementary therapies that you want to review with the physician as soon as possible and you are told, "Sorry, the doctor only can only spend 10 minutes with each patient," you may need to assess whether this is the best source for your primary care.

It is helpful for people to write a list of health concerns or questions that they want to discuss before the appointment. They can include information about symptoms, such as when the symptoms started, what they feel like, whether they are constant, or whether they come and go. Records that are kept, such as blood sugar levels, daily blood pressure, dietary history, and mood changes, should be taken to the appointment. It is important for people to tell their practitioners personal information—even if it feels uncomfortable to do so. Any allergies or reactions experienced in the past to any medicines or foods need to be discussed. Health practitioners need to know as much as possible about their clients.

It is helpful for people to compile a written list of the medicines and supplements that are taken, including those that do not require a prescription, such as over-the-counter pain relievers, laxatives, vitamins, herbs, flower essences, or eye drops. Also helpful is a list of the names of other healthcare practitioners visited, including physicians, chiropractors, acupuncturists,

---

KEY POINT

People need to prepare for visits to their healthcare practitioners by writing down health history, symptoms, medications and complementary and alternative therapies used, and other pertinent information that can be shared with the practitioner.

massage therapists, nutritionists, and herbalists. People should supply each of their healthcare practitioners with this information. It is useful for people to be open about their reasons for seeing other healthcare practitioners and the care they are receiving from them. If a healthcare practitioner does not agree with what is being done, it is useful for people to listen to the reasons. All of the facts can be evaluated and decisions made to stay with this practitioner or change to one who is more supportive. Practitioners need to be given permission to contact each other for clarification when necessary.

People should take a notepad to appointments and take notes when given instructions. Most individuals only remember a portion of what they are told, and if anxious, they could forget most of the information. In some circumstances, it could be helpful to bring along someone to help ask questions, clarify, and remember responses.

KEY POINT

It is wise for people to keep their own health file that records their medicines, treatments, surgeries, hospitalizations, and visits to all practitioners.

When Diagnosed With a Medical Condition

If diagnosed with a medical condition, people need to ask about the causes. Often, improvements can be made if the reasons are understood. It is beneficial for people to ask for guidelines as to symptoms that should be reported. For example, if the temperature is over what degree? If the discharge is what color? If the stools are loose for how long? If the coughing persists for how many days? If the pain is not relieved within what time period? Healthcare practitioners may have printed information about conditions (e.g., diabetes fact sheets) that can be given to take home; if not, people should ask where more information can be obtained. People need to be knowledgeable about their conditions and informed about the tests that are ordered (see **Exhibit 15-1**).

KEY POINT

It is useful for copies of diagnostic tests to be requested and kept in a personal health record.

EXHIBIT 15-1   QUESTIONS TO ASK WHEN TESTS ARE RECOMMENDED

Why is the test needed?

Does the test require any special preparation?

What are the side effects and risks?

When can the results be expected and how are they obtained?

Will additional testing be required at a later date?

Does my insurance cover this test?

What are the alternatives to this test?

EXHIBIT 15-2   QUESTIONS TO ASK WHEN TREATMENTS ARE ORDERED

When should treatment start?

How long will the treatment be needed?

What are the benefits of the treatment, and how successful is it usually?

What are the risks and side effects associated with the treatment?

What are the alternatives to this treatment?

What is the cost of the treatment?

It is important for people to learn about the available treatments for their conditions. Some of the questions that need to be answered when treatments are ordered are listed in **Exhibit 15-2**. People also can check out information in their libraries or on the Internet. Discussing the issue with friends or relatives who are nurses or other healthcare practitioners is beneficial.

After the visit, follow-up may be needed. Questions should be phoned or emailed to the healthcare practitioner. If symptoms worsen or people have problems with what they are asked to do, they need to contact the practitioner. If tests are ordered, arrangements should be made at the laboratory or other offices to get them done; when the tests are completed and people have not been informed of the outcome, they need to call for test results. Some healthcare practitioners' offices call patients to see how everything is going, but in most instances, it is up to the individual to do the follow-up.

REFLECTION

How much responsibility do you assume for your personal health care? Do you ask the right questions, demand answers, take action when you think you are not being heard, and follow up as needed? If you do, what has caused you to develop these skills? If you do not, what prohibits you from doing so?

When Hospitalized

Everyone needs to know the physician in charge of their care. This may seem like a ridiculous statement; however, this may be an issue when people go into the hospital for a surgical procedure and have an existing medical problem or several health-related issues. Different doctors may be responsible for different aspects of care. It can be particularly difficult in a teaching hospital where there are several medical residents writing orders. People need to make sure all health professionals know their health history. It is smart for people to keep a brief summary (just a short page) of their condition and health history at the bedside. It cannot be assumed that each doctor or resident knows everything about the history. Very likely, they do not.

Protection from errors during hospitalization can be increased when a summary of personal health history is kept by the bedside.

When patients have questions about procedures or a medicine and the answer is merely, "Your doctor ordered it," the patient needs to ask which doctor. If they do not understand the purpose or did not receive an adequate explanation, they are wise to politely refuse. They are not being difficult or uncooperative patients, but, rather, informed and responsible patients participating in their own care. They are within their rights to ask their primary doctor to give an explanation. If illness prevents someone from taking on this total responsibility, he or she can ask a family member or friend to be an advocate (see **Exhibit 15-3**).

It is helpful for people to learn the names of the hospital staff who do anything to or for them. By having their notebooks handy, they can write down staff names and positions, whether it is the nursing assistant who is responsible for bathing assistance or the anesthesiologist who assists with surgery.

When Surgery Is Recommended

There is information that needs to be considered in order to make an informed decision about surgery. The purpose of the procedure needs to be understood, along with benefits, risks, and alternatives. Sometimes surgery is not the only answer to a medical problem. Nonsurgical treatments, such as a change in diet, special exercises, acupuncture treatments, herbs, or other nonconventional treatments might help just as well or better. Individuals can research these on their own.

There should be a clear understanding of how the proposed operation relates to the diagnosis. It is reasonable for people to ask why a particular surgeon was chosen. When they meet with the surgeon, people should ask for an explanation of the surgical procedure in detail. They should find out if there are different ways of doing the operation and why the surgeon has chosen

---

EXHIBIT 15-3 "THE DOCTOR ORDERED IT...."

An example of what can happen when the only response to your question is "the doctor ordered it" is what happened to a patient who had just gone through a surgical procedure in the hospital to prepare her for kidney dialysis treatments. Early one morning the nurse came in to do a finger stick for a drop of blood. The patient accepted this, assuming that blood tests may be needed. The nurse came in again before lunch to repeat the procedure, and the patient asked, "Why are you sticking me again?" The reply was simply, "The doctor ordered it."

Her daughter, a registered nurse, came to visit later in the day just as the nurse was about to do yet another finger stick. Knowing that this type of testing was typically done for persons with diabetes and that her mother was not diabetic, the daughter asked, "Why are you checking her blood sugar? Is it high?"

"The doctor ordered it," came the reply.

"Well, before you do it, please tell me her blood-sugar level."

With this, the patient said. "Blood sugar! I'm not a diabetic, I don't want you to do this."

The family later learned that one day her blood-sugar level was 147 mg, a little high, but not alarming considering that she was receiving intravenous fluids with dextrose (sugar water). An intern had ordered the test.

When the patient told her primary medical doctor he remarked, "The intern was being overly cautious. I'll write an order that they need to check with me before ordering any more tests."

This situation caused the patient unnecessary worry, discomfort, and expense.

---

the particular procedure. One way may require more extensive surgery than another one and a longer recovery period.

It is important that people know what they will gain by having the operation and how long the benefits are likely to last. For some procedures, the benefits may last only a short time with a need for a second operation at a later date. For other procedures, the benefits may last a lifetime. Published information about the outcomes of the procedure can be reviewed. Because all operations carry some risk, people need to weigh the benefits of the operation against the risks of complications or side effects.

> **KEY POINT**
>
> One approach when surgery is recommended is watchful waiting, in which the doctor and patient monitor the problem to see if it gets better or worse. If it gets worse, surgery may be needed fairly soon. If it gets better, surgery may be able to be postponed, perhaps indefinitely.

Pain almost always occurs with surgery. People should ask how much pain can be expected and what the doctors and nurses will do to reduce the pain. Individuals may want to discuss how staff will respond if they want to use nonconventional, alternative methods of pain control. Controlling the pain helps people to be more comfortable while they heal, helps them get well faster, and improves the results of their operation.

It is useful for individuals to find out whether the surgeon agrees to their own treatment plans during and after the surgery. For example, what does the physician think about a plan to take higher doses of vitamins and minerals before surgery to promote healing afterward? Will the hospital permit music or guided imagery tapes during the surgery? Can a therapeutic touch practitioner offer treatments after surgery? All desires and suggestions should be openly discussed.

Getting other opinions from a different surgeon and healthcare practitioner of choice is a very good way to make sure having the operation is the best alternative. When people seek second opinions, they can spare themselves unnecessary duplicative testing by obtaining a copy of their records from the first doctor.

Anesthesia is used so that surgery can be performed without unnecessary pain. People should ask to meet with the anesthesiologist to learn about his or her qualifications. The anesthesiologist can explain whether the operation calls for local, regional, or general anesthesia and why this form of anesthesia is recommended for the procedure. Questions about expected side effects and risks of having anesthesia should be asked. Some hospitals are using acupuncture as an adjunct to anesthesia, and patients should ask if it is an appropriate treatment.

If surgery is agreed on, people need to ask if the operation will be done in the hospital or in an outpatient setting. If the doctor recommends inpatient surgery for a procedure that is usually done as outpatient surgery, or just the opposite, the reason should be examined because it is important to be in the most appropriate place for the operation. Until recently, most surgery was performed on an inpatient basis and patients stayed in the hospital for 1 or

> **KEY POINT**
>
> Surgeons should be asked about possible complications and side effects of the surgery. Complications are unplanned and undesirable events, such as infection, too much bleeding, reaction to anesthesia, or accidental injury. Some people have an increased risk of complications because of other medical conditions. Side effects such as swelling and some soreness at the incision site are anticipated occurrences.

more days. Today, many surgeries are done on an outpatient basis in a doctor's office, a special surgical center, or a day surgery unit of a hospital.

People should ask how long they will be in the hospital. The surgeon can describe how people can expect to feel and what they will be able to do or not do the first few days, weeks, or months after surgery. Before discharge, individuals should find out what kinds of supplies, equipment, and any other help they will need when they go home. Knowing what to expect can help them better prepare and cope with recovery. They should ask when they can start regular exercise again and return to work. People should not want to do anything that will slow down the recovery process. Lifting a 10-pound bag of potatoes may not seem to be too much a week after surgery, but it could be.

Choosing a Health Insurance Plan

Selecting a health plan to cover themselves and their families is an important decision many people make every year. Health insurance varies greatly, and thus, to avoid surprises, it is important for people to find out exactly what their insurance plan covers. Does it allow for any alternative approaches? Many policies now cover acupuncture, and some cover nutrition counseling, massage, and other forms of body work. People need to be informed. Some questions worth asking are offered in **Exhibit 15-4.**

Health insurance coverage for surgery can vary, and there may be some costs individuals will have to pay. Many health insurance plans require patients to get a second opinion before they have certain nonemergency or elective operations. Even if a plan does not require a second opinion, people may still ask to have one. Before having a test, procedure, or operation, it is advisable for people to call their insurance plan to find out how much of these costs it will pay and how much they will have to pay themselves. They also may be billed by the hospital for inpatient or outpatient care, any doctor who visited, the anesthesiologist, and others providing care related to the operation.

EXHIBIT 15-4   QUESTIONS TO ASK OF A HEALTH INSURANCE PLAN

- What benefits and services are covered?
- Is my current doctor in the network?
- How does the plan work?
- How much will it cost me?
- Are there copays?
- Does the plan have programs in place to assist me in managing chronic conditions?
- Is it evident that qualified healthcare professionals make decisions about medical treatments and services provided to plan members?
- How will I get needed emergency services and procedures that ensure I will get the level of care needed?
- Is the information about member services, benefits, rights, and responsibilities clearly stated?
- Is there easy access to primary care and behavioral health care (mental health services)?
- Does the plan include alternative approaches?
- How promptly are decisions made about coverage of medical treatments and services?
- Are there clear communications from the health plan to members and doctors about reasons for denying medical treatments or services and about the process for appealing decisions to deny treatment or services?

The Agency for Health Care Policy and Research, within the Department of Health and Human Services, has excellent information on choosing quality healthcare coverage. The National Committee for Quality Assurance, a non-profit accrediting agency, has a health plan report card that shows consumers how well managed health plans are doing (see the resources at the end of the chapter).

KEY POINT

People should do their homework to assure that they are not surprised by unexpected bills.

How to Get Quality Care

From the Primary Healthcare Practitioner

People must decide what they want and need in a healthcare practitioner. What is most important to them in working in partnership? Internists and family physicians are the two largest groups of primary healthcare practitioners for adults. Many women see obstetricians/gynecologists for some, or all, of their primary care needs. Pediatricians and family practitioners are primary healthcare practitioners for many children.

REFLECTION

What do you look for in a healthcare practitioner?

Nurse practitioners, certified nurse midwives, and physician assistants are trained to deliver many aspects of primary care. Physician assistants must practice in partnership with doctors. Nurse practitioners and certified nurse midwives can work independently in some states, but not others. Some people may choose an acupuncturist, chiropractor, or naturopath for their health maintenance and consult with the medical doctor when hospitalization or special tests are needed.

There are minimum requirements to look for in any practitioner (see **Exhibit 15-5**). Is the primary healthcare practitioner listening with full attention and not distracted with other things going on? Does the practitioner answer questions without impatience? When a client calls the office with a concern, do the staff members respond, or do they refer it to the practitioner? Does the practitioner return calls? Does the entire staff treat clients with respect? Does the receptionist or billing person answer all questions and assist in a courteous way if help is needed?

What role does prevention have in the plan for health care? Does the practitioner offer advice about a healthful diet, exercise, adequate sleep, addictive behaviors, and how to prevent minor illnesses?

Does the practitioner have additional certification and/or experience (e.g., is he or she board certified and licensed)? There are organizations that certify

---

KEY POINT

Staff members are an extension of the primary healthcare practitioner, and their attitude and actions often reflect those of their employer.

EXHIBIT 15-5   CHARACTERISTICS TO LOOK FOR IN A PRIMARY HEALTHCARE PRACTITIONER

Look for someone who:

- Listens to you
- Explains things clearly
- Encourages you to ask questions
- Treats you with respect
- Understands the language that you are most comfortable speaking (or has someone in the office who does)
- Takes steps to help you prevent illness
- Is rated or certified to give quality care
- Has the training and experience that meets your needs
- Has privileges at the hospital of your choice
- Participates in your health plan, unless you choose to pay out of pocket

---

people in many different specialties and modalities. Is the practitioner knowledgeable about alternative treatments or willing to listen to the patient or search for alternative/complementary treatments? Does the practitioner have privileges at the hospital that the patient wishes to use?

If they are already enrolled in a health plan, people's choices may be limited to doctors and healthcare practitioners who participate in the plan. If they have a choice of plans, individuals may want to first think about whom they would like to see. Then they may be able to choose a plan that fits their preferences.

When Surgery Is Recommended

It is important for people to check their surgeons' qualifications. One way to reduce the risks associated with surgery is to choose a surgeon who has been thoroughly trained to do the procedure and has plenty of experience doing it. Surgeons can be asked about their recent record of successes and complications with a procedure. If they are more comfortable, people can discuss the topic of a surgeon's qualifications with their primary healthcare practitioner or do their own research.

People undergoing surgery will want to know that their surgeon is experienced and qualified to perform the operation. Many surgeons have taken special training and passed exams given by a national board of surgeons. People should ask whether their surgeon is board certified in surgery. Some surgeons also have the letters *FACS* after their name, which means they are fellows of the American College of Surgeons and have passed a review by surgeons of their surgical practices. To check out qualifications, people can contact the American Board of Medical Specialties or Administrators in Medicine (see the resources at the end of the chapter).

Working with a Complementary/Alternative Modality Practitioner

Complementary/alternative modality (CAM) covers a broad range of healing philosophies, approaches, and therapies that conventional medicine does not commonly use or make available. People use CAM treatments in a variety of ways. Therapies may be used alone as an alternative to conventional treatment. They may also be used in addition to, or in combination with, conventional methods, in what is referred to as an integrative approach. Many CAM therapies are called holistic, which generally means they consider the whole person, including physical, mental, emotional, and spiritual aspects. Many are reviewed in this book. Some useful sources of information are the National Institutes of Health, National Center for Complementary and Alternative Medicine, American Holistic Nurses Association, and Natural Healers (see the resources at the end of the chapter).

Often people learn about these modalities and the therapists after referrals from friends. Today, people can go on the Internet, search for the particular modality, and find most anything they want to know. (Caution is needed, however, as claims made regarding a practitioner or modality on the Internet may not be supported by fact.) Schools often have a listing of their graduates, which can serve as a referral source. The telephone book yellow pages can be explored for alternative medicine. Primary healthcare practitioners can be good referral sources, too. Many chiropractors are knowledgeable about alternative treatments. Friends who are nurses can also be asked for ideas.

How do people know whether these therapies and practitioners are appropriate for them? They can check to see whether the practitioner has had training and experience. Practitioners can be asked for referrals of other clients. People should look for the same qualities as they would for any healthcare practitioner—and more. Are they holistic in their practice? Do they consider mind, body, and spirit when planning treatment? Do they explain what they are doing? Have they trained at a licensed school? Do they have certification or state licensing for what they do? It is useful for people to check the credentials of practitioners. Practitioners can be asked whether their therapy has been used to treat conditions similar to the ones for which treatment is sought, whether there are any side effects or cautions in using a particular therapy, and whether they can they share research to support the treatment or product they use.

Working in partnership with their healthcare practitioners gives people the tools that they need to get the high quality health care they deserve. It takes work and being alert and proactive, not just accepting everything that is said by a practitioner. By being active, informed participants, people increase their chances of getting the best health care available.

Summary

When meeting with a healthcare practitioner, consumers have responsibilities, including preparing for the visit, informing the practitioner of medications and supplements used and other practitioners seen, and remembering instructions in detail. When diagnosed with a medical condition, consumers need to ask about the cause, symptoms, and guidelines for treatment.

Asking questions about procedures, medications, and the staff are important during a hospitalization. When facing surgery, consumers need to ask about alternatives, benefits, and risks of surgery, as well as expected pain, where the surgery will be done, and anticipated time for recovery.

A variety of questions need to be asked when considering a health insurance plan, including benefits covered, how the plan works, cost, how decisions are made, and whether one's current practitioners are in the network.

When evaluating a primary healthcare practitioner, consumers should consider whether the practitioner listens, explains things clearly, encourages questions, treats them with respect, speaks the same language, takes steps to prevent illness, is certified, has privileges at the hospital of their choice, and participates in their insurance plan.

By being active participants and informed consumers of their health care, people can maximize the quality of care they receive and reduce risks.

Suggested Reading

Benner, P. (2003). Enhancing patient advocacy and social ethics. *American Journal of Critical Care, 12*(4), 374–375.

Clark, C. C. (2003). *American Holistic Nurses Association guide to common chronic conditions: Self-care options to complement your doctor's advice.* Hoboken, NJ: John Wiley & Sons.

Ekegren, K. (2000). We are all advocates. *Journal of the Society of Pediatric Nursing, 5*(2), 100–102.

Ford, S., Schofield, T., & Hope, T. (2003). What are the ingredients for a successful evidence-based patient choice consultation? A qualitative study. *Society of Science and Medicine, 56*(3), 589–602.

Greggs-McQuilkin, D. (2002). Nurses have the power to be advocates. *Medical Surgical Nursing, 11*(6), 265, 309.

Henderson, S. (2003). Power imbalance between nurses and patients: A potential inhibitor of partnership in care. *Journal of Clinical Nursing, 12*(4), 501–508.

Hyland, D. (2002). An exploration of the relationship between patient autonomy and patient advocacy: Implications for nursing practice. *Nursing Ethics, 9*(5), 472–482.

Penson, R. T. (2001). Complementary, alternative, integrative, or unconventional medicine? *Oncologist, 6*(5), 463–473.

Roter, D. L., & Hall, J. A. (2006). *Doctors talking with patients/patients talking with doctors: Improving communication in medical visits* (2nd ed.). Westport, CT: Praeger.

Strax, T. E. (2003). Consumer, advocate, provider: A paradox requiring a new identity paradigm. *Archives of Physical Medicine and Rehabilitation, 84*(7), 943–945.

Resources

Administrators in Medicine

www.docboard.org

Information on doctors in many states is available from state medical board directors.

Agency for Healthcare Research and Quality

2101 E. Jefferson St., Suite 501

Rockville, MD 20852

301-594-6662

www.ahrq.gov

Your Guide to Choosing Quality Health Care is based on research about the information people want and need when making decisions about health plans, doctors, treatments, hospitals, and long-term care.

American Board of Medical Specialties

800-776-2378

www.certifieddoctor.org

This site can tell you whether a doctor is board certified. Certified means that the doctor has completed a training program in a specialty and has passed an exam (board) to assess his or her knowledge, skills, and experience to provide quality patient care in that specialty.

American Holistic Nurses Association

800-278-2462

www.ahna.org

The American Holistic Nurses Association has a directory of members with additional training in various modalities.

American Medical Association

American Medical Association Chicago Headquarters

515 N. State Street

Chicago, IL 60610

312-464-5000

800-665-2882

www.ama-assn.org

"Physician Select" information is available on training, specialties, and board certification about many licensed doctors in the United States.

Joint Commission

One Renaissance Boulevard

Oakbrook Terrace, IL 60181

630-792-5000

www.jcaho.org

The Joint Commission evaluates and accredits nearly 20,000 healthcare organizations and programs in the United States. An independent, not-for-profit organization, the Joint Commission is the nation's predominant standards-setting and accrediting body in health care.

National Center for Complementary and Alternative Medicine

P.O. Box 8218

Silver Spring, MD 20907-8218

888-644-6226

www.nccam.nih.gov

Focuses on evaluating the safety and efficacy of widely used natural products, such as herbal remedies and nutritional and food supplements; supporting pharmacological studies to determine the potential interactive effects of CAM products with standard treatment medications; and evaluating CAM practices.

The National Committee for Quality Assurance

Health Plan Accreditation

2000 L Street, NW, Suite 500

Washington, DC 20036

202-955-3500

www.ncqa.org

This is an accrediting agency that gives standardized, objective information about the quality of managed-care organizations, managed behavioral healthcare organizations, credentials verification organizations, and physician organizations.

National Institutes of Health

Bethesda, MD 20892

301-496-4000

www.nih.gov/niams/healthinfo/library

This is a place to start to find information on any medical topic, resources, research, and self-help groups.

National Library of Medicine Medline Plus

http://www.nlm.nih.gov/medlineplus

This site has useful topics on a variety of issues affecting patients.

Natural Healers

www.naturalhealers.com

This website describes many complementary and alternative therapies and gives a list of schools in the United States and Canada that teach the various disciplines. Schools often have lists of graduates who are practitioners in various areas.

U.S. Department of Health and Human Services

200 Independence Ave.

Washington, DC 20201

877-696-6775

www.healthfinder.gov

This site has free guides to reliable health information.

Menopause: Time of the Wise Woman

OBJECTIVES

This chapter should enable you to

- Describe the three seasons of a woman's life
- Discuss how menopause is a sacred journey
- Define menopause
- List the three types of estrogen produced by the ovaries
- Describe what is meant by a natural hormone
- List a multistep approach to assisting women in menopausal transition years
- Describe at least six factors that can trigger hot flashes

Imagine a time long, long ago, when cycles of life were celebrated. There were ceremonies celebrating the cycles of the sun and the moon, the cycles of planting and harvesting, the cycles of seasonal change, and the cycles of birth and death, marriages, and rites of passage. The cycles of our lives were honored and celebrated as necessary transitions and initiations. People knew how to stay connected to their spirit and to that which had meaning in their lives.

REFLECTION

How do you connect with your spirit and to that which has meaning for you?

Menopause is the time that helps women reflect on life. The physiological, emotional, and spiritual changes that take place direct women's hearts, souls, and spirits. If they do not pay attention, their bodies and emotions call out to them—sometimes whispering gently, sometimes yelling to get noticed.

KEY POINT

Women are reminded to honor the different seasons of their lives:

The time of the maiden: of innocence, joy, playfulness, passions
The time of the mother: of unconditional loving, giving, and creativity
The time of the crone: of achieving the crowning glory of the wisdom of age

All of these seasons evoke celebration in honor of these transitions.

We must get back to remembering our roots. Remembering means to reconnect, to put our members back together, and to put all the pieces of ourselves back together again. Remembering our roots reminds us to reconnect to what keeps us centered and balanced in our lives. We can also reconnect to the plants that have nurtured women for centuries.

Many cultures have rites of puberty for males and females, honoring the transition into adulthood. There are trials, challenges, and initiations that evoke honor, integrity, courage, and moral character. Individuals then are welcomed into society as important and contributing members, often being given new names that honor special characteristics or individual traits. They are launched into their lives with a supportive send-off and expected to live as responsible members of the community. As their lives progress, there is another initiation when these people become elders. Respect is shown to these individuals, honoring the wisdom that they have gleaned from living life and life's experiences. The wisdom is honored and cherished, for it is passed down through generations, ensuring the future survival of the community. It is considered imperative to pass one's knowledge to others. It is not customary to keep it to oneself.

Many of the experiences that build wisdom were gathered from mistakes along the way—wisdom does not come easy. To take what life gives you and turn it into wisdom requires courage and fortitude. As with any initiation, this does not occur overnight. It requires patience, endurance, and intention and a willingness to step outside your comfort zone. The courage to forge ahead in the midst of bodily changes can bring a new relationship with your body, that of a much deeper connection.

KEY POINT

If ever there is a time for a woman to tune into her spiritual nature, it is during menopause. If by that time she has not discovered her spiritual essence, her spirit will call upon her to start listening.

Menopause as a Sacred Journey

How does this idealistic philosophy translate into our Western culture? For the most part, gray-haired men become more distinguished with age. Gray-haired women are encouraged to dye their hair, use wrinkle-reducing creams, consider cosmetic surgery, and hide any signs of aging whatsoever. Many women approach aging with dread, feeling like they are unappreciated and fearing they will be useless and discarded.

A more holistic approach to menopause is to view this as a sacred journey on one of life's rivers. It is an initiation into wisdom and creativity that takes time, preparedness, courage, fortitude, patience, and support. The river must be approached with respect, for you cannot predict what nature has in store for you throughout the journey. As you embark with anticipation, you do not know what will be waiting for you at the mouth of the river. You may start in peaceful waters but may face big rapids ahead. Sometimes you get caught in an eddy and need to sit in the stillness awhile. At times, you have to weather the storm on terra firma.

Because menopause is such an important and vital transition in a woman's life, it is meaningful to approach it as a time of celebration rather than a time of dread. Perhaps it is this dreaded anticipation that causes women to approach the menopausal transition with more fear and trepidation, which causes more intense symptoms. That is unknown; however, now, more than ever, women have many more options available to them so that they can embark on the menopausal journey with knowledge and many hormonal and herbal preparations that will facilitate this wondrous and sacred passage.

Taking Charge

The first thing a menopausal woman must consider is her goals. Are they to relieve symptoms, such as hot flashes, night sweats, insomnia, or vaginal dryness? Are they to have future protection against osteoporosis and heart disease? A combination? She must reflect deeply on this to make her decision. She can use her wise-woman intuition combined with true medical facts and then choose her intention. After she decides, she needs to surround herself with a healing team. One aspect of this is finding health professionals who will support her decisions and provide information to keep her informed of new options. She needs to learn about nutritious foods to nurture her body. Keeping her body balanced with massage therapies or energy therapies could prove helpful, as could seeking a spiritual mentor. She will find it beneficial, if she is not already doing so, to exercise with a friend. She can start her own circle of women who can gather monthly to support one another. Keeping company with people who energize, not drain her, will prove to be therapeutic. With a healing team, a woman can realize that she does not have to walk this path alone.

REFLECTION

As you go through life passages, who will make up your healing team to offer support, guidance, and assistance?

The Energetics of Healing

Most indigenous cultures have a word in their language that means energy flow through the body. The Chinese call it *chi*. The Japanese call it *ki*. In India it is called *prana*. The English language, unfortunately, lacks a word that translates into energy flow. Most other cultures are aware of the energy centers in our bodies and how to keep the energy flowing to maintain harmony and balance and to prevent disease. There are several energy systems in the body, including energy fields, energy meridians, and energy centers called chakras. You can learn how to look at your body in an energetic way and keep your energy centers strong (see Healing Touch International, listed under the resources section, for classes pertaining to energy healing). Menopause is a perfect time for a woman to start observing her own energy flow and to commit to a way of life that promotes balanced energy in her body and mind. Everyone has a personal and individual energy makeup or blueprint. That is why certain things work for some people and do not work for others. Each woman needs to begin looking at things energetically. She needs to learn to trust her body's inner wisdom to find medicines, herbs, and approaches that work for her individually. Finding her own balance is another example of the wise-woman way.

The Art of Mindfulness

There is a spiritual practice termed *mindfulness*, which teaches to be always present, centered, and observant in each thought that you have and in each action that you take. Mindfulness is a perfect practice during the transition of menopause. If a woman can be mindful of what is happening with each different emotion or symptom that she experiences, then she can observe the effects of different substances and situations that may trigger symptoms. When she knows what triggers a symptom, then she can modify it. It is important for a woman to learn to be mindful of her body during this phase. She can start tuning in to her body's natural rhythms so that she will know when it needs rest, when it needs nourishing foods, when it needs quiet, when it needs play and laughter. If she believes that menopause directs her toward self-growth, then if she is mindful, she can see the connections more clearly and gain wonderful insights.

The Art of Healing

Healing is different from curing. The word *healing* means *wholeness*. Each individual has a unique energy matrix that defines wholeness. Menopause is a time for discovering and connecting to that which brings a woman closer to wholeness. Only her own inner wisdom ultimately knows what that is. During menopause, she is remembering—bringing mind, body, and spirit together to be more whole. This can be done with the use of herbs, foods, friends, exercise, quiet time, and nature. By reflecting in the quiet and the stillness, a woman will know what she needs for her own personal wholeness and healing.

> **KEY POINT**
>
> A new wholeness of mind, body, and spirit can be realized during menopause.

Defining Menopause

The medical definition of *menopause* is the absence of menstrual periods for 12 months in a row. Some experts define it as no periods for 6 months in a row, whereas others say if you are having menopausal symptoms unrelated to other medical pathology, then that constitutes menopause. The average age to experience menopause in the United States is age 51, although the onset of menopause can occur in the 40s through the late 50s.

Some blood tests can be used to confirm menopause, especially thyroid-stimulating hormone (TSH) and follicle-stimulating hormone (FSH) levels

> **KEY POINT**
>
> Sixty million women will be experiencing the menopausal journey by the year 2020. Women's health programs that address the holistic needs of this population will need to expand.

EXHIBIT 16-1   FOLLICLE-STIMULATING HORMONE

FSH is produced in the pituitary gland. Through a feedback loop system, it sends a message to the ovaries each month to stimulate ovulation. In the perimenopausal and menopausal years, there is a gradual decline in estrogen production. When the pituitary gland notices that there is less estrogen being produced in the ovaries, it produces more FSH in an effort to get the ovaries to respond by producing more estrogen. As this feedback loop continues, FSH levels start rising.

A normal FSH level for women who are menstruating ranges between 4.7 and 21.5 mIU/ml. For postmenopausal women, the usual range is 25.8–123.8 mIU/ml. (Normal values can differ among laboratories doing the testing.)

(see **Exhibit 16-1**). It is important to rule out underlying thyroid disease by getting a TSH level, as some of the symptoms can be similar to menopause. FSH is elevated in menopause, and thus, measuring the level can be a useful diagnostic measure.

Estrogen

Estrogen, which is produced in the ovaries, adrenal glands, and fat cells, is involved in many functions. Increasingly, research reveals new insights into the effects of estrogen; researchers are finding more estrogen receptors in the body than were originally considered. There are estrogen receptors in the skin, the brain, the heart, the bones, the genitourinary tract, and the intestines. When estrogen levels decline in menopause, it affects all these organ systems. Through much research since the 1930s and through medical evaluation, doctors began advising estrogen replacement in the 1950s. It was believed that after menopause, estrogen replacement helped to protect against heart disease, osteoporosis, colon cancer, vaginal atrophy, and urethral and bladder atrophy. It was suggested that estrogen helped to keep skin supple, keep our teeth strong, and help our brains stay healthy, even preventing Alzheimer's disease; however, the benefits of estrogen replacement were found to be overshadowed by the risks when the results of the Women's Health Initiative study began to show that women who took oral estrogen with progesterone therapy showed increased risks for breast cancer, coronary heart disease, and venous thromboembolism (Writing Group for the Women's Health Initiative Investigators, 2002). Many women, frightened by these findings, ceased using

KEY POINT

Five times as many women die of heart disease and osteoporosis than breast cancer. Heart disease causes as many deaths each year as the next eight leading causes combined. Approximately 240,000 women die of heart attacks each year. Women are more likely to die of a first heart attack than men. The American Heart Association claims that a woman's lifetime risk of heart attack is one in two (see **Exhibit 16-2**).

EXHIBIT 16-2   HEART DISEASE AND WOMEN

RISKS FOR HEART DISEASE
- Natural menopause occurring before age 40
- Menopause induced by surgery or illness before age 45
- Previous diagnosis of heart disease or hypertension
- Family history of heart disease especially heart attack prior to age 50
- Bulk of body fat is in upper body/waist
- Smoking
- Poor diet—low in nutrients, high fat, frequent fast foods
- Physically inactive
- No passion for life
- Inability to express anger; unresolved anger and grief
- Social isolation; limited community support

HOW TO STRENGTHEN YOUR HEART
- Exercise 3 to 5 times per week for 30–40 minutes
- Eat a high-nutrient diet, rich in whole grains, green leafy vegetables, low-fat protein, low sugar
- Keep your heart emotionally strong by expressing feelings
- Learn the art of forgiveness (yourself and others)
- Practice the art of the four-chambered heart
    1. Be full hearted. Do not do anything halfheartedly.
    2. Be openhearted. Open yourself to receiving as well as giving.
    3. Be strong hearted. Let go of fear.
    4. Be clear hearted. Set a clear intention.

estrogen. Although estrogen is no longer seen as the cure for all of the negative consequences of aging, it now is viewed as having a place for short-term
use in the relief of menopausal symptoms.

Over 35 million American women over age 50 have osteoporosis or are at
risk of developing the disease. Approximately 40% of these women will sustain a fracture as a result of their fragile bones. A woman's risk of developing
a hip fracture is equal to her combined risk of breast, uterine, and cervical
cancer. The increased risk for developing osteoporosis caused by declining
estrogen levels caused many women to consider hormonal replacement
therapy, but there are safer options for women to consider (see **Exhibit 16-3**).

The ovaries produce three types of estrogen—estrone, estradiol, and
estriol. Estrone (E1) was the first estrogen discovered and is very potent.
This potency is attributed to many intense side effects when taking it, such
as breast tenderness, nausea, and headache. Estrone is possibly the estrogen
most linked to a greater risk for breast cancer. Estradiol (E2) is the primary
estrogen produced by the ovary prior to menopause. Estriol (E3) is the principal estrogen produced in pregnancy. It is a weak form of estrogen and has
not been used much in the past because it was more difficult to formulate a

---

EXHIBIT 16-3   OSTEOPOROSIS: IDENTIFYING AND REDUCING RISKS

RISKS FOR OSTEOPOROSIS
- Surgical or abrupt menopause
- Menopause
- Strong family history of osteoporosis
- Caucasian or Asian heritage
- Small body frame/slender build
- High caffeine intake
- Smoking
- High intake of phosphates from cola drinks
- Carbonated beverages
- Frequent antacids
- Low-calcium diet
- Alcoholic beverages
- High exercise/low body fat ratio/infrequent periods as seen in athletes
- Infrequent periods
- Premature gray hair

high enough dose in the laboratory. During perimenopause, estrone, which is produced in the ovaries and body fat, is the predominant type of estrogen.

Other Hormones

Progesterone falls in perimenopause but not in direct proportion to estrogen. A disproportionate amount of estrogen to progesterone is referred to as *estrogen dominance*. Some of the symptoms of estrogen dominance include breast swelling and tenderness, heavy irregular periods, water retention, cold hands and feet, and symptoms associated with premenstrual syndrome and menopause. Phytoestrogens—the estrogens obtained from plants and food—can increase this imbalance.

Women produce *testosterone* in their ovaries and adrenal glands. This hormone serves an important role in our interest and enjoyment of sex. Testosterone declines by as much as 50% by midlife, resulting in reductions in sex drive, sensitivity to nipple stimulation, sexual arousal, and capacity for orgasm. The decline in this hormone also causes women to have less energy and to experience a thinning and loss of pubic hair.

Thyroid hormone has significant functions in our bodies, too. It regulates the body's metabolism; therefore, altered levels of thyroid hormone can have profound and far-reaching effects. A conservative estimate is that by age 50, 20% of women have hypothyroidism—low thyroid function. This can cause symptoms that often mimic low estrogen levels, such as:

- Weight gain
- Breathing difficulties
- Heavy periods
- Low energy
- Depression
- Reduced immunity
- Muscle weakness
- Fibromyalgia
- Constipation
- Memory problems
- Night sweats
- Sleep disturbances
- Dry skin and hair
- Low body temperature
- Intolerance to cold
- Decreased libido

When thyroid levels in the blood are low, a greater amount of thyroid stimulating hormone (TSH) is produced, which is one way that this change can be detected through blood tests. (Other blood tests of thyroid function include measurements of levels of thyroxine, triiodothyronine, and thyroid antibodies.) Although there is a relationship between the decline of estrogen and thyroid hormones with age, at this point the way in which a reduction in one influences a reduction in the other is uncertain. Estrogen dominance can block the action of thyroid hormone.

The adrenal glands have an effect on the various hormones in our body. This is a consideration in regard to stress management as stress can affect the function of the adrenals leading to hormonal imbalances. Reduced adrenal function is characterized by:

- Moodiness
- Muscle weakness
- Allergies
- Low energy, particularly during late afternoon
- Not feeling refreshed when waking

Although menopause is thought to be associated with a reduction in estrogen and progesterone, you can see that menopause is a much more complicated hormonal process than that. Reliable testing of hormonal levels is done through analysis of blood or saliva.

Hormonal Replacement

After a woman is determined to be in menopause or has severe menopause-related symptoms that she wants to treat, she must then decide on a regimen. If she still has an intact uterus, adding estrogen can cause a buildup of the uterine lining (or endometrial lining). This buildup is called hyperplasia and can lead to an increased risk of uterine cancer. Adding a progestin component (a natural or synthetic product that mimics the action of natural progesterone) prevents uterine hyperplasia; therefore, if a woman takes estrogen and has a uterus, it is advised that she also take a progesterone supplement. The progestin component often causes more intolerable side effects, such as headaches, mood swings, and irritability, which cause some women to stop taking hormone supplements.

Estrogen is used for the treatment of moderate to severe menopause symptoms and for the prevention of osteoporosis. When used, it is prescribed at the lowest effective dosage for the shortest period of time. It should not be used by women with breast cancer, a history of breast cancer, suspected or known estrogen-sensitive cancers, coronary artery disease, untreated hypertension, active liver disease, pulmonary embolism, undiagnosed vaginal bleeding, or high sensitivity to hormone therapy. Women with a uterus should have progestogen prescribed with the estrogen to counteract the increased risk of endometrial cancer; women without a uterus should not use progestogen. Women 60 and older who have never taken hormone therapy shouldn't start unless there is a strong reason to do so; these women need to be monitored closely by their physicians for cardiovascular risks. When estrogen is used only to treat vaginal symptoms, local administration rather than oral is recommended. The use of bioidentical custom-compounded hormones is not recommended.

REFLECTION

Some women are willing to assume the increased risk for cancer and heart disease in order to obtain the benefits from hormonal replacement therapy. They claim that they are more concerned with feeling and looking their best today than with the possible risks they may face in the future. How does this coincide with your beliefs and values? What has influenced your personal beliefs and values concerning this type of view?

Multistep Approach

A multistep approach can prove helpful to women in the menopausal transition years, consisting of:

- Herbs
- Physical exercise
- Essential oils and aromatherapy
- Breathing techniques, meditation, journaling, self-care techniques

Herbs

The wise-woman approach focuses on a nourishing and nurturing approach. The phase of menopause can be used by a woman to become more in tune with her body, her surroundings, and her attitude on how she wants to live the next phase of her life. The focus is on finding a balance in her life that supports health, vitality, harmony, and joy. Menopause is seen as a celebration. The work of childbearing and child rearing is approaching its close, lending time for joyful expression of a woman's true essence. As her children are growing and becoming adults, a woman can direct her energies to things that support her own continued growth.

> **KEY POINT**
>
> Menopause can move a woman toward more outward expression in the world and, at the same time, toward becoming more introspective.

Although few herbs have been proven to have significant benefit in managing menopausal symptoms, they can provide a means for a woman to rediscover the ways of nourishing, nurturing, and replenishing her body and spirit. It can be very gratifying for a woman to begin to learn about these wondrous plants that have been growing alongside us for centuries. Herbs can be important allies for her as she takes time to observe the plants, smell their fragrances, sit with them to discover their medicine, bask in the beauty of their colors and textures. She will find certain plants call to her and resonate with her own energy vibration. These will be her healing herbs. Classes on herbs and herbal medicine can help one to become more knowledgeable so that herbs can be used wisely.

The National Center for Complementary and Alternative Medicine has studied herbs promoted for the treatment of menopausal symptoms, and their findings reveal the following:

- Black cohosh (*Actaea racemosa, Cimicifuga racemosa*) has been promoted for its estrogenic effect in relieving menopausal symptoms; however, studies of its effectiveness in reducing hot flashes have had mixed results. A study funded by National Center on Complementary and Alternative Medicine and the National Institute on Aging found that black cohosh, whether used alone or with other botanicals, failed to relieve hot flashes and night sweats in postmenopausal women or those approaching menopause. Other research suggests that black cohosh does not act like estrogen, as once was thought. It is advised that this herb not be used by people with liver disorders as it can have very serious consequences.

- Chaste berry (*Vitex agnus-castus*) is believed to affect pituitary function in increasing LH and reducing FSH, which results in increased progesterone; it can take several months to work.

- Dong quai (*Angelica sinensis*) has been reported by some women to reduce hot flashes and enhance energy, although research shows it to be no more effective than a placebo. Dong quai is known to interact with, and increase the activity in the body of, the blood-thinning medicine warfarin. This can lead to bleeding complications in women who take this medicine.

- Red clover (*Trifolium pratense*) provides isoflavones, which have estrogenic effects. There is no evidence that red clover has any benefit in reducing menopausal symptoms. Some studies have raised concerns that red clover, which contains phytoestrogens, might have harmful effects on hormone-sensitive tissue (for example, in the breast and uterus).
- Ginseng (*Panax ginseng or Panax quinquefolius*) may help with some menopausal symptoms, such as mood symptoms and sleep disturbances, and with one's overall sense of well-being. Ginseng has not been found helpful for hot flashes.
- Green tea strengthens bones and is a general tonic for good health. It is not advised for use if you have a clotting disorder or are taking an anticoagulant.
- Ginkgo biloba has been suggested to enhance brain function, but studies have not supported this.
- St. John's wort can improve mild symptoms of depression but has not been proven to be helpful for moderate to severe depression; it can increase the risk of bleeding.
- Valerian is believed to have a calming effect and promote sleep. Studies by the National Center for Complementary and Alternative Medicine suggest that valerian is generally safe to use for short periods of time (e.g., 4–6 weeks).

It is important to remember that herbs can take several weeks or months to work, so patience is needed. Also, knowledge about dosage, effects, and risks is limited because herbs have not been extensively studied. Although they are natural plant substances, herbs need to be used with the same discretion as drugs.

Physical Exercise

Do not underestimate the importance of exercise. Frequent physical exercise is imperative to moving through this transition with ease and health. Because

---

KEY POINT

Tea making can be used as a ritual for the menopausal woman. She can choose a beautiful teapot and mug that reflects her unique personality. As she drinks the tea, she can pay attention to the color, the aroma, the flavor, and the temperature, visualizing that it is nourishing all the organs of her body as she drinks it, bringing her vitality and health.

people are no longer working in the fields, hauling wood and water, or being active as part of a day's work, physical exercise is needed to maintain hormonal balance and produce endorphins that help with well-being.

A woman should find some kind of movement that she enjoys and commit to moving at least 4 times a week for at least 30–40 minutes. It takes at least 3 weeks to feel the effects of beginning to exercise again. It will not work to binge exercise. A woman should not go more than 72 hours without moving. It is useful for a woman to find the kind of movement that she enjoys, such as dancing, gardening, walking, or swimming. Yoga and T'ai Chi are excellent. Almost all community recreation centers have adult exercise classes, including water classes if there is a pool. There are lots of hiking or cycling clubs that offer all types of activities. Walking is free and easy. The woman can take mindful meditation walks where she will notice all of her surroundings and how they are changing with the seasons. This is a good way to smell the flowers and destress.

REFLECTION

Do you engage in some kind of movement for at least 30 to 40 minutes, four to five times a week? If not, what prevents you from doing so, and what can you do to change this?

A woman will notice immense differences in how she feels if she keeps her commitment to exercise. If she does not enjoy any type of physical activity, she could find benefit in joining a movement class that interests her at a community education center and commit to it for at least 6 weeks. Often, she will end up liking it, even if she did not think she would. Exercising with a buddy can help. Of course, any weight-bearing exercise will be of additional value in helping to keep bones strong, but initially, the woman should be advised to start with something that interests her. To prevent osteoporosis, a weight-bearing exercise program must be added to the program.

Essential Oils

Many different essential oils used in aromatherapy work very well to help emotional well-being. A woman should find ones that seem to call to her and experiment with ones that suit her best. Lavender is always good for calming. She can place it on her pillow to help her sleep or put a few drops on her forehead or temples to help ease tension. It is nice to place lavender drops in the bath, too.

Clary sage is another good oil during menopause (consider the name—the saging of the wise woman). Orange, jasmine, and ylang-ylang can evoke beauty and creativity and can also be uplifting.

Breathing Techniques, Meditation, Journaling, Self-Care

Another important step is for each woman to add a disciplined, self-care program that helps her stay centered, focused, and destressed. Learning how to switch to deep, abdominal breathing instead of shallow chest breathing helps oxygenate the body to give her more energy.

Meditation techniques are good life tools that will be invaluable throughout the rest of an individual's life. Simple meditation techniques can be done in an instant, and they energize for hours. Meditation can be simply looking at a sunset or a flower or washing the dishes in a mindful manner—it does not necessarily have to be a formal meditation practice. Many books and classes are available to help in developing techniques that work for each lifestyle (see "The Art of Self-Care," later in this chapter).

Managing Specific Symptoms

Using a multistep approach as the basic foundation, specific symptoms associated with menopause can be managed. Some specific suggestions for common problems are offered on the following pages.

Insomnia

Women should identify potential triggers to insomnia, which include caffeine, alcohol, sugar, NutraSweet (aspartame), lack of exercise, emotional stress, and worry. Some women have reported problems with insomnia if they have any sugar or caffeine after 3:00 P.M. (especially ice cream, candy, or sweet desserts).

Journaling is a good tool to combat insomnia. It is helpful for a woman to journal what she did during the day, what her thoughts and feelings were, any frustrations or joys, and so forth. Journaling is a good daily practice for self-nurturing (see **Exhibit 16-4**). Several good books are listed in the "Suggested Reading" section at the end of the chapter to offer some good journaling techniques, and it could also be useful to take a class on journaling.

Calming essential oils and hot baths help relax nervous tension and anxiety. Women can try lavender, geranium, ylang-ylang, and clary sage as essential oils to place externally on the skin, on pillows or sheets, or in bath water.

EXHIBIT 16-4   FREE ASSOCIATION WRITING: GUIDELINES TO OFFER WOMEN

Thirty minutes before bedtime, drink a relaxing herbal tea and then sit down with a pen, paper, and a timer. Set the timer for 5 minutes. Begin writing all of the things on your mind. Write down words or phrases pertaining to what you are thinking. Think of all the things you did that day and all the things you have to do tomorrow or later in the week. Free associate all the words that come to you. For instance, you may think that tomorrow you have a meeting where you need to bring a potluck item. That reminds you that you need to go to the grocery store, which reminds you that you should buy a card for your friend who is in the hospital while you are in the store. That reminds you that you should call your friend's mother to see how she is doing and whether she needs any help. You also remember that you have to find your notes from the last meeting and so forth. Keep writing all the things you think of until the timer goes off. Ideally, you should be done writing at least 30 seconds before the timer goes off, and at that point you should have to think really hard to find other things on your mind to write about. If you are still writing furiously when the timer goes off, then extend the timer to 7, 8, or even 10 minutes. This exercise helps clear your mind of extraneous thoughts and worries, which in turn helps your mind relax. This relaxation helps prevent the mind chatter that sometimes makes it difficult to fall asleep or that awakens you in the night.

The spiritual state could affect a woman's sleep patterns. She should ask herself questions such as these: What is my spiritual connection? How do I create mindfulness and centeredness? How do I nurture myself? She should be encouraged to be open to asking some hard questions. Insomnia may not only be an adjustment to hormonal and physical changes, but also a signal that her spirit is calling out to her. When she does not listen to the stillness and misses her spirit calling, her spirit may awaken her at night to get her attention.

Night sweats can cause sleep interruptions. Some of the triggers for night sweats include stress, sugar, alcohol, lack of exercise, and spicy foods. Reducing these can improve night sweats.

Do not forget that daily physical exercise is important in promoting good sleep patterns.

Decreased Libido

During menopause, the fluctuating hormone levels can cause changes in libido. Some women experience an increased sex drive, whereas others experience a decrease. When women are under stress, libido is one of the first things affected. In our fast-paced society and with all the many things women do in a day, making love is sometimes the last thing on their minds.

Some herbs will enhance libido, and some women report good results with hormonal therapies as well; however, other factors need to be examined. In a quiet and reflective state, the woman can ask her body's inner wisdom to speak its truth. She needs to be willing to hear the truth and answer some tough questions. Is she truly in a relationship that supports her as the person she truly is? Does the love she gives out return to her in the way she desires? Is she around people who love, support, and energize her, or does she feel drained by them? Does she get the companionship that she desires? Does she receive the communication from her partner that she wishes for? Is she too fatigued? Does her body need a deep rest?

High stress can cause the adrenal glands to work overtime. This leaves a person feeling depleted, with no extra energy for expressing sexuality. Some herbs (e.g., American ginseng) help to strengthen the adrenal glands, which in turn may increase libido; however, herbs do not take the place of rest, sleep, and stress-reduction strategies that offer the adrenal glands a chance to recover.

REFLECTION
What do you notice about the way your energy and stress levels affect your libido?

Vaginal Dryness
Vaginal dryness can be attributed to the decreased estrogen effects on vaginal and vulvar tissue. The vulvar and vaginal tissue becomes thinner and atrophies. This can cause irritation and discomfort, painful intercourse, and contribute to urinary incontinence. This can be treated by oral hormone replacement and/or prescription estrogen cream that is applied to the vulvar and vaginal tissue. The estrogen cream is available as a synthetic hormone or as a bioidentical hormone.

The herbal options include creams and salves that are soothing and healing to vaginal tissue and mucous membranes. There are also moisture-enhancing teas and tinctures available.

Triggers to Hot Flashes

You have probably already heard this expression: "It's not a hot flash. It's a power surge." This may arise from the energy principle that hot flashes

> **KEY POINT**
>
> In the United States, approximately 60%–70% of women experience hot flashes. In other countries, there is much less reported experience of hot flashes—30% in Canadian women, 9% in Japanese women, and 0% in Mayan women. It is not certain whether the hot flashes are not reported or whether they are actually not experienced or not deemed uncomfortable. Many indigenous cultures, such as the Mayans, do not have a term in their language that translates into hot flashes.

signify a rewiring of our nervous system toward intuition. Some energy theorists believe that hot flashes are an actual energy release. It may be a release of toxins or a type of cleansing and clearing. If a woman notices that hot flashes usually occur after an exposure to an intense substance, then this theory makes sense.

Declining estrogen levels cause vasomotor symptoms, such as hot flashes, but there may be other triggers. If a woman tunes in to her body and identifies these triggers, sometimes she can modify her experience of hot flashes. It is good for her to observe keenly and to keep a diary of her symptoms to identify her own personal triggers. Common triggers to hot flashes are sugar, spicy or hot foods, alcohol, aspartame, stress, chronic sleep deprivation, anxiety and worry, and anger.

Sugar

The effects that sugar and refined carbohydrates have on causing hot flashes and premenstrual syndrome, such as symptoms of mood swings, headaches, fatigue and low energy, anxiety and restlessness, insomnia, and breast tenderness cannot be emphasized enough. If a woman became a Sherlock Holmes for a few weeks and analyzed her sugar intake and her body's response, chances are that she would notice a connection. One way to identify this is to eliminate all sugar and wheat for 3 weeks and then gradually add them back in, in moderate amounts. This can be an extremely difficult undertaking for many people. It takes a lot of preparation, but the rewards at the end are worth it. People who have done this often notice an unbelievable increase in energy, as if a cloud or veil has lifted. They have increased mental clarity and alertness. There also tends to be some weight loss, which is a benefit that many would welcome!

Sometimes a little sugar or wheat products, especially ingested in the afternoon or later, can cause night sweats and insomnia. It is very common to experience hot flashes and night sweats after a special evening going out to dinner, where a woman may have a little wine, pasta, dessert, or richer foods that she may not normally eat.

Spicy or Hot Foods

Some spicy foods, such as curry, hot peppers, or cayenne can trigger hot flashes or night sweats. Women need to observe their own personal responses to hot, spicy foods.

Alcohol

Although a glass of wine or beer may be relaxing and calming, alcohol breaks down into sugars and also can be a trigger for hot flashes. Each woman's personal response to alcohol needs to be observed.

Aspartame

It is important to read labels, as this chemical sugar substitute can be found in more products than realized and can worsen menopausal symptoms.

Stress

Stressors are different in everyone. Each woman should identify her own stressors and pay attention to them. Are they signaling that one needs to honor her wise-woman ways and her intuition? Are they signaling her to be gentler with herself or more nurturing? Oftentimes, it is the frequent little things that cause the most stress, not the big disasters. One might reflect on the hot flash she just had. Was there a message in it for her?

Chronic Sleep Deprivation

New studies are showing that Americans are chronically sleep deprived. Adults actually need about 8 hours of sleep each night, especially in our fast-paced society. A lack of sleep places extra stress on the adrenal glands that are already depleted during menopause. This is also a paradox, because during menopause women are awakened at night due to hot flashes or night sweats, and their sleep is disrupted often; therefore, it is difficult to obtain a restful night's sleep. Keeping a consistent sleep schedule by going to sleep and awakening at the same times each day, together with some herbs that help promote sleep, will sometimes help establish a better sleep pattern.

Anxiety and Worry

It is difficult to let go of worries and trust and go with the flow. If a woman has a worry, she can either take action to relieve it or let it go. She must consider whether it is worth worrying about. A woman can allow a hot flash to help her keep in tune with what worries are important to her. For example, if a woman is worried about a medical symptom, she must recognize that action is needed and get it evaluated. In anxiety-producing situations, deep-breathing techniques can help her to get through, as can learning techniques to help her stay centered, calm, and detached from the outcome.

Anger

The relative increase in testosterone in relation to the declining levels of estrogen can cause more aggression or assertiveness, which can make a woman feel out of control. Women say they become quick to anger, and they do not like that feeling. Some women say that after they have reflected on the situation they realized that their anger may have been legitimate, but that it was out of proportion, and they wish they could have controlled it better. A woman may benefit by asking herself this: Is this anger so intense because I have been silent for too long, and now cannot hold it in any longer? Do I really need something to help me temper this anger so that I can be more balanced and not so out of control? Often, when women begin a nourishing self-care program, including herbs or hormones, the anger and ensuing hot flashes diminish.

The Art of Self-Care

Women's nature is to give and they tend to give a lot; therefore, it is imperative that women also learn how to receive. The more a woman gives, the more it is necessary to receive also, for then she can be replenished to give again. This is the true art of recycling.

REFLECTION

The more you give, the more you need to receive in order to pass the giving around the circle. Does what you give balance with what you receive?

When someone offers something to us, it is a gift to them for us to receive it. This may need to be reinforced to women as they are encouraged to receive as part of their own nurturing. Women who have been other oriented for most of their lives may find it hard to receive and care for themselves. They may be receptive if it is put in the perspective that they can help others much more if they are energized and not depleted.

Each individual must develop her own self-care model that energizes and supports her and then make a disciplined commitment to integrate it into her daily life. Some suggestions that can be offered to women are presented in **Exhibit 16-5**.

The Celebration

It is important to honor the transitions in life and mark one's changing seasons with ritual. A woman can create her own ritual or ceremony to acknowledge

EXHIBIT 16-5   SELF-CARE HINTS TO NOURISH AND ENERGIZE YOU

- Give gratitude to your body every day. While you are taking a shower, allow the water to flow over every body part, mindfully acknowledging how it serves you, and be grateful that this day you are healthy.

- Scan your body energetically and notice areas of tension or discomfort that may need your nurturing attention.

- Embrace humor. Add levity every day. Smile and laugh often.

- Have music in your life in some way.

- Allow your creativity to flow: Sing a song, bake a cake, or paint a picture. Start by doing easy and simple things first.

- Search to find your passion and allow it to unfold.

- Find a connection to your spiritual essence, whatever that may be.

- Allow quiet time often (for reflection and basking in the stillness). Discover how much quiet time you need and honor that. Create a healing space in your home or outdoors where you can go to be quiet.

- Spend time with nature, mindfully drinking in your surroundings to nurture and replenish you.

- Create rituals to mark the passing of time or the special events in your life.

- Share a connectedness with others. Surround yourself with friends that are loving, supportive, stimulating, and energizing.

- Learn to care without rescuing or enabling.

- Honor your own body rhythms for rest, work, and play. Practice self-acceptance.

- Be open to possibilities and be willing to step outside your comfort zone to try new things or new attitudes. A trusted friend can help lend clarity if you are unsure.

- Be open to receiving and be willing to ask for help. Human nature causes you to want to give and help others. There are people around you that want to reach out and all they need is to be asked. Your gift back to them is allowing yourself to receive from them.

- Check in. If someone or something fires up your emotions negatively sit with it and ask yourself, is the problem the issue or is it the way I am reacting to the issue?

- Be open to looking at your own issues and past patterns, and be willing to make changes that may move you toward a higher level of awareness and communication. Menopause is a time to claim your power and sometimes stand your ground over things that you have tolerated for way too long. However, remember that being in power is walking the path of the heart.

and celebrate the wise woman she has become. She can invite a small circle of women or include special family, friends, and children, give gratitude, sing a song, light candles, share stories, and give blessings. A shared feast can follow the ceremony. A woman can use her own inner wisdom, intuition, and creativity to make her ceremony a special one.

Indigenous cultures look at life in a circle. We are all equal because we are all equidistant from the center or the source. We all travel the circle through the seasons of our lives, and we all pass in each other's footsteps at some point on the circle. We must remember that we do not travel the circle alone. We are all connected, all related. Women need to reach out to each other, knowing that they all share the energy of their heart connections. When they have times of doubt or despair, they can call on the energy of the worldwide circle of women that will uplift their spirits and give them hope. When they have moments of joy and bliss, they should send it out to others. At midlife, women can learn to walk the path of the heart, using the gifts of the wise woman—compassion, understanding, and unconditional love.

Summary

Menopause is defined as the absence of menstruation for 12 consecutive months. This is a significant event for women, not only because of the physiological challenges that are present, but because it happens at a time when they can take stock of many aspects of their life. From this standpoint, menopause is viewed as a sacred journey into wisdom and creativity.

A multistep approach is useful for women as they face the menopause transition. Useful strategies include the use of herbs, essential oils, and exercise and self-care techniques. Women should try to identify triggers for hot flashes, which could include sugar, spicy foods, alcohol, artificial sweeteners, stress, chronic sleep deprivation, anxiety, and anger. Individualized plans are essential to address each woman's unique needs.

Menopause can be used as a time to launch new self-care behaviors that allow women to nurture themselves and receive from others. New experiences and opportunities are possible.

References

Writing Group for the Women's Health Initiative Investigators. (2002). Risks and benefits of estrogen plus progestin in healthy postmenopausal women: Principal results from the Women's Health Initiative Randomized Controlled Trial. *Journal of the American Medical Association, 288,* 321–333.

Suggested Reading

Crooks, R. L., & Baur, K. (2007). *Our sexuality.* New York, NY: Wadsworth Publishing.

Eliopoulos, C. (2012). *Women afire! Living with meaning and purpose midlife and beyond.* Charleston, SC:Amazon.com.

Esposito, N. (2005). Agenda dissonance: Immigrant Hispanic women's and providers' assumptions and expectations for menopause healthcare. *Clinical Nursing Research, 14*(1), 32–56.

Hall, L., Callister, L. C., & Matsumura, G. (2007). Meanings of menopause: Cultural influences on perception and management of menopause. *Journal of Holistic Nursing, 25*(2), 106–118.

Jones, M. L., Eichenwald, T., & Hall, N. W. (2011). *Menopause for dummies.* Hoboken, NJ: Wiley Publishing.

Klaiber, E. L. (2001). *Hormones and the mind. A woman's guide to enhancing mood, memory, and sexual vitality.* New York, NY: HarperCollins Publishers.

Lesser, J., Hughes, S., & Kumar, S. (2007). Sexual dysfunction in the older woman: complex medical, psychiatric illnesses should be considered in evaluation and management. *Geriatrics, 60*(8), 18–21.

Lock, M. (2005). Cross-cultural vasomotor symptom reporting: Conceptual and methodological issues. *Menopause: The Journal of the North American Menopause Society, 12,* 239–241.

McCall, K., & Meston, C. (2007). Differences between pre- and postmenopausal women in cues for sexual desire. *Journal of Sexual Medicine, 4*(2), 364–371.

Northrup, C. (2012). *The wisdom of menopause: Creating physical and emotional health and healing during the change.* New York, NY: Bantam books.

Writing Group for the Women's Health Initiative Investigators. (2002). Risks and benefits of estrogen plus progestin in healthy postmenopausal women. *Journal of the American Medical Association, 288,* 321–333.

Resources

Healing Touch International
www.healingtouchinternational.org

National Cancer Institute
www.cancer.gov/clinicaltrials/digest-postmenopausal-hormone-use

National Institutes of Health
http://health.nih.gov/topic/Menopause

North American Menopause Society
www.menopause.org

Women's Health Initiative
www.nhlbi.nih.gov/whi

Living Fully With Chronic Conditions

17

OBJECTIVES

This chapter should enable you to:

- Describe the meaning of healing as it applies to chronic conditions
- List potential reactions to being diagnosed with a chronic condition
- List actions that can empower people to live fully with chronic conditions

Suppose you plan to meet friends for dinner and an unbearable headache sends you to bed instead. A case of the flu cancels plans for a long weekend getaway. An important meeting is missed because you're unable to leave your home due to pulled back muscles. The effects from the medications taken to control cold symptoms causes you to feel like doing nothing but crashing in bed. These types of situations can detour plans and make us feel miserable. Fortunately, they usually are short term and we recover, returning to life as usual. However, there are conditions that cause discomfort and disruption to life that are not temporary—chronic conditions (**Exhibit 17-1**). Once they develop, these conditions remain for life. Their impact can range from the inconvenience of having to take a daily medication or having to use a cane to a major impairment in function, such as being blind or unable to walk.

Approximately half of all Americans have at least one chronic disease and more than one fourth have two or more; the number is growing. Women are more likely than men to have more than one chronic condition. The likelihood of having a chronic condition increases with age, with three out of four people over age 65 being affected by these diseases. Hypertension is the leading chronic condition among adults. Arthritis is the leading cause of disability among the chronic conditions. About half the deaths each year are the result of conditions with chronic effects—heart disease, cancer, and stroke.

EXHIBIT 17-1   MAJOR CHRONIC CONDITIONS

- Arthritis
- Autoimmune diseases, such as ulcerative colitis, lupus erythematosus, Crohn's disease
- Blindness
- Cancer
- Cardiovascular diseases
- Cholesterol disorders
- Chronic fatigue syndrome
- Chronic hepatitis
- Chronic pain syndromes
- Chronic renal failure
- Deafness and hearing impairment
- Diabetes mellitus
- Epilepsy
- Hypertension
- Osteoporosis
- Respiratory diseases including asthma and chronic obstructive pulmonary disease (COPD)
- Sickle cell anemia and other hemoglobin disorders
- Stroke

As chronic conditions will be companions for life, it is important that people affected by them understand the conditions and their related care so that they can achieve the highest possible quality of life.

A Healing Approach

Much of the U.S. healthcare system is based on the medical model in which the emphasis is on diagnosing, treating, and curing a health problem. Success often is based on the patient having the problem treated and eliminated and on returning to normal. Although appropriate diagnosis and treatment are important components of the care of someone with a chronic condition, the

KEY POINT

Growing numbers of people are living with one or more chronic conditions.

focus will differ from that of acute care as cure is not a realistic goal. Instead, the emphasis is on *healing*.

Many people consider healing as being the same as cure, and in some situations it is. However, in the realm of chronic conditions, healing implies living in harmony with the condition, achieved by:

- Mobilizing the body, mind, and spirit to control symptoms.
- Maintaining or improving self-care efforts.
- Managing the disease effectively (i.e., taking medications and doing treatments as recommended).
- Preventing complications.
- Delaying decline in status and function.
- Using the disease to learn about other aspects of self and life.
- Achieving the highest possible quality of life.

Although there may be imperfections in the body's structure or function, healing implies that a good quality of life can be achieved. Rather than being victim of a disease and defining oneself by it (e.g., "I'm a diabetic," or "I can't do that because I've got a heart condition."), the person who adopts a healing approach incorporates the condition into life without being defined by it. For example, a person with diabetes establishes a desired lifestyle and then adjusts diet and medication administration to accommodate that lifestyle rather than significantly changing his or her lifestyle to follow a strict plan. Likewise, the person with a newly diagnosed heart condition who enjoys eating at fine restaurants will learn how to make wise selections from any menu rather than forfeit dining out.

Most people affected by chronic conditions have the ability to choose if they will adopt a healing approach or become a victim to their condition. A healing approach is fostered by becoming an informed consumer. This entails obtaining thorough explanations for healthcare providers, asking questions for clarity, and independently learning as much about the condition as possible. A wealth of information can be found through Internet searches of the condition. Local support groups also can be great sources of information. Honest discussions with family and friends are useful in erasing any misconceptions they may have, obtaining support, and establishing ground rules as to minimizing the impact of the condition on relationships.

REFLECTION

What factors do you think influence how you would react to being diagnosed with a chronic disease?

Reactions to a Chronic Condition

Learning that one has the diagnosis of hypertension, diabetes, osteoporosis, or another condition can be difficult. A chronic condition can affect every aspect of life (**Figure 17-1**). In addition to physical symptoms, psychological state, finances, and socialization can be impacted. Roles and responsibilities may need to be changed or forfeited to accommodate the effects and demands of the disease. One's vulnerability and mortality may be seriously considered for the first time. Meeting the demands of taking care of a family, supporting oneself, and maintaining relationships can become challenging. The knowledge of the negative impact a similar diagnosis has had on someone else can trigger anxiety. Even in the absence of symptoms that limit function, an individual may find that others view him or her differently based on stereotypes or misinformation. Understandably, defense mechanisms may kick in when the situation is too much to cope with, which can include:

- *Denial.* Actions that are inconsistent with the realities of the condition may be taken. For example, the person with diabetes may claim that he doesn't need to use insulin as he doesn't believe his body really

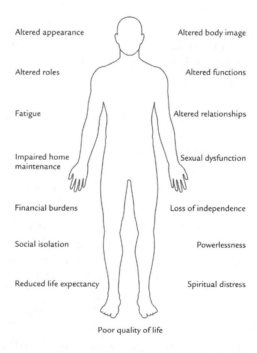

Altered appearance

Altered body image

Altered roles

Altered functions

Fatigue

Altered relationships

Impaired home maintenance

Sexual dysfunction

Financial burdens

Loss of independence

Social isolation

Powerlessness

Reduced life expectancy

Spiritual distress

Poor quality of life

Figure 17-1   Chronic conditions potentially affect every aspect of a person's life

needs it; a mountain climbing vacation may be planned by someone with degenerative joint disease.

- *Anger.* The person may resent that he or she has been diagnosed with the disease and displace feelings on family and friends.
- *Guilt.* A person may perceive that the disease resulted from years of not following good health practices or as punishment. For example, a man who has become visually impaired may associate this with viewing pornography.
- *Depression.* When the potential impact of a chronic disease is considered, a person can become overwhelmed and depressed. Plans may have to be altered. The realization that one is not in complete control will be faced.

In addition to the individuals with chronic conditions, the significant others in their lives may experience reactions as well. Shocked at his spouse's new diagnosis and at a loss of what to do to help, a husband may tell his wife, "Don't worry, these doctors don't always know what they're talking about." An adult child who has to care for a parent impaired with a heart condition may be angry that the parent didn't listen to advice to stop smoking and now not only the parent, but also the caregiver child must suffer the consequences. A man who has a child with a disability may feel that this is his punishment for having been unfaithful to his wife. Dashed hopes for the vision they had of a dynamic life may lead to depression in family members who now have to make adjustments to accommodate the needs of a member with a chronic disease.

In time and with support people should work through their reactions and reach a place in which they view the condition realistically, engage in appropriate care measures, and make necessary adjustments to their lifestyle. It is important that necessary, but not unnecessary adjustments be made. For example, a person with diabetes may need to incorporate blood testing and insulin administration into his lifestyle as necessary adjustments; however, it is unnecessary for him to decline social invitations because he can only eat certain foods or give up his interest in sailing because he may get hypoglycemic while on the water. Accepting one's chronic condition doesn't mean being defined by it.

Goals

As mentioned earlier, chronic conditions differ from acute ones; therefore, goals of care will be different and include effectively managing the condition; stimulating the body's healing abilities; preventing complications; maximizing

> **KEY POINT**
>
> It is important for people with chronic conditions to avoid labeling them-selves with the disease (e.g., a diabetic, a heart patient, an arthritis victim) and instead, consider themselves people who happen to have the condition (e.g., a person with diabetes/heart condition/arthritis). This mind-set can be empowering in that people are controlling their conditions rather than being controlled by them.

quality of life; dying with peace, comfort, and dignity; and empowering one to live effectively with the condition.

Effectively Managing the Condition

Learning to live with a chronic condition often requires that new knowledge and skills be gained. This includes a review of the disease, conventional as well as complementary and alternative therapies that are used, medications and their effects, procedures for performing treatments (e.g., dressings, irri-gations), care of equipment and special devices, and any necessary diet or lifestyle modifications.

Stimulating the Body's Healing Abilities

The body has tremendous abilities to fight disease, control symptoms, and prevent complications. Learning to stimulate these activities can assist in avoiding complications and reducing the need for medications and other treat-ments that carry risks. In some circumstances, complementary and alternative approaches (e.g., biofeedback, acupuncture, herbal medicine) can substitute for conventional treatments, as when relaxation exercises are used instead of a tranquilizer.

Preventing Complications

Chronic diseases and many of the conventional treatments used to manage them can increase the risk for infections, injuries, and other complications. For example, a person with emphysema can easily develop pneumonia from the secretions that pool in the lungs; medications used to treat hypertension can cause the person to have dizziness leading to falls; and poor vision can cause a person to misread instructions on a medication label and administer the drug incorrectly. Risks that a person may have based on his or her unique status and function need to be considered, and actions must be taken to mini-mize them.

Maximizing Quality of Life

Adhering to the ideal diet, taking medications exactly at the prescribed time, and avoiding experiences that could pose risk may keep the disease under control but at the expense of a decent quality of life. Although lifestyle modifications may be required to keep a chronic condition under control, it is important that the person control the condition rather than being controlled by the condition. Quality of life is an extremely important consideration.

Dying with Peace, Comfort, and Dignity

Decline and death will accompany some conditions, despite the best care, such as late-stage renal failure and cancer. For some persons, the time between diagnosis and death will be brief; for others there could be years of steady decline. Pain control, preservation of energy, comfort, and assistance in meeting basic needs become crucial. Emotional and spiritual support becomes important, also. Describing one's wishes during the last stage of life in an advance directive helps to assure preferences and wishes are understood.

REFLECTION

How much sacrifice of normality and quality of life in the present would you be willing to make to add several years to your life?

Empowering One to Live Effectively With a Chronic Condition

Being proactive can assist individuals with chronic conditions to live in harmony with their conditions and enjoy high-quality lives. There are several actions that can empower people to achieve this positive state, such as selecting the right provider, becoming informed, getting support, and developing a positive mind-set.

Select the right provider. Chronic conditions require regular, and sometimes frequent, contact with a healthcare provider; therefore, it is crucial that the right provider(s) be chosen. In addition to having expertise in the area, the provider should:

- Allocate adequate time for office visits and telephone consultation.
- Communicate in a style and language that are appropriate for the patient.
- Thoroughly explain and educate the patient about the condition and its management.
- Welcome and encourage the patient's active participation in his or her care.

- Be sensitive to the needs of the entire family unit.
- Be open to complementary and alternative therapies.
- Demonstrate hope and optimism.

Become informed. Although a good healthcare provider will explain and educate, each person needs to take responsibility for equipping himself or herself with knowledge. Investing time reading about the condition, prescribed treatments, treatment options, medications used, and resources is essential. If a person is unable to conduct this research independently, a family member or friend should be recruited to assist. Maintaining a file that contains information obtained, summaries of office visits with the healthcare provider, drug fact sheets, resources, and other information can prove to be highly beneficial.

Get support. Living with a chronic condition is hardly a smooth journey. There are bumps in the road and detours from anticipated plans. Understandably, this can be a challenging experience that depletes physical, emotional, social, and economic resources. Eliciting support can help a person to carry the burdens faced. Support can be recruited from family, friends, faith communities, and support groups. Each person should identify the specific needs for which support and assistance are needed, such as accompaniment to doctors' visits, doing Internet searches for treatment options, or visiting weekly to check in and offer encouragement.

Develop a positive mind-set. Rather than drowning in the demands imposed by a chronic condition and becoming overwhelmed and discouraged, each person can break down the demands into pieces and recognize his or her ability to address the pieces. By reflecting on his or her life in totality and identifying the challenges that have been successfully met, confidence can be strengthened. In addition, taking stock of capabilities, support systems, and other things to be thankful for can promote a positive mind-set.

REFLECTION
What do you have to be thankful for?

Summary

Growing numbers of people are being diagnosed with chronic conditions. Living with a chronic condition adds new challenges and responsibilities to one's life that affects body, mind, and spirit. Fortunately, we live in a time when people can survive and live high-quality lives with these conditions.

Being equipped with knowledge, good support, and a positive mind-set enables a person to live effectively with a chronic condition.

Suggested Reading

Anderson, G. (2004). *Chronic conditions: Making the case for ongoing care.* Baltimore, MD: Johns Hopkins University.

Baan, R., Straif, K., Grosse, Y., Secretan, B., El Ghissassi, F., Bouvard, V., . . . Cogliano, V.; WHO International Agency for Research on Cancer Monograph Working Group. (2007). Carcinogenicity of alcoholic beverages. *Lancet Oncology, 8,* 292–293.

Centers for Disease Control and Prevention. (2006). Prevalence of doctor-diagnosed arthritis and arthritis-attributable activity limitation—United States, 2003–2005. *MMWR, 55,* 1089–1092. Available from http://www.cdc.gov/mmwr/preview/mmwrhtml/mm5540a2.htm

Centers for Disease Control and Prevention. (2008). *National diabetes fact sheet, 2007.* Atlanta, GA: U.S. Department of Health and Human Services. Available from http://www.cdc.gov/Diabetes/pubs/fact sheet07.htm

Centers for Disease Control and Prevention. (2008). Prevalence of self-reported physically active adults—United States, 2007. *MMWR, 57,* 1297–1300. Available from http://www.cdc.gov/mmwr/preview /mmwrhtml/mm5748a1.htm

Centers for Disease Control and Prevention. (2008). Youth risk behavior surveillance—United States, 2007. *MMWR, 57*(SS-04), 1–131. Available from http://www.cdc.gov/mmwr/preview/mmwrhtml/ss5704a1.htm

Kung, H. C., Hoyert, D. L., Xu, J. Q., & Murphy, S. L. (2008). Deaths: Final data for 2005. *National Vital Statistics Reports, 56*(10). Available from http://www.cdc.gov/nchs/data/nvsr/nvsr56/nvsr56_10.pdf

Lorig, K., Holman, H., Sobel, D., & Laurent, D. (2007). *Living a healthy life with chronic conditions* (3rd ed.). Boulder, CO: Bull Publishing.

National Center for Health Statistics. (2007). *Health, United States, 2007. With chartbook on trends in the health of Americans.* Hyattsville, MD: National Center for Health Statistics; 2007. Available from http://www.cdc.gov/nchs/data/hus/hus07.pdf

Addiction: Diseases of Fear, Shame, and Guilt

OBJECTIVES

This chapter should enable you to

- Describe what is meant by an addiction
- List at least five types of addictions
- Discuss at least five significant areas that can be affected by addictions
- Describe at least three characteristics of persons with addictions
- Describe four measures that can aid in healing the body of the addict

---

Addictions are among the most common diseases in the United States. An estimated 20% of our population lives with the pain of being addicted to something. Each person who is an addict affects about 10 other people who experience pain, stress, fear, shame, and guilt just as the addict does. Although we often think of alcohol and drugs when considering addictions, other addictions, such as gambling or exercise, appear more subtle but cause many of the same problems.

Scenarios of Addictions

Addictions are played out in many ways, such as in these scenarios in which people:

- Are compelled to gamble no matter what the costs. A loss of home, maxed out credit cards, risky loans, and threats on their lives do not stop them.
- Are driven to the next new sexual conquest in spite of the risk of HIV and AIDS, hepatitis, and loss of their marriages and children.

- Must eat and vomit even though they weigh 90 pounds and have painful dental problems from stomach acid damaging their teeth when they vomit. Additionally, they just got out of the hospital because of cardiac problems and electrolyte imbalances in their blood.
- Are unable to stop themselves from eating the whole container of food (usually not vegetables, but ice cream, cookies, chips, etc.). It happens in secret especially if their weight is more than 250 pounds.
- Need the cocaine to complete some important work task, but cannot get the sustained high anymore. Their heart hurts while their nose runs and bleeds.
- Drive drunk for the seventh time since their last DUI and lose their driver's licenses.
- Continue to grow their own pot, although their best friends died in an automobile accident last year after they used a lot of drugs and marijuana.
- Worked 90 hours again this week and missed their youngest child's birthday. They just have to get that promotion.
- Can't get control of the collection of newspapers that has become a fire hazard or the 17 cats that have overtaken their home, despite the fact that things are getting moldy and smelling and they are getting sick.
- Must run another 15 miles today even though they have not menstruated for more than 12 months. It does not matter that it is 10 P.M. If they hurry, they can do it before midnight.

These scenarios depict some extreme late-stage addictive behaviors. Before people get to this stage, a nonjudgmental health risk appraisal with hopeful feedback can start the healing process.

REFLECTION

Was there anyone in your family who had an addiction? How did that impact that person and the rest of the family? How did the family address this issue with the person?

Definition of Addictions

Addictions are behaviors people do repeatedly that result in problems in one or more major areas of their lives. All addictions are the result of attempts to relieve or control feelings of anxiety, vulnerability, and anger. They are related

> **KEY POINT**
>
> Significant areas that potentially are affected by addictions include the following:
>
> - Physical and mental health (e.g., causing accidents, leading to depression, etc.)
> - Spirituality, including loss of hope, belief, and peace; isolation from supports; anger at God, creator, supreme being
> - Relationships, including those with family, mate (partner), parents, children, and friends
> - Work, including poor performance; overachievement; poor working relationships; attendance problems; workaholism
> - Legal, including litigation, DUI, violence, stealing, tax evasion, sexually acting out
> - Financial, including loss of income/savings; living in guilt, fear, and shame over spending

to problems with impulse control and self-esteem. The repetitious behaviors that constitute an addiction include misuse, overuse, and abuse of alcohol, food, work, sex, shopping, smoking, money, Internet, power, relationships, exercise, drugs (prescription, nonprescription, and street drugs), and collections of stuff (papers, memorabilia, guns, etc.). You probably could add a few more to the list of things that are used addictively to escape facing a feeling.

Prognosis for Persons With Addictions and Their Families

The prognosis may look bleak and appalling, but it is not. Early identification, education, and intervention can lead to prevention of part or all of the problems noted previously. Intermediate and late interventions can reduce the severest symptoms of the addiction-related diseases and problems. Later in this chapter, assessment and intervention are addressed. These techniques help to reduce the inevitable destruction. Thus, the prognosis can be excellent when specific steps are followed and there is support for all who are directly involved.

Who Are Persons with Addictions?

At the very least, 1 of every 10 people suffers with some type of addiction. People with addictions are represented in all strata of society. Everyone is at risk. Persons with addictions are presidents, teenagers, generals, nuns, home-less individuals, grandmothers, rabbis, doctors, nurses, accountants, skid row people, pilots, school dropouts, valedictorians, and beauty queens. The types of people who become addicted are not determined by social status, income, education, ethnicity, gender, or any other demographic characteristic.

People with addictions are not necessarily bad people, although they sometimes do very bad things. Most persons with addictions do not know why they repeat actions that keep hurting themselves and others. At first, many do not really think they have a problem. They believe they can cut back on their behaviors whenever they desire. They repeatedly try to stop the addiction and often succeed in exercising some control although the problem remains. As the behavior begins to bother others, they may make promises to stop; however, there comes a point when they cannot control the outcomes of their addictive behaviors. Unfortunately, they lack the ability to break the pattern.

Next, people with addictions may feel fear, shame, and guilt. They want to stop, but the urges are out of their control. Some may not want to stop because they cannot conceive living without the habit that they believe allows them to function. If they muster a superhuman effort and manage to stop acting out the addiction, they suffer severe anxiety and pain. With drugs, alcohol, gambling, and food addictions, there can be tremendous physical pain, as well as anxiety and emotional devastation.

Finally, people with addictions lose all hope, suffer guilt, lie in their shame, and hide themselves in constant fear (beginning paranoia). Prison, chronic illnesses, financial ruin, suicide attempts, and death are frequently the final results of untreated addictions.

Those with an addiction cannot imagine life without the escape and feel-ings of relief that their addiction gives them. Most of the people with addic-tions have a need to escape from some emotion that is not at a conscious level.

Persons who have loved ones who have addictions suffer many feelings, too. Families, friends, and the people with addictions themselves need com-passion, love, support, and information to recover their lives. It is possible. People need to know they are not alone and that there is hope. People with an addiction are not bad, worthless, and hopeless; rather, they are sick. Since 1957, the American Medical Association recognized alcoholism as a disease.

KEY POINT

Withdrawal symptoms from some addictions can include insomnia, head-aches, extreme irritability, anxiety, upset stomach, sweating, and chills.

The *Diagnostic and Statistical Manual IV*, a guide for identifying mental illnesses and their characteristics, lists various addictions as diseases. These listings are helpful because they acknowledge many addictions as treatable conditions.

Commonalities Among Addictions

There are similarities among many addictions. Most people with an addiction contributed significantly to the development of their problems, which can cause society to have limited sympathy for them; however, just as many people with heart disease have contributed to the development of their illnesses because of lifestyle choices they have made. People who stray from the special diets they should follow knowing they are putting their lives in jeopardy contribute to their medical problems just as much as persons who abuse alcohol and drugs. Nevertheless, society considers people with traditional medical diagnoses as legitimately needing help and care despite their responsibility for causing their problems. There is less support and sensitivity for those who are addicted.

> ### KEY POINT
>
> People with addictions deserve reasonable help and support just as any person with a disease. They are sick people who need help and healing desperately. People in their lives also need healing because loving and living with an addict makes people close to them hurt and sick, too.

Help, Hope, Healing, and Health

How can you help persons with addictions and yourself if you have relationships with them? There are risk factors and risk patterns associated with the development of addictions. Identification of those patterns and the persons at risk is ideal. You can use a health risk appraisal questionnaire such as the Efinger Addictions Risk Survey (EARS) to identify persons at risk (see **Exhibit 18-1**). Please feel free to make copies for future use.

The EARS asks questions that are nonoffensive to most people. The survey includes topics that are common experiences and assigns a score to the degree the experience affects the person. It does not appear as invasive or judgmental like many other surveys and screening tests. For example, the traditional alcohol surveys ask questions about alcohol use that are likely

EXHIBIT 18-1   EARS

Efinger Addictions Risk Survey (EARS)

1. Various stresses and losses occur at all stages of our lives. Please check Column A if any of these losses or stresses occurred or are occurring in your life. Please check Column B if the item checked in Column A still causes you emotional pain.

Column A	Stresses, losses, and feelings	Column B
Yes, I had or have this stress/loss.	Check the left column if you have or had these stresses or losses. Check the right column if it still causes you pain	Yes, still causes me some pain.
	Feelings of shyness as a child	
	Felt other children were favored	
	Frequent illnesses as a child	
	Feelings of aggression	
	Addicted mate	
	More than two "panic" or anxiety attacks	
	Few friends I can count on for help	
	Bothered or worried about addicted parent	
	Can't count on spiritual advisor; minister, priest, rabbi, or other	
	When I need help, I'm uncomfortable asking for it	
	Insomnia	
	Death of a pet	
	Guilt in some areas of life: religion, sexual activities, parenting	
Total:	Add column A and B	Total:

Total of Columns A and B_____

Indicator: Early risk   10–13
              Intermediate risk   14–19
              Late risk   20–23
              Probable addictions   24–26

2. When you have leisure time, how often do you have these feelings or reactions? Please check the box in the column that applies to you. Answer the questions as you would in a normal week. For example, do not respond as you would when you are on vacation.

Feelings or reactions to your leisure time	2 Never	2 Almost never	3 Sometimes	4 Often	5 Most of the time
Tired					
Do not want to exercise					
Bored					
Anxious or "uptight"					
Want to shop					
Want to gamble					
Want to make love					
Want to have a drink					
Lonely					
Column totals:	X 1	X 2	X 3	X 4	X 5
Multiply by:					
TOTALS					

to cause an alcoholic to deny or outright lie. In typical nursing and health assessments, the questions evoke an underestimation of drinking, smoking, and sexual behaviors. These surveys encourage the people with addictions to lie to health professionals and to feel guilt and fear.

People should not be made to feel guilty about their addictions; it is the only way they know how to cope with life. With a risk assessment of behaviors and thought patterns, they can be helped to identify the patterns of behaving and thinking that are leading them to addictions that are hurting, damaging, and eventually going to destroy their lives. They can be shown new patterns of living, behaving, and thinking that will help them to live more comfortably

---

EXHIBIT 18-1   EARS (CONTINUED)

Total of:
Column 1 _____
Column 2 _____
Column 3 _____
Column 4 _____
Column 5 _____

TOTAL of 5 columns _____

Total scores
Addiction indicators:
Early risk   24–27
Intermediate risk   28–34
Late or high risk   35–39
Probable addiction 40–45

Please check the column that most applies to the frequency you have these feelings or do the
following behaviors.

Behavior or feelings you may experience.   Almost always   Frequently   Sometimes   Almost never
3. How often do you wear a seat belt? 1 2 3 4
4. How often do you feel overwhelmed? 4 3 2 1
5. How often do you feel the need for some help to change a low or sad mood? 4 3 2 1
6. How often do you do a creative activity? 1 2 3 4
7. How often do you worry about your family or being alone? 4 3 2 1
8. How often do you worry about your past? 4 3 2 1
9. How often do you worry about criticism by others? 4 3 2 1
10. How often do you feel guilty about some behavior? 4 3 2 1
11. How often are you annoyed by criticism of what you do? 4 3 2 1
12. How often do you feel grief over a loss or death of someone? 4 3 2 1
13. How often do you wish you could tell your parents, mate, or family what they've done to hurt you? 4 3 2 1
14. How often are you fearful about the results of something you've done? 4 3 2 1
15. How often have you had legal problems? 4 3 2 1
16. How often do you feel compelled to do something you really wish you didn't do? 4 3 2 1
17. How often are you lonely? 4 3 2 1
18. How often have you had a problem enjoying sexual activity? 4 3 2 1
19. How often do you feel compelled to control you anger or resentments? 4 3 2 1
20. How often are you concerned about your job performance or other responsibilities? 4 3 2 1

Add the numbers in the boxes you checked.

Totals of the numbers in the boxes you checked.

Add each of the columns' totals

_____

_____

TOTAL _____

Copyright 1998. Efinger.

---

with less anger, victimization, and fear. They can be given hope that they can live happier and more serene, courageous lives without their addictions and avoidance of feelings. The new patterns can be applied to dealing with all addictions because they all have some common sources of feelings that result in the damaging behaviors. It is clear that the early addictive behaviors are often effective in coping with anxiety, anger, and other uncomfortable events. The intermediate addictive behaviors can have both effective and painful, damaging results. The late behaviors have ineffective results. There is no relief at this stage for those with an addiction. They are obsessed prisoners of their addictions.

Recognizing these stages of early, intermediate, and late addictive behaviors provides hope that help can be given at earlier stages. Earlier help and hope lead to quicker healing.

Denial

Denial is a strong barrier to seeking help, hope, and healing. Denial exists for the addict, as well as for the loved ones because of the fear, shame, and guilt associated with every addiction. When communicating with an addict, gentleness, acceptance, promotion of self-esteem, and recognition of pain help to reduce denial. When family members or healers provide compassion to an addict, they enhance treatment effectiveness and provide an opening to reduce denial.

> **KEY POINT**
>
> The addict's denial must be confronted.

The need for denial causes an addict to try to manipulate others. Open, compassionate confrontations and interventions are necessary in helping the addict consider beginning any type of healing and recovery process. Confronting and risking the anger of the addict may save his or her life. At times it is hard for family and traditional healers to be compassionate. Many stereotypes that influence feelings toward someone who has caused turmoil and pain to self and others by his or her addictions exist.

At this stage, a noncaring attitude directed toward the addict by family and friends is common. Some parents privately wish their drug-addicted child was dead. Some people who have lost a child to addiction or an accident grieve and start their own addictions to sexual activities, excessive shopping, work, controlling another child, religion, and secret things that they often feel too ashamed to share.

EARS

The EARS (Exhibit 18-1) is a health risk appraisal survey for addictions. It can help identify those at risk for addictions at an early stage of their disease. When the risks are identified, there can be interventions, education, motivation, and reduction of the horrendous costs of addictions to the persons, their families, and society. Risk identification provides information about where the problems may be. It gives a focus for the most effective counseling, education, and support interventions to reduce the progression of the risks to addictions. It appears that if the major areas of pain, fear, shame, and guilt vulnerability can be addressed first and ameliorated, a person with an addiction may be more amenable to change and to attending therapy, and to be motivated to work on his or her problems.

Discussion of Risk Scores

The EARS scores are divided into early, intermediate, late, and probable addiction risk scores. There is flexibility in the scoring. The scores serve as a general framework to assess addictions risk.

Early Risk Indicators

Early risk indicators are events or predictors that occur in childhood or adolescence that reflect behaviors and attitudes characteristic of persons who later became people with addictions. Feelings of shyness as a child are one of the earliest indicators. This could be reflected through low self-esteem or other uneasy feelings when dealing with people. Some of these children have been abused and are fearful and insecure. They often have experienced a nonsupportive environment at home or school. Frequent illnesses experienced by a child set a pattern of thinking that medications or drugs can fix everything. The child may feel different from classmates and friends, and they become isolated and lonely. A lack of coping skills to deal with loss can be reflected through unresolved feelings about the death of a pet. In fact, any of the responses in question 1 that are checked as still causing pain are risk indicators. The more items that continue to cause pain, the greater the risk.

Question 3, regarding frequency of wearing seat belts, is an early risk indicator. Perhaps an answer of "almost never" reflects the rebellion a potential addict feels toward society. This could indicate lack of self-care or a subtle, self-destructive attitude. It also represents an effort to be in control of how rules are applied in the life of a person with an addiction. Research supports this behavior as an early risk. The number of close relatives who have a serious addiction problem should also be analyzed. The genetic component of addictions and depression are major early warning risks.

Intermediate Risk Indicators

Intermediate risk indicators are predictors that occur in childhood, adolescence, and early adulthood that reflect behaviors and attitudes that are characteristic of persons who later became people with addictions. Being very reluctant to ask for help, rarely asking for help, and having few people to

> **KEY POINT**
>
> People with addictions often feel anxious or uptight, frequently bored, tired, lazy, and lonely when faced with leisure time. The need to avoid the quiet time alone reflects their discomfort with inner feelings and thoughts.

count on are intermediate risk indicators. Not viewing leisure time positively is a precursor to serious risk.

Feelings of isolation are a common experience. Persistent guilt, insomnia, panic attacks, and not feeling spiritually connected are serious patterns of intermediate risks. Unresolved pain related to the loss or death of a person is not an unusual experience when people receive no help with grieving; however, when this feeling is added to the other patterns of feelings and behaviors, the person is at risk.

Late Risk Indicators

Late risk indicators are predictors and stresses that affect healthy functioning of the person's spiritual, mental, physical, and economic life. All of the risk indicators already mentioned can be found to a greater degree in the late risk indicators. When questions 4, 7, 8, 9, and 10 to 20 are primarily marked "almost always," late risk for addiction is strongly present. The degree of risk is related to the higher scores. For example, when there is a lack of any creative activity, and the "almost never" column is marked, the risk score increases.

Driving while under the influence of drugs or alcohol is a serious late risk that leads to legal problems. That behavior and other problems in legal, financial, work, or intimate relations are patterns that approach the strongest late risk indicator category.

Probable Diagnosis of an Addiction

The scores for many of the risk indicators listed previously will be higher when there is a probable diagnosis of an addiction. In addition, there may be arrests, serious illnesses, admissions to the hospital (greater than two related to the addiction), and suicidal feelings. These require immediate attention by their healthcare providers and family members.

Treatment Modalities and an Essential Paradigm Shift

The medical model of addictions management has not had a highly successful record of arresting and healing these diseases. A paradigm (worldview) shift

KEY POINT

Medications, therapy, supervision, and support groups, such as the anonymous groups for narcotics (NA), alcoholics (AA), gamblers (GA), overeaters (OA), and so forth, have a role in holistic healing, but treatment methods must go beyond those common resources. Holistic treatment addresses the mind, body, and spirit.

toward a holistic approach is essential to access the healing tools and energy to begin effective treatment and healing. We only need a small window of hope to access a spiritual connection to the vast energy of love, compassion, and acceptance to start the healing for the addict and those involved.

Holistic Approach to Healing Using Feedback on Risk Factors

The areas of risk identified by the EARS can be individually addressed using a holistic approach. Most of the stresses and fears (concerning work, relationships, etc.) can be decreased through use of imagery and progressive relaxation.

All transitions and changes are stressful. Planning relaxing activities for the period of time between the completion of employment activities and the start of home responsibilities can reduce stress. This is a time when people feel anxious, and addictive thoughts strongly enter their minds. Simple strategies, such as clarification of job description and responsibilities, can reduce work-related stress. Reducing interruptions at work and home when involved in tasks requiring concentration can decrease irritability and stress.

Anxiety (panic) attacks are instant triggers for addictions. The thoughts with the emotional and bodily responses during an anxiety attack feel life threatening. Seeking help is essential.

KEY POINT

A strategy that can help control an anxiety attack is to slow down breathing, look up, focus on an object, and repeat, "These feelings are terrifying, but they are not dangerous. I will not be harmed." This action seems very simple, yet it works to get people through the panic. A thought, even a prayer, can interrupt the body's responses.

Support groups for losses, illnesses, and painful life events can be healing experiences for many people. People with addictions feel acceptance and hope with the knowledge that they are not alone as they experience the group support. Sometimes the groups may be spiritual or religious in orientation.

Coping skills, assertiveness, conflict resolution, and crisis-management techniques increase the comfort of daily living. With these capabilities, people can interrupt the cycle of victimization and vulnerability leading to addictions.

The identified risk factors lead to specific treatment strategies. The education, interventions, and treatments include integration of mind, body, and spirit approaches to healing.

KEY POINT

A neurotransmitter can be thought of as a brain messenger.

Healing the Mind

Addictions create alterations in optimal brain chemistry. Abnormal neurotransmitters (chemicals that help send impulses through the nerves) of the brain can cause and perpetuate addictions. Healing the mind is essential to permanently heal addictions and restore normal brain chemistry.

A diet rich in vitamins, amino acids (proteins), essential fatty acids, and minerals needs to be integrated into daily living. Some food supplements help the brain manufacture chemicals like the neurotransmitter norepinephrine that boosts energy and mood; examples of these supplements include:

- *Glutamine.* This has been successfully used to reduce alcohol cravings in early recovery.
- *Tyrosine.* Tyrosine is an important building block for norepinephrine.
- *Tryptophan.* Tryptophan is a precursor for serotonin. Foods high in tryptophan are popcorn, turkey, chicken, and dairy products.
- *Serotonin.* Serotonin is a neurotransmitter that creates a calm, relaxed state and promotes sleep. The absence of sufficient serotonin is a trigger for addictions.

Hypoglycemia (low blood sugar) can cause depression, anxiety, panic attacks, and mood swings leading to increased addictions. Another neurotransmitter, histamine, regulates mood and energy. High levels of histamine cause the mind to race and lead to obsessive-compulsive thoughts and behaviors. Methionine, an amino acid, reduces the impact of histamine on the brain.

The mind has habits of thought that are responses to physical imbalances. Blocked energy (of qi, or life energy) causes imbalances in the brain. The blockages occur throughout the body and affect the mind in harmful ways. Disturbances in brain chemistry can be reduced by helping a person to alter negative thoughts through therapy, prayer, meditation, and other modalities. Next, the body must be integrated into the holistic approach to healing.

Healing the Body

The body has many systems that affect all other aspects of healing. In turn, the body is affected by the mind, spirit, and the energies within and without. Touch and movement have key roles in healing addictions.

Massage

Touch is a gesture of support. When someone touches, respects, and cares for the body of a person with an addiction, the addict is aided to reconnect physically and center emotionally—important healing steps because avoiding problems and dissociating from the body are common among people with addictions. The person is more connected to his or her body and able to discuss and come to terms with his or her addiction.

Massage has a powerful impact on the body, releasing endorphins, substances that have a mood-enhancing effect. Self-massage of hands and ears helps to reduce cravings. The ear has pressure points that stimulate the body's natural pain relievers.

Full-body massage releases tension and blocked energy. Loosening the tight muscles sends the body messages to cut down on the production of stress hormones. Massage also moves lymphatic fluid through the body. This movement of fluid assists in the body's natural cleansing processes and enhances immune response. Acupressure can help people deal with stress, depression, anger, and the issues underlying their addictions.

Hatha Yoga

Hatha yoga simulates the parasympathetic nervous system and removes tension from all major muscle groups. Yoga helps the addict to become more physically aware, leading to greater mental self-awareness. Eventually, as peaceful body responses are developed, the addict becomes sensitive to how his or her behavior can change. The person experiences the ability to feel better without addictions. The connection between mind and body heals in an integrated way.

Acupuncture

The use of acupuncture reduces cravings and unblocks energy (qi). The body feels better, and there is a release of the healing neurotransmitters. It restores harmony in the body. These results are related to the skill of the practitioner.

Other Body-Healing Therapies

Dance therapy, shiatsu, Ta'i Chi Ch'uan, reflexology, aromatherapy, chiropractic, acupuncture, qigong, therapeutic touch, and healing touch are major complementary therapies. Some other treatments include homeopathy, hypnotherapy, herbals, biofeedback, and music therapy.

> **KEY POINT**
>
> People with addictions need support and a spiritual awakening to acknowledge that total healing is possible for them.

Healing the Spirit

Spiritual healing encompasses the removal of grudges, negative views of self, and compassion for self and others. Forgiveness, making amends, and acceptance of self and others promote spiritual love and healing. These practices also remove the fear, shame, and guilt that are characteristic of addictions.

Healing the spirit is a wonderful experience for people with addictions. People with addictions have lost focus and are unaware of their own identities and purposes. They become disconnected from the purpose of their creative energy, their inherent goodness, and ability to love and be loved. With support and guidance through the process, people with addictions may experience relief from their compulsions in a brief time frame. The willingness to heal opens the channels for healing to occur. The process is a personal, empowering journey. From its deepest perspective, an unresolved longing causes addictions. This awareness is a message that opens the heart and soul of a person with an addiction to see that he or she has lost his or her way. All addictions have a common source. If only the obvious addiction is addressed, the cause will remain, and there is the risk that when one addiction pattern starts to diminish, another will start to take over. This happens because the sources of the addictions have not been identified, addressed, and healed. The new addiction's role is the same as the old addiction's role; therefore, if patterns of behavior and the needs for the patterns are identified, then intervention, education, support, and prevention can be implemented.

REFLECTION

In *For Whom the Bell Tolls,* Ernest Hemingway stated that, "We become stronger at our broken places." How have you seen this occur in your own life?

Summary

Addictions are behaviors people do repeatedly that result in problems in one or more areas of their lives. People can be addicted to gambling, drugs, alcohol, sex, eating, work, collecting, or exercise. Health, spiritual, relationship, work, legal, and financial problems can result from addictions. Those with an addiction may not believe that they have a problem, feel shame, live in fear, experience guilt, and lose hope. The body of the addict can be healed using measures such as a good diet, supplements, massage, yoga, acupuncture, and other body-healing therapies. Forgiveness, making amends, and acceptance of self and others are important aspects of healing the spirit of the person who has an addiction.

References

Federation of American Hospitals. (1981). Alcoholism and the increasing role of the hospital. *Federation of American Hospitals Review, 14,* 4.

Suggested Reading

Beattie, M. (2001). *Codependent no more and beyond codependency.* New York, NY: MJF Books.

Carnes, P. (2001). *Out of the shadows: Understanding sexual addiction* (3rd ed.). Center City, MN: Hazelden.

Carnes, P. (2005). *Facing the shadow: Starting sexual and relationship recovery* (2nd ed.). Carefree, AZ: Gentle Path Press.

Dodes, L. M. (2011). *Breaking addiction: A 7-step handbook for ending any addiction.* New York, NY: HarperCollins.

Durham, M. (2003). *Painkillers and tranquilizers.* Chicago, IL: Heinemann Library.

Hausenblas, H., & Fallon, E. (2002). Relationship among body image, exercise behavior, and exercise dependence symptoms. *International Journal of Eating Disorders, 32*(2), 179–185.

Hunt, J. (2011). *How to defeat harmful habits: Freedom from six addictive behaviors.* Eugene, OR: Harvest House.

Lee, S. G. (2000). *Light in the darkness: A guide to recovery from addiction: A physician talks openly about his own addiction to sex.* Newport News, VA: Five Star Publications.

Mate, G. (2010). *In the realm of hungry ghosts: Close encounters with addiction.* Berkeley, CA: North Atlantic Books.

Schwartz, J., & Beyette, B. (2005). *Brain lock: Free yourself from obsessive-compulsive behavior; A four-step self-treatment method to change your brain chemistry.* New York, NY: HarperCollins.

Urcchel, H. (2009). *Healing the addicted brain: The revolutionary, science-based alcoholism and addiction recovery program.* Naperville, IL: Sourcebooks.

West, M. (2000). *An investigation of pattern manifestations in substance abuse–impaired nurses.* Chester, PA: Widener University.

West, M. (2002). Early risk indicators of substance abuse among nurses. *Journal of Nursing Scholarship, 34*(2), 187–193.

Wright, J. (2003). *There must be more to this: Finding more life, love, and meaning by overcoming your soft addictions.* New York, NY: Broadway Books.

Resources

AddictionSearch.com
www.addictionsearch.com

Alcoholics Anonymous
www.aa.org

Eating Disorder Recovery Center
www.addictions.net

National Institute on Alcohol Abuse and Alcoholism
www.niaaa.nih.gov

National Institute on Chemical Dependency
http://nationalinstituteonchemicaldepend.blogspot.com

National Institute on Drug Abuse
www.drugabuse.gov/

Online Gamers Anonymous
www.olganon.org

Web of Addictions
www.well.com/user/woa

Symptoms and Chakras

OBJECTIVES

This chapter should enable you to

- Describe the characteristics and purposes of chakras
- List the names and locations of the seven main chakras of the body
- Describe a symptom of imbalance or dysfunction for each of the seven chakras

Chakras

Chakra is a Sanskrit word for spinning vortices, or wheels, of energy. Many ancient systems of belief have this understanding of energy in and around the body with entry points located at distinct places along the body. The chakras are like the input valves for the body to receive nourishing energy from the universe around it and output valves to release or transfer energy. The energy received is transformed into useable energy for the body; however, the chakras also extend outward to affect everything and everyone. It is thought to be vital to the health of all aspects of the body—physical, emotional, mental, and spiritual—that the chakras be open, flowing, and healthy.

From a physical standpoint, it is believed that the energy received through the chakras affect the hormones, the organ systems, and the cellular functions of the body. Each chakra is associated with a particular nerve plexus (group of nerve endings) or endocrine system (organs of the body that secrete necessary hormones for health and well-being). The chakras receive energy from outside of the body and transform it internally to stimulate some form of hormonal gland response or nervous system response, which in turn affects the entire body (Gerber, 2001).

> **KEY POINT**
>
> The chakras are vortices of energy that receive nourishing energy from the universe and send out energy from the body. They have distinct names, often described for the area of the body over which they reside.

There seems to be a general acceptance of the existence of seven main chakras of the body (see **Figure 19-1**):

First: The root chakra comes downward from the base of the spine.

Second: Sacral chakras are a pair of chakras located just below the umbilicus (belly button) on the front and lower back portion of the body.

Third: The solar plexus chakras are a pair of chakras located where the ribs form a *V* on the front of the body and the middle of the back of the body.

Fourth: The heart chakras are a pair of chakras located between the breasts and between the shoulder blades.

Fifth: The throat chakras are a pair of chakras located at the point of the Adam's apple in the neck and the curve of the neck on the back of the body.

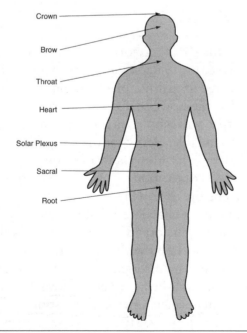

Figure 19-1   Chakras

Sixth: The brow chakras are a pair of chakras located in the middle of the forehead on the face and directly behind on the back of the head.

Seventh: The crown chakra is a single chakra located on the top of the head.

These seven make up the chakras of the physical body. Some believe there are more chakras that affect the nonphysical portion of our existence (Gerber, 2001, p. 128).

The crown chakra and the root chakra are the only two without a corresponding pair. The five central chakras exist on the front of the body and the back of the body. The main channel of energy exists from the top of the head, or crown, to the base of the spine. All chakras meet in this channel and connect with each other through nadi. Nadi are fine threads of energy that help connect the pairs of chakras to the main channel of energy through the body and to each other. In this way, all of the chakras are connected and can affect each other.

The Root Chakra

In the yogic tradition, the root chakra is called *muladhara*, or the manifestation of life energy. The center relates to the lower portion of the body—hips, legs, and feet—and to the functions of movement and elimination.

In addition to the physical aspect, this chakra relates to feeling secure in the world, feeling grounded to the earth, and having the will to survive. It deals with the sense of joy and vitality. The root chakra governs the adrenal glands, which are responsible for secreting hormones that, in frightening situations, help one to be more alert and escape danger if necessary. If the chakra is not balanced or unhealthy from prolonged stress, problems such as high blood pressure and anxiety can be seen. One often feels a sense of lacking presence, of not being here, or of being low on physical energy. In persons with HIV/AIDS and cancer, this chakra is said to be severely damaged or nonfunctioning.

KEY POINT

Energy can be seen through color and heard through sound. The vibrational energy of the root chakra is associated with the color red (the slowest vibration on the color spectrum) and the sound of bumble bees. It can also be related to the first sound, do, in the scale, *do, re, mi, fa, sol, la, ti, do.*

The Sacral Chakra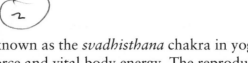

The sacral chakra is known as the *svadhisthana* chakra in yogic tradition and is related to the life force and vital body energy. The reproductive organs and the lower abdomen are affected by this chakra, especially the gonads (endocrine organs of sexuality and reproduction). Nutrients and the absorption of fluids through the intestine are also related to this chakra. The color vibration that relates to this chakra is orange, and the sound is the one of a flute or the sound, *re*.

Psychologically, this chakra deals with feelings and emotions. It also relates to the expression of sexuality and desires, pleasures, and feelings about reproduction. Relationships will occur easily and fall into place when the sacral chakra works properly.

When imbalanced or nonfunctioning, the sacral chakra will cause problems with emotions, feelings of sexuality, or the ability to reproduce or create. If energy is deficient to this chakra, one may feel a lack of emotions or flat. There may be a tendency to avoid emotions altogether. Some people have problems feeling passionate about anything or anyone. If the chakra is overactive, it is believed that one may experience mood swings or be very emotional. Another possibility is the need for constant pleasurable stimulation.

The Solar Plexus

The third chakra, the solar plexus, is also known as the *manipura* and is associated with one's own power, strength, and ego identity. It also centers around the issues of the will to think, control, exert authority, and display aggression and warmth, as well as the physical components of digestion and metabolism. With this chakra intact, one has the ability to start things and see them through to completion. The color associated with this chakra is lemon yellow and the sound is of the stringed instruments, such as the violin, or the sound, *mi*.

Physically, imbalances or malfunctioning of this chakra can lead to obesity, diabetes, hypoglycemia, heartburn, gallbladder problems (stones or infection), and stomach ulcers. Most of us have felt we have been hit in the pit of the stomach when someone verbally or physically attacks us. In fact, one

---

KEY POINT

Creativity or the desire to create another human being or other things is greatly affected by the sacral chakra.

may feel nauseous or even vomit. Our stomach gets into "knots" when we get nervous or fearful, and things are out of our control.

Emotionally, one feels timid or dominating when there is an imbalance. There can be feelings of anger and rage that sit within this chakra. Some persons who are passive aggressive (act disinterested, but are really angry inside) have imbalances in this chakra. A lack of self-worth or a hidden sense of shame is often found within this chakra. A healthy outlet is one of assertiveness or being able to state your needs and feelings freely and clearly while respecting others and yourself.

The Heart Chakra

The heart chakra, in the Indian tradition, is called the *anahata* and is the center of love. It has the qualities of loving self and others, harmony, compassion, openness, giving, peace, and grace. The heart chakra is called the transformation center because at the heart level one begins to deal with his or her spiritual side. Whereas the first three chakras dealt with the physical, emotional, and mental components of the self, the heart center speaks to the soul. It is through this chakra we experience the feeling of unconditional love or the deep love for others. This love is of a pure or spiritual kind of caring. The heart chakra is thought to be open when one cries tears of happiness or joy or in those instances in which the heart strings are pulled. The color associated with the heart is emerald green, and the sound is of bells or the *fa* sound.

The heart, circulation, and respiratory systems are related to this center on the physical plane. Imbalances can be seen through physical problems, such as heart disease, high blood pressure, poor circulation, angina (chest pain), asthma, emphysema, bronchitis, or pneumonia. Without a properly functioning heart and lungs, life is not possible and death will result.

Psychologically, imbalances in the heart can leave one feeling self-hatred or lacking self-esteem. One may avoid interpersonal relationships for fear of being hurt or unloved. Some choose isolation to avoid this pain. Others may have suffered from neglect, abandonment, or abuse. Grief weighs heavy on the heart and, if left unresolved, can create many problems. Some choose to give their hearts away, leaving very little for themselves, therefore feeling very empty and lonely.

Forgiveness is not necessarily letting others off the hook but releasing the anger and hurt one experienced from the action. Forgiveness of the self is one of the hardest things to do because we tend not to feel worthy of it. By showing compassion to ourselves, we are able to show compassion to others better. Forgiveness and the release of old grudges help the heart energy to expand.

REFLECTION

Is there anything that you have done for which you are unable to forgive yourself? What prevents you from offering or receiving this forgiveness from yourself?

The Throat Chakra

The center of creative energy, the throat chakra, is also called the *vishuddha*. It deals with issues surrounding communication and expression. By using the voice, one is able to create sounds of speech or song to express oneself. This connection to others by sound allows us to reveal who we are. The throat chakra relates to the neck, esophagus, throat, mouth, and ears. The sound of wind blowing through trees or the note, *sol*, is associated with this chakra. It resonates to the vibration color of turquoise or light blue. Some people describe it as an electric blue.

Physically, a healthy throat chakra is one revealing a clear voice that communicates thoughts, feelings, and ideas concisely. If the chakra is unbalanced, one may speak softly or with a hoarseness. Having a "frog in the throat" or "choking" on words are descriptions of the free flow of speech being impeded. Other symptoms of imbalance are knots in the neck muscles, tight throats, and clenched jaws.

Emotionally, people need to express themselves and their talents, knowledge, and ideas. Hindering this expression may reflect a poor self-esteem or identify family patterns that suppressed the individual's need to create or communicate. Grief can become locked here when emotions are not released through words or tears. Anger and stress can inflame the temporomandibular joint (in front of the ears) with teeth gnashing and grinding at night.

One way of opening this chakra is through sound—speaking, singing, chanting, or toning. By opening one's mouth and emitting sound, the vibration clears the throat and it becomes stronger. Additionally, the throat chakra is the communicator from the heart to the soul. It allows us to sing to our heart's content and allow music to soothe the soul. It lets us speak from our heart and make sounds reflecting our emotions. Silence can be used to actively listen to others or their surroundings. By actively listening, one can learn to communicate more effectively by hearing exactly what the other is saying.

The Brow Chakra

The brow chakra, known as the *ajna* center, is also called the third eye. It is associated with the pituitary gland, which controls most of the hormonal functions of the body. The brow chakra relates to memory, dreams, intuition, imagination, visualization, insight, and psychic perception. This chakra is all about seeing, within and without. The color associated with the brow chakra is deep indigo blue or deep purple. The sound *la* or of waves crashing on the beach help with this center.

Sinusitis, headaches, and poor vision may be seen as imbalances in this chakra. From an emotional standpoint, the brow chakra deals with putting oneself in another's shoes, being sensitive to the needs of others, and having a sense of humor. Mentally, recognizing patterns helps people predict that which will occur next in addition to providing insight into their behaviors. Remembering recent events, long-term events, and even past lives is encoded within this chakra. Stored images that can be seen like a movie in one's head allow one to view the same scenario again or replay it, making changes to produce a different outcome. Dreams are another mental component of the brow chakra, whether they are daydreams or those occurring in sleep. The symbology of dreams provides a great deal of insight into feelings, thoughts, and desires.

> ### KEY POINT
>
> Physically, the brow chakra relates to the eyes, brain, nose, and head. Any imbalance will result in problems with these parts.

Imbalances with the brow chakra may be seen with hallucinations, delusions, nightmares, and misguided thinking. Some people will be insensitive to others' needs either deliberately or simply by not having observed them. These are the people who just do not seem to take the hint for whatever you are trying to tell them. Others are not able to see what is in front of them; they are blinded by inaccurate thinking.

The Crown Chakra 7

The seventh chakra, or crown chakra, is called the *sahasrara* center in yogic tradition. It is often called the spiritual center or center of connection to the oneness of the universe. The crown chakra relates to the divinity of self and our connection to a higher source however one defines it. The crown chakra also relates to intelligence, thought, consciousness, and information. An alignment to a higher power is seen through this center. It is associated with the color white or lavender and the sound *om* or *ti*. The pineal gland is an endocrine gland that is not well understood. Its purpose is to deal with the body's rhythms and timing.

> **KEY POINT**
>
> The crown chakra is the one with the highest energy vibration of the physical body.

There does not seem to be much dysfunction with this center on the physical level. Some people with imbalances experience a sense of disconnection, a holier-than-thou thinking, or a rigid belief system of thinking. These people can be described as having their heads in the clouds or being spacey.

The crown chakra is the center of our connection to the oneness of the universe. It allows people to define who they are in relation to the universe and to experience their higher selves as whole. It is where they connect to the higher power and experience the oneness with all of creation in peace and grace.

Putting It All Together

An understanding of what the various chakras represent and symptoms associated with their imbalances can provide important information as to the status of the mind, body, and spirit. It can be helpful for people to consider whether certain physical symptoms that they experience on a regular basis are related to emotions that are blocking a chakra. For example, problems with weight control or chronic upset stomach could be associated with feeling a sense of low self-worth or unexpressed anger. Learning to be assertive in expressing feelings in a healthy manner could improve the physical symptoms. Meditating and thinking of the color associated with the specific chakra that is expressing itself as imbalanced could prove useful. Also, healers who work with the energy fields, such as healing touch practitioners, can assist in balancing the chakras.

Summary

Chakra is a Sanskrit word for spinning vortices of energy. Chakras transform energy from the universe into useable energy for the body. In turn, chakras send out energy from the body to the universe. The seven main chakras of the body are the crown, brow, throat, heart, solar plexus, sacral, and root. The names often refer to the area of the body over which they reside. Specific symptoms can be manifested when there is imbalance or dysfunction of a chakra. Changing thought patterns, working with an energy therapist, and using specific sounds and meditating on colors associated with the chakra are some of the ways that chakras can be balanced and symptoms improved.

References

Gerber, R. (2001). *A practical guide to vibrational medicine: Energy healing and spiritual transformation.* New York, NY: HarperCollins.

Suggested Reading

Buhlman, W. (1999). *Chakra technique and the vibrational technique.* Chatsworth, CA: Spiritual Adventures.

Chopra, B. (2010). *The chakra diaries.* Pahoa, HI: Creative Counsel.

Lee, I. (2002). *Healing chakra: Light to awaken my soul.* Mesa, AZ: Healing Society, Inc.

Mercier, P. (2007). *The chakra bible: The definitive guide to chakra energy.* New York, NY: Sterling Publishing Co.

Pond, D. (1999). *Chakras for beginners: A guide to balancing your chakra energy.* St. Paul, MN: Llewellyn Publications.

Saradananda, S. (2008). *Chakra meditation: Discover energy, creativity, focus, love, communication, wisdom, and spirit.* New York, NY: Duncan Baird.

Saradananda, S. (2011). *The essential guide to chakras: Discover the healing power of chakras for mind, body, and spirit.* London, UK: Watkins Books.

Wauters, A. (1997). *Chakras and their archetypes: Uniting energy awareness and spiritual growth.* Berkeley, CA: Crossing Press.

Wauters, A. (2002). *The book of chakras: Discover the hidden forces within you.* Haupauge, NY: Barrons Educational Series.

Medication Wisdom

OBJECTIVES

This chapter should enable you to

- Describe medication-related information that is significant for health-care providers to review and discuss with consumers
- Distinguish between medication side effects and adverse drug reactions
- List eight common undesirable consumer behaviors with prescription medications
- Describe specific interventions the healthcare provider can use to help patients resolve issues related to obtaining prescription medications

Throughout history, people have exerted self-control over the selection and use of healing agents, including medicines. The interest in selection and use of medications continues to be important to consumers. People want to know specifically about the medications they are taking, how medications work, their potential side effects, possible and potential interactions with other medications, and how the medications will impact their functional status.

KEY POINT

Medication use has increased over the past decade with more than 80% of all Americans consuming over-the-counter or prescription drugs on a regular basis. Approximately half of all Americans take at least one prescription drug with more than 10% using five or more drugs (Gu, Dillon, & Burt, 2010).

Consumer Issues Related to Prescription Medications

The ultimate goal of taking medication is to enhance one's health state or level of functioning. The desired outcome is to obtain the maximum benefit with minimal side effects and little or no toxicity. Each person has a unique and individual body composition. Because of each person's unique biochemical composition, no two individuals will respond to a medication in exactly the same way. **Table 20-1** depicts general, essential information the consumer needs to know when taking any medication.

Name of the Medication

Medications have both a generic and a trade name. This information is essential for people to know for each of their prescribed medications. The trade (brand) name of the medication is the drug name as it is available from pharmaceutical manufacturers. Pharmacists, unless otherwise indicated, can substitute an equivalent generic drug without prescriber approval, and in many cases, this will pass on significant cost savings to consumers. In some cases, however, only the trade (brand) name medication can be dispensed if the prescription specifies, "Dispense as written." Consumers need to be educated that although generic products are considered to be pharmaceutically equivalent to the trade (brand) name counterparts, not all are therapeutically the same. This could ultimately result in failed therapy because of ineffectiveness of a medication.

Use of the Medication

It is important to know the purpose for each medication used (e.g., to treat inner ear infection, lower blood pressure, eliminate fluids). This can be particularly

---

TABLE 20-1   ESSENTIAL INFORMATION TO KNOW WHEN USING A MEDICATION

- Name (brand and generic)
- Use
- How to administer
- Side effects
- Adverse reactions
- Storage and handling
- Special precautions
- Interactions with other medications, foods, and supplements
- What to do if there are missed doses

important, as there can be situations where the pharmacist cannot accurately read the handwriting on the prescription. Many medication names are spelled very similarly, and there could be serious consequences if the wrong medication is given. (An example of two medications that easily could be confused is Celebrex [celecoxib], used to treat musculoskeletal conditions, and Celexa [citalopram], used for treatment of depressive disorders.) By knowing the name and purpose of their prescribed drugs, consumers can detect errors and spare themselves complications.

REFLECTION

Do you know the names, purposes, side effects, and adverse effects for every medication you use?

How to Administer the Medication

Healthcare providers need to educate consumers to know the correct dosage of their medications, routes of administration, and any other important information. There are several routes for administration of prescribed medications. The most common routes for home medication administration are oral, rectal, topical, otic (eardrops), ophthalmic (eye drops), subcutaneous (sub-Q) injection, intranasal, inhalation, and intramuscular (IM) injections. Failure to take prescribed medication via the correct route could result in outcomes ranging from insignificant consequence to death. An example of incorrectly administering a medication could be instilling eye drops in the ear. In this case, the outcome will most likely not be life threatening (although not receiving the needed medication could have effects for the eye). An example of serious potential harm might be administering a medication intended for topical (skin) administration orally.

Administration of the correct dosage at the specified intervals is important. Not taking enough of the drug or skipping doses could reduce the benefit and leave the health problem with inadequate treatment. On the other hand, taking doses that exceed the prescribed amount or taking drugs more frequently than prescribed could cause damage to body organs and serious complications. Consideration also must be given to the existence of conditions, such as liver or kidney disease, that could slow the metabolism and excretion of drugs. Dosage adjustments would be warranted in these situations.

Side Effects of the Medication

Medication side effects are responses that can occur with medications. Although side effects result in discomfort or cause concern, they usually are not life threatening. An example of a side effect would be the onset of nausea after taking an antibiotic on an empty stomach.

Side effects may or may not require that a medication be stopped. People must be knowledgeable of potential and actual side effects for every medication they use in the event that they occur. Although healthcare providers need to inform consumers of the side effects that should be reported, people should read the drug information provided with their prescriptions, search the Internet, and use other means to become informed about their medications, also. Together, the healthcare provider and consumer can discuss the situation and make an informed decision about whether to stop the medication and whether the side effects warrant intervention.

Adverse Reactions to the Medication

In addition to medication side effects, consumers must be aware of potential adverse drug reactions (ADRs). ADRs have been studied in hospitalized patients, and it has been estimated that there are over 2 million ADRs annually in the United States, resulting in ADRs being the fourth leading cause of death (U.S. FDA, 2008). ADRs can be severe adverse reactions and toxic effects that can result in loss of life or permanent impairment and loss of functioning. ADRs require prompt intervention by healthcare providers, because without emergent care some ADRs could result in death within minutes.

Storage and Handling of the Medication

People need specific information on medication storage and handling for every drug they use. Improper storage or handling decomposition of biochemical ingredients can result in decreased potency or efficacy. Detailed, specific medication storage and handling instructions should be provided with each medication prescribed. Examples of frequently asked questions regarding storage and handling are as follows: "Does this medication need to be refrigerated?" "Can I travel with this medication?" "How long is the medication good if I store it?" Medication labels contain specific details on storage and handling, but they can be difficult to read for some older or sight-impaired

---

KEY POINT

Adverse drug reactions demand speedy attention. A person taking a new medication who develops tightness in the throat or difficulty breathing, indicative of a life-threatening ADR known as anaphylaxis, could die within a short time without intervention.

persons. Special considerations would be in order in these situations to assist consumers with needed storage and handling information.

Special Medication Precautions

Nearly every prescription medication has some unique precautions, making it essential for people to be aware of times when a medication should or should not be used. These special precautions also help to alert the healthcare provider that the person may need close surveillance when using the drug. Healthcare providers need to discuss the risks and benefits of taking a medication with consumers prior to prescribing any medication.

Older adults, young children, and pregnant women are at increased risk for harmful medication effects as a result of their physiological differences. Children's medication dosages must be carefully and accurately calculated according to age, weight, level of growth and development, and height. Older adults have an increased risk of ADRs because they do not metabolize nor excrete medications as easily as younger people because of age-related changes. Pregnant women are at increased risk of teratogenicity or harm to the unborn fetus with many medications. Healthcare providers should discuss risks and benefits of medications prescribed during pregnancy with the obstetrician and document the outcome of the conversation in the health record.

Interactions With Other Medications

Drug-to-drug interactions are a significant concern. People are using more and more over-the-counter medications, as well as herbal and nontraditional medicinal products. There can be interactions with or contraindications for certain foods with over-the-counter medications, herbal, and nontraditional therapies that they use. The likelihood of drug-to-drug interaction occurrence is increased because of the complex biochemical composition of the many products now available. As the number of medications a person takes increases, there is a commensurate increase in the likelihood of an interaction. Drug-to-drug interactions can range from relatively benign or not harmful to life threatening. An example of a possible food contraindication would be taking warfarin (a blood thinner) and consumption of green, leafy vegetables (such as lettuce, cabbage, or brussels sprouts), as these vegetables counter the desired effects of the drug.

Missed Dose of Medication

It is always advisable for people to call the healthcare provider if a dose of medication is missed. This is a situation that may occur because the person simply forgot to take his or her medicine, or perhaps he or she was too ill to take it.

An example would be a person with diabetes who has intractable nausea and vomiting and does not know whether or not to take his or her morning insulin. The healthcare provider can review the situation and advise accordingly.

Common Consumer Medication Mistakes and How to Avoid Them

Medications should be taken precisely as prescribed. People can increase medication safety and efficacy by following five fundamental principles of medication administration each time a medication is taken:

1. Right drug
2. Right dose
3. Right route
4. Right time
5. Right person

People often engage in several common undesirable behaviors with respect to medications. These behaviors are quite common and need to be addressed by the healthcare provider on each patient encounter. **Table 20-2** depicts eight common undesirable consumer behaviors that frequently occur with prescribed medications.

Saving Unused Medication for Oneself or Others for Future Illness

Using medications that have remained from a date far in the past or prescribed for someone else is never advised. Medications are prescribed in a dosage that

---

TABLE 20-2   COMMON UNDESIRABLE BEHAVIORS TO AVOID WHEN TAKING MEDICATIONS

1. Saving unused medication for oneself or others for future illness
2. Forgetting to get the prescription refilled
3. Not finishing medications as prescribed because symptoms are relieved
4. Forgetting to take the medicine at the prescribed times
5. Not taking the medication as directed
6. Stopping or not filling the prescription because of cost
7. Sharing medications
8. Taking over-the-counter or nonconventional remedies without informing the healthcare provider

is necessary to treat a specific condition for a particular person at the given time. Prescriptions for the treatment of acute conditions, like infections, need to be fully consumed until the condition is completely resolved. People can develop a flare-up of an inadequately treated infection or resistance to certain pathogens by not completing the medication as prescribed, subjecting themselves to new risks.

Forgetting to Get the Prescription Refilled

It has become common for people to get medications filled via direct mail order or the Internet. This can add to the problem of forgetting to refill a prescription or to refill it in time to avoid missed medication doses. It is not advisable to miss even one medication dose, as this can cause potentially serious or life-threatening harm. Many medications are used for the control of long-term, chronic conditions, such as heart disease, hypertension, or diabetes, and missing even one dose can result in undesirable or harmful effects. An example would be a person who takes the medication, Lanoxin (digoxin). Digoxin affects the contractility of the heart. Missing even one dose of digoxin can result in heart failure, a potentially life-threatening situation.

KEY POINT

When using mail-order or Internet services to fill prescriptions, be careful to allow adequate time for ordering refills to assure that the existing supply of the medication does not run out before the new supply arrives.

It is especially important to give the mail-order or Internet pharmacy ample time to process the order. In the event that a person runs out of a drug, it could prove useful to check with the healthcare provider to see whether any medication samples are available to use until the shipment arrives. If the provider does not have samples, then the person must request a written prescription for a few doses, enough to last until his or her shipment arrives. In the majority of cases, the individual will most likely have to pay for these interim prescriptions, as many insurers do not. It is also important to note that many third-party insurers often will not pay for more than 90 days on a prescription, requiring the consumer to return to the primary healthcare practitioner to get a new prescription.

Not Finishing Medications as Prescribed Because a Person Feels Better

Prematurely discontinuing a medication is a common behavior that can result in potentially life-threatening situations later on. For example, in the case of

-hemolytic streptococcus, the most frequent cause of strep throat, the bacteria may not be completely eradicated even though a person feels better and the symptoms are essentially resolved. These medications are usually prescribed for a full 10 days of therapy in order to eradicate the causative organism completely. When antibiotic therapy is suboptimal, the organism can become resistant to the treatment for future events, or the organism can grow in other areas of the body, such as the heart valves. This organism could then harm the heart valves, resulting in abnormal leaking of blood back into the heart chambers and impaired circulation or damaged heart valve leaflets.

Forgetting to Take the Medication at Designated Times

Frequently, a person forgets to take a dose of medication. In some cases, the dose may be taken as soon as one remembers, if there are no special indications, such as taking it with or between meals. If the missed dose is close to the time of the next scheduled dose, the healthcare provider most likely will advise the person to take the missed dose and resume the usual schedule. The consumer should not double up or take an extra dose, as this can result in a variety of potential outcomes, depending on the medication and potential side effects. An example is a missed a dose of penicillin; attempting to double the dose to make up the missed dose often results in gastrointestinal upset, such as nausea, diarrhea, or severe stomach cramping.

Not Taking the Medication as Directed

Medications need to be taken specifically as prescribed. This is a very important consideration because many medications are designed biochemically to work in certain environments. For example, if the medication is to be taken 1 hour before meals and is consumed with a meal, its absorption could be affected and the drug may not achieve its intended purpose. Many medications should not be taken with milk or other dairy products. If this is the situation, it will be specified on the medication administration sheet and on the label of the medication container. There can be quite serious, even life-threatening consequences if directions are not adhered to as specified.

Stopping or Not Filling a Medication Prescription Due to Lack of Funds

There are reasons for medications to be prescribed; therefore, not filling prescriptions has consequences. When the healthcare practitioner prescribes the medication, it is always helpful for people to ask for an estimate of what the prescription will cost. Although the provider may not be able to give exact figures, some information regarding the approximate cost of the medication should be available. In other cases, the healthcare provider may ask that a pharmacy be called to obtain the price. If a prescription calls for a new medication, especially if it is a very new and expensive medication where there is no generic, less expensive equivalent, it is useful to ask the healthcare practitioner whether she or he has samples available; this can spare one the cost of filling the full prescription only to discover that one cannot tolerate the drug. It is very frustrating to spend a large amount of money on a medication that does not agree with the person and produces so many side effects that he or she cannot take it. If the provider wants the person to have the medication but has no samples, and the person cannot afford the medication, he or she can try to contact a hospital social worker or community resource person for assistance. There may be a community agency that assists in these purchases—and there are also some medication programs for those who cannot pay through some of the pharmaceutical companies that cover some medication prescriptions.

The bottom line is that people should never do without a medication if they cannot afford it. Healthcare providers need to consider this issue when recommending medications.

REFLECTION

Some people complain about the cost of medications and other healthcare expenses while not hesitating to spend money on expensive restaurant dinners, manicures, theater tickets, and other nonessentials. Why do you think they hold such attitudes?

Sharing Medications

Giving medication to anyone other than the person for whom it was prescribed unfortunately does occur, and this practice carries many significant risks. Using someone else's medications can mask symptoms, resulting in improper diagnosis and treatment of life-threatening disorders. It also can result in microbial resistance to organisms.

Taking Over-the-Counter or Nonconventional Products Without Informing the Healthcare Provider

As previously stated, there can be potentially lethal or life-threatening interactions between medications and food substances. It is crucial that people share all information regarding over-the-counter, herbal, or nonconventional (complementary or alternative) therapies with their healthcare providers at each visit. In turn, providers need to make a practice of inquiring about the addition of an over-the-counter or nontraditional treatment with each client visit.

Health records should reflect this information to provide continuity of care by all providers. An example of the importance of sharing this information involves the use of green tea preparations, which can enhance the potency of warfarin, an agent used to thin the blood. If the herbal substance enhances the effect of the medication, a person may experience a life-threatening hemorrhage, such as in the brain, and die as a result of cerebral hemorrhage.

Suggested Times for Taking Scheduled Medications

All prescribed medications include specificity regarding the frequency of administration. **Table 20-3** depicts common abbreviations and meanings for medication administration, suggested times for the common medication administration schedules, and special considerations. These abbreviations are of Latin origin.

Strategies to Assess Medications' Effectiveness

Most pharmacies now provide informational pamphlets with prescriptions. This information usually provides detailed written information ranging from medication dosing to potential side effects. Consumers should be certain to obtain this information on any prescription that is filled. In addition, many pharmacists now provide individual counseling for each consumer. If people opt not to have medication counseling, they may be asked to sign a waiver to that effect. Medication information provided with each prescription includes medication purpose, potential side effects and ADRs, special precautions when taking the medication, and food–drug interactions. Medication information provided by pharmacies also includes specific tips to insure the prescription medication will be effective (see **Table 20-4**).

Summary

More than 100,000 nonprescription and 40,000 prescription drugs are used annually. All of these products, although beneficial in many ways, can carry serious risks if used improperly. Consumers and their healthcare providers need to be in partnership in assuring safe drug use.

People need to be informed users of all medications to assure safety. They need to be knowledgeable about the generic and brand names, intended use, administration, side effects, adverse reactions, storage, handling, precautions, and interactions related to each drug they use.

TABLE 20-3 COMMON ABBREVIATIONS AND RELATED INFORMATION FOR MEDICATIONS

Common Abbreviation	Meaning	Suggested Times for Taking the Medication	Special
ac	Must be taken before meals in order for it to be effective and to do what it is supposed to do	Take the medication 30 minutes before the scheduled or planned mealtime.	If the medication is skipped, take it 30 minutes before the meal; wait at least 2 hours after the meal to take the medication.
bid	Must be taken two times a day	Suggested times for taking the medication should reflect a person's lifestyle for ease of compliance. For example, if one arises at 6 A.M., the medication can be taken at 7 A.M. and 4 P.M. The two times a day dosing is usually spaced approximately 8 hours apart.	If one dose is forgotten, take it as soon as remembered and resume the regular schedule the following day. Do not double up and take two doses at one time.
tid	Must be taken three times a day	Try to incorporate these three times into individual lifestyle if possible to take the medication on time. Space the medications across the day so that doses are not taken close together. For example, if one awakens at 6 A.M., the drug could be taken at 7 A.M., 3 P.M., and 10 P.M. (These sample times are approximately 8 hours apart.)	If one dose is forgotten, take it as soon as remembered and resume the regular schedule the following day. Do not double up and take two doses at one time.
qid	Must be taken four times a day	Suggested times for a four times a day dosing would be 9 A.M., 1 P.M., 6 P.M., and 9 P.M. There can be individual variance and adjustment of this suggested schedule to fit individual lifestyles. It is not recommended to double doses to catch up in the event of a missed dose.	If one dose is forgotten, take it as soon as remembered and resume the regular schedule the following day. Do not double up and take two doses at one time.

(continues)

TABLE 20-3   COMMON ABBREVIATIONS AND RELATED INFORMATION FOR MEDICATIONS (CONTINUED)

Common Abbreviation	Meaning	Suggested Times for Taking the Medication	Special
hs	Must be taken at bedtime	The suggested time for a bedtime dose is 9 or 10 P.M.; however, if one retires earlier or later individual adjustment can be made.	None
qd	Must be taken once a day	This dosing regimen is usually flexible and accommodates the person's individual lifestyle. Most persons take the medication for daily dosing at 8 or 9 A.M.	It is important to take the medication at the same time each day in order not to forget to take it or to mistakenly repeat administration. Choose a consistent time that will increase the likelihood of remembering, such as with breakfast.

TABLE 20-4   SPECIFIC TIPS TO INSURE MEDICATION EFFECTIVENESS AND OPTIMIZE THERAPEUTIC EFFECTS

- Always take medication specifically as prescribed
- Take the medication at the time it is prescribed
- Take the medication in the dose that is prescribed
- Do not skip medication doses
- Do not double up on the medication if a dose is missed
- Notify the primary care provider who prescribed the medication of any symptoms experienced whether they are suspected or known to be related to the medication
- Immediately report to the healthcare provider any unusual feelings or events immediately, such as swelling in any part of the body, joint aching, heart palpitations, or unusual fatigue

Common mistakes that must be avoided include saving unused medications for future use, using someone else's drugs, running out of medications because of a failure to get refills in time, discontinuing the medication prematurely, forgetting to take the drug, not taking the drug as directed, and using other drugs or over-the-counter remedies without the healthcare provider's knowledge.

Informed medication use is an important part of self-care. People can seek information from pharmacists who fill prescriptions to enhance their knowledge.

References

Gu, Q., Dillon, C. F., & Burt, V. L. (2010, September). *Prescription drug use continues to increase*. Data Brief No. 42. Atlanta, GA: National Center for Health Statistics.

U.S. Food and Drug Administration (FDA), Center for Drug Evaluation and Research. (2008). (*Preventable adverse drug reactions*. Retrieved from http://www.fda.gov/Drugs/DevelopmentApprovalProcess /DevelopmentResources/DrugInteractionsLabeling/ucm110632.htm

Suggested Reading

Griffith, H. W., & Moore, S. (2008). *Complete guide to prescription and nonprescription drugs*. New York, NY: Penguin Group.

Karch, A. M. (2008). *Lippincott's nursing drug guide*. Philadelphia, PA: Lippincott.

Lippincott. (2008). *Drug facts and comparisons*. Philadelphia, PA: Lippincott.

Mancano, M. A., & Gallagher, J. C. (2012). *Frequently prescribed medications*. Burlington, MA: Jones & Bartlett Learning.

Medwatch. Retrieved from http://www.drugintel.com/public/medwatch/

Silverman, H. M. (2008). *The pill book*. New York, NY: Bantam.

Resources

Drugs.com
www.drugs.com
Includes information on drugs, supplements, and herbal remedies.

Medline Plus
http://www.nlm.nih.gov/medlineplus/druginformation.html

RxList
www.rxlist.com/script/main/hp.asp

Surviving Caregiving

OBJECTIVES

This chapter should enable you to

- Define the terms *sandwich generation* and *club sandwiches*
- Discuss the many roles caregivers can fill
- List at least two major instructions provided in an advance directive
- Discuss the difference between a delirium and a dementia
- Describe at least two ways that caregivers' burdens can be relieved

There has been a tremendous increase in the number of aging Americans, also known as a graying of the population. The fact that growing numbers of people can expect to survive to later years than their grandparents is wonderful news, but it does create major challenges. These challenges can affect people very directly, either in terms of the human and financial resources needed to meet their own needs as they age or the care they will have to provide to the older adults in their lives.

> **KEY POINT**
>
> In the early 1900s, life expectancy was age 47 and only about 4% of the population lived to age 65 or older. Now in the 21st century, life expectancy is over 77 years, with nearly 13% of the population 65 years or older. The fastest growing segment within the older age group is those over the age of 85 years. It will become more commonplace to see people reaching their 100th birthday.

Sandwich Generation

In the early 1990s, the term *sandwich generation* was coined. This term appropriately described the way that it feels to be caught in the middle—responsible for the care of both parents and children at the same period of time. For centuries, family members took care of each other, and thus, this certainly is not a recent development. The important differences now, however, are that so many more people are surviving much longer, primarily because of new medications and advances of technology in health care; rather than a few years, care of an elder relative can span several decades. Also noteworthy is the fact that few families have the luxury of having a stay-at-home parent, as most women are employed outside the home. The combined stressors, both internal and external, often contribute to feelings of alienation within families instead of enjoyment in relationships.

Increasingly, what can be termed *club sandwiches* are being added to the menu of family profiles. A club sandwich refers to a caregiver who is providing support and assistance to a parent, a child, and a grandchild. There are, however, a variety of club sandwiches. Sometimes the grandparent may become the primary care provider. The reasons for this are many. With approximately 50% to 60% of marriages ending in divorce, some grandparents are faced with assisting their single-parent children with the demands of raising children. Others are handed their new obligations because their adult children became involved in drugs, alcohol abuse, or other unhealthy lifestyles that prevented them from raising their own children. Whatever the forerunner, the fact remains that many families are struggling with caregiver issues and need information and assistance.

KEY POINT

Caregiving wears many faces and includes a variety of tasks and responsibilities.

Many Faces of Caregiving

Providing daily personal care or supervision to a parent can be demanding and overwhelming; however, caregiving often entails more than physical care. One's presence—being there for a parent—can be a significant factor in preserving emotional well-being. An example of a situation that may require intervention is the vulnerable time when one's parents are attempting to make

> **KEY POINT**
>
> If you consider the fact that a person may have worked in a job for decades and then one day is no longer required to perform that role, it is easy to understand how emotional upheaval may occur in retirement, especially if the individual has not prepared for this change.

a sometimes difficult adjustment to retirement. The opportunities of retirement, although perhaps envied by many, are often accompanied by loss and role confusion.

Adult children may assist their parents to recognize their continued value during the time of transition to retirement. It is also a time when children can be instrumental in assisting their parents to navigate the healthcare system. Older parents may be very concerned about their ability to pay for their health care with a limited income. The complexity of Medicare billing also is worrisome to many. There have been so many changes in the healthcare delivery system that clarification and advocacy are often the most supportive activities that children can provide for their parents. Being available to accompany them to doctors' visits and helping them to make some very difficult healthcare decisions can be valuable ways children can help their parents in late life.

As one ages, chronic health problems become more common. These chronic conditions (e.g., hypertension and other cardiac diseases, arthritis, diabetes, and gastrointestinal disturbances) can interfere with the quality of life and ultimately disrupt a person's function. Often these conditions may necessitate more frequent visits to healthcare providers. Many older people have never challenged a physician's diagnosis or requested a second opinion. They may be fearful of offending their doctors by asking questions. Adult children can help their parents to be assertive in this most important aspect of decision making. Frequently, making the right choice means preventing unnecessary discomfort and decline in function, which could accompany inappropriate treatments.

> **KEY POINT**
>
> Any declines in function that can be prevented will help to preserve independence and enhance quality of life, as well as reduce caregiving responsibilities.

Advance Directives

To help assure that appropriate care is provided and that informed decisions are made, discussions and thoughtful planning should be done well in advance of when needed. These issues can be addressed in a document called an *advance directive* or *living will*.

KEY POINT

An advance directive provides specific instructions on the treatment people want or do not want to have in the event they are not capable of expressing their opinions. This document also names a person who is authorized to make healthcare decisions for matters not specifically described in the document.

Preparation of an advance directive is extremely beneficial. This is frequently a topic of discussion at senior meetings and senior centers; however, if one's parents have not yet completed this document, they should be encouraged or assisted in this process. A copy of an advance directive is usually available from a healthcare provider or area agency on aging office.

The decision of naming a healthcare representative—someone who can make decisions regarding treatments if an individual is unable to do so for himself or herself—is sometimes a difficult one for an older adult to make. It is vital that permission be obtained from the person who will be named as proxy healthcare representative. Also, the physician, hospital, perhaps an attorney, and other interested family members should have a copy of the advance directive, in addition to the named representative. The original should remain in a safe place with important papers.

Financial Considerations

Assisting parents or other older relatives to establish a financial system to prevent financial chaos or hardship is another useful action. Often older adults are the victims of scams. In their attempt to handle situations independently, they frequently become easy prey to unscrupulous predators. After they have sustained significant loss, they may conceal the facts and bear their shame with a mask of depression. They may be fearful that they will be considered incompetent; therefore, they may never come forward or even share the information with people who could assist them to recoup their losses. Putting some financial safety checks in place early on often helps to prevent loss and provides a good record-keeping mechanism for future accounting. These records may be necessary to justify financial eligibility for a state or federal assistance program or to verify appropriate distribution of funds, which sometimes becomes a significant issue even within close-knit families.

> **KEY POINT**
>
> It is possible that early caregiving interventions can be so natural and subtle that you may not even realize that you are doing anything that resembles taking care of another.

Financial issues that may be difficult to discuss should not be avoided but rather brought to awareness and addressed before a crisis or disability occurs. Critical topics, those of financial consequence as well as those of advance directives, end-of-life issues, healthcare proxy, and guardianships are too important to wait until a court makes the decisions for individuals and their loved ones.

REFLECTION

Have you and your older relatives discussed issues pertaining to their finances and health care in the event that they are incapacitated by illness or disability? If you have, what were some of the emotional issues that you had to confront? If you have not, what are the issues preventing this discussion and how can you address them?

Age-Related Changes

Many changes occur with age. People develop into more complex physical, psychological, and spiritual beings with a greater understanding of life, having experienced many unique situations. On the down side, aging usually brings negative changes in all of the senses. Vision is frequently compromised with age by such conditions as macular degeneration, cataracts, or *presbyopia* (blurring of vision up close due to changes in the lens). Hearing may deteriorate and limit the tones that can be heard, with the lower tones being the most audible. This condition, known as *presbycusis*, causes people to miss large portions of conversation because certain letters sound muffled. Often older adults hesitate to obtain or use hearing aids because the aids can be distracting, difficult to adjust to, or emphasize the fact that they are getting old.

> **KEY POINT**
>
> Sometimes presbycusis can cause people to give inappropriate responses because they misinterpret speech. Others mistakenly may view them as being confused or having early dementia.

Diminished sense of smell and taste may lead to nutritional problems, social isolation, and unsafe conditions in the home. Common reasons for visits to the emergency room for older adults are dehydration, electrolyte imbalance, malnutrition, and falls. These conditions sometimes can be prevented by frequent contact with family, friends, and others who can make visits and identify problems early. In some communities, a friendly visitor service is provided by church groups or gatekeeper programs. The postal service as well as utility companies (gas/electric/telephone/cable) often train their employees to look for signs that indicate a customer may need assistance. In many communities, older persons can receive Meals on Wheels or daily lunches through senior centers. With funding from the Older Americans Act, seniors can be transported to local sites where nutritious meals, social stimulation, and interaction await them. Coordinators of these programs are usually aware of significant changes in their members and can intervene before unsafe conditions develop or provide rapid assistance through Adult Protective Services personnel if a senior becomes unable to make independent decisions any longer.

Changes in tactile (touch) sensations cause difficulty feeling heat, cold, pain, and pressure, which increases the risk of personal injury. Sometimes fear of being injured by temperature extremes, the discomfort of feeling chilled, and the fear of falling in the bathtub or shower interfere with bathing and personal care needs. This can be most distressing to family or others who attempt to provide care; however, in the older person, a process in their skin causes it to become drier than it did when they were younger. Unless there is incontinence or it is necessary to remove other irritants from the skin, a shower or bath twice a week will suffice for hygienic needs; however, even this is not essential as a thorough cleansing with a washcloth and basin can achieve cleanliness without having to enter the tub or shower.

Perhaps the most distressing alterations are the cognitive changes, which are increasingly common with advanced age. Age-related changes in cerebral function can result in slower reaction time and increased time to learn new information, as well as greater potential for attention to be distracted. These events, coupled with the sensory changes already mentioned, and the stress of everyday living, can create memory difficulties. Memory loss, significant personality changes, confusion, or changes in intellectual abilities are not normal

> **KEY POINT**
>
> A *delirium* is a change in mental function with a rapid onset that usually can be treated and corrected. Some possible causes are infections, low blood pressure, dehydration, and side effects from medications. A *dementia* is a deterioration in mental function that has a slow, subtle onset. Some possible causes include Alzheimer's disease, AIDS, and alcoholism. The losses sustained cannot be regained.

consequences of aging. If an older adult is experiencing what is thought to be unusual changes in cognitive function, a comprehensive geriatric assessment is a wonderful means to identify the source of the problem. Some causes of memory problems that can be detected in a comprehensive geriatric assessment include delirium, depression, and dementia.

Although a dementia is not able to be reversed, there have been tremendous strides in new medications that slow the progression of the disease. These medications are the most successful when begun at an early stage of the disease. The longer function can be maintained, the longer one will be able to remain independent. When memory impairment or other cognitive changes are present, it is essential that the individual obtain a comprehensive assessment. Typically, this assessment consists of a series of visits to a geriatrician, an advanced practice geriatric nurse, and a geriatric social worker. Included in the examination are blood tests that assess complete blood count, folic acid, thyroid function, and electrolyte imbalances. There also is radiological imaging of the brain. Sufficient testing is done to assure that the memory changes are not the result of a severe depression. After all other disease processes are ruled out, a diagnosis can be established and appropriate treatment prescribed. The goal of this therapy is to help the individual maintain function as long as possible.

REFLECTION

How would your life be affected if a parent developed a dementia? What plans for that person would you believe to be necessary?

Function Versus Diagnosis

Despite having their diagnoses, most people with chronic conditions are able to function quite well. Function is a key indicator for a caregiver to use to assess how the person is doing, which includes the present state of wellness/

illness and the potential for the person to increase or decrease independence. Changes in function can be clues of changes in the status of a disease or the presence of new factors that require intervention. For example, if your mother was walking 100 feet unassisted last week and today she is huffing and puffing at 20 feet, something has occurred that should be immediately addressed with her healthcare provider.

Because of the numerous changes that take place throughout the body with age, physical diseases can be more difficult to detect. It is very common for an older person to have an unusual response to illness. Signs and symptoms may be diminished, absent, or vary significantly from what one would observe in a younger individual. An example of this is delirium (acute onset of confusion), which may occur in older persons when they develop infections, such as urinary tract infections. In younger persons, there usually are signs and symptoms that would indicate the urinary tract infection, such as pain on urination, fever, and frequent urination. These symptoms may be absent in older adults, and instead, the infection may be displayed through fatigue, reduced appetite, or a change in mental status.

> ### KEY POINT
>
> It is important for older adults and their caregivers to establish relationships with healthcare providers who are knowledgeable about geriatric medicine, sensitive to older adults, and aware of caregiver issues.

Medications

Medications afford many people not only more years to their lives, but add a higher quality of life to those years; however, medications can create a new host of problems. Every drug carries a risk of side effects and interactions with other drugs, herbs, and food. There is the chance that a medication may be indiscriminately used to manage a symptom when a nonpharmacologic approach could suffice. Side effects of drugs can be missed or attributed to another problem; worse still, another drug can be ordered to manage the side effects of the drug.

It is beneficial for caregivers to maintain a list of all prescriptions, over-the-counter medications, and dietary supplements that the individual is using. This list, including dosages, should accompany the individual to all visits to healthcare providers.

To assist an individual with self-administration of medications, the caregiver can develop a charting system whereby the person checks off the times medications are taken or use a multiday container in which several days' supply of medications can be prepoured (these are available at pharmacies or medical supply stores).

Becoming a Caregiver

Families have their own unique manner of selecting who will be the designated caregiver. Statistically, it will be a woman (about 75%)—a wife, daughter, or daughter-in-law.

Although many people perceive a caregiving responsibility as a burden, it can prove to be a special experience. Much can be discovered through the caregiving process when it is manifested as an act of love. It allows the caregiver the opportunity for intimacy and to heal relationships. Caregivers also can learn much about themselves—their talents, strengths, and weaknesses. Sometimes they can gain new direction in their lives or at least awareness of the relationship between themselves and others, especially those in their families. Positive family relationships and traditions can be strengthened. The act of caregiving can serve as an important lesson on values to younger members of the family.

This is not to minimize the stress and sacrifice involved with caregiving. There is usually significant effort and frustrations that may be endured while providing care. Caregiving can result in financial strain, fewer social opportunities, inconvenience in living arrangements, and reduced personal space. Career opportunities are frequently lost by caregivers, and this can have great impact on someone who is torn between climbing that corporate ladder and being the dutiful daughter or son. If a caregiver attempts to perform all caregiving responsibilities unassisted, it can be a very trying and difficult road.

Finding Help

When care is needed for a short time, as with a terminal illness, there may be sufficient help available among family and friends; however, when an illness is chronic and the need for caregiving extends for years, support and assistance may wane. Alternative arrangements for help from outside the family become crucial. There are organizations that assist people who are in need of caregiving and their caregivers that can be found through local information and referral services. Churches often have ministries that can provide assistance through volunteer caregivers. Nonprofit organizations that assist people with specific conditions (e.g., Alzheimer's Disease and Related Disorders Association, Multiple Sclerosis Society) can also provide valuable assistance.

> **KEY POINT**
>
> The local office of the Area Agency on Aging can provide assistance in locating resources within your community. You can contact your local library for information and referral assistance.

Support Groups

A wonderful resource available for caregivers is special support groups that address issues of people dealing with caregiving. Information as well as successful techniques in handling complex situations are often shared at these informal meetings. There is a special bond that develops between members who share their unique stories that hold aspects to which many others in the group can relate.

An important message that surfaces in support groups is that caregivers have rights, too. Some caregivers feel they need to be given permission to take care of themselves. Somewhere along the road they may have put their own self-care needs on the back burner. They may need to be encouraged to maintain their self-esteem and emotional balance.

There can be valuable exchanges between those who are experiencing the daily challenges of caring for loved ones who may be difficult or deteriorating very rapidly. Suggestions on the most workable method of handling a situation can save another group member hours of aggravation and frustration.

> **KEY POINT**
>
> Sometimes there is a problem finding someone to stay with a dependent individual while the caregiver goes to a support group meeting. The group facilitator may have suggestions to help with this.

Caregiver support groups are an efficient, economical, user-friendly way of gaining access to a very complicated social and healthcare services maze. To obtain information about a local caregiver support group, meeting time, and place, call a nearby hospital's community education department or the local Agency on Aging. If they do not host the meetings themselves, they will be able to offer assistance in finding them. Searching the Internet for the condition a person has (e.g., Alzheimer's disease, multiple sclerosis) often leads to information about local support groups and resources (see the resources section at the end of this chapter).

Planning

To provide caregiving duties efficiently, a plan should be established with the input of the person who is receiving the care when possible. The plan will identify what the person needs and details on the caregiving arrangements (who will do what and when). Sometimes the person who is the recipient of the care may be a very private person or very demanding and have a specific idea of the assistance desired; the desired conditions may be unrealistic or unreasonable. The caregiver needs to be kind yet firm about setting limits. Too often, guilt or fear of displeasing the sick person leaves the caregiver never saying "no," which can lead to caregiver being a victim of burnout.

> **KEY POINT**
>
> Respite is a break from caregiving duties. It can be for several hours a week or several weeks a year. Respite can be achieved from a home health aide companion coming into the home or the ill person attending adult day care services or briefly staying at an assisted-living residence, at a nursing facility, or with another relative. The important point is that the caregiver receives a break from daily caregiving or duties and is able to return to responsibilities refreshed, renewed, healthy, and in good spirits. When burnout is not addressed and respite care is not arranged, the caregiver is at risk for illness.

Caregiving can be a full-time responsibility, yet often it must be combined with many other household responsibilities and/or a paid job. It is important that the caregiver does not exceed realistic limits, but reaches to family, friends, professionals, or community resources for help. If other family members are unable to provide hands-on assistance, they can be asked to contribute financially to the purchase of services that can ease the caregiving burden. Respite from caregiving responsibilities is something that should be incorporated into the plan. Often a professional can assist the caregiver in obtaining respite services.

Stresses of Caregiving

Caregiver stress is not only something that one experiences in the day-to-day, hands-on care of a loved one. This stress can exceed all boundaries and distance. Long-distance caregiving can be extremely difficult and challenging. With many families spread across the nation and around the world, caregiving has taken on a whole new appearance. Just a few generations ago, families remained in close proximity, and most women did not work outside the home. The care of young children and the aged was considered a normal family responsibility. There were few "old people's homes," as they were called, in which to place an older person. The conditions of most of those facilities were very poor, causing them to be dreaded by older persons. Many families committed never to put their loved ones in such a place. Although the conditions of today's long-term care facilities are significantly improved over those of the past, many families are still reluctant to seek institutional care for a loved one and experience tremendous guilt when there is no alternative but to do so.

Long-Distance Caregiving

Family lifestyles have changed, and often both parents work outside the home. With some families separated by thousands of miles, family visits may be rare. Telephone calls, pictures, and even emails have taken the place of Sunday and holiday gatherings for many extended families. Although only about 5% of the senior population lives in nursing homes, many older adults have moved to a more supportive environment. Assisted-living residences, continuing-care retirement communities, and adult communities have become increasingly common residences for elders. Most older adults no longer expect their children to provide for their daily care.

Many adult children who do not live in close proximity to their parents may not know that there is a problem with their parents until they receive a crisis telephone call; they may be ill prepared to handle that emergency. They may have never discussed advance directives, healthcare proxy, and other important issues with their parents, yet they may be facing some difficult decisions that have to be made. Decisions may need to be based on secondhand information (e.g., what a parent's friend recalls her desires to be regarding life-sustaining treatments).

KEY POINT

Every state has local Area Agency on Aging offices provided under the Older Americans Act, Title III. They frequently are excellent providers of information and referral and are able to make the task of long distance caregiving easier. In some states, the Area Agency on Aging is the center for one-stop shopping for all issues that involve the older adult.

There is another group of professionals who have assisted many long-distance caregivers—geriatric care managers. This group of professionals is located throughout the United States. Their role is complex and varied. They are advocates for their clients, liaisons to clients' families, and supporters of their client's well-being. They perform assessments, develop care plans, arrange for services, supervise auxiliary personnel, and serve as surrogate families. An additional function that is provided by many geriatric care managers is that of family or individual counseling. Many geriatric care managers are self-employed professionals, usually nurses or social workers. Some of them are part of larger corporations or affiliations, such as hospitals, Catholic Charities, or Jewish Family Services. There is a fee for their services, usually determined on an hourly basis. The cost of services varies and can be quite expensive, but these care managers are able to deliver high-quality service and peace of mind that can be invaluable for long-distance caregivers. More information about care managers can be obtained from their national office at the National Association of Geriatric Care Managers (www.caremanager.org).

KEY POINT

The Family Medical Leave Act entitles employees to as much as 12 weeks of unpaid leave per year for the care of a family member.

Being Proactive

Caregivers can face significant change and loss. There is an alteration of roles, and frequently their entire lives are greatly affected. They may find themselves in the position of parenting their parents. Such lifestyle changes may result in them, as the caregivers, displaying adaptive behaviors that could include denial, excessive physical complaints, rigidity or stubbornness, an overly critical attitude, selective memory, or regression. They also could experience exhaustion, loneliness, depression, anger, or guilt. These behavioral manifestations sometimes cause them to experience the additional loss of friends, coworkers, and other support systems at a time when they most need this support. It is important for caregivers to recognize negative feelings toward their responsibilities and perhaps toward the individuals for whom they are providing care. It could prove very helpful for them to take the caregiver assessment in **Exhibit 21-1** to help identify red flags for problems.

After completing the assessment in Exhibit 21-1, caregivers should consider discussing the assessment with family and friends who form their support system and asking for advice and assistance in finding ways to change the situation as needed. Options could include these:

- Calling a family meeting to assign or encourage additional family support
- Negotiating expectations and limitations with the care recipient
- Contacting outside community resources to come into the home and assist with care
- Finding someone to listen
- Scheduling time off
- Joining a support group
- Seeking counseling
- Exploring assisted living or nursing home care

It is far wiser to develop strategies to avoid having caregiver stress lead to negative consequences for all parties involved.

Summary

As people are living longer, their risk of becoming disabled and dependent on others increases. For most of these people, care will be provided by family members. This can present many challenges for the care recipients and the caregivers.

EXHIBIT 21-1   SELF-ASSESSMENT OF CAREGIVER STRESS

___ There are many days that I wonder if my life is worth living.

___ The care recipient requires care and attention that exceed my abilities.

___ I am often overwhelmed by my responsibilities.

___ My spouse and children feel that I am neglecting their needs.

___ I often do not sleep or eat properly.

___ There is little to no opportunity for me to exercise or engage in recreation.

___ I have postponed taking care of my own health needs because of the caregiving demands.

___ I often argue or have conflicts with the care recipient.

___ Sometimes I have thoughts of harming or abandoning the care recipient.

___ I worry about what will happen to the care recipient if something happens to me.

___ My health is suffering.

___ There is little time or space for me to have personal private time.

___ It has become difficult for me to have a social life.

___ I have become resentful of other family members who do not provide assistance.

___ The care recipient takes me for granted.

___ Caregiving is creating financial hardship for me and/or my family.

___ I'm feeling guilty that my best does not seem good enough.

___ I feel trapped in the situation with no end in sight.

Caregivers can fill many roles, such as providing supervision, personal care, guidance in navigating the healthcare system, assistance with decision making, and support.

Often, caregivers have to juggle caregiving responsibilities with other family and work demands. Many people are caught between providing care for their children and for their parents, leading to them being referred to as the sandwich generation. The term club sandwich is being added to family profiles to include caregivers who are providing care to their parents, children, and grandchildren.

It is important for caregivers to help their family members consider financial and health plans in the event that the dependent family members are incapable of making decisions and performing functions independently. An advance directive is one means to plan for future events; it provides instructions on treatment desired in the event people cannot express their opinions at a later date and the person authorized to make healthcare decisions for them.

A variety of age-related changes can affect older individuals' abilities to function independently and engage in self-care. Furthermore, aging changes can alter the presentation of symptoms. It is important for caregivers to understand age-related changes to differentiate normal from abnormal findings.

Caregivers can lighten their burdens by planning care, seeking help with caregiving, participating in support groups, and preventing and managing stress.

Suggested Reading

Acton, G. J., & Miller, E. W. (2003). Spirituality in caregivers of family members with dementia. *Journal of Holistic Nursing, 21*(2), 117–130.

Blair, P. D. (2005). *The next fifty years. A guide for women midlife and beyond.* Charlottesville, VA: Hampton Roads Publishing Company.

Callone, P., & Kudlacek, C. (2010). *The Alzheimer's caregiving puzzle: Putting together the pieces.* New York, NY: Demos Medical Publishing.

Calo-oy, S. (2004). *The caregiver's guide to dealing with guilt.* San Antonio, TX: Orchard Publications.

Dodds, M. (2006). *A Catholic guide to caring for your older parent.* Chicago, IL: Loyola Press.

Farran, C. J. (2001). Family caregiver intervention research: Where have we been? Where are we going? *Journal of Gerontological Nursing, 27*(7), 38–45.

Larrimore, K. L. (2003). Alzheimer disease support group characteristics: A comparison of caregivers. *Geriatric Nursing, 24*(1), 32–35.

Lerner, H. (2002). *The dance of connection.* New York, NY: Quill.

Lilly, M. L., Richards, B. S., & Buckwater, K. C. (2003). Friends and social support in dementia caregiving: Assessment and intervention. *Journal of Gerontological Nursing, 29*(1), 29–36.

Logue, R. M. (2003). Maintaining family connectedness in long-term care: An advanced practice approach to family-centered nursing homes. *Journal of Gerontological Nursing, 29*(6), 24–31.

McCall, J. B. (2000). *Grief education for caregivers of the elderly.* New York, NY: Haworth Pastoral Press.

McCullough, D. (2008). *My mother, your mother: Embracing "slow medicine," the compassionate approach to caring for your aging loved ones.* New York, NY: HarperCollins.

Meyer, M. M., & Derr, P. (2007). *The comfort of home: A step-by-step guide for caregivers.* Portland, OR: CareTrust Publications LLC.

Moore, S. L., Metcalf, B., & Schow, E. (2000). Aging and meaning in life: Examining the concept. *Geriatric Nursing, 21*(1), 27–29.

Nouwen, H. J. M. (2011). *A spirituality of caregiving.* Nashville, TN: Upper Room Books.

Ostwald, S. K., Hepburn, K. W., & Burns, T. (2003). Training family caregivers of patients with dementia: A structured workshop approach. *Journal of Gerontological Nursing, 29*(1), 37–44.

Rackner, V. (2009). *Caregiving without regrets: 3 steps to avoid burnout and manage disappointment, guilt, and anger.* Mercer Island, WA: Medical Bridges.

Sheehy, G. (2010). *Passages in caregiving: Turning chaos into confidence.* New York, NY: Harper.

Winters, S. (2003). Alzheimer disease from a child's perspective. *Geriatric Nursing, 24*(1), 36–39.

Resources

Family Caregiver Alliance
690 Market Street, Dept. P, Suite 600
San Francisco, CA 94104
www.caregiver.org

National Council on Family Relations
3989 Central Avenue NE, Suite 550
Minneapolis, MN 55420
612-781-9331
www.ncfr.com

National Eldercare Locator
1112 16th Street NW, Suite 100
Washington, DC 20036
800-677-1116
www.eldercare.gov

PART IV

Safe Use of Complementary and Alternative Therapies

Navigating the Use of Complementary and Alternative Therapies

OBJECTIVES

This chapter should enable you to

- List three factors that have stimulated Americans' interest in complementary therapies
- Discuss at least three philosophical differences between holistic and conventional approaches
- Describe the role of self-awareness in health and healing
- List at least five questions that should be asked to guide the decision to use a complementary therapy
- Describe the five major categories of complementary therapies

Complementary therapies have been used for centuries throughout the world; however, it has been primarily since the 1990s that their use has soared in the United States. It was during that decade that the landmark study by David Eisenberg and his colleagues, published in the prestigious *New England Journal of Medicine*, revealed that one third of Americans were using alternative therapies—a figure that has continued to grow since that study (Eisenberg, 1998). By the turn of the century, Americans were spending over $27 billion annually for complementary and alternative therapies, most of which was out of pocket. The 2007 National Health Interview Survey (NHIS), which included a comprehensive survey of complementary and alternative medicine (CAM) use by Americans, showed that approximately 38% of American adults use complementary and alternative therapies. To say this caught the medical community's attention would be an understatement!

Many factors have contributed to the growing use of complementary therapies. The interest in preventive health has stimulated individuals to explore practices and products that they can use independently, and many complementary therapies, such as meditation and dietary modifications, fit the bill. Growing reports of adverse drug reactions and other complications from conventional medical care have led people to explore natural means to manage illness. The personalized attention provided by practitioners of complementary therapies offers people a superior experience to the abbreviated and often impersonal office visits and hospital stays. Furthermore, research proving the benefits of complementary therapies is growing by the day. *NO!*

The heightened attention to complementary therapies also has created much confusion regarding the safety of using many of these therapies. Questions such as these emerge: Why and when is it appropriate to use these therapies? Are these therapies consistent with my health beliefs? My spiritual beliefs? How do I choose a practitioner or a therapy? Above all, are these therapies right for me?

Mystery and what seems to be a strange language surround many complementary treatments. Even the terminology used to describe these types of therapies is confusing. These therapies were first called (and continue to be called) *alternative therapies*. This term does not fit for many because it creates an either/or choice situation. Because it is felt by many that these therapies can be used in conjunction with Western medical treatments, the word *complementary* conveyed a clearer meaning. Clarifying even more the joint use of several methods of treatment at the same time, use of the term *integrative* has evolved.

Terms such as *holistic* and *natural* are frequently used to describe complementary therapies. *Allopathic* is a word that describes Western conventional medicine. Some feel that using the term *traditional medicine* to describe today's practice of medicine is incorrect. It is felt by many that *traditional* indicates the use of a treatment since the beginning of time and, therefore, *conventional* better describes the present use of Western medical treatments.

KEY POINT

The evolution of terms used in describing complementary therapies is most evident in the changed name of the division of the National Institutes of Health (NIH) that investigates and researches these types of practices and products. First established as the Office of Alternative Therapies, in 1998, it became the National Center for Complementary and Alternative Medicine. Along with the name change has been an increase in its annual budget for research of these therapies.

History

Many complementary therapies originated from ancient and non-Western healing traditions, many of which have their roots in spiritually based health-care systems. These systems use such measures as prayer, meditation, drumming, storytelling, and mythology to help people in their search for wholeness by allowing them to experience sacred moments in their lives. Spirituality is not in itself religion, but it underlies and enhances all world religions. It is also seen as a drive to become a complete, balanced person and is believed by many to be related to intuition, creativity, and motivation (Fontaine, 2000). Most ancient and non-Western cultures express healing as being in balance and harmony.

Holistic Health Beliefs

Today the word *holistic* is being used freely, and yet many of the basic concepts that underlie its true meaning are not understood or fully embraced by those using it. The most basic of these concepts is the concept of wholeness. Many ancient healing traditions have as a belief that wellness exists when there is balance of the physical, emotional, mental, and spiritual components of our being. Your physical body has an innate physical tendency to work toward equilibrium or homeostasis. It has a built-in potential to maintain physical health or optimal function of all body systems and a complex natural ability to repair itself and overcome illness. In the quest for balance emotionally, you strive to feel and express your entire range of human emotions freely and appropriately. Mentally, you seek a sense of self-worth, accomplishment, and positive self-identity. Spiritually, you seek a connectedness to others and to a higher power or divine source. This balanced state is seen as wellness in a holistic approach to health care. Illness is considered an imbalance of these components.

Healing is seen as an ongoing, lifelong process, and when viewed in a positive manner, it is seen as a continual journey of self-discovery. All healing is self-healing, and conventional (Western) medical treatments and complementary treatments help by creating an environment that supports your attempt to balance the components of your beingness. Nevertheless, some philosophical

---

KEY POINT

Holistic health is the balance among your physical, emotional, mental, and spiritual components.

6 Pillars   Phys   Emotional
             Environ   Social
             Mental   Spiritual

differences exist between the holistic and conventional (Western) medical approaches. These include:

- The holistic approach to healing seeks the root cause of the problem (the source of the imbalance) and tries to reestablish a balance of mind, body, and spirit, whereas conventional medical approaches are more disease oriented and focus more on removing or managing signs and symptoms of physical illness.

- In Western medical approaches, people have been conditioned to turn over the responsibility of healing to the healthcare provider; however, participation in one's own healing process is a key component of a holistic approach to health care. Being a passive recipient or expecting others to fix a problem independent of your efforts is inconsistent with the holistic caring process. This belief in seeking balance or healing is carried out in a partnership. You and your care provider have complementary responsibilities. Learning how to partner is a very important concept that conventional medicine, until recently, has not strongly emphasized.

- Another element of the holistic approach to health care is the requirement of self-care. The care and nurturing of all aspects of one's self support a healthier balance and result in more productivity and a fuller participation in the life experience. Western medicine tends to focus on diagnosing symptoms and treating illnesses with minimal attention to teaching and supporting individuals in practices that promote the health.

- A holistic orientation to health recognizes the interconnectedness of the mind, body, and spirit. If any aspect of yourself is not attended to and nurtured, it cannot function to its capacity; this will have a detrimental effect on the other components of your health. Imbalances are identified and addressed before they become disease processes. Although Western practitioners may appreciate the relationship of body, mind, and spirit, they tend to focus on restoring and promoting normal function of the aspect of the body that relates to their area of practice. For example, a cardiologist may prescribe treatments to restore a normal heart rate but not explore the spiritual distress that is burdening the patient.

Self-awareness then becomes paramount in this dance between balance and imbalance of the mind, body, and spirit. This self-awareness is supported and/or learned through many of the complementary therapies. Many of the therapies can precipitate an awareness of emotions as well as a physical awareness. It may have been the physical complaint that directs a person to seek the therapy, but the awareness of an emotional component may surface in the process. Knowing this is of importance when seeking complementary therapies. This is also the reason that these therapies are referred to as holistic; they involve all aspects of the person. Recognizing and choosing to act on this self-awareness is part of the personal healing process.

REFLECTION

How do you react when you experience headaches, indigestion, and other symptoms? Do you quickly try to eliminate them with medications, or do you first take time to understand the underlying causes so that you can prevent them in the future?

Acute and Chronic Illness

Pharmacological and technological advancements have equipped conventional medicine to handle acute conditions effectively and efficiently. Heart attacks can be halted, shattered bones mended, and infections eliminated. Unfortunately, as medical technology has increased, the caring components of medical care seem to have shrunk. Hospital stays and office visits are shorter; healthcare providers seldom have ample time to learn about the whole person or to teach and empower the person for self-care. For these reasons, conventional medicine is less successful at managing chronic conditions than at treating acute illnesses.

Persons with chronic conditions are increasingly integrating complementary therapies into their medical care. They find that complementary practitioners invest more time in getting to understand their clients, encourage an active provider–client partnership, empower clients for self-care, promote healthy lifestyle practices, and tend to see the whole person rather than merely treat the symptoms (Eliopoulos, 1999).

KEY POINT

Less than 25% of the healthcare dollar in the United States today is spent on prevention and acute care; the balance is spent on the management of chronic conditions.

85%

Self-Awareness: Body Wisdom

The journey toward self-awareness, like anything else, begins with the first step, and a beginning point is physical body awareness. Physical body awareness appears to be a foreign concept in Western society today, as most people have been taught to deny, ignore, or push through early signs the body may express. Self-awareness is a major developmental pearl that provides key information that influences choices in making informed decisions when choosing a therapy that is most appropriate and/or safe at any given point in time.

> **KEY POINT**
>
> Self-awareness enables one to make informed decisions.

As body awareness develops, awareness of feelings and thoughts follows. This knowledge and understanding leads to inward focusing. Inward focus and awareness protect and guide in many ways. Most of us have been directly taught by our parents, teachers, and others how to get around our weaknesses and imbalances. We tend to ignore body messages until they scream so loudly that we finally are incapacitated and forced to stop and take notice.

Befriending and listening to the body is a lifelong process that continues to be refined as more attention is paid to it. Paying attention leads to learning the body's wisdom versus overriding it. Learning how to recognize this wisdom increases personal knowledge, and with this knowledge comes personal power—power to be in greater control of knowing the best choice to help the body regain or move toward balanced health.

Self-awareness feeds into self-responsibility in a holistic approach to maintaining wellness. Rather than following the pattern of expecting others to provide information of what is best, self-awareness provides the ingredient of self-involvement in illness prevention. This journey of exploration and self-discovery can be fascinating and removes the image of the body as foreign territory. This voyage begins with small steps and requires notation of results obtained. This tuning into the body, instead of tuning out, allows the body to be worked with in a cooperative way.

Self-awareness better equips a person in making an initial choice of which complementary therapy to experience and provides information after receiving the therapy. Paying attention to the physical, mental, and emotional responses after receiving a treatment is invaluable in making future treatment decisions. Were the reasons for choosing a particular therapy satisfied? In what way? For how long? What new information was obtained about the body and the mind? This information allows for the choice to continue an old pattern of ignoring the messages received, or paying attention, learning, and deciding to make new informed choices.

> **KEY POINT**
>
> Paying attention to their physical, mental, and emotional responses helps people to determine what is best for their health rather than relying on the decisions of others.

The continued evolution of this self-awareness can become very powerful and self-empowering. Becoming more involved allows for greater participation in decision making in health care or wellness choices. This gentler, friendlier approach to body–mind maintenance allows for less faultfinding with the body, less rushing ahead without willpower, and less blindness to personal weaknesses. Self-awareness and self-acceptance are major components that propel a person on the journey toward a more balanced mind, body, and spirit and establish a more personal feeling of control of life in general.

Selecting a Therapy

Some common categories of complementary and alternative therapies are as follows:

- Alternative systems of medical practice, including traditional Chinese medicine, acupuncture, homeopathy, ayurveda, and naturopathy
- Manipulative and body-based practices, including chiropractic, osteopathy/craniosacral therapy, energy therapy/energy medicine, bodywork therapy, such as massage therapy, movement therapy (the Trager approach, the Feldenkrais method, and Alexander therapy)
- Mind–body therapies, including meditation/relaxation, biofeedback, guided imagery, hypnotherapy, yoga, and T'ai Chi
- Natural products, including herbs, aromatherapy, vitamins, minerals, other supplements

Some therapies overlap into several categories. For example, T'ai Chi and yoga can be categorized under both movement and mind–body therapies. Specifics about these therapies can be found in other chapters in this text.

> **KEY POINT**
>
> If a person desires to use CATs, he or she must be committed to being an active participant in regaining and/or maintaining health.

Deciding to use a complementary/alternative therapy (CAT) requires much consideration and forethought. In order for a CAT to be safe and appropriate, many questions need to be asked.

A major consideration is how to work effectively with a primary healthcare provider. Is conventional medical care required to monitor a particular medical condition? If so, developing a partnership with the conventional healthcare provider is crucial. Cooperation by the provider and the person seeking CAT is essential. For proper assessment of treatment effects, if combining conventional medicine and CAT, the conventional healthcare provider needs to be informed. Clear baseline assessment information of current symptoms or concerns is needed. The person must be prepared to keep a diary of information to assist in evaluating CAT results. Decisions need to be made regarding the safety of combined therapies or the impact of postponing conventional medical care while using CAT. Agreements need to be reached regarding follow-up visits with the Western medical care provider, if necessary, during or after receiving CAT. Some suggestions for using CAT safely are listed in **Exhibit 22-1**.

Choosing a Practitioner

Deciding which therapy to receive is part of the process, but deciding which provider will deliver that therapy requires just as much serious thought and investigation. There are some important steps in selecting a treatment specialist.

First, people need to be urged to take their time. They should gather names of practitioners by contacting professional organizations, asking for recommendations from people they respect, and checking local directories. After a few practitioners are identified, they can be called and asked questions regarding education, experience, and credentials/certification. Some therapies require degrees, whereas others require training with specific criteria. People need to beware of any pressure or claims about cures. Brochures can be requested, as can the names of some of the practitioners' clients who can be contacted for references. State licensing boards can provide information about standards that practitioners need to meet.

EXHIBIT 22-1   SUGGESTIONS FOR SAFELY USING CATS

- Know your reason for choosing CAT. Is it out of frustration with experiences of Western conventional medicine? Increasing awareness of other cultures' approaches to health and illness? A desire for wellness or to explore a wholeness approach to health? Friends' positive experiences that have aroused your curiosity?
- Use CAT under the supervision of a qualified doctor if you have a serious medical illness.
- Understand and establish your goals in using CAT.
- Beware of mixing and matching conventional and CATs on your own.
- Do your homework and gather information; do not rely merely on testimonies of persons who have used a therapy.
- Avoid the more is better fallacy.
- If something sounds too good to be true, it generally is.
- Keep an open mind, and use the best of conventional medicine and CAT.
- Ask questions of practitioners before selecting their therapy, such as the following:
  - What is it?
  - How does it work?
  - What health conditions respond best to this CAT?
  - When is it best to use it/not use it?
  - What should I expect?
  - What are the possible harmful effects?
  - What is the cost and length of a session?
  - How long must I use the therapy/receive treatment?
- What information and resources are available to help me learn about this CAT?
- Do you need to be licensed or certified to practice this CAT? If so, are you?
- What professional organizations can be contacted to get information about practitioners of this CAT?

The information collected needs to be reviewed. People need to assess their own comfort level with the qualifications (educational background/licensing/certification), or lack of, and the practitioner's manners in interacting with them. They may want a consultation visit. During this visit, they should be clear about their goals, review the goals with the practitioner, discuss organizing a treatment plan, and inquire about side effects and/or adverse reactions. Also, people need to remember to budget their time and expenses appropriately to attain their goals.

After an initial session, reevaluating is helpful. People need to understand that it is their decision to continue the sessions or not; they are in charge. They should consider the practitioner's professionalism, willingness to answer questions, ability to listen, understanding of concerns, and their overall feelings of ease with the provider. People need to trust their judgment and confidence in the practitioner's skills and degree of healing relationship established.

REFLECTION

Have you ever received care from a healthcare practitioner who did not treat you with respect or listen to your concerns? How did this influence your ability to feel in charge of your care? How can you avoid behaving in the same manner when you care for others?

Today there are many options available in the quest to maintain or regain health and wellness. Being prepared to make appropriate individual choices regarding these options requires considerable thought and a thorough investigation. This investigation leads to an understanding of the concepts of wholeness (the inseparableness of mind, body, and spirit) and insights into personal health beliefs. This investigation also leads to greater knowledge of self-awareness, self-responsibility, and self-nurturance and provides a clearer vision of personal needs and more personal power in developing a partnership with the healthcare provider.

A broader view of health allows for the realization that optimum health is a lifelong journey rather than an ultimate state of being. The goal of this journey is the maintenance of balance between health and illness. The attempt at maintaining this balance is thwarted with many challenges and choices. What is the extent of your personal knowledge of your health and wellness status? Is there a need to consult with a conventional medical healthcare provider if you choose to integrate a complementary therapy in your journey? What do you know about the origin, today's use, and safety of a specific non-Western medical therapy? Is it safe to combine the use of conventional and complementary healthcare therapies? Is the practitioner of choice knowledgeable and appropriately educated and trained? Will there be communication between your providers if necessary? In directing your own health care, these are but a few of the questions that need to be addressed. Selection of an appropriate complementary therapy requires much consideration, and the information provided here is offered as beginning guidance and assistance along the journey of self-discovery and balanced health.

Other chapters offer detailed discussions of specific categories of complementary and alternative therapies.

Summary

Americans have shown a growing interest in complementary therapies because they are concerned about preventive health, adverse drug reactions, and personalized care offered by complementary practitioners. Although new to many Americans, many complementary therapies originated from ancient and non-Western healing traditions that have been used in other parts of the world for centuries.

Holistic care is not the same as complementary or alternative therapies. Holistic care implies a balance and harmony among mind, body, and spirit. It assumes that individuals take an active role in achieving maximum wellness. Although people can use complementary or alternative therapies as part of holistic care, the use of these therapies does not guarantee holism.

Complementary therapies need to be used wisely. To use complementary therapies safely, people need to gather information to base decisions on facts, establish goals, and keep an open mind. They need to ask questions before using a complementary therapy, including these: What conditions respond best to it? How does it work? What should be expected? How long must it be used, and what are the harmful effects?

References

Eisenberg, D. (1998). Advising patients who seek alternative medical therapies. *The Integrated Medical Consult, 1*(1), 4–5.

Eliopoulos, C. (1999). *Integrating conventional and alternative therapies: Holistic care for chronic conditions.* St. Louis, MO: Mosby, Inc.

Fontaine, K. L. (2000). *Healing practices: Alternative therapies for nursing.* Upper Saddle River, New Jersey: Prentice Hall.

Suggested Reading

American Cancer Society. (2009). *American Cancer Society complete guide to complementary and alternative cancer therapies.* Atlanta, GA: American Cancer Society.

Carlson, L. K. (2002). Reimbursement of complementary and alternative medicine by managed care and insurance providers. *Alternative Therapies in Health and Medicine, 8*(1), 38–49.

Cerrato, P. L. (2001). Complementary therapies update. *RN, 61*(6), 549–552.

Chopra, D. (2001). *Perfect health: The complete mind/body guide.* New York, NY: Random House.

Decker, G. (1999). *An introduction to complementary and alternative therapies*. Pittsburgh, PA: Oncology Nursing Press, Inc.

Earthlink, Inc. (2000, June/July). Alternative healthcare: Is it the right alternative for you? *Blink*, p. 27.

Eisenberg, D. M. (1997). Advising patients who seek alternative medical therapies. *Annals of Internal Medicine, 127*(1), 61–69.

Freeman, L. (2008). *Mosby's complementary and alternative medicine. A research-based approach* (3rd ed.). St. Louis, MO: Mosby.

Huebscher, R., & Shuler, P. A. (2003). *Natural, alternative, and complementary health care practices*. St. Louis, MO: Mosby.

Jesson, L. E., & Tovino, S. A. (2010). *Complementary and alternative medicine and the law*. Durham, NC: Carolina Press.

Kirskey, K. M., Goodroad, B. K., Kemppainen, J. K., Holzemer, W. L., Bunch, E. H., Corless, I. B., et al. (2002). Complementary therapy use in persons with HIV/AIDS. *Journal of Holistic Nursing, 20*(3), 250–263.

Krohn, J., & Taylor, F. A. (2002). *Finding the right treatment. Modern and alternative medicine: A comprehensive reference guide that will help you get the best of both worlds*. Point Roberts, WA: Hartley & Marks Publishers.

Micozzi, M. (2010). *Fundamentals of complementary and alternative medicine*. St. Louis, MO: Saunders.

Strovier, A. L., & Carpenter, J. E. (2008). *Introduction to alternative and complementary therapies*. Philadelphia, PA: Haworth Press.

Trivieri, L., & Anderson, J. W. (Eds.). (2002). *Alternative medicine: The definitive guide* (2nd ed.). Berkeley, CA: Celestial Arts.

Young, J. (2007). *Complementary medicine for dummies*. New York, NY: IDG Books Worldwide, Inc.

Resource

National Center for Complementary and Alternative Medicine

Offers fact sheets, clinical practice guidelines, and other useful resources. http://nccam.nih.gov/

Alternative Medical Systems

CHAPTER

23

OBJECTIVES

This chapter should enable you to:

- Describe the alternative systems of medical practice of traditional Chinese medicine, acupuncture, homeopathy, ayurvedic medicine, and naturopathy

Traditional Chinese Medicine

What It Is

Traditional Chinese medicine (TCM) is a complex and sophisticated ancient system of health rooted in Chinese culture that has been passed down from generation to generation. TCM approaches the person as a whole and sees mind, body, and spirit as interrelated elements that are intertwined with nature and the universe. It embraces many theories, methods, and approaches and emphasizes prevention. Some of the basic principles inherent in TCM are listed in **Exhibit 23-1**.

How It Works

Diagnostic approaches are based on identifying patterns of disharmony or imbalance and consist of the following:

- Examination or assessment of the voice, pulse, and respiration
- Observation of overall appearance, eyes, skin, and tongue
- Questioning about functions of the whole person (both physical and emotional)

> **EXHIBIT 23-1   BASIC PRINCIPLES INHERENT IN TRADITIONAL CHINESE MEDICINE**
>
> - Qi (chi) is believed to be the vital life force or invisible flow of energy that circulates through specific pathways in plants, animals, and people and are called meridians, which are necessary to maintain life.
> - Yin and yang, which is the interaction of opposing forces and seen as complementary aspects.
> - Five phases theory of fire, earth, metal, water, and wood.
> - Five seasons of summer, autumn, winter, spring, and late summer.
> - View of mind, body, and spirit as the three vital treasures.

The patterns determined by this investigative process are the basis on which the treatment plan is made. Treatments then are focused on restoring or maintaining harmony and balance of the qi's flow.

Various therapies may be prescribed to treat patterns of disharmony. Some of these therapies are diet; herbal remedies; massage; acupuncture (insertion of very fine needles at specific points on the body to stimulate and balance energy flow); acupressure, consisting of the application of finger pressure over specific points to also stimulate the flow of qi; moxibustion, or the burning of a special herb (mugwort) over acupuncture points in order to stimulate the energy flow through the energy channels; and qigong (pronounced chee-gong), involving low-impact stretches, abdominal breathing, meditation, and visualization, and/or possibly T'ai Chi (pronounced tie-chee), which is a form of movement meditation designed to unite body and mind, improve muscle tone, and encourage relaxation. Individualized plans determine the number of sessions prescribed for any of these forms of treatment.

Acupuncture

What It Is

As already mentioned, acupuncture is one of the treatment methods used in TCM. Acupuncture is used to unblock the pathways (meridians or channels) at specific juncture points through which the qi or life force flows. The theory is that disease or illness results from the blockage or obstructions of the pathways, therefore clearing the blockages allows the qi to run freely through the meridians and restore health.

Archeologists have traced the use of acupuncture as far back as the Stone Age. Americans began to take note in 1972, after President Richard Nixon visited China. It was during that trip that a news reporter developed appendicitis, and acupuncture was used during his emergency appendectomy. The reporter wrote of this event and stimulated much interest and many visits by Western physicians to observe the use of acupuncture in China. The National Institutes of Health has demonstrated the effectiveness of acupuncture for pain management after dental surgery and for controlling nausea and vomiting in pregnancy, after chemotherapy, and after surgery.

How It Works

It has been determined that acupuncture stimulates physical responses, such as changing brain activity, blood chemistry, endocrine functions, blood pressure, heart rate, and immune system response. More specifically, medical research has shown that acupuncture can regulate blood cell counts, trigger endorphin production, and control blood pressure.

Acupuncture needles are sometimes inserted and removed immediately, and at other times, they are allowed to remain in place for a period of time. Sensations described as rushing, warmth, and tingling are experienced. As the immune system is stimulated, a sense of well-being is experienced. Acupuncture is used both to maintain health and to treat illness or pain.

What It Helps

The World Health Organization (WHO) has described more than 100 different health conditions that acupuncture can treat. Much positive research has supported the use of acupuncture in the treatment of alcohol and drug addiction.

Words of Wisdom/Cautions

One of the biggest advantages of the use of acupuncture is the lack of harmful side effects. A feeling of light-headedness or euphoria after a treatment has been reported, which is stabilized by a few minutes of rest.

KEY POINT

Acupuncture has been one of the most heavily researched complementary and alternative therapies. Evidence supporting its effectiveness for many conditions exists.

Homeopathy

What It Is

Homeopathy is a system of medical practice that is based on the principle that like cures like. It is a therapeutic system that assists self-healing by giving small doses of remedies prepared from plant, animal, and mineral substances. This system uses diluted (the more diluted the better) portions or remedies to cure symptoms of disease. The remedies encourage the body to eliminate symptoms by encouraging the symptoms to run their course instead of suppressing them. Symptoms are considered signals as the body works to restore natural balance. During this process of self-healing, the immune system is stimulated, healing is accelerated, and the body is strengthened.

How It Works

The effects of homeopathy on the healing process within the human body were identified by Constantine Hering (1800–1880), who became known as the father of homeopathy in the United States and who was also the founder of the first schools and hospitals in which homeopathy was taught throughout the country. He established the laws of cure, which are often referred to as "Hering's laws" (see **Exhibit 23-2**). These were based on his observations of how healing occurs.

According to Hering's laws, a person's health seems to get worse before it improves. In those cases in which the individuals may experience an initial worsening of their symptoms, their status will be followed by improvement and relief. This so-called worsening is often known as a healing crisis that signals the body's increased activity toward healing and usually passes quickly. The homeopathic practitioner refers to this worsening of symptoms as an aggravation. It is viewed as a sign that the remedy is working and has affected the vital force.

> **KEY POINT**
>
> The basic premise underlying the science of homeopathy is that the body's own healing process is activated to cure illnesses naturally. Specifically, the remedies stimulate and increase the vital force (often referred to as the life force) and restore balance to it, facilitating the body's innate healing ability. The vital force is the energy responsible for the health status of the body and for coordinating its defense against illness. If this vital force is disturbed because of environmental factors, lifestyle, inadequate nutrition, or lack of exercise, illness or unexpected, undesirable symptoms can occur.

EXHIBIT 23-2   HERING'S LAWS OF CURE

Healing takes place from the top to the bottom. Any symptom or ailment associated with the head area heals first, before the symptoms in the feet. Likewise, the symptoms disappear first from the shoulders, then the elbow, and then down the arm.

Healing takes place from inside to outside. This law refers to the fact that the symptoms will be relieved from the more centrally located organs before the extremities are improved.

Healing occurs from the most important organs to the least important organs. The symptoms move from the major or vital organs to the less vital or minor ones. Symptoms associated with any conditions of the heart will disappear first before symptoms associated with the intestines.

Symptoms disappear in reverse order of their appearance, with emotions improving first and then the physical symptoms. An example of this law is the instance in which a person who has been struggling with chronic fatigue for several months develops flu or a cold. The symptoms related to the flu or cold will disappear before the chronic fatigue symptoms clear. The healing process occurs in the reverse order to the onset of the symptoms.

In essence, it is essential that a person who uses homeopathic remedies understands the action of the remedies. The homeopathic effects on the body are different than what one expects from the use of traditional medicines, especially prescription drugs.

Remedies of healing compounds are made through a process of serial dilution. One single drop of a plant substance is mixed with 100 drops of water and shaken. This mixture is called C. One drop of that solution is then mixed with another 100 drops and shaken. This is repeated, and the number of repetitions indicates the number placed before the C. For instance, the process repeated 30 times would be called a 30C dilution. During this process, each dilution actually becomes higher in potency.

Oral homeopathic remedies are available in several forms. The most common ones are pellets (most preferably those formed as tiny beads), tablets

KEY POINT

The more dilute a homeopathic remedy is, the higher is its potency.

of soft consistency, and liquids. Some individuals prefer pellets and tablets to liquids because the dose in the tablets is predetermined and uniform. Other types of homeopathic remedies are suppositories, ointments, creams, and gels.

It is not clearly understood how homeopathic remedies work, but there are a number of theories. One theory is the hologram theory—meaning no matter how many dilutions occur, a complete essence of the substance remains. Another theory is that the water has memory—meaning the original substance leaves an imprint of itself on the water molecules.

Homeopaths believe everyone expresses illness and heals in unique ways. This differs greatly from conventional Western medicine, which lumps symptoms into categories and believes that everyone with the same disease can be treated in the same way.

Diagnostic methods in homeopathy work toward obtaining a composite picture of a person, taking into consideration the physical, emotional, and mental aspects. A person is encouraged to tell his or her story while the homeopath observes many things, including dress, posture, tone of voice, and rate of speech. Homeopathic practitioners refer clients to conventional medical physicians for drugs and surgery when appropriate. The focus is on how a person is expressing a particular problem or condition so that an individualized plan can be made for treatment. Homeopathic practitioners will determine the remedy that can most closely mimic the sick person's pattern of symptoms.

REFLECTION

Do the laws of cure make sense to you? Can you think of illnesses you experienced that followed these laws?

There are two types of homeopathic approaches for treatment—classical and nonclassical. A classical homeopath will prescribe a single remedy for a specific problem after determining a symptom picture and matching it to an individual's constitution. The nonclassical approach would be trying to match a symptom to a remedy without having the detailed individualized information about the person with the symptom(s).

What It Helps

Homeopathic remedies benefit those with arthritis, pain, anxiety, muscular aches and pains, asthma, sinusitis, allergies, headaches, acute infection, skin disorders, circulatory disorders, infant and childhood illnesses, digestive problems, endocrine imbalances, and cardiovascular problems, as well as pregnant and lactating women.

Words of Wisdom/Cautions

There is the potential for using remedies to treat serious conditions instead of seeking appropriate conventional care. This is especially true if someone is trying to self-treat an illness.

Because symptoms can get worse before they get better, it may be hard to determine whether a remedy is working or whether a side effect requiring immediate Western medical attention is occurring.

Dosages are different from conventional Western medicines; therefore, it would be better to seek the advice of a qualified homeopath rather than using an over-the-counter remedy.

Homeopathy can be used for minor acute care, but it is best to become informed by seeking a group study program instead of trying over-the-counter remedies or reading related literature.

Guidelines for using homeopathic remedies are provided in **Exhibit 23-3**.

Ayurvedic Medicine

What It Is

Ayurveda is a sophisticated ancient healing system derived from Hindu and Indian culture and practiced in India for 4,000 years. It is a way of life, a philosophy of living, which supports the belief in the interconnectedness of mind, body, and spirit and of the individual to the environment. It teaches and emphasizes individual responsibility in becoming an active participant in maintaining healthy body systems. The focus is on prevention and regaining good health.

Ayurveda views individuals as composed of five elements: earth, water, fire, air, and space. Ayurvedic philosophy holds that there are three basic operating principles or doshas that govern the function of health. It is believed that a person is born with a particular dosha (body type) or combination of doshas and that the basic constitution is expressed through this body type. There is also a belief that the mind has a powerful influence on the body; therefore, a major role for a person is to become aware of the positive and negative thought patterns that support or destroy health.

> **KEY POINT**
>
> Ayurveda means science of life. In Sanskrit, ayur means life, and veda means knowledge.

EXHIBIT 23-3   GUIDELINES FOR USING HOMEOPATHIC REMEDIES

Whenever a person takes homeopathic remedies, certain guidelines must be followed for effectiveness of these natural products in alleviating symptoms and various health conditions.

STORING THE REMEDIES

Improper storage may interfere with the effectiveness of homeopathic substances, so it is strongly recommended that you store them as outlined here. If these remedies are properly stored and handled, they maintain their strength for years.

- Keep homeopathic remedies in their original containers. Avoid transferring the remedies to another bottle as this helps prevent any contamination.
- Secure the container tops of the remedies tightly; this prevents moisture from forming.
- Keep homeopathic remedies in a cool, dry, dark place away from humidity and out of direct sunlight and extremes of temperature (e.g., higher than 100°F). These factors may cause the remedies to lose their potency.
- Store the remedies away from any strong, pungent-smelling substances, such as perfume, camphor products, mothballs, and strong aromatic compounds found in various products, including mint foods and aromatic oils. It is believed that all of these substances act as an antidote and interfere with the effectiveness of the homeopathic therapy.
- Store homeopathic remedies out of the reach of children.

HANDLING THE REMEDIES

Careful handling of homeopathic remedies, particularly those in liquid or tablet form, is very important.

- The homeopathic remedies (pellets and tablets) or the bottle dropper of the liquid remedies should be handled as little as possible because this is a source of contamination that may reduce the effectiveness of the product.
- When you pour the remedy, gently tip the pellets or tablets into the lid of the bottle or onto a clean, dry teaspoon. If any of the tablets have fallen out onto the floor or anywhere else or are unused, do not put the tablets back into the container as this will contaminate the stock. Discard the tablets.

ADMINISTERING THE REMEDIES

Following specific recommendations for taking homeopathic remedies will provide the best results of these natural-healing substances.

- The environment of the mouth should be in its natural condition. That is, the homeopathic remedies should not be taken within 20 minutes before or after eating, brushing one's teeth, or drinking anything other than water. It is important to take the solid dosage forms of homeopathic remedies when

---

EXHIBIT 23-3 GUIDELINES FOR USING HOMEOPATHIC REMEDIES (CONTINUED)

---

the mouth is empty and clear of any food or beverages because these other substances in the mouth may decrease the absorption and effectiveness of the remedy.

· When using homeopathic remedies, it is best to eliminate the use of certain spicy foods and strong-smelling foods, including garlic, any form of caffeine, camphor, mints, toiletries, and medications (aspirin, laxatives, etc.) during this time, as these substances may counteract the effects of the homeopathic remedies.

· The use of tobacco should be avoided. Nicotine may alter the response of the body to homeopathic remedies.

· Read the label on the specific homeopathic remedy. Pay attention to what is stated. Note the ingredients and amounts.

· Follow the dosage instructions on the container or as instructed by your healthcare practitioner.

· Take the homeopathic dose on an empty stomach. This is essential because gastric juices and digestive processes can destroy or inactivate the remedy.

· Place the solid forms (pellets or tablets) under the tongue (sublingually) and dissolve gradually without chewing or any tongue movement. The remedies are absorbed into the buccal lining of the mouth. This, in turn, allows the remedies to go directly to the point of action, bypassing the stomach, intestines, and liver.

· Place the drops (liquid) under the tongue in amounts ranging from 1 to 10 drops. (Follow the dosage on the label or use as suggested by the healthcare practitioner.)

· A general rule would be to start with 5 drops and then after 1 week increase to 10 drops; however, there is an exception to this rule. In the case of sensitivities, the dosage amount is reduced. The healthcare practitioner may recommend starting with 1 or 2 drops and increasing the dosage by a drop each week until 10 drops is reached.

· Finally, the homeopathic remedies should be taken only for as long as one needs them. As soon as positive results are observed, the homeopathic remedies should be discontinued.

How It Works

By assessing an individual's physical and emotional makeup, food and environmental preferences, and lifestyle, a particular dosha or body type is determined. Knowing the specific elements that make up this dosha along with basic patterns, 24-hour cycles, seasons, and stages of life associated with each of the doshas, a particular health plan is developed by the practitioner.

Some diagnostic methods used are detailed history taking through questioning (family, interpersonal relationships, and job situation), pulse diagnosis, tongue diagnosis, and other observational skills (eyes, nails, and urine).

A treatment plan's goal is to help a person arrive at a lifestyle that results in a balance of body and mind for optimum health. Treatment suggestions may include a combination of the following:

- Nutritional suggestions, including the six tastes that are important to include in every meal (sweet, sour, salty, pungent, bitter, and astringent)
- Herbs (which are classified according to the six tastes)
- Exercise (specific for each dosha/body type)
- Breathing exercises (called pranayama)
- Meditation exercises (to help develop moment-to-moment awareness and cleansing of the body)
- Massage (marma therapy)
- Aromatherapy (to help balance body functions, emotions, and memories)
- Music (certain tones or rhythms to be used at certain times of the day)
- Purification technique called panchakarma (that includes five procedures or therapies to be experienced over the course of 1 week)

What It Helps

In addition to being appropriate for a wide range of physical and emotional illnesses, ayurveda benefits anyone who is interested in optimum health.

Words of Wisdom/Cautions

Physical side effects are rare, although there may be occasional side effects from certain ayurvedic herbs. It is always best to seek advice from a practitioner who is experienced or who has been educated in the ayurvedic principles.

Naturopathy

What It Is

Naturopathic medicine grew out of the 19th century medical system and was given its name by Dr. John Scheel in 1895 and was formalized by Benedict Lust in 1902. It was popular in the early 1900s, but with the development of antibiotics and vaccines in the 1940s and 1950s, that popularity declined. It was not until the 1970s that interest was renewed.

Naturopathy is a way of life, viewing health as more than the absence of disease. Naturopaths look at symptoms as signs of the body eliminating toxins and believe that a person should be treated as a whole, looking at physical, psychological, emotional, and genetic factors. A naturopath's emphasis is on finding and treating the cause (not just symptoms), self-responsibility, education, health maintenance, and disease prevention. The basic belief is that the body has the ability to heal itself and an innate ability to maintain health, and thus, any treatment focuses on a combination of natural healing methods that strengthen the body's natural abilities. Techniques and approaches that do no harm and support and restore harmony within the body are used.

A practice is generally built around two or more therapeutic approaches such as traditional Chinese medicine or acupuncture, clinical nutrition, counseling, dietary and lifestyle modifications, exercise, herbal medicine, homeopathy, hydrotherapy, or osteopathy.

Training for a doctor of naturopathic medicine (ND) degree varies from a 4-year graduate level education within a naturopathic medical college to training from a correspondence school.

How It Works

As the primary role of a naturopath is as an educator, recommendations and encouragement for self-responsibility for health are offered, and a person seeking naturopathic help must make a commitment to change. The first visit involves a medical history with detailed assessment of lifestyle habits, diet, occupation, family dynamics, emotional state, and environmental and genetic influences. Diagnostic procedures, such as laboratory testing and X-rays may be used and if specialized care, surgery, or hospitalization is warranted, referral is made to other healthcare professionals as appropriate. Naturopaths prefer the least invasive intervention and do not do emergency care, although some may practice natural childbirth. Naturopaths pay attention to a person's individuality and susceptibility to disease.

What It Helps

There is unlimited potential for most health conditions to respond to this type of approach.

Words of Wisdom/Cautions

Natural therapies are less likely to cause complications, but choose the practitioner who knows his or her limitations with any of the approaches practiced.

Summary

Traditional Chinese medicine, acupuncture, homeopathy, ayurveda, and naturopathy are considered alternative medical systems because they fall outside the scope of medical practice as it is known in the United States. However, these systems have a long history and are used successfully and commonly in other parts of the world.

Traditional Chinese medicine approaches the person as a whole and sees mind, body, and spirit as interrelated elements that are intertwined with nature and the universe. Acupuncture, part of traditional Chinese medicine, is based on the theory that disease or illness results from the blockage or obstructions of the energy pathways (meridians) and by clearing the blockages allow qi to run freely through the meridians and health can be restored. Homeopathy is a system of medical practice that is based on the principle that like cures like using remedies of healing compounds made through a process of serial dilution. Ayurveda views individuals as composed of the five elements of earth, water, fire, air, and space, and is based on the belief that there are three basic doshas (body types) that influence health. Naturopathy holds the basic belief that the body has the ability to heal itself and an innate ability to maintain health, and naturopaths promote treatments that focus on a combination of natural healing methods that strengthen the body's natural abilities.

Any of these methods could hold the potential for benefit health and healing. It is a sound principle to seek medical advice to identify health conditions so that the type of health issue being addressed is thoroughly understood prior to using these approaches. Ideally, these systems can be integrated with conventional practices to use the best of both worlds.

Suggested Reading

Bing, Z., & Hongcai, W. (2010). *Basic theories of traditional Chinese medicine*. Philadelphia, PA: People's Military Medical Press.

Cummings, S., & Ullman, D. (2004). *Everybody's guide to homeopathic medicines*. New York, NY: Penguin.

Hammond, C. (1995). *The complete family guide to homeopathy*. New York, NY: Penguin Studio.

Herhoff, A. (2000). *Homeopathic remedies*. Garden City, NY: Avery Publishing Group.

Laxarides, L. (2010). *A textbook of modern naturopathy*. London, UK: BCM Waterfall.

Lockie, A. (1993). *The family guide to homeopathy: Symptoms and natural solutions*. New York, NY: Fireside.

Lockie, A. (2001). *Homeopathy handbook*. New York, NY: Dorling Kindersley.

McCabe, V. (2000). *Practical homeopathy*. New York, NY: St. Martin's Griffin.

Monte, T. (1997). *The complete guide to natural healing*. New York, NY: Perigee Book.

Ninivaggi, J. (2010). *Ayurveda: A comprehensive guide to traditional Indian medicine for the West*. Westport, CT: Praeger Publishing.

Reichenberg-Ullman, J. (1994). *The patients' guide to homeopathic medicine*. Edmonds, WA: Picnic Point Press.

Reichenberg-Ullman, J. (2000). *Whole woman homeopathy*. Roseville, CA: Prima Health.

Reiter, R. (2003). *Healing without medication*. North Bergen, NJ: Basic Health Publications, Inc.

Schmukler, A. (2006). *Homeopathy: An A to Z home handbook*. Woodbury, MN: Llewellyn Press.

Shalts, E. (2006). *Easy homeopathy*. New York, NY: McGraw Hill.

Skinner, S. (2001). *An introduction to homeopathic medicine in primary care*. Gaithersburg, MD: Aspen Publications.

Sollars, D. (2001). *The complete idiot's guide to homeopathy*. Indianapolis, IN: Alpha Books.

Strovier, A. L., & Carpenter, J. E. (2006). *Introduction to alternative and complementary therapies*. Philadelphia, PA: Haworth Press.

Trivieri, L., & Anderson, J. W. (Eds.). (2002). *Alternative medicine: The definitive guide* (2nd ed.). Berkeley, CA: Celestial Arts.

Ullman, D. (2002). *The consumer's guide to homeopathy*. New York, NY: Jeremy P. Tarcher/Putnam.

Wardie, J. (2010). *Clinical naturopathy: An evidence-based guide to practice*. Chatswood, Australia: Elsevier Press.

Welch, C. (2011). *Balance your hormones, balance your life: Achieving optimal health and wellness through ayurveda, Chinese medicine, and Western science*. Cambridge, MA: Da Capo Press.

Young, J. (2007). *Complementary medicine for dummies*. New York, NY: IDG Books Worldwide, Inc.

Resources

Acupuncture

Acupuncture and Oriental Medicine Alliance
6405 43rd Avenue Ct., NW, Suite B
Greg Harbor, WA 98335
253-851-6896
www.aomalliance.org

American Academy of Medical Acupuncture
4929 Wilshire Boulevard., Suite 428
Los Angeles, CA 90036
800-721-2177
www.medicalacupuncture.org

American Association of Acupuncture and Oriental Medicine
5530 Wisconsin Avenue, Suite 1210
Chevy Chase, MD 20815
888-500-7999
www.aaaomonline.org/

Ayurvedic Medicine

Ayurvedic Institute
11311 Menaul NE, Suite A
Albuquerque, NM 87112
505-291-9698
www.ayurveda.com

Homeopathy

National Center for Homeopathy
801 N. Fairfax Street, Suite 306
Alexandria, VA 22314
703-548-7790
www.homeopathic.org

North American Society of Homeopathy
1122 E. Pike Street, Suite 1122
Seattle, WA 98122
206-720-7000
www.homeopathy.org

Naturopathy

American Association of Naturopathic Physicians
3201 New Mexico Avenue NW, Suite 350
Washington, DC 20016
866-538-2267
www.naturopathic.org

Nutritional Supplements

OBJECTIVES

This chapter should enable you to

- Describe the difference between recommended daily allowances and dietary reference intakes
- Discuss the development of the nutritional supplement industry
- List factors to consider in assessing the need for supplements
- Describe facts that are listed on supplement labels
- List the antioxidants and their sources
- Define the term *phytochemical*
- Discuss health conditions for which nutritional supplements can be beneficial

---

The importance of proper and sensible nutrition, as stated in a previous chapter, cannot be emphasized enough. In its *Healthy People 2010* report, the U.S. government's Office of Disease Prevention and Health Promotion states that what people eat, especially when they exercise regularly and do not smoke and/or drink excessively, is the most significant controllable risk factor affecting their long-term health (U.S. DHHS, n.d.). Basic dietary principles supporting optimum health included the following:

- Eating a wide variety of foods that provide adequate nutrients, including plenty of fresh fruits and vegetables, complex carbohydrates, plant protein, and fiber.
- Keeping consumption of sugary foods, caffeinated beverages, and alcohol to a minimum.
- Eating only a small amount of animal protein and fat.
- Drinking plenty of water.

It was once thought that anyone who followed these principles was considered properly nourished and did not need supplementation; however, the dietary habits of most Americans lead them to be overweight and undernourished.

Views on Nutritional Supplements

There has been much debate regarding vitamins and minerals as nutritional supplements and how (or if) they should be taken daily. The early 1900s focused largely on deficiencies of vitamins and minerals that caused diseases such as scurvy and beriberi. Vitamins began to be added to food—a process called fortification. Through the 1950s and 1960s, a number of foods were fortified, such as breads, cereals, and milk. The addition of calcium has been among the recent fortification of many food products.

By the mid-1970s, however, the focus was less on vitamin deficiency and more on the value of vitamin and mineral supplementation for the prevention of illness and disease. Vitamin C was thought to prevent and alleviate symptoms of the common cold. Vitamin E was said to help keep the heart healthy. A low-fat, high-fiber diet was the order of the decade. The late 1970s saw the response of the U.S. Senate Select Committee on Nutrition that listed the number one public health problem in this country as poor nutrition. Experts believed Americans consumed too much food of too little nutritive value and that this was a contributing factor to poor quality of life and increased disease (U.S. Senate, 1977).

KEY POINT

Over the past decade, there has been an explosion in the use of nutritional supplements, which has caused the government to begin to regulate the supplement industry.

How the Government Is Involved

Recommended Dietary Allowances

Since the 1940s, the Food and Nutrition Board of the National Academy of Sciences has made recommendations for nutrient intake. These recommendations have been termed *recommended dietary allowances (RDAs)* and represent the standards that should meet the needs of most healthy people in the United States. The RDAs address energy, protein, and most vitamins and minerals. Fats and carbohydrates standards were set in 2002 (National Academies of Science, 2002); they had never been set before that time because the experts assumed that if the average person met the energy requirements for protein then the demand for fats and carbohydrates would also be met.

There has been much confusion surrounding RDAs. Most consumers have had some misunderstanding of how to use RDAs and what food choices best meet these recommendations. The misconception exists that RDAs are *optimum* daily requirements rather than recommended *minimum* intakes as a standard for healthy individuals. The development of daily values (DV) and dietary reference intakes (DRIs) is phasing out RDAs.

KEY POINT

The RDAs reflect the minimum, not optimum, daily requirements for nutrients.

Dietary Reference Intakes

In an attempt to continue focusing on the benefits of healthy eating, dietary reference intakes (DRIs) were developed to update RDAs. These new guidelines represent the latest understanding of nutrient requirements for optimum health. The first set focused on nutrients related to bone health and fluoride (National Academies of Science, 2004a); in 1998, folate, the B vitamins, and choline were added (National Academies of Science, 1998). Vitamin C, vitamin E, selenium, and carotenoids came in 2000 (National Academies of Science, 2001); in 2001 recommendations were published for vitamin A, vitamin K, arsenic, boron, chromium, copper, iodine, iron, manganese, molybdenum, nickel, silicon, vanadium, and zinc (micronutrients) (Russell, 2001). Additional nutrients and electrolytes are anticipated to be added in the future.

The need to update and change the standards for intake of nutrients remains a challenge. The Food and Nutrition Board of the National Academy of Sciences is updating these standards on a regular basis. The newest work through these agencies is called daily values (DVs). DVs are divided into two groups—reference daily intakes (RDIs) and daily reference values (DRVs). RDIs are to be used in reference to vitamins, minerals, and proteins.

DRIs are divided into the following four subcategories:

1. Estimated average requirements (EARs) (IOM, n.d.)
2. RDAs, continued from 1989
3. Adequate intakes
4. Tolerable upper intake levels

> ### KEY POINT
>
> In 1994, an office was created within the National Institutes of Health (NIH) for overseeing research on dietary supplements through the Dietary Supplement Health and Education Act (DSHEA). This requires manufacturers of dietary supplements to include the words *dietary supplement* on product labels.

EARs would satisfy 50% of requirements for men and women for specific age groups and are intended for use by nutritional professionals. If calculations of EARs are not available, adequate intakes are used instead of RDAs. RDAs continue to be considered as sufficient amounts of nutrients to meet nearly all needs. Tolerable upper intake levels indicate the largest amount of a nutrient that someone can ingest without adverse effect (National Academies of Science, 2004b).

The Food and Drug Administration (FDA) describes acceptable claims that can be made for relationships between a nutrient and the risk of a disease or health-related condition. These claims must be clear as to the relationship of the nutrient to the disease and be understandable by the general public. The claims can be made in several ways—through third-party references (such as the National Cancer Institute), symbols (such as a heart), and vignettes or descriptions (see **Exhibit 24-1**).

The FDA does not allow claims for healing, treatment, or cure of specific medical conditions on the labels or advertisements of nutritional supplements. This would put supplements in the category of drugs. The DSHEA allows only three types of claims to be used with supplements—nutrient content, disease, and nutrition support claims. The nutrient content explains how much of a nutrient is in a supplement. Claims regarding disease must have a basis in scientific evidence and refer to health-related conditions or diseases and a particular nutrient. Nutrition support claims (which may be used without FDA approval, but not without notification to that agency) are set up to explain how a deficiency could develop if the diet was deficient in that nutrient. These claims are accompanied by an FDA disclaimer on the label of the supplement and are therefore easy to determine. In March 1999, the DSHEA required that all nutritional supplements carry a supplement facts panel (see **Figure 24-1**).

REFLECTION

What motivates you to use supplements? Do you carefully evaluate claims about them?

EXHIBIT 24-1   STATUS OF HEALTH CLAIMS

APPROVED HEALTH CLAIMS FOR DIETARY SUPPLEMENTS AND CONVENTIONAL FOODS

- Calcium and osteoporosis
- Folate and neural tube defects
- Soluble fiber from whole oats and coronary heart disease
- Soluble fiber from psyllium husks and coronary heart disease
- Sugar alcohols and dental caries

APPROVED HEALTH CLAIMS FOR CONVENTIONAL FOODS ONLY

- Dietary lipids and cancer
- Dietary saturated fat and cholesterol and coronary heart disease
- Fiber-containing grain products, fruits, and vegetables and cancer
- Fruits and vegetables and cancer (for foods that are naturally a "good source" of vitamin A, vitamin C, or dietary fiber)
- Fruits, vegetables, and grain products that contain fiber, particularly soluble fiber, and coronary heart disease
- Sodium and hypertension

HEALTH CLAIMS NOT AUTHORIZED

- Antioxidant vitamins and cancer
- Dietary fiber and cancer
- Dietary fiber and cardiovascular disease
- Omega-3 fatty acids and coronary heart disease
- Zinc and immune function in older individuals

Source: The Commission on Dietary Supplementation

A Supplement Extravaganza

Vitamins and minerals were what traditionally made up typical nutritional supplements. Today the definition of nutritional supplements is expanded to include "vitamins, minerals, herbs, botanicals, and other plant-derived substances; and amino acids, concentrates, metabolites, constituents, and extracts of these substances" (NIH, n.d.).

Consumers can easily become confused by the burgeoning of scientific studies on nutrients; the immediate release of single studies relating to nutrition, diet, and health; and supplement advertisements that use health claims

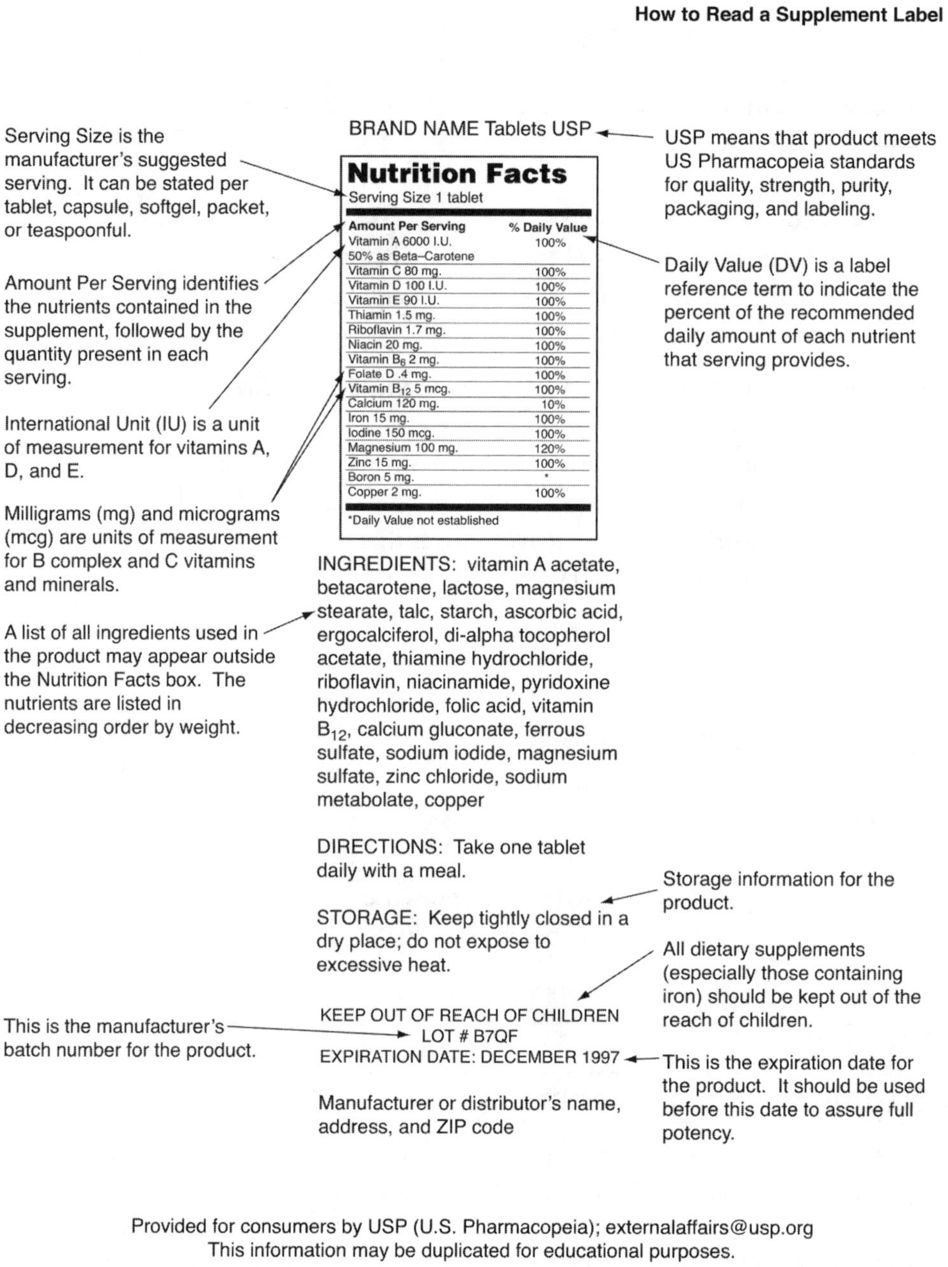

Serving Size is the manufacturer's suggested serving. It can be stated per tablet, capsule, softgel, packet, or teaspoonful.

Amount Per Serving identifies the nutrients contained in the supplement, followed by the quantity present in each serving.

International Unit (IU) is a unit of measurement for vitamins A, D, and E.

Milligrams (mg) and micrograms (mcg) are units of measurement for B complex and C vitamins and minerals.

A list of all ingredients used in the product may appear outside the Nutrition Facts box. The nutrients are listed in decreasing order by weight.

BRAND NAME Tablets USP

Nutrition Facts
Serving Size 1 tablet

Amount Per Serving	% Daily Value
Vitamin A 6000 I.U.	100%
50% as Beta–Carotene	
Vitamin C 80 mg.	100%
Vitamin D 100 I.U.	100%
Vitamin E 90 I.U.	100%
Thiamin 1.5 mg.	100%
Riboflavin 1.7 mg.	100%
Niacin 20 mg.	100%
Vitamin B_6 2 mg.	100%
Folate D .4 mg.	100%
Vitamin B_{12} 5 mcg.	100%
Calcium 120 mg.	10%
Iron 15 mg.	100%
Iodine 150 mcg.	100%
Magnesium 100 mg.	120%
Zinc 15 mg.	100%
Boron 5 mg.	*
Copper 2 mg.	100%

*Daily Value not established

INGREDIENTS: vitamin A acetate, betacarotene, lactose, magnesium stearate, talc, starch, ascorbic acid, ergocalciferol, di-alpha tocopherol acetate, thiamine hydrochloride, riboflavin, niacinamide, pyridoxine hydrochloride, folic acid, vitamin B_{12}, calcium gluconate, ferrous sulfate, sodium iodide, magnesium sulfate, zinc chloride, sodium metabolate, copper

DIRECTIONS: Take one tablet daily with a meal.

STORAGE: Keep tightly closed in a dry place; do not expose to excessive heat.

KEEP OUT OF REACH OF CHILDREN
LOT # B7QF
EXPIRATION DATE: DECEMBER 1997

Manufacturer or distributor's name, address, and ZIP code

USP means that product meets US Pharmacopeia standards for quality, strength, purity, packaging, and labeling.

Daily Value (DV) is a label reference term to indicate the percent of the recommended daily amount of each nutrient that serving provides.

Storage information for the product.

All dietary supplements (especially those containing iron) should be kept out of the reach of children.

This is the expiration date for the product. It should be used before this date to assure full potency.

This is the manufacturer's batch number for the product.

Provided for consumers by USP (U.S. Pharmacopeia); externalaffairs@usp.org
This information may be duplicated for educational purposes.

Figure 24-1   Supplement Label

Source: Provided for consumers by USP (U.S. Pharmacopeia) external affairs.

to increase sales. In many cases, one report contradicts another—what was touted as good yesterday is considered harmful today.

The new millennium was accompanied by buzzwords, such as antioxidants, phytochemicals, functional foods, and nutriceuticals. It is yet to be determined whether these new compounds deserve the onslaught of press, print, and manufacture they so readily receive. When a new study has been completed that suggests a benefit from a nutrient, it is released immediately. Americans rush to purchase the latest combination of vitamins, minerals, and nutritive supplements—spending billions of dollars on dietary supplements annually. Are the supplements worth taking, and are the claims made by manufacturers true? Nutritional supplements may be helpful, but they may also be harmful. Taking supplements without knowledge of their actions and interactions could lead to imbalances of other nutrients and, potentially, toxicity; however, if taken properly and with forethought, supplements can increase general health and ward off some diseases.

When making the decision to take nutritional supplements, evaluation of the information available should be done in a carefully planned manner as part of a total nutrition program. If people hear of studies that seem to relate to their circumstances, it is important for them to do some investigating on their own. Studies need to be assessed in the context they were performed and their relationship to a person's particular situation. Health professionals can assist individuals in understanding what studies reveal. In any case, taking a supplement based on what has been read or heard in the news does not guarantee that a given person will have the same outcome.

The nutritional supplement industry is still in its infancy and grows exponentially each day. There continues to be an increasing variety of supplements readily available, which can be bought over the Internet, by mail order, at supermarkets, in drug stores, and in other types of stores. Supplements are no longer the domain of natural food stores. The FDA regulates and oversees manufacturing, product information, and safety; and the Federal Trade Commission regulates advertising of supplements; however, decisions regarding whether supplements should be taken, in what form, and how much, still remain a personal choice.

There is increasing interest among Americans to use nutritional supplements for optimum health. Some experts are also indicating a definite need for supplementation as part of a nutritional plan. Relying solely on diet to meet the body's needs for vitamins and minerals is fast becoming a thing of the past. We no longer eat food picked fresh from the local garden and cooked within hours. Additionally, many food preparation practices decrease the nutritive value of food.

Vitamins and minerals play an important role in optimum health and in reducing the risk of chronic diseases. Is the diet a person is consuming, however, giving him or her enough nutrients and meeting individual needs? If so, then all that may be needed is a good quality multivitamin and mineral; however, if a person has poor nutritional habits, is under a great deal of stress, is pregnant or planning a pregnancy, or has a health problem, supplementation is a must. The supplements that are needed, how much of each, and for how long they need to be taken then becomes the issue. The research is promising, but for many nutrients, the results are still not definite.

KEY POINT

Supplements can never substitute for a healthy, sensible nutrition program where whole foods, containing hundreds of substances, working together synergistically, are consumed. Poor food choices and eating habits cannot be banished by supplementation.

Age, nutritional lifestyle, quality and quantity of food, gender, life stage, environment, family history, personal history, diet, exercise, and rest patterns should be considered in determining a person's need for nutritional supplementation.

As with all holistic approaches, knowledge, self-care, balance, and using what nature provides are the keys. Individual nutritional needs can be determined by looking at the responses made on your nutritional lifestyle survey (earlier in this book) and by considering several factors, which include the following:

Age. The need for nutrient supplementation increases with age. There are some nutrients that are not absorbed as well as a person ages, even if they are consumed in good quantity. Older adults may be affected by poor lifelong nutrition habits, social isolation, and chronic diseases that require special diets or affect food intake. Children need a balanced diet, along with a good quality multivitamin and limited sweets and fats to promote their growth and development.

Chronic health problems. Some chronic conditions create special nutritional needs, whereas others produce symptoms that threaten a healthy nutritional status. If a person has a chronic health problem, is taking medication, and wants to take nutritional supplements, a thorough knowledge of drug–supplement interactions is necessary. Nurses and other health practitioners can help in this area.

Female issues. Women who are pregnant, planning to become pregnant, breastfeeding, menopausal, or postmenopausal need added nutrients and/or supplementation. Different supplements are needed for different life cycles. Within the past decade, the National Academy of Sciences has increased DRIs for folate and calcium for women (Barr, Murphy, & Poos, 2002; Eldridge, 2004). These are of great import for women.

Lifestyle choices. Cigarette and alcohol use reduces the levels of certain nutrients and predisposes the body to diseases for which added protection is helpful. A stressful lifestyle can deplete nutrients. Deficiencies also can arise from the high consumption of caffeine and sugar.

If people presently take multivitamin/mineral supplements and believe that they may benefit from taking additional supplements, they should first examine the labels of their food and supplements. They need to check the DVs of each vitamin and mineral listed. They should be receiving at least 100% of all nutrients from all food and supplements. If they are lower than 100%, they first need to determine whether they can meet this need through a change in diet. If they cannot meet the requirements through diet, then supplements of specific nutrients may be necessary. If the DV is over 100% for some or all nutrients, they may want to cut back on that supplement. The guideline is to keep below 300% DV of any nutrient. A nutritional analysis can be used, also. A registered dietician, nutritionist, or nutrition-knowledgeable health-care professional can help with an analysis.

KEY POINT

When using supplements, always read the labels. Know what you are taking, how easily it is absorbed, if it has United States Pharmacopeia (USP) initials, and when it expires; and by all means, if you have any adverse effects—stop taking it.

Supplement Labels

Supplements come in many forms—tablets, capsules, softgels, powders, and liquids. The USP is the agency that oversees drug products and sets the standards for dietary supplements. All supplements now require labels, which are regulated by the USP. The USP has certain standards that must be met for single vitamins and those in combination, as well as dietary supplements, and botanical and herbal preparations. Figure 24-1 shows the information that is contained on a supplement label.

One facet of the dietary supplement label that is important to read is the DV column. The DV is the percentage of the recommended daily amount of the nutrient that a serving gives. According to the USP, intake should be between 50% and 100% of the DV of each nutrient. One nutrient that would not provide 100% of the DV is calcium. If 100% of calcium were added to any supplement, it would be too big to swallow. Calcium should be taken in divided doses throughout the day.

The bioavailability of a particular nutrient is the amount of a nutrient that enters the bloodstream and actually reaches the various organs, tissues, and cells of the body. Nutrients have greater bioavailability when they are taken with compounds that help their absorption.

Fruits and vegetables contain many compounds that, when eaten, allow for synergy to take place in the body, increasing bioavailability. Fruits and vegetables are especially affected by mode of storage and preparation. Those that have been exposed to heat, light, or air have lost some or all of their nutrients; their bioavailability is lowered.

Two important factors when considering the bioavailability of a supplement are ease of absorption and the benefit of taking it in combination with another supplement. Disintegration (how quickly a tablet/supplement breaks apart) and dissolution (how fast the supplement dissolves in the intestinal tract) (U.S. Pharmacopeia, 2008) are two additional factors directly related to bioavailability. Vitamin C assists with the absorption of iron into the body. If taking iron as a supplement, people also should consider their vitamin C intake. Taking an iron supplement with a glass of orange juice would increase the bioavailability of the iron. On the other hand, calcium inhibits iron and magnesium absorption. This is important to remember when adding supplements to a nutritional plan, especially if iron, calcium, and vitamin C individually or in a multivitamin are being taken. More research is needed to determine and document supplement interactions. The NIH issued a call for research, mandated by Congress, "to explore the current state of our knowledge about the important issues related to bioavailability of nutrients and other bioactive components from dietary supplements" (Scalli, Zaid, & Soulamaymani, 2007, p.680).

Birth Defects

One of the first vitamins addressed by the advisory committee on dietary supplements was folic acid. The original aim was to reduce neural tube birth defects, especially spina bifida. The RDA for folic acid was doubled from 200 to 400 mcg/day. The Centers for Disease Control, the USP, the FDA, and the March of Dimes have all recommended that women of childbearing age consume 400 mcg per day of folic acid, either through diet or supplements (CDC, 2002) (see **Figure 24-2**). Women who are planning to become pregnant and who are of childbearing age should eat a varied diet and also take folic acid through supplementation. Sufficient folate (folic acid) is critical from conception through the first four to six weeks of pregnancy when the neural tube is formed. This means adequate diet and supplement use should begin well before pregnancy occurs.

Antioxidants

Antioxidants are compounds that naturally protect the body from free radicals and help to depress the effects of metabolic by-products that cause degenerative changes related to aging. Mounting evidence shows that antioxidants play a role in preventing or delaying the onset of diseases such as cancer, stroke, arthritis, heart disease, immune problems, and neurological problems (Rahman, 2007).

Free radicals have a helpful function in the body, but in higher levels, they can damage cells and tissues. They are produced by the body's own metabolism and are generated from exposure to environmental factors and toxins. Antioxidant nutrients are said to neutralize the harmful free radicals that occur in the body constantly and arise from improper nutrition, eating fatty foods; smoking; drinking alcohol; taking drugs; and exposure to environmental pollutants (such as herbicides and pesticides), toxins, carcinogens, iron, smog, and radiation.

KEY POINT

The neural tube is a type of membrane that grows into the spinal cord and brain in utero (during pregnancy). Neural tube defects are problems in the development of the brain and spinal cord that arise during pregnancy. This is now believed to be a result of folate deficiency of the mother in the weeks before and in the early weeks of pregnancy. It is estimated that nearly 50% of all neural tube defects that occur can be arrested by adequate folate intake.

Folic acid is a B vitamin that everyone needs to help cells grow and divide. It is especially important for women who may become pregnant.

Why is Folic Acid So Important for Women?

Taken before and during early pregnancy, folic acid, also known as folate, reduces the chances of having a baby born with birth defects of the spine and brain (spina bifida and anencephaly). If you have already had a baby with spina bifida or anencephaly, see your doctor if you are considering another pregnancy.

For women of childbearing years USP, the CDC, the FDA, and the March of Dimes all recommend 400 mcg of folic acid daily.

You should not take more than 1000 mcg daily unless your healthcare professional tells you to do so.

How Can I Get an Adequate Amount of Folic Acid?

It is possible to get folic acid by eating foods such as green leafy vegetables; cereal and cereal products; and citrus fruits and juices. Certain fully fortified breakfast cereals have 100% of the recommended daily amount of folic acid. Many women, however, prefer to take a multivitamin supplement.

What About a Dietary Supplement?

Folic acid may be taken as a tablet or as part of a multivitamin supplement. Make certain that the letters "USP" appear on the label to ensure that your vitamin/mineral product meets established standards for strength, quality, and purity.

Where Can I Find Additional Information About Pregnancy and Folic Acid?

Centers for Disease Control and Prevention
Mail Stop F 45
4770 Buford Highway N.E.
Atlanta, GA 30341-3724
770/488-7160
e-mail: pgm5@cdc.gov

Food and Drug Administration
HFE-88
5600 Fishers Lane
Rockville, MD 20857
800/332-4010
http://www.fda.gov

March of Dimes
1275 Mamaroneck Avenue
White Plains, NY 10605
888/663-4637
e-mail: resourcecenter@modimes.org

Spina Bifida Association of America
4590 MacArthur Blvd.
Suite 250
Washington, D.C. 20007
202/944-3285
e-mail: sbaa@sbaa.org

United States Pharmacopeia
12601 Twinbrook Parkway
Rockville, MD 20852
301/816-8223
e-mail: externalaffairs@usp.org

Ask your healthcare professional to help you determine your need for folic acid supplements.

Provided for consumers by USP (U.S. Pharmacopeia); externalaffairs@usp.org
This information may be duplicated for educational purposes.

Figure 24-2   Folic Acid
Source: Provided for consumers by USP (U.S. Pharmacopeia) external affairs.

The body has natural antioxidant enzymes that regulate the effects of free radicals. These enzymes are catalase, superoxide dismutase, and glutathione peroxidase. Vitamins A (as beta-carotene), C, and E and selenium assist the enzymes in the body to fight free radical damage.

Benefits Versus Risks of Antioxidant Supplementation

A diet low in fats, sweets, and animal protein that includes at least five fruits and vegetables per day, in variety, is a much safer way to obtain antioxidant protection than through supplementation. A greater chance of imbalances and toxicity exists when supplements are being used (**Tables 24-1** and **24-2** show guidelines for information on vitamins and minerals); however, there continues to be increased evidence supporting antioxidant supplements as part of nutritional lifestyle.

Vitamin A

Vitamin A is necessary for good immune function, tissue repair, healthy skin and hair, bone formation, and vision. Fat-soluble vitamin A can cause toxicity and even be fatal in amounts higher than 10,000 IUs per day. There may be greater risk of birth defects for babies whose mothers take preformed vitamin A during pregnancy—especially during the first trimester (Strobel, Tinz, & Bielsalski, 2007).

Beta-carotene, one of the family of carotenoids and a precursor to vitamin A, is a natural antioxidant that enhances the immune system and may protect against certain cancers, cataracts, and heart disease. Beta-carotene is converted in the intestines and liver into preformed vitamin A; its sources are bright, orange-yellow fruits and vegetables.

Lycopene is another carotenoid and powerful antioxidant shown to reduce the risk of prostate and other cancers. The sources are fruits and vegetables of deep red to pink color. Some orange fruits and vegetables, some green leafy vegetables, and broccoli contain lutein, another carotenoid that has been found to arrest the development of macular degeneration and help protect the eyes from other diseases (VERIS, 2002).

Vitamin C

Vitamin C is a popular supplement often consumed in large amounts. Recent research may be causing consumers of this supplement to rethink their intake. Between 100 and 200 mg of vitamin C is considered an optimum dose (Levine, Wang, Padayatty, & Morrow, 2001). High doses can easily upset bowel function, causing diarrhea. The dose for each person is highly individualized. Dr. Andrew Weil suggests a trial-and-error method for finding your optimum dose, which he believes to be 3,000 to 6,000 mg per day (Weil, 2006, p. 57). It is easy to have a 500-mg intake from eating at least six fruits and vegetables a day.

TABLE 24-1 VITAMINS

Vitamin	Deficiency May Cause	How It Works	Excess May Cause	Best Foods to Eat
A	Night blindness; skin problems; dry, inflamed eyes	Bone and teeth growth; vision; keeps cells, skin, and tissues working properly	Birth defects, bone fragility, vision and liver problems	Eggs, dark green and deep orange fruits and vegetables, liver, whole milk
B-1	Tiredness, weakness, loss of appetite, emotional upset, nerve damage to legs (late sign)	Helps release food nutrients and energy, appetite control, helps nervous system and digestive tract	Headache, rapid pulse, irritability, trembling, insomnia, interference with B2, B6	Whole grains and enriched breads, cereals, dried beans, pork, most vegetables, nuts, peas
B-2 Riboflavin	Cracks at corners of mouth, sensitivity to light, eye problems, inflamed mouth	Helps enzymes in releasing energy from cells, promotes growth, cell oxidation	No known toxic effect Some antibiotics can interfere with B2 being absorbed	whole grains, enriched bread and cereals, leafy green vegetables, dairy, eggs, yogurt
Niacin	General fatigue, digestive disorders, irritability, loss of appetite, skin disorders	Fat, carbohydrate, and protein metabolism; good skin, tongue, and digestive system; circulation	Flushing, stomach pain, nausea, eye damage, can lead to heart and liver damage	Whole wheat, poultry, milk, cheese, nuts, potatoes, tuna, eggs
B-6 Pyridoxine	Dermatitis, weakness, convulsions in infants, insomnia, poor immune response, sore tongue, confusion, irritability	Necessary protein metabolism, nervous system functions, formation of red blood cells, immune system function	Reversible nerve injury, difficulty walking, numbness, impaired senses	Wheat and rice bran, fish, lean meats, whole grains, sunflower seeds, corn, spinach, bananas
Biotin	Rarely seen since it can be made in body if not consumed. Flaky skin, loss of appetite, nausea.	Cofactor with enzymes for metabolism of macronutrients, formation of fatty acids, helps other B vitamins be utilized	Symptoms similar to vitamin B_1 overdose	Egg yolks, organ meats, vegetables, fish, nuts, seeds, also made in intestines by normal bacteria there
Folic Acid	Anemia, diarrhea, digestive upset, bleeding gums,	Red blood cell formation, healthy pregnancy, metabolism of proteins	Excessive intake can mask B_{12} deficiency and interfere with zinc absorption	Green leafy vegetables, organ meats, dried beans

TABLE 24-1   VITAMINS (CONTINUED)

B$_{12}$ Cobalamin	Elderly, vegetarians, or those with malabsorption disorder are at risk of deficiency-pernicious anemia, nerve damage	Necessary to form blood cells, proper nerve function, metabolism of carbohydrates and fats, builds genetic material	None known except those born with defect to absorb	Liver, salmon, fish, lean meats, milk, all animal products
Pantothenic Acid	Not usually seen; vomiting, cramps, diarrhea, fatigue, tingling hands and feet, difficult coordination	Needed for many processes in the body, converts nutrients into energy, formation of some fats, vitamin utilization, making hormones	Rare	Lean meats, whole grains, legumes
C Ascorbic Acid	Bleeding gums, slow healing, poor immune response, aching joints, nose bleeds, anemia	Helps heal wounds, collagen maintenance, resistance to infection, formation of brain chemicals	Diarrhea, kidney stones, blood problems, urinary problems	Most fruits, especially citrus fruits, melon, berries, and vegetables
D	Poor bone growth, rickets, osteoporosis, bone softening, muscle twitches	Calcium and phosphorus, metabolism and absorption, bone and teeth formation	Headache, fragile bones, high blood pressure, increased cholesterol, calcium deposits	Egg yolks, organ meats, fortified milk, also made in skin when exposed to sun
E	Not usually seen, after prolonged impairment of fat absorption, neurological abnormalities	Maintains cell membranes, assists as antioxidant, red blood cell formation		Vegetable oils and margarine, wheat germ, nuts, dark green vegetables, whole grains
K	Tendency to hemorrhage, liver damage	Needed for prothrombin, blood clotting, works with Vitamin D in bone growth	Jaundice (yellow skin) with synthetic form, flushing & sweating	Green vegetables, oats, rye, dairy

TABLE 2-5   MINERALS

MINERALS	DEFICIENCY MAY CAUSE	HOW IT WORKS	EXCESS MAY CAUSE	BEST FOODS TO EAT
Calcium	Rickets, soft bones, osteoporosis, cramps, numbness and tingling in arms and legs	Strong bones, teeth, muscle and nerve function, blood clotting	Confusion, lethargy, blocks iron absorption, deposits in body	Dairy, salmon and small bony fish, tofu
Phosphorus	Weakness and bone pain, otherwise rare	Works with calcium; helps with nerve, muscle, and heart function	Proper balance needed with calcium	Meat, poultry, fish, eggs, dairy, dried beans, whole grains
Magnesium	Muscle weakness, twitching, cardiac problems, tremors, confusion, formation of blood clots	Needed for other minerals and enzymes to work, helps bone growth, muscle contraction	Proper balance needed with calcium, phosphorus, and vitamin D	Nuts, soybeans, dried beans, green vegetables
Potassium	Lethargy, weakness, abnormal heart rhythm, nervous disorders	Fluid balance; controls heart, muscle, nerve, and digestive function	Vomiting, muscle weakness	Vegetables, fruits, dried beans, milk
Iron	Anemia, weakness, fatigue, pallor, poor immune response	Forms of hemoglobin and myoglobin supplies oxygen to cells, muscles	Increased need for antioxidants, heart disorders	Red meats, fish, poultry, dried beans, eggs, leafy vegetables
Iodine	Goiter, weight gain, increased risk of breast cancer	Helps metabolize fat, thyroid function	Thyroid-decreased activity, enlargement	Seafood, iodized salt, kelp, lima beans
Zinc	Poor growth, poor wound healing, loss of taste, poor sexual development	Works with many enzymes for metabolism and digestion; immune support, wound healing, reproductive development	Digestive problems, fever, dizziness, anemia, kidney problems	Lean meats, fish, poultry, yogurt
Manganese	Nerve damage, dizziness, hearing problems	Enzyme cofactor for metabolism; controls blood sugar, nervous, and immune functions	High doses affect iron absorption	Nuts, whole grains, avocados
Copper	Rare, but can cause anemia and growth problems in children	Enzyme activation, skin pigment, needed to form nerve and muscle fibers, red blood cells	Usually by taking supplements. Liver problems, diarrhea.	Nuts, organ meats, seafood
Chromium	Impaired glucose tolerance in low blood sugar and diabetes	Glucose metabolism	Most Americans have low intakes	Brewer's yeast, whole grains, peanuts, clams
Selenium	Heart muscle abnormalities, infections, digestive disturbances	Antioxidant with vitamin E, protects against cancer, helps maintain healthy heart	Nail, hair, and digestive problems, fatigue, garlic odor of breath	Meat and grains, dependent on soil in which they were raised
Molybdenium	Unknown	Element of enzymes needed for metabolism, helps store iron	Gout, joint pains, copper deficiency	Beans, grains, peas, dark green vegetables

Vitamin E and Selenium

Vitamin E, discovered in the early 1920s, has antioxidant proprieties. It has been shown to decrease the risk of cardiovascular disease and some cancers and may offer protection from Parkinson's disease and slow the progression of Alzheimer's disease.

Vitamin E together with selenium offers powerful antioxidant properties. Selenium has anticancer properties. Plants grown in selenium-rich soil are the best source. A combination of vitamin E and selenium has been found to reduce some cancers by 37%, and research has shown that the death rate from cancer was reduced by 50% in groups that took selenium supplements (Peters et al., 2007).

Vitamin E comes in two forms—natural and synthetic. The natural type is preferred and usually is listed on the label as d-alpha tocopherol (d-alpha tocopheryl) acetate. Synthetic vitamin E is listed on the label as dl-alpha tocopherol (tocopheryl) acetate.

There is little evidence of vitamin E toxicity, but at high levels, it can increase the effect of anticoagulant (blood thinning) medications and may also interfere with vitamin K's action in the body (blood clotting).

Vitamin E has been shown to help immune response, keep low-density lipoprotein cholesterol levels in check, and assist other antioxidants to be more available for use against free radicals. A diet that includes whole grains, wheat germ, nuts, sunflower oil, and corn oil will meet vitamin E RDAs. If you wish to consume 400 to 800 IUs per day, supplementation is necessary.

Phytochemicals

Phytochemicals (plant chemicals) are compounds that exist naturally in all plant foods and give them their color, flavor, and scent. They are the nonnutritive substances of plants and are not vitamins or minerals; nevertheless, phytochemicals have been associated with assisting the immune system, working as antioxidants, and fighting cancer (Beliveau & Gingras, 2007). Foods that have been identified as having these health benefits are fruits, vegetables, legumes, grains, seeds, soy, licorice, and green tea. Researchers have discovered many classes of phytochemicals in food. Isoflavones (phytoestrogens) and lignins (soy), lycopene (tomatoes), anthocyanins and proanthocyanidins (grapes, blueberries, cherries, and other red crops), saponins (whole grains and legumes), flavonoids (cherries, tea, and parsley), and isothiocyanates and indoles (broccoli, cauliflower, and cabbage), have antioxidant properties that may lower LDL (bad cholesterol levels) and curb growth of tumors (Johnson, 2007).

One example of a vegetable that has gotten much press over the last several years is broccoli. Broccoli is in the cruciferous family, which includes cauliflower, cabbage, kale, brussels sprouts, bok choy, and Swiss chard. Cruciferous vegetables are excellent sources of fiber, beta-carotene, vitamin C, and other vitamins and minerals. Their cross-shaped flowers give them their name. The phytochemicals that have been found in these vegetables are indoles, isothiocyanates, and sulforaphane, which assist the body in triggering the formation of enzymes that block hormones and may protect cells against damage from certain carcinogens. Research is promising with regard to these and other phytochemicals and cancer; however, more studies are needed.

Benefits and Risks of Phytonutrient Supplementation

The benefits of soy products have been clearly established in studies showing that soy fights cancer and lowers cholesterol levels; however, the use of soy products, especially in women who are vegetarians, has been linked to iron deficiency. One reason cited is that a vegetarian diet uses soy to replace the meat of conventional diets; therefore, supplementing the diet with vitamin C to enhance iron absorption is recommended (Venderley & Campbell, 2006).

Optimum levels for phytochemicals have not yet been determined. Individual foods contain different phytochemicals in varying amounts, and experts believe that it is the combination of these compounds that may make the difference. Some scientists believe there are thousands of phytochemicals in a single food. Because researchers have advanced the most active phytonutrients found in fruits and vegetables over the last few years, it was inevitable that these components, in supplement form, would soon follow.

Although this provides an extremely convenient way of receiving the benefits of phytonutrients, supplements contain only isolated components and not the entire compound as it is found in the whole-food state. Currently, it would be best to consume phytochemicals by eating a variety of fruits, vegetables, grains, and legumes. Supplementation with isolated phytochemicals is discouraged until more information is obtained.

Cardiovascular Health

Heart disease is the number one killer of Americans, although it is largely preventable. The main issue leading to cardiovascular problems is arteriosclerosis (buildup of fatty deposits on the inner wall of arteries) more commonly known as hardening of the arteries. There has been a preponderance of literature explaining many factors affecting a person's risk of heart disease. Some factors include diet, stress, heredity, and lifestyle. Over the last several years,

---

EXHIBIT 24-2   RISKS WITH VITAMIN B GROUP SUPPLEMENTATION

There are risks associated with excess doses of certain vitamin B group vitamins, particularly niacin, B_6, folate, and choline. Vitamin B_6, when taken in amounts in excess of 100 mg per day can cause neuropathy (a disorder of the nerves), which could lead to weakness, pain, and numbness of the limbs. Niacin's upper level intake is 35 mg per day. Symptoms of overdose with niacin include flushing, itching, and warm sensation. The folate upper level limit is set at 1000 mcg (1 mg), whereas choline is set at 3.5 grams per day. Excess choline may lead to low blood pressure.

Thiamine, riboflavin, B_{12}, pantothenic acid, and biotin do not have upper limits set because of the lack of evidence suggesting adverse effects from high intakes of these B vitamins; however, excessive consumption of these B vitamins is not wise.

---

researchers have indicated the important role that some B vitamins—especially B_6, B_{12}, and folic acid—play in cardiovascular health by lowering blood levels of homocysteine.

Homocysteine, an amino acid, forms in the blood vessels and can accumulate there as a result of the breakdown of protein in foods (such as meats and dairy). High levels of homocysteine may be a leading cause of atherosclerosis. Second, research shows that people over the age of 50 have been found to have lower levels of vitamin B_{12}. Although older adults may get sufficient vitamin B_{12} in their food, between 10% and 30% no longer have the ability to adequately absorb the naturally occurring form of B_{12}; therefore, the National Academies of Science recommends those over age 50 should consume foods fortified with B_{12} and add supplements if necessary (Food and Nutrition Board, 2008). B vitamins often work synergistically with each other as well as with other body processes, such as enzymes being activated; B vitamins in combination are the best way to supplement (see **Exhibit 24-2**).

Strong evidence exists for the benefits of fish oil supplements in promoting cardiovascular health and improving outcomes in persons with cardiovascular disease. Vitamin A, as beta-carotene, vitamin C, calcium, magnesium, selenium, vitamin E, manganese, potassium, bioflavonoids, choline, and essential fatty acids (EFAs) are some nutrients that also have been linked to optimum cardiovascular health. These nutrients are needed to repair, protect, and prevent degeneration of the blood vessels.

Hypertension

High blood pressure responds very well to lifestyle changes. Increased fiber, low sodium intake, relaxation techniques, and exercise are helpful in lowering and maintaining blood pressure. The DASH diet (see **Figure 24-3**) suggests foods that tend to be rich in potassium, magnesium, and calcium; these are the minerals experts believe to be important for controlling blood pressure. Foods rich in potassium include bananas, apricots, grapes, oranges, spinach, lentils, and almonds.

When diuretics (fluid pills) are used, the addition of magnesium as a supplement may be needed. Diuretics cause fluid loss, and when fluid is lost from the body it takes potassium and magnesium with it. Magnesium assists potassium to keep blood pressure levels optimum.

Diabetes Mellitus

Diabetes is a chronic disease that is one of the major causes of blindness in the United States. It is caused from either a defect in or insufficiency of insulin that does not allow for the management of appropriate blood glucose levels.

Food is broken down and absorbed by the body when you eat. Enzymes—chemicals made by the body with the help of nutrients—turn protein into amino acids and starches and sugars into their simple sugars. Fats are broken down into fatty acids. After this happens, there is usually a rise in blood sugar, leading to the hormone insulin being secreted by the pancreas (a gland located behind the stomach). Insulin assists in the movement of nutrients from the bloodstream into the muscles and fat tissues and also the liver. This allows the liver to stop producing glucose (blood sugar). If there is not enough insulin being excreted or if what is excreted is unable to be used, then diabetes develops.

Chromium was first found to have a relationship with insulin control in the body in the mid-1950s, and in the late 1970s, it was finally accepted as a nutrient. Chromium is a mineral, essential to the body, which acts cooperatively with other substances that control metabolism. It is a component of the glucose tolerance factor used to help with fat metabolism by transporting glucose to the cells (and being metabolized to produce energy) and activates certain enzymes. The recommended daily allowance of chromium is 50 to 200 mcg. It is believed that only 10% of the population of the United States receives enough chromium in their diet. Deficiency or inadequacy of chromium effectively blocks insulin function, resulting in elevated glucose levels. Supplementation decreases fasting glucose levels and insulin levels, improves glucose tolerance, and suppresses cholesterol and triglyceride levels.

Following The DASH Diet

The DASH eating plan shown below is based on 2000 calories a day. The number of daily servings in a food group may vary from those listed, depending on your caloric needs.

Use this chart to help plan your menus, or take it with you when you go to the store.

Food group	Daily servings (except as noted)	Serving sizes	Examples and notes	Significance of each food group to the DASH eating plan
Grains and grain products	7–8	1 slice bread 1/2 cup dry cereal* 1/2 cup cooked rice, pasta, or cereal	whole-wheat bread, English muffin, pita bread, bagel, cereals, grits, oatmeal, crackers, unsalted pretzels, popcorn	major sources of energy and fiber
Vegetables	4–5	1 cup raw leafy vegetable 1/2 cup cooked vegetable 6 oz vegetable juice	tomatoes, potatoes, carrots, green peas, squash, broccoli, turnip greens, collards, kale, spinach, artichokes, green beans, lima beans, sweet potatoes	rich sources of potassium, magnesium, and fiber
Fruits	4–5	6 oz fruit juice 1 medium fruit 1/4 cup dried fruit 1/2 cup fresh, frozen, or canned fruit	apricots, bananas, dates, grapes, oranges, orange juice, grapefruit, grapefruit juice, mangoes, melons, peaches, pineapples, prunes, raisins, strawberries, tangerines	important sources of potassium, magnesium, and fiber
Low-fat or fat-free dairy foods	2–3	8 oz milk 1 cup yogurt 1 1/2 oz cheese	fat-free (skim) or low-fat (1%) milk, fat-free or low-fat buttermilk, fat-free or low-fat regular or frozen yogurt, low-fat and fat-free cheese	major sources of calcium and protein
Meats, poultry, and fish	2 or less	3 oz cooked meats, poultry, or fish	select only lean; trim away visible fats; broil, roast, or boil, instead of frying; remove skin from poultry	rich sources of protein and magnesium
Nuts, seeds, and dry beans	4–5 per week	1/3 cup or 1 1/2 oz nuts 2 tbsps or 1/2 oz seeds 1/2 cup cooked dry beans	almonds, filberts, mixed nuts, peanuts, walnuts, sunflower seeds, kidney beans, lentils, and peas	rich sources of energy, magnesium, potassium, protein, and fiber
Fats and oils**	2–3	1 tsp soft margarine 1 tbsp low-fat mayonnaise 2 tbsps light salad dressing 1 tsp vegetable oil	soft margarine, low-fat mayonnaise, light salad dressing, vegetable oil (such as olive, corn, canola, or safflower)	besides watching fats added to foods, choose foods that contain less fat
Sweets	5 per week	1 tbsp sugar 1 tbsp jelly or jam 1/2 oz jelly beans 8 oz lemonade	maple syrup, sugar, jelly, jam, fruit-flavored gelatin, jelly beans, hard candy, fruit punch, sorbet, ices	sweets should be low in fat

* Serving sizes vary between 1/2 and 1 1/4 cups. Check the product's nutrition label.
** Fat content changes the serving sizes for fats and oils. For example, 1 tbsp of regular salad dressing equals 1 serving; 1 tbsp of a low-fat dressing equals 1/2 serving; 1 tbsp of a fat-free dressing equals 0 servings.

Figure 24-3   The Dash Diet
Source: DASH.

> **KEY POINT**
>
> Omega-3 oils are found in fish oils and flaxseed; the omega-6 oil sources are borage (an herb), evening primrose, and black currant oils.

Essential Fatty Acids

The two essential fatty acids necessary for life and the proper function of the body are omega-3 (alpha-linoleic) and omega-6 (linoleic). EFAs enhance immune function, protect the lining of the gastrointestinal system, increase kidney blood flow, reduce inflammation, and inhibit platelet aggregation (cells of the blood sticking together). These fatty acids become converted with the help of eicosanoids, which are produced by enzymes. Prostaglandins are probably the most commonly known eicosanoids. One to two tablespoons of flaxseed ground fresh and sprinkled on food will provide an adequate amount of omega-3 oil for the average person. Cold-pressed and fresh canola, sunflower, and safflower oils are excellent sources of EFAs. These can be easily incorporated into your daily nutritional plan. Supplementation should be considered if a person has cardiovascular risks or conditions or does not include these nutrients in the diet.

Osteoporosis

Calcium is an important mineral and may help prevent osteoporosis. When bones are calcium rich, they are less susceptible to fractures. DRIs suggest that Americans take between 1,000 and 1,300 mg of calcium per day. After menopause, some women may require up to 1,500 mg of calcium per day. The upper tolerable limit for calcium is set at 2,500 mg per day. Excess calcium can cause muscle cramps, kidney stones, high blood calcium, or poor absorption of iron, zinc, or magnesium.

When considering calcium intake, it is important to know that vitamin D is essential in metabolizing and absorbing the calcium that you are ingesting. Magnesium and phosphorus are minerals that work together with calcium. Meeting calcium requirements means consuming adequate quantities of calcium and vitamin D. If people depend on dietary sources for these nutrients they may have to eat a very large quantity of calcium-rich foods in addition to consuming foods that include vitamin D, phosphorus, and magnesium in the proper ratios. There are several forms of calcium available in supplements; the most bioavailable form is calcium citrate; the least absorbable is calcium carbonate.

Sorting It All Out

The concept that people must be constantly on guard and change course to follow the most recent scientific study, not really knowing if they are heading in the right direction, is ominous. Nutritional supplementation can seem like a daunting task and quickly becomes overwhelming, unless the basics are implemented. Rather than becoming overloaded with facts regarding optimum nutritional supplementation, people should focus on eating basic, well-balanced diets. When a balanced diet is eaten, many compounds that help with the protection, breakdown, absorption, and integration of all that is ingested are consumed. Nature provides what is needed with the proper ingredients in the right amounts for the body—if a nutritious and healthful food plan is followed. If people are confused or have any doubt regarding what or how much they should be taking, consulting a healthcare practitioner who is knowledgeable in nutrition and nutrition supplementation for guidance is beneficial.

Summary

RDAs represent minimum standards of food intake to meet the needs of an average person, whereas DRIs go beyond RDAs to describe nutrient requirements for optimal health.

In 1994, the Dietary Supplement Health and Education Act required dietary supplement manufacturers to include the words *dietary supplement* on product labels. It also established an oversight office within the NIH.

It is important to read supplement labels. Factors to consider when contemplating the use of supplements include age, the presence of chronic health problems, gender, and lifestyle habits. Supplements can never replace a healthy diet.

Antioxidants include vitamins A, C, E, and selenium. Phytochemicals are compounds that exist naturally in all plant foods.

Supplements can be beneficial in the treatment of many chronic conditions; however, they must be used wisely. People need to learn about the actions, interactions, and risks associated with the specific supplements they use.

References

Barr, S. I., Murphy, S. P., & Poos, M. I. (2002). Interpreting and using the dietary reference intakes in dietary assessment of individuals and groups. *Journal of the American Dietetic Association, 102,* 6.

Beliveau, R., & Gingras, D. (2007). Role of nutrition in preventing cancer. *Canadian Family Physician, 53*(11), 1905–1911.

Centers for Disease Control and Prevention. (2002). *Folic acid. PHS recommendations.* Retrieved from http://www.cdc.gov/mmwr/PDF/rr/rr5113.pdf

Eldridge, A. L. (2004). Comparison of 1989 RDAs and DRIs for water-soluble vitamins. *Nutrition Today, 39*(2), 88–93.

Food and Nutrition Board, National Institutes of Medicine. (2008). *Dietary reference intakes: A risk assessment model for establishing upper intake levels for nutrients.* Retrieved from http://www.nap.edu/catalog.php?record_id=6432#toc

Institute of Medicine (IOM). (n.d.). *Dietary reference intakes.* Retrieved from http://www.iom.edu/Activities/Nutrition/SummaryDRIs/DRI-Tables.aspx

Johnson, I. T. (2007). Phytochemicals and cancer. *Proceedings of the Nutrition Society, 66*(2), 207–215.

Levine, M. Wang, Y., Padayatty, S., & Morrow, J. (2001, August 14). New recommendations for vitamin C. *Proceedings of the National Academy of Sciences, 98,* 9842–9846.

National Academies of Science. (1998, April 7). *Dietary reference intakes for thiamin, riboflavin, niacin, vitamin B6, folate, vitamin B12, pantothenic acid, biotin, and choline* (Press release).

National Academies of Science. (2001, April 10). *Antioxidants' role in chronic disease prevention still uncertain; huge doses considered risky* (Press release).

National Academies of Science. (2002, September 5). *New eating and physical activity targets to reduce chronic disease risk* (Press release).

National Academies of Science. (2004a). *Dietary reference intakes (DRIs).* Retrieved from http://fnic.nal.usda.gov/dietary-guidance/dietary-reference-intakes/dri-tables

National Academies of Science. (2004b). *A report of the Panel of Micronutrient, Subcommittee on Upper Reference Levels of Nutrients and of Interpretations and Uses of Dietary Reference Intakes, and the Standing Committee on Scientific Evaluation of Dietary Reference Intakes.* Washington, DC: Food and Nutrition Board, IOM, and National Academy of Sciences Press. Retrieved from http://www.nap.edu/openbook.php?record_id=10490&page=793

National Institutes of Health, Office of Dietary Supplements. (n.d.). *What are dietary supplements?* Retrieved from www.ods.od.nih.gov

Peters, U., Littman, A. J., Kristal, A. R., Patterson, R. E., Potter, J. D., & White, E. (2007). Vitamin E and selenium supplementation and risk of prostate cancer in the vitamins and lifestyle (VITAL) study cohort. *Cancer Causes and Control, 18*(8), 1131–1140.

Rahman, K. (2007). Studies on free radicals, antioxidants, and co-factors. *Clinical Interventions in Aging, 2*(2), 219–236.

Russell, R. (2001, January 9). *Dietary reference intakes for vitamin A, vitamin K, arsenic, boron, chromium, copper, iodine, iron, manganese, molybdenum, nickel, silicon, vanadium, and zinc* (Opening statement, Institute of Medicine public briefing). Washington, DC.

Scalli, S., Zaid, A., & Soulamaymani, R. (2007). Drug interactions with herbal medicines. *Therapeutic Drug Monitor, 29*(6), 679–686.

Strobel, M., Tinz, J., & Bielsalski, H. K. (2007). The importance of beta-carotene as a source of vitamin A with special regard to pregnant and breastfeeding women. *European Journal of Nutrition, 46*(Suppl 1), I1–I20.

U.S. Department of Health and Human Services, Office of Disease Prevention and Health Promotion Healthy People 2010. (n.d.). Retrieved from http://www.healthypeople.gov/

U.S. Pharmacopeia. (2008). *Dietary supplements lexicon.* Retrieved from www.usp.org

U.S. Senate (the McGovern Report) Select Committee on Nutrition and Human Needs. (1977). *Dietary goals for the U.S.* (2nd ed.). Washington, DC: U.S. Government Printing Office.

Venderley, A. M., & Campbell, W. W. (2006). Vegetarian diets: Nutritional considerations for athletes. *Sports Medicine, 36*(4), 293–305.

VERIS. (2002). *Carotenoids and eye health.* LaGrange, IL: VERIS Research Information Services.

Weil, A. (2006). *8 weeks to optimum health: A proven program for taking full advantage of your body's natural healing power.* New York, NY: Alfred A. Knopf Publishers.

Suggested Reading

Balch, P. A. (2010). *Prescription for nutritional healing. A to Z guide to supplements.* New York, NY: Penguin Putnam.

Balentine, R. (2007). *Diet and nutrition: A holistic approach.* Honesdale, PA: Himalyan Press.

Caballero, B. (2010). *Guide to nutritional supplements.* Oxford, England: Elsevier.

Igloe, R. S., & Hui, Y. H. (2001*). Dictionary of food ingredients* (4th ed.). New York, NY: Chapman and Hall.

Lieberman, S., & Bruning, N. P. (2003). *The real vitamin and mineral book.* New York, NY: Penguin.

MacWilliam, L. (2007). *Comparative guide to nutritional supplements* (3rd ed.). Oswego, NY: Northern Dimensions Publishing.

Mielke, T. (2009). *The book of supplement secrets: A beginner's guide to nutritional supplements.* Bloomington, IN: Author House.

Mindell, E. (2004). *Earl Mindell's vitamin bible for the 21st century.* New York, NY: Time Warner Books.

Murray, M. (2001). *Encyclopedia of nutritional supplements* (2nd ed.). Rocklin, CA: Prima Publishing.

PDR Staff. (2008). *PDR for nonprescription drugs, dietary supplements, and herbs, 2008 physicians' desk reference (PDR) for nonprescription drugs and dietary supplements.* Montvale, NJ: Thomson.

Ulene, A. (2000). *Complete guide to vitamins, minerals, and herbs.* New York, NY: Avery Books.

Weil, A. (2006. *8 weeks to optimum health: A proven program of taking full advantage of your body's natural healing power.* New York, NY: Alfred A. Knopf Publishers.

Resources

American Council on Science and Health

1995 Broadway

New York, NY 10023

212-362-7044

www.acsh.org

This site covers many issues regarding nutrition and health with links to other sites.

Ask Dr. Weil Bulletin

www.drweil.com

The online information features Dr. Andrew Weil, founder of the Integrative Medicine Program, University of Arizona Health Sciences Center, Tucson, Arizona. Offers many topics on integrative medicine and health.

Facts About the DASH Eating Plan

www.nhlbi.nih.gov

This site offers the DASH diet plan with recipe suggestions for downloading or to order by mail.

Healthfinder Webpage

www.healthfinder.gov

This is an online consumer health information service provided by U.S. Department of Health and Human Services. It contains links to medical journals and databases with special resources on health.

National Heart, Lung, and Blood Institute

P.O. Box 30105

Bethesda, MD 20824-0105

301-592-8573

National Institutes of Health

Office of Dietary Supplements

9000 Rockville Pike

Bethesda, MD 20892

www.ods.od.nih.gov

The International Bibliographic Information on Dietary Supplements
database provides access to bibliographic citations and abstracts from
published international, scientific literature on dietary supplements.
This database was set up to help consumers, healthcare providers,
educators, and researchers.

National Library of Medicine

8600 Rockville Pike

Bethesda, MD 20894

www.nlm.nih.gov

**Nutrient Data Laboratory USDA Agricultural Research Service,
Beltsville Human Nutrition Research Center**

4700 River Road, Unit 89

Riverdale, MD 20737

301-734-8491

ndb.nal.usda.gov/

This database gives facts about the composition of food, a glossary of
terms, and links to other agencies.

PlaneTree Health Library

@ Cupertino Library

10800 Torre Avenue

Cupertino, CA 95014

408-446-1677 ext. 3350

www.planetree-sccl.org

This is a consumer health and medical library that is free and open to
the public with the aim of providing access to information to make
informed decisions about health. There is a range of information from
professional/technical to easy-to-understand materials for all areas
of medical treatment. They also offer some materials in Spanish and
Vietnamese.

Herbal Medicine

25

OBJECTIVES

This chapter should enable you to

- Discuss the actions of a tonic, an adaptogen, and an immune stimulant
- List at least eight forms in which an herb can be used
- Describe three precautions to observe when wildcrafting herbs
- Discuss precautions when using herbs with children
- Describe the common use and cautions of at least five popular herbs

Advances in medical technology have given us antibiotics, laser surgery, and organ transplants that have changed the face of health care; however, this mushrooming of technology has come at a cost, including new risks and the insidious belief that healthcare professionals and technology are the only sources of health and healing.

We have not always looked to technology for solutions to health problems. There was a time when our ancestors were very aware of and connected to the healing power of nature. It was a natural part of human existence. Unfortunately, much of this information has been lost or ignored. Throughout the last century, in our quest for modern technology, we thought that we might be able to improve on nature.

A change is taking place, however, in which we are rediscovering that one needs to venture no further than the kitchen spice cabinet, backyard garden, or nearest woods to discover the abundance of herbs that can be readily used to influence human health and well-being. Indeed, we are literally surrounded by a bounty of medicinally charged leaves, flowers, seeds, barks, and roots. We are learning of the healing power of nature. Common garden weeds, such as St. John's wort, and garden perennials, such as echinacea, are offering natural ways to improve health and treat illnesses. Many benefits can be found in developing a relationship with plants.

Employing the use of herbs from a holistic perspective is the best way to maximize their healing potential. This means using herbs in a way that addresses the whole person—mind, body, and spirit—within the dynamic environment as opposed to just trying to control, suppress, or alleviate symptoms. It is important to incorporate the appropriate herbs into a larger effort of care that includes other lifestyle decisions and factors, such as nutrition, exercise, and stress reduction (much of which has been addressed in other chapters of this book).

KEY POINT

Rather than control specific symptoms, herbal therapy, when used holistically, considers the needs of the whole person—mind, body, and spirit.

Historical Uses of Herbs and Folklore

Let us take a quick look at where herbs have fit into the history of medicine. In the United States, an untold amount of information from millennia of cultural plant medicine use by indigenous people has been lost in just a few generations (see **Exhibit 25-1**). Fortunately, other countries and cultures have, to

EXHIBIT 25-1   WISDOM OF THE ANCIENTS

One particularly interesting remnant of an ancient system of herbal use is referred to as the Doctrine of Signatures. This system suggested that the physical characteristics of the plant indicated its function or action. In other words, the color, shape, or appearance of the plant signified how it could be used. For example, a plant with a thick yellow root would indicate use for liver problems or red stems for blood conditions. This explanation is an oversimplification of a very intricate, insightful system of which too little information has survived to make it relevant or safe for use today; however, it is evidence of the tremendous depth of understanding and relationship people the world over once had with their environment.

help rebalance hormones and alleviate anxiety. Thus, depending on how deep a relationship you would like, becoming familiar with some of the known plant phytochemicals and their actions would give you an added advantage in choosing the herbs that would be most effective in a given situation and avoid possible side effects or conflicts with other herbs or treatments.

Getting Started

In getting started, it is important to familiarize yourself with the different forms in which herbs are available and their advantages and disadvantages. From the holistic perspective, the form of herb is just as important a consideration as the type of herb. If it is a tea and the person does not "do teas," it is not of much use. Most of us are used to taking our medicines in the form of a pill; however, this is not the only way, nor is it always the best or most effective method. Remember that a pill is a form that must first be digested before the medicine it contains can be absorbed, assimilated, and used. This is a process that is often compromised when someone is dealing with an illness. Teas and tinctures are in liquid form, and thus, they are more readily available to the body to absorb. Also, there is an entire science involving the solubility of phytochemicals in different solutions, some releasing their properties more readily in alcohol, others in water.

Most herbal remedies at some point in their creation, unless you just eat them fresh or dried in an unaltered state, involve an extraction process. Extracts are made by separating the active constituents (phytochemicals) from the inactive ones (which may include sugars, starch, etc.) with the use of a solvent. This concentrates the active ingredients, which can then be kept as fluid (known as a liquid extract) or condensed and dried into a powder (referred to as powdered extract, which can then be put into capsules or made into tablets). A more potent form of the herb is created by these processes and is generally more effective for dealing with health imbalances than simply eating fresh or raw dried herbs.

Tea is a liquid extract, with water being the solvent. This a great way to get your daily medicine, particularly tonic herbs, that nourish the body's

---

KEY POINT

Herbal remedies can be used in many forms, such as extracts, teas, infusions, decoctions, tinctures, capsules, compresses, poultices, liniments, salves, or ointments.

various tissues and systems, especially when taken over a period of weeks or months—it also does not seem much like medicine. A cup of tea is a thoughtful thing to do for yourself or someone else, providing you the opportunity to add your own caring intention to the preparation. Relatively speaking, it is often the least expensive way to take your medicine.

There are two methods of tea making: infusion and decoction. An infusion is a gentle form of preparation designed to preserve the valuable nutrients and essential oils and is used to prepare the more delicate parts of plants—leaves, flowers, and fresh berries. Generally speaking, the herbs are placed in a covered container and steeped in freshly boiled water for 10–30 minutes. The proportions used will vary depending whether the herbs are fresh or dried; usually 1 teaspoon dried or 2 teaspoons fresh to 1 cup of water or 1 ounce of dried herb (roughly 2 ounces of fresh) to 1 quart of water is used. The second method, decoction, is used for harder parts of the plant (roots, bark, and seeds), which are gently simmered for 15 minutes to an hour. There are some wonderful books that extol the virtues of medicinal teas that include tried and true recipes to make it easy. Supplies are readily available at most health food stores in ready-to-steep bags or as loose bulk herbs to custom blend to your needs and desires. For those who are interested, there is little that can be more gratifying than growing, picking, and using some herbs of your own, and it is fairly easy; however, teas may not be for everyone. They do take time and space, can be cumbersome to travel with, and may not appeal to some taste buds, although most herbalists will tell you that the tasting of the herb can be a very important part of the healing relationship. Additionally, there are some phytochemicals that are not water soluble (meaning that they are not amenable to water extraction).

There are health situations when forms other than teas are more appropriate. In these cases, other liquid extracts such as tinctures can be used, which will vary depending on the solvent used—generally alcohol, vinegar, or glycerin. Again, these preparations are made by the active ingredients literally being pulled out of the plant into the solution; the solvent then acts as a preservative. Of the three types of solvents, food-grade alcohol is most often used as it is the strongest and provides the longest shelf life (3 years or longer); however, regardless of the type of solvent used, all tinctures are easy to carry and are readily absorbed by the body's digestive process. Some people object to the taste of alcohol tinctures (even though most can be concealed in juice or tea), and there are situations when even small amounts of alcohol are undesirable. The extracts made with glycerin are nonalcoholic and sweet tasting (and thus child friendly), but are considerably weaker in potency than the alcohol extracts. Likewise, vinegar extracts are not as potent as alcohol tinctures and not readily available commercially, but there are those who feel that vinegar, especially apple cider vinegar, adds healing properties of its own. When using tinctures, one should follow specific manufacturers' recommendations for dosage.

Capsules may contain crushed, dried herbs or the more concentrated powdered extract. The size of the capsule will determine the dose of the unaltered dried herb, with the standard double 00 capsules holding 500 mg. If the contents are powdered extract, the strength will be higher. This is one compelling reason why it is so important to develop the habit of reading labels so you know as much as possible what exactly you are getting (see **Table 25-1**).

External forms of herbal medicines also are extremely effective, and with many possible variations in form and content, they can be useful in a wide array of situations. For example, an infusion or decoction of echinacea or goldenseal can be used as a gargle to ease the inflammation of a sore throat. A compress is made by soaking a soft cloth in a warm or cool tea made with the appropriate herb and applied to an injury, sore, or wound to speed healing. A poultice is similar, but the herb itself is applied to the skin. This could be as quick and simple as crushing a fresh plantain leaf and pressing it to a bug bite or sting to bring relief, or as specific as mixing a combination of herbs, say garlic, mustard, and onions, wrapping them in gauze and securing them over the chest to break up the congestion of a cold. Herbal tea bags make a handy poultice (wet thoroughly and bandage where needed). Yet another very effective way to use a tea is as an herbal steam inhalation, breathing in the medicinal steam to lessen the inflammation, irritation, and discomfort of sinus infections or head colds. This also is a great idea to add to your routine to prevent illness, as many herbs that are used this way are antiviral and antibacterial, with the steam helping to keep the mucous membranes in top condition to fend off infection-causing bugs.

Liniments are tinctures that are used only externally. Because the alcohol that they contain is usually the isopropyl (rubbing) type, a liniment absorbs quickly on the skin, carrying the medicine into the tissues. Salves and ointments are semisolid preparations designed for application to the skin using oil-based substances or beeswax with dried herbs. They are not meant to blend into the skin but to form a protective outer layer that holds the medicine in place and prolongs the time that the herbs remain moist. (Included in the "Suggested Reading" list are books that go into great detail about making your own remedies, some providing extremely helpful illustrations of each step and pictures of commonly used herbs.)

TABLE 25-1   BASIC DAILY FOOD PLAN

STANDARDIZED EXTRACT	WHOLE PLANT PRODUCTS
• Highly purified standard amount of specific constituents • Nonsynthetic powerful medicine • Some herbs have organ-specific activities and indications • Insures proper identification of plants • Promotes more allopathic approach of treating symptoms • Clinical testing and research data are available • Evidence of increased incidence of side effects • May lose all other activity, but targeted effect • Solvents such as hexane are involved in the extraction process; solvent residues can be liver toxins • Takes plant constituents out of context; perpetuates the idea that we can outsmart or improve on nature	Can confirm active constituents present at a certain level • Does not interfere with the natural synergistic balance of nature's intent • Promotes traditional or holistic approach; prevention and nutrition • Some variations in concentration of components depending on growing conditions • Record of thousands of years of use and efficacy • Less expensive than standardized extracts • Solvents and preservatives typically include alcohol, vinegar, water, and glycerin

Note: It stands to (holistic) reason to determine what will serve the individual best in each situation. Perhaps employing standardized extract in more acute cases and relying on whole plants for the majority of health needs involving nutrition, prevention, and tonification (tonics). The important issue is having access to and being able to use what works best for the individual in a given scenario.

Cultivating Herbal Wisdom

After you have made the decision to incorporate herbal medicine as part of your holistic health approach, some general guidelines and tips can help make your relationship with plants rewarding and safe. Become familiar with a few herbs. A single herb can offer multiple health benefits, and thus, you may find most of your herbal needs can be met from a few plants. Attend one of the increasingly available classes and workshops. To locate courses and events, check with local health food stores, wellness centers, and universities. Also, the American Herbalists Guild and the American Botanical Council (listed in the "Resources" section) are great organizations to check with regarding educational programs.

When deciding where to buy herbs, try to purchase locally and organically grown produce when possible. Become attuned to looking for sustainably harvested products (plants that are collected in an ecologically and environmentally conscious manner). This may require a little investigation but is well worth the effort. Search for reputable companies, particularly those with the reputation of being in the business for more reasons than just making a profit. The quality of the medicine will be enhanced by the social consciousness and good intentions of the company producing it.

Some precautions are needed when gathering your own herbs from the wild (*wildcrafting*). First, be absolutely sure that you properly identify the plant, as there are many examples of different plants that resemble each other very closely, some of them being very toxic. Be aware of the potential for chemical/pesticide contamination where you pick. For instance, do not pick along busy roads where the plants are exposed to many different pollutants from auto exhaust to detergent-laden runoff. Also, know plants well enough to avoid picking them if they are endangered species. All of these precautions can be disregarded if you are in a position to grow some herbs on your own. There are some books listed at the end of this chapter to help you do that.

For best results when using herbs, be consistent and take them in the recommended amounts and with the recommended frequency. If taking something long term (as a tonic), it is a good idea to omit the herb at regular intervals; for instance, 5 days on/2 days off or 1 month on/1 week off—the individual situation and the herb used will help determine the schedule.

When treating a specific condition, avoid starting herbs that are generally promoted as good for the condition without some assessment. Consider how you feel and what may have contributed to or precipitated the situation. Do not ignore symptoms or delay seeking the most appropriate care or therapy. If you have a known problem, consult with a healthcare practitioner, especially if you are taking other medications. The ideal approach in these situations is to work with an experienced professional herbalist. If you do not know of one, try contacting the American Herbalists Guild to see whether there is a member in your area. In lieu of an herb-savvy healthcare professional, the safest way to proceed is with one herb at a time.

KEY POINT

Remember that the term natural does not necessarily mean safe!

You will do well to remember that herbs are medicines and need to be used appropriately and correctly. It is important to know the dosage range and understand that a higher dose does not necessarily mean greater effectiveness. Also, in regard to safety, you must consider individual allergies and sensitivities, particularly if taking an herb for the first time, start with a low dose (perhaps half the recommended dose) and work up to a standard dose over a couple days.

KEY POINT

When using herbs with children, particular attention must be paid to dose. A general rule is to use half the adult dose for children ages 7 to 12 years and one-quarter of the dose for children less than 7 years of age. Of course, the same guidelines of safety must be applied, as are considerations for the specific situation and individual child. Using herbs with infants should only be done with the guidance of a healthcare practitioner.

Be aware that there are definitely situations when the use of specific herbs is inappropriate and contraindicated (should be avoided), as they may give rise to side effects or complicate the situation. This is particularly true during pregnancy and breastfeeding. There seems to be information released daily regarding drug–drug and food–drug interactions. To date, relatively little is known about interactions between synthetic pharmaceuticals and herbs. Until more research is available, use common sense and check several reliable resources, including the practitioner who prescribed the medication. Healthcare providers need to be kept aware of all medicines, herbs, remedies, and supplements that are being taken by their patients. A growing number of pharmacists are becoming more knowledgeable about herbs, and many have access to computer databases that can help alert people to possible undesirable interactions to avoid side effects. The literature shows that most problems related to using herbs arise from misuse, allergic reactions, or improper combining with pharmaceuticals, all of which can be avoided by taking the responsibility to use herbs wisely, thoughtfully, and with respect for the abundance of health that they offer.

The following pages list some of the more popular and useful herbs readily available, along with a combination of information gathered from time-tested experience and what modern science has validated. This information is by no means all encompassing but is offered as a place to begin.

Review of Some Popular Herbs

Burdock (*Arctium lappa*)

Family: Compositae

Other names: Beggar's buttons, cockle buttons, cocklebur, burr seed

Habitat: Open fields, roadsides, and waste places

Parts used: Root and seed

Common uses: Blood cleanser/purifier (alternative)

Cautions: Do not take during pregnancy. It may cause dermatitis in sensitive individuals. It can affect blood sugar in people with diabetes and produces a diuretic effect in some people. Antibiotics and medications taken to treat gout, cancer, or HIV may interact with burdock.

Burdock is a large biannual plant growing up to 6 feet tall with huge leaves and a deep taproot that can reach 30 feet in length. Small purple flowers appear at the top of a single stalk in late spring of the second year and mature into seed heads, which readily stick to almost anything they touch. These are also known as beggar's buttons because they were once used to fasten clothing together. The concept of hook and loop fasteners is based on the sticky nature of burdock seed heads.

In Japan, the long taproot is known as gobo and is used for food. Fresh, young roots can be sliced and added to a stir-fry or soup to make a nutritious meal. In general, it is an important tonic herb, which is considered gentle and nourishing. Traditionally, burdock root has been used as a blood purifier because of its ability to support the body's function in elimination of waste products via the liver. Herbalists use the root as a mild diuretic (increasing fluid elimination through the kidneys) and to promote sweating during some illnesses. A decoction of the root or tincture of the seeds can be used for dry skin disorders, such as eczema and psoriasis. Additionally, some cases of acne respond to treatment with burdock.

A root poultice or oil infusion of leaves can be applied to skin sores and leg ulcers. A compress made with a strong decoction will help treat topical fungal infections.

Black Cohosh (*Cimicifuga racemosa*)

Family: Ranunculaceae

Other names: Black snake root, fairy candles, bugbane

Habitat: Densely shaded, deciduous woods

Parts used: Dried root and rhizome

Common uses: General anti-inflammatory, menopausal symptoms, menstrual cramps, antispasmodic, sedative

Cautions: Large doses (over 2 teaspoons) may cause headache. Do not use during pregnancy or while nursing. Take with food to avoid stomach irritation with long-term use.

Black cohosh is a perennial plant that often reaches 6 feet in height and produces a rather showy spike of white flowers in midsummer. It is a spectacular sight when the sunlight filters down though the woodland canopy and strikes individual "candles," setting them aglow. This herb was greatly valued by Native Americans as a remedy for joint pain. It is often used as part of a formula (in combination with other herbs) to reduce inflammation of joints and soft tissues. This is an herb people have found useful in easing the discomfort of some forms of arthritis, bursitis, and fibromyalgia. It is also has been thought to have a role in normalizing to the female reproductive system, providing relief of uterine pain and decreasing menopausal symptoms, such as hot flashes and anxiety; however, research results concerning its effectiveness are mixed. Because black cohosh can reduce spasms, it can be an aid in treating whooping cough and relaxing tense muscles. Black cohosh can also be used as a tincture or a decoction.

Calendula (*Calendula officinalis*)

Family: Compositae

Other names: Pot marigold

Habitat: Mediterranean area; however, it can be cultivated in any good garden soil.

Parts used: Flower petals

Common uses: Topically for healing skin and mucous membranes, internally for stomach ulcers, fevers, and menstrual cramps

Cautions: Do not take during pregnancy or when breastfeeding.

This bright yellow member of the marigold family is native to the Mediterranean region and lacks the strong smell of its more familiar nonmedicinal cousin. Almost anyone can easily grow calendula in a sunny location.

Calendula's anti-inflammatory and wound-healing properties make it a very useful herb. As an ointment, it can be applied to bruises, cuts, and scrapes. In the form of a tincture, it is a wonderful mouth rinse for red, irritated gums, gingivitis, and pyorrhea. A tea can be used as an aid in healing mouth tissues after oral surgery, as well as treating a sore throat or mouth ulcers. Just gargle and rinse. A poultice or compress can be applied to varicose veins and bruises. A tea or glycerite tincture used topically can help with healing bedsores. Although calendula is considered a mild remedy, it is effective as a tea to soothe unpleasant conditions like stomach ulcers. Externally, it can often reduce the effects of eruptions, such as measles or shingles. For day-to-day use, calendula cream makes skin feel soft and silky. In the form of a lotion, it is an excellent beauty aid for cleansing and soothing the skin.

Cayenne (*Capsicum annum*)

Family: Solanaceae
Other names: Hot pepper, red pepper
Habitat: Tender annual can be cultivated in any good garden soil
Common uses: Stimulate circulation, aid in nerve pain, anti-inflammatory
Parts used: Dry, ground pods without seeds
Cautions: Do not use the seeds (as they can be too irritating). Use with caution during pregnancy. Do not use on broken or injured skin. Avoid getting capsicum in the eyes. Some individuals may develop sensitivity to both internal and external applications. It is not recommended to take capsicum for more than 2 days at a time.

Cayenne is popular as a condiment for food, especially in Asian, Mexican, and Indian cuisines. Recently, it has gained popularity in contemporary medicine as a topical cream to reduce nerve pain and to reduce itching and inflammation associated with psoriasis. It has been shown to be effective in treating mild frostbite, muscle tension, rheumatism, and chronic lumbago (lower back pain) by increasing circulation. Administering capsaicin via the nose seems to help relieve cluster headaches; this should be done by a trained professional.

Internal uses include stimulating the appetite and the prevention of atherosclerosis (plaque buildup in the arteries). Cayenne is available in capsule form, which helps in avoiding its hot, spicy sensation. Many topical ointments are available over the counter. Always follow the manufacturer's directions.

Chasteberry (*Vitex agnus-castus*)

Family: Verbenaceae
Other names: Chaste tree, monk's pepper, hemp tree
Habitat: Mediterranean region of Asia

Parts used: Fruit

Common uses: Female tonic, kidney tonic, and thyroid tonic

Cautions: Chasteberry may affect certain hormone levels. Women who are pregnant or taking birth control pills or who have a hormone-sensitive condition (such as breast cancer) should not use chasteberry. Do not take while breastfeeding, either. It may cause urticaria (itching). Because chasteberry may affect the dopamine system in the brain, people taking dopamine-related medications, such as selegiline, amantadine, and levodopa should avoid using chasteberry. This herb can cause gastrointestinal problems, rashes, and dizziness.

Chasteberry is a deciduous shrub growing up to 10 feet high with flower spikes made up of dense, showy clusters of pale, lilac blue flowers. In folklore, the plant was given the name of monk's pepper because of the alleged use of the fruit in monasteries for its ability to reduce male libido.

As a female tonic, chasteberry has been used to reduce common symptoms associated with imbalances of the menstrual cycle and menopause. It is believed to work through the female pituitary gland, which is responsible for the secretion of the hormones that regulate the ovaries. Because of this mechanism of action, it is a primary herb in helping menopausal symptoms, such as hot flashes and mood swings. Although some small studies support the benefit of this herb, additional scientific studies are needed to draw conclusive results. Chasteberry can be safely taken for months at a time with intermittent breaks to check if it is still needed.

Chasteberry can be used as a tea or a tincture.

Cinnamon (*Cinnamomum zeylanicum*)

Family: Lauraceae

Other names: Cassia

Habitat: Tropical Asia

Common uses: Antiviral, antibacterial, analgesic (pain relieving), mild digestive disorders and intestinal cramping in children and adults, flatulence, circulatory stimulant

Parts used: Bark

Cautions: Do not use with active stomach ulcers or during pregnancy. Some individuals may be sensitive and develop contact irritation.

Cinnamon is an evergreen with dense, leathery leaves that grows 30 to 40 feet tall. It is a tropical tree native to China. Cinnamon is used as both a food and a medicine. Although widely known for adding a pleasing, mellow flavor to desserts and ethnic foods, it is very safe as a medicine for children and adults. As a tea for nausea, vomiting, and motion sickness, it is pleasant and soothing.

Ground cinnamon can reduce diarrhea, especially if mixed in applesauce, because applesauce contains pectin, which helps bind the bowels. Cinnamon also possesses antibacterial properties.

Cinnamon can be ingested as fresh spice with foods or in capsules. It can be taken for amounts up to 6 grams daily for 6 weeks or less.

Dandelion (*Taraxacum officinale*)

Family: Compositae
Other names: Piss-a-bed, teeth of the lion, Dent de' Leon
Habitat: Lawns, meadows, and roadsides
Parts used: Whole plant, leaves, flowers, roots, stem
Common uses: Blood tonic, diuretic, digestive bitter, stimulates the liver
Cautions: Do not collect from sprayed lawns or roadsides.

This ubiquitous little weed, which is the bane of many homeowners, is indeed a wonderful tonic and medicine, with every part having uses. The leaves are a rich source of calcium, magnesium, sodium, zinc, manganese, copper, iron, phosphorus, and vitamins C and D. It is so nutritious that it made the top of the list in a Japanese vegetable survey of the world's most nutrient-dense plants! The golden yellow flowers are high in flavonoids and antioxidants, and a good way to use them is in savory spring biscuits or salads. The tender young leaves are also good steamed, sautéed, or raw in salads. The leaves as a medicine are a gentle, potassium-sparing (will not deplete the body of vital potassium) diuretic, making it useful in some types of congestive heart failure, high blood pressure, and water retention related to premenstrual syndrome. It is believed to support and strengthen liver function while reducing liver congestion and enhancing the flow of bile and that dandelion's bitter action stimulates digestion, absorption of nutrients, and elimination of wastes. These are some of the reasons that the dandelion was usually one of the plants used to make traditional spring tonics. Despite the long-term use of the herb, there is limited evidence supporting this. Seek the advice of your practitioner or medical doctor if your symptoms include pain or the whites of your eyes are yellow.

Dong Quai (*Angelica sinensis*)

Family: Umbelliferae
Other names: Tang kwei (there are a variety of spellings and
    pronunciations)

Habitat: China

Common uses: Premenstrual syndrome, menopause, balancing female
hormones, anemia, heart and circulatory tonic, antispasmodic

Part used: Root

Cautions: Women who have midcycle spotting or menstrual flooding
should not take dong quai. Do not take during pregnancy.

This is a small, fern-leafed, aromatic plant that is native to China. Dong
quai is a much revered traditional Chinese tonic herb and can be found in
many Asian grocery stores. The roots, which are the parts used, are often
sliced and incorporated in soups and stews. In the West, it is used as a circulatory stimulant and as a laxative in older people. It has been promoted to help
nourish women with long menstrual cycles, bloating, and heavy bleeding with
associated weakness, mild anemia (because of its significant iron content),
and menopausal symptoms of hot flashes, skin crawling, and vaginal dryness.
Although dong quai has many historical and theoretical uses based on animal
studies, there is little human evidence supporting the effects of it for any condition. Most of the available clinical studies have either been poorly designed
or reported insignificant results. Also, most have examined combination formulas containing multiple ingredients in addition to dong quai, making it difficult to determine which ingredient may cause certain effects.

Dong quai can be added to any soup by placing the roots in a cheesecloth
bag and removing it before serving. This herb can also be taken as a tea or
tincture.

Echinacea (*Echinacea angustifolia, purpurea,* or *pallida*)

Family: Compositae

Other names: Purple coneflower

Habitat: Prairies, meadows; is easily cultivated

Parts used: Root, whole flowering head

Common uses: Immune system stimulant, anti-inflammatory,
antibacterial

Cautions: Do not use with autoimmune diseases, such as lupus, or
AIDS. Because it stimulates the immune system, it has the potential of
causing a flare-up. People with allergies to ragweed, chrysanthemums,
marigolds, and daisies may have allergic reactions to echinacea.

Echinacea once grew in abundance on the Great Plains. Native Americans
used this plant for medicinal purposes long before White settlers arrived on
the shores of North America. It is now grown commercially with tons being

exported to Europe annually. It has also become a common garden perennial, growing well in most sunny, dry locations and attracting butterflies.

Many people know echinacea as the immune herb. It has been used to prevent a cold or flu or shorten the duration and severity of symptoms, although research supporting this has yielded mixed views. A tincture is an effective way to take this herb for this purpose. Be aware that a good quality preparation will make the inside of the mouth tingle for a short period of time. The most appropriate and effective use is on exposure to a bacterial or viral infection or at the first signs of the same. The strategy most people find effective is to use the tincture, taking one or two droppers full every 2 to 3 hours as long as symptoms exist (or a couple of days), four times a day for another 2 to 3 days.

A less well-known use for echinacea is as a topical treatment for skin infections, such as boils, carbuncles, and bug bites. For this purpose, a strong tea or tincture can be applied as a compress. A tea may also serve as a mouthwash or gargle for gingivitis or inflamed sore throat.

Garlic (*Allium sativum*)

Family: Allium

Other names: Stinking rose

Habitat: Any good garden soil

Parts used: Individual cloves from the bulb

Common uses: Heart tonic, blood thinner, lung infections, to lower cholesterol, to lower blood pressure

Cautions: Do not take medicinal amounts of garlic if you are on blood thinners (such as warfarin) or high daily doses of vitamin E without medical advice. Discontinue at least 1 week before any surgical procedure, to avoid prolonged bleeding. Garlic has been found to interfere with the effectiveness of saquinavir, a drug used to treat HIV infection. Its effect on other drugs has not been well studied.

Garlic was cultivated over 5,000 years ago. It is sometimes known as the stinking rose because of its acrid smell when sliced or chopped. Garlic is rich in germanium, which is a powerful antioxidant, and sulfur, which can reduce the risk of stomach, lung, and bowel cancers. In World War I and World War II, garlic was used as a wound dressing because of its strong antibacterial and antiviral properties.

Garlic is both a food and a medicine as many herbs are—adding to the wisdom, "Let your medicine be your food and your food be your medicine." The activity of garlic makes it good for preventing atherosclerosis (buildup of plaque in the veins and arteries); although some evidence indicates that taking garlic can slightly lower blood cholesterol levels, studies done by the National Center for Complementary and Alternative Medicine have not drawn the same conclusions. It also acts as a mild blood thinner and can help lower blood lipids (cholesterol).

Garlic possesses potent antimicrobial (antibacterial, antiviral, and antifungal) activity, but it is best used fresh and uncooked for this purpose. Because the medicinal volatile oils are excreted through the lungs, it is useful for respiratory infections, such as colds and bronchitis. The suggested dose is two to three raw, crushed cloves (the small sections of the garlic bulb) four times a day.

Because garlic can be unpleasant if taken straight, mixing it with a little honey, yogurt, or applesauce is very helpful. Try chopping the clove and placing it on a spoon. Do not chew it, but wash it down with water (like a pill). This method can help reduce the taste and residual odor. Also, raw garlic can be delicious and medicinal eaten in the form of pesto or grated over pasta. Because of the high levels of volatile oil compounds, garlic can be irritating to the stomach lining despite taking it with food. If irritation occurs, discontinue use for a period of time and then restart at a small dose.

A great variety of commercial products are available, manufactured to minimize the odor and other less desirable effects. Some of these have been the subject of research for their effects on cholesterol and blood pressure and are quite effective; however, for the antimicrobial action and cost effectiveness, fresh organically grown garlic is still the best bet.

Ginkgo (*Ginkgo biloba*)

Family: Ginkgoaceae

Other names: Maidenhair tree

Habitat: Native to China

Common uses: Improve circulation

Parts used: Leaves

Cautions: Some individuals will have allergic reactions. Ginkgo dilates blood vessels; therefore, individuals who have fragile blood vessels and a tendency to bleed easily should not take ginkgo. Those with a history of stroke-related aneurysm (bleeding as opposed to blood clot) should avoid ginkgo as should those on blood-thinning therapies, such as warfarin. Uncooked ginkgo seeds contain a chemical known as ginkgotoxin, which can cause seizures. Consuming large quantities of seeds over time can cause death.

Ginkgo trees are among the oldest living plants in the world. Their survival is partly explained by the fact that they were considered sacred trees by the Chinese and therefore protected. Recently, cultivated trees have proven to be one of the finest specimen trees for inner cities, thriving undeterred by pests and pollution. These are usually the trees you see growing out of cracks in the sidewalk.

Ginkgo is one of the most researched herbs in the world. It has many uses, but most people became familiar with ginkgo as an herb that potentially could boost memory. To date, research has not supported this use for the herb.

The National Center for Complementary and Alternative Medicine is funding research that includes studies on ginkgo for other conditions for which it has been used, including asthma, symptoms of multiple sclerosis, vascular function (intermittent claudication), cognitive decline, sexual dysfunction caused by antidepressants, and insulin resistance.

Take tinctures as recommended by the manufacturer for up to 3 months before evaluating improvement. Because of the blood-thinning potential of ginkgo, consult your practitioner or herbalist before self-treatment.

Ginseng (American ginseng [*Panax quinquefolius* L], Asian/Chinese/ Korean Ginseng [*Panax ginseng*], Siberian Ginseng [*Eleutherococcus senticosus*])

Family: Araliaceae

Other names: Ginseng

Habitat: Siberia, China, Northern Korea

Part used: Root bark

Common uses: Normalize body systems, help in adapting to stress (adaptogen), lower blood glucose

Cautions: Breast tenderness in some normally menstruating women. Some individuals may develop high blood pressure and should discontinue use; individuals with hypertension should not use. Occasionally, headaches, insomnia, and nervousness have been reported with ginseng use.

Siberian ginseng is a relative of American and Chinese ginsengs. It was first studied in Russia for its effects on productivity of factory workers and was shown to increase productivity and reduce the incidence of disease. In studies among athletes, endurance, speed, and stamina were increased, and recovery time was shorter.

Siberian ginseng is used as a tonic remedy for people who are stressed out, overworked, and burning the candle at both ends. It is milder and less stimulating than American ginseng and greatly valued for its ability to help the body adapt to and handle stress.

Ginseng appears to have antioxidant effects that may benefit people with heart disorders. Although additional research is needed in this area, some studies suggest that ginseng also reduces oxidation of low-density lipoprotein (LDL or bad) cholesterol and brain tissue. Several studies suggest ginseng may lower blood sugar levels in patients with type 2 diabetes before and after meals.

It can be taken as a tea or tincture for up to 3 months at a time. Then one should take a break and reevaluate how he or she is feeling.

Goldenseal (*Hydrastis canadensis*)

Family: Ranunculaceae

Other names: Yellow root

Habitat: Deciduous woodlands (endangered species)

Common uses: Antibacterial, anti-inflammatory, antifungal

Parts used: Rhizome and root

Cautions: Should not be used by women who are pregnant or breast-feeding or with infants and small children. Do not exceed the recommended dose. It is not meant for long-term use.

Goldenseal is native to deciduous woodlands of North America. Overcollection and misuse have made it an endangered species. Although it can be cultivated, it is a slow and tricky process. Fortunately, there are other herbs that contain some of the same powerful constituents as goldenseal.

Clinical studies on a compound found in goldenseal, berberine, suggest that the compound may be beneficial for certain infections—such as those that cause some types of diarrhea, as well as some eye infections; however, goldenseal preparations contain only a small amount of berberine, and thus, it is difficult to extend the evidence about the effectiveness of berberine to goldenseal. Goldenseal may be wasted on systemic (distributed by the bloodstream) diseases. It will not help with general malaise, fever, or aches and pains. Other herbs, such as echinacea, are better suited for helping to fight the flu or a cold. Suggested and appropriate uses for goldenseal include urinary tract infections, gastritis, and athlete's foot. It is a very strong herb, and a little goes a long way.

Use a tea for nasal wash, eyewash (sterile tea solution) for conjunctivitis, and as a mouth rinse for gum disease, infection, or sore throat. When preparing eyewash, carefully strain the tea through a coffee filter, then reheat to sterilize. Cool to room temperature for use.

Lavender (*Lavandula angustifolia*)

Family: Lamiaceae

Other names: English lavender, garden lavender

Habitat: Originally found in France and the western Mediterranean region; now easily cultivated in any garden that has good sunlight

Common uses: Relaxant, sedative

Parts used: Flower, essential oil

Cautions: Lavender oil can be poisonous if taken by mouth. Sedative effects can be compounded if taken with medications that have sedative effects. Applying lavender oil to the skin can cause irritation.

Lavender's most popular use is as an essential oil in aromatherapy to promote relaxation. Other health claims for the use of this herb have not been proven. The essential oil can be diluted with another oil to apply to the skin or placed in bath water.

Lemon Balm (*Melissa officinalis*)

Family: Lamiaceae

Other names: Sweet Mary, honey plant, cure-all, dropsy plant, Melissa

Habitat: Native to the Mediterranean region and western Asia. Lemon balm will grow vigorously in average soil in temperate climates; it is a common garden herb.

Common uses: Antibacterial, antiviral, antidepressant, nervine (calms nervousness), insomnia

Parts used: Fresh leaves are preferred; dry leaves can be used.

Cautions: Few studies have investigated the safety and effectiveness of lemon balm alone, except for topical use.

Lemon balm is a mild, aromatic, tasty, and effective remedy. It can be safely used to settle digestive problems. Several studies have found that lemon balm combined with other calming herbs (such as valerian, hops, and chamomile) helps reduce anxiety and promote sleep. It has been known for centuries as the gladdening herb. Just sniffing fresh lemon balm can lift one's spirits. The crushed leaves, when rubbed on the skin, are used as a repellant for mosquitoes. A poultice or compress (made by soaking a cloth in a strong tea) can be used to ease the discomfort of herpes lesions or shingles.

For cold sores, mix a few drops of lemon balm essential oil with 2 to 3 tablespoons of glycerin and dab on the sore.

Licorice Root (*Glycyrrhiza glabra*)

Family: Leguminosae
Other names: Sweet root
Habitat: Southeastern Europe and western Asia
Common uses: Gastric irritation, tonic, expectorant (helps remove secretions from the chest), anti-inflammatory
Part used: Root
Cautions: Licorice should be avoided by individuals with high blood pressure, kidney disease, and edema. Large amounts over time can cause sodium retention and potassium depletion. When taken in large amounts, licorice can affect the body's level of the hormone cortisol and steroid drugs. Pregnant women should avoid using licorice as a supplement or consuming large amounts of licorice as food, as some research suggests it could increase the risk of preterm labor.

This perennial member of the pea family has long been cultivated for its flavorful root. Licorice has been a popular ingredient in candy and to disguise the unpleasant taste of other medicine. It is an integral part of traditional Chinese medicine and is used to balance other herbs used in a formula. Although traditionally used to relieve gastric irritation and stomach ulcers, research supporting this use is inconclusive at this time.

Milk Thistle (*Silybum marianum*)

Family: Compositae
Other names: Mary thistle, silymarin
Habitat: Originated in Europe but will grow in any temperate climate (can become a noxious weed)
Common uses: Liver tonic, liver protectant, stimulating breast milk production
Parts used: Leaves, seeds
Cautions: Do not take during pregnancy. In large doses, can cause mild diarrhea and allergic reactions, especially in persons with allergies to plants in the same family (e.g., ragweed, daisy, marigold)

Milk thistle is a stout, hardy, invasive, annual plant. It can grow to up to 3 feet high, sporting dark green, scallop-edged spiny leaves with white streaks. The petals of the solitary purple flowers end in sharp spines. Although the leaves and seeds both have medicinal value, the seed contains the highest amount of the active component silybin, which is credited with the ability to protect the liver from the damage caused by many drugs, including chemotherapy. There have been some studies of milk thistle on liver disease in humans, but these have been small. Some promising data have been reported, but study results at this time are mixed. The National Center for Complementary and Alternative Medicine is studying milk thistle's benefits for chronic hepatitis C and nonalcoholic steatohepatitis (liver disease that occurs in people who drink little or no alcohol). The National Cancer Institute and the National Institute of Nursing Research are also studying milk thistle for cancer prevention and to treat complications in HIV patients.

Silybum marianum was named milk thistle because of the traditional use of a tea from the leaves to stimulate milk production in nursing mothers. The leaves also enhance digestion.

For therapeutic purposes, standardized extracts (tinctures and freeze-dried extracts) are probably most appropriate because of the high concentration of active constituents that are most soluble in alcohol.

Plantain (*Plantago major, P. lanceolata*)

Family: Plantaginaceae

Other names: White man's foot

Habitat: Common weed of lawns, gardens, and meadows

Common uses: Topical and internal antibacterial, anti-inflammatory, demulcent (soothing) coughs, wound healing, insect bites and stings; seed can be used as a bulk laxative

Parts used: Leaves, roots, seeds

Cautions: Do not collect plantain from contaminated areas or sprayed lawns. Some individuals may be allergic to plantain.

The Native Americans named plantain White man's foot because it appeared to sprout up in the footsteps of the white settlers as they moved west. It is now common throughout most of the United States. The dark green, glossy, ribbed leaves radiate from the ground. Beneath the earth are the short, dense, radiating, brown roots. The flowers and seeds form at the top of tall stalks.

A poultice of the leaves of the plant has been used effectively for insect or spider bites and bee stings. In an emergency, plantain can be gathered from a lawn or meadow, chewed up, and applied directly onto the bite or sting; it is then covered and kept in place for 1–2 hours. The pain and swelling will quickly diminish. This remedy often works better than over-the-counter pharmaceuticals. A poultice can also be applied to cuts, scrapes, and burns to aid in healing. Plantain leaf tea or juice (combined with tomato, carrot, or vegetable juice) is an effective way to soothe the symptoms of gastritis, irritable bowel, or colitis and to relieve the discomfort from urinary tract infections. A tea of the leaf or root is also a mild, soothing expectorant (facilitates removal of secretions from the lungs), which makes it useful for treating bronchitis and lung congestion. The seeds are a rich source of zinc and psyllium, which is a popular bulk laxative.

Rosemary (*Rosmarinus officinalis*)

Family: Labiatae

Other names: Dew of the sea

Habitat: Native to the Mediterranean region, will grow in any average garden soil.

Common uses: Antimicrobial, dyspepsia, rheumatism, moth repellant, some types of headache, memory aid, antioxidant, digestive aid

Parts used: Leaves

Cautions: Do not use in medicinal amounts during pregnancy. Rosemary leaves are quite safe, but the essential oil should be used with caution because of its potency.

Rosemary is a native of the Middle East and around the Mediterranean Sea. From afar, it looks like green sea foam on the face of the cliffs by the sea, hence its name, *dew of the sea*. It is easy to grow in average garden soil but is not winter hardy. It can be grown indoors but it is temperamental and does not like to dry out.

This herb is excellent on roasted potatoes and with lamb and other foods, but as a medicine, the crushed leaves possess potent antimicrobial activity (kills bacteria and virus), which is due to the high content of volatile oils. During World War II, rosemary leaves and juniper berries were burned in hospitals as a disinfectant.

For gas, nausea, and biliousness, take as a tea or tincture. To stimulate circulation, soothe aches, and relieve rheumatic pain, make a strong tea and add it to bath water, or make a warm compress and apply over affected areas. This is a good herb to use in steam inhalations for prevention or treatment of colds.

Saw Palmetto (*Serenoa repens, Sabal serrulata*)

Family: Palmaceae

Other names: Cabbage palm, American dwarf palm tree

Habitat: Subtropical sandy soil

Common uses: Tonic for male and female reproductive organs, respiratory system, irritable bladder, enlarged prostate gland

Part used: Berry

Cautions: No known serious side effects when used as recommended; some people may experience mild stomach discomfort.

Saw palmetto is also known as Spanish sword because its long slender leaves, which radiate from the ground, have sharp, serrated edges that can rip clothing and skin. This can make collecting the berries a challenge. Writings suggest that the berries smell and taste like rotten cheese. Because of this and the added fact that many of its active components are not released in water, its use as a tea is undesirable.

Numerous human trials report that saw palmetto improves symptoms of benign prostatic hypertrophy such as nighttime urination, urinary flow, and overall quality of life, although it may not greatly reduce the size of the prostate. It does not affect readings of prostate-specific antigen (PSA) levels. It is best taken as a liquid or powdered alcohol extract as the active properties are not water soluble.

St. John's Wort (*Hypericum perforatum*)

Family: Guttiferae
Other names: Hardhay, amber, goat weed, Klamath weed, Tipton weed
Habitat: Open fields, roadsides
Common uses: Antidepressant, antianxiety
Parts used: Flowers and buds
Cautions: The most common side effects of St. John's wort include dry mouth, dizziness, diarrhea, nausea, increased sensitivity to sunlight, and fatigue. Combining St. John's wort with pharmaceutical antidepressants can lead to increased serotonin-related side effects, which could potentially be serious. A commonsense approach is to check with your healthcare practitioner before combining any herb and drug.

This stout little plant has come to be known as the depression herb, and it has been the subject of substantial research and clinical trials. Studies suggest that St. John's wort is of no greater benefit in treating major depression than a placebo (National Center for Complementary and Alternative Medicine, 2012; Rapaport, Nierenberg, Howland, Dording, Schettler, & Mischoulon, 2011).

Sage (*Salvia officinalis, Salvia lavandulaefolia*)

Family: Laminaceae
Other names: Black sage, common sage, broad-leafed sage
Habitat: Mediterranean region; can be grown in average garden soil

Common uses: Antiseptic, astringent, antispasmodic, antioxidant, antiviral, antibacterial

Parts used: Leaves

Caution: Do not take during pregnancy.

This beautiful, woody perennial makes a nice addition to any herb garden. It likes a sheltered, sunny location and will withstand moderately cold, snowy winters.

Historically, sage was associated with fertility. Native Americans used it topically for skin conditions and to stop wounds from bleeding. The astringent, antiseptic qualities of sage make it an ideal gargle in the form of a tea for sore throats, gingivitis, or bleeding gums. A tea is also a good way to make a digestive tonic and to relieve night sweats during menopause or reduce excessive perspiration.

Some recent small studies have suggested that sage may improve memory and mental performance (Natural Medicines Comprehensive Database, 2012; Scholey et al., 2008).

Brew a strong tea for making a compress to soothe slow-healing wounds. Add honey to an infusion for sore throat or cough and take over 1–3 days. For colds and sinus congestion, use a steam inhalation to dry up excessive secretions and postnasal drip.

Thyme (*Thymus vulgaris*)

Family: Laminaceae

Other names: Garden thyme

Habitat: Native to the Mediterranean region, northern Africa, and parts of Asia. It can be grown in average garden soil in a sunny location.

Common uses: Antibacterial, antiviral, expectorant (helping to remove secretions) of colds and bronchitis, antifungal

Parts used: Leaves

Caution: Avoid large amounts with hypothyroidism.

Although research on its therapeutic value is scant, thyme has been widely used for centuries as both a culinary and medicinal herb. This perennial shrub can be easily grown in a sheltered spot in the garden.

A strong tea of thyme should be considered to help eliminate mucus congestion, coughs, or sore throat associated with a cold or flu. A soothing cough medicine can be made by steeping dried thyme in honey. Tea can also be used as a gargle to ease or prevent a sore throat. Add a strong tea to bath water to soothe and deodorize the skin. A steam inhalation is effective for sinus congestion.

Yarrow (*Achillea millefolium*)

Family: Compositae

Other names: Soldier's wound wort, thousand weed, staunchweed, sanguinary, milfoil

Habitat: Temperate regions of North America and Europe

Common uses: Styptic, anti-inflammatory

Parts used: Flower heads

Cautions: Some individuals' skin may be sensitive; avoid during pregnancy.

Yarrow is a hardy, rampant grower and easily crowds out more delicate plants. Thus, you may want to confine it to its own section of the garden. It likes a hot, sunny location and is not fussy about rich soil, but will not tolerate wet roots.

Archeologists have identified fossils of yarrow pollen in Neanderthal burial caves of 60,000 years ago. It was used as a styptic 3,000 years ago to stop bleeding from wounds suffered in the Trojan War. Native American tribes used this herb for skin sores and wounds, and it was included in the medical supplies issued during the American Civil War.

Yarrow, applied as a poultice, has been used for its ability to stop bleeding and to reduce inflammation; however, there is no research supporting this claim to date.

Summary

Phytochemicals are chemicals that occur in plants. Herbs can have therapeutic effects in the body, some of which include tonics, which are nourishing; adaptogens, which help the body regain normal function in the presence of stress; and immune stimulants, which enhance the immune system's ability to fight an illness.

Herbal remedies can be used in the form of an extract, tea, infusion, decoction, tincture, capsule, compress, poultice, liniment, salve, or ointment.

When wildcrafting herbs, caution must be taken to identify the plant properly, assure that there is no contamination from pesticides or other chemicals, and avoid picking endangered species.

Children require lower doses of herbs. Herbs should not be used with infants unless guided by a healthcare practitioner.

Each herb has unique uses and cautions. It is important that individuals become knowledgeable of herbs they intend to use to determine appropriateness for the given condition, dosage, and safety issues.

References

National Center for Complementary and Alternative Medicine. (2012*). St. John's wort and depression.* Retrieved from nccam.nih.gov/health/stjohnswort/sjw-and-depression.htm

Natural Medicines Comprehensive Database. (2012). *Sage.* Retrieved from www.naturaldatabase.com

Rapaport, M. H., Nierenberg, A. A., Howland, R., Dording C., Schettler P. J., & Mischoulon, D. (2011). The treatment of minor depression with St. John's wort or citalopram: Failure to show benefit over placebo. *Journal of Psychiatric Research;45*, 931–941.

Scholey, Tildesley, Ballard, Wesnes, Tasker, Perry, et al. (2008). An extract of *Salvia* (sage) with anticholinesterase properties improves memory and attention in healthy older volunteers. *Psychopharmacology, 198*(1), 127–139.

Suggested Reading

Avins, A. L., & Bent, S. (2006). Saw palmetto and lower urinary tract symptoms: What is the latest evidence? *Current Urology Report, 7*(4), 260–265.

Basch, E., Bent, S., Foppa, I., Haskmi, S., Kroll, D., Mele, M., et al. (2006). Marigold (*Calendula officinalis*): An evidence-based systematic review by the Natural Standard Research Collaboration. *Journal of Herbal Pharmacotherapy, 6*(3–4), 135–159.

Bent, S., Kane, C., Shinohara, K., Neuhaus, J., Hudes, E.S., Goldberg, H., & Avins, A.L.(2006). Saw palmetto for benign prostatic hyperplasia. *New England Journal of Medicine, 354*(6), 557–566.

Dog, T. L., Johnson, R., Foster, S., & Kiefer, D. (2012). *National Geographic guide to medicinal herbs: The world's most effective healing plants.* Margate, FL: National Geographic Books.

Duke, J. A. (2003). *The green pharmacy.* Emmaus, PA: Rodale Books.

Gaby, A. R. (2006). Natural remedies for herpes simplex. *Alternative Medicine Review, 11*(2), 93–101.

Gagnier, J. J., van Tulder, M., Berman, B., & Bombardier, C. (2006). Herbal medicine for low back pain. *Cochrane Database System of Reviews, 19*(2), CD004504.

Gardner, C. D., Lawson, L. D., Block, E., Chatterjee, L. M., Kiazand, A., Balise, R. R. , et al. (2007). Effect of raw garlic vs. commercial garlic supplements on plasma lipid concentrations in adults with moderate hypercholesterolemia: A randomized clinical trial. *Archives of Internal Medicine, 167*(4), 346–353.

Gruenwald, J., Brendler, T., & Jaenicke, C. (Eds.). (2007). *PDR for herbal medicines.* Montvale, NJ: Medical Economics Company.

Hobbs, C. (2006). *Herbal remedies for dummies.* Foster City, CA: IDG Books Worldwide, Inc.

Kains, M. G., & Frank, R. A. (2011). *Culinary herbs: Their cultivation, harvesting, curing, and uses.* CreateSpace.

Kasper, S., Anghelescu, I. G., Szegedi, A., Dienel, A., & Kieser, M.(2006). Superior efficacy of St. John's wort extract WS 5570 compared to placebo in patients with major depression: A randomized, double-blind, placebo-controlled, multi-center trial. *BMC Medicine, 4*(1), 14.

Kennedy, D. O., Little, W., Haskell, C. F., & Scholey, A. B. (2006). Anxiolytic effects of a combination of *Melissa officinalis* and *Valeriana officinalis* during laboratory induced stress. *Phytotherapy Research, 20*(2), 96–102.

Ladas, E., Kroll, D. J., & Kelly, K. M. (2005). Milk thistle (*Silybum marianum*). In P. Coates, M. Blackman, G. Cragg, M. Levine, J. Moss, & J. White (Eds.). *Encyclopedia of dietary supplements* (pp. 467–482). New York, NY: Marcel Dekker.

Ody, P. (2000). *The complete medicinal herbal.* New York, NY: DK Publishing, Inc.

Winston, D. (2003). *Herbal therapeutics: Specific indications for herbs and herbal formulas.* Broadway, NJ: Herbal Therapeutics Research Library.

Resources

American Botanical Council
P.O. Box 144345
Austin, TX 78714-4345
Phone: 512-926-4900; fax: 512-926-2345
www.herbalgram.org

American Herbalists Guild
14 Waverly Court
Asheville, NC 28805
Phone: 617-520-4372
www.americanherbalistsguild.com

Herb Research Foundation
4140 15th Street
Boulder, CO 80304
Phone 303-449-2265; fax 303-449-7849
www.herbs.org

Natural Medicines Comprehensive Database
www.naturaldatabase.com

Office of Dietary Supplements, National Institutes of Health
Bethesda, MD 20892
http://ods.od.nih.gov

Aromatherapy: Common Scents

OBJECTIVES

This chapter should enable you to

- Define *aromatherapy*
- Describe the method of extraction that makes an essential oil pure
- List the four basic types of aromatherapy
- Describe four methods of using aromatherapy
- Describe the use of aromatherapy for at least five different health problems
- List at least 10 precautions or contraindications with the use of aromatherapy

---

Aromatherapy refers to the therapeutic use of essential oils. This branch of herbal medicine is often misunderstood and maligned; even its name is a bit of a misnomer. Contrary to popular belief, aromatherapy is not a new therapy, but part of one of the oldest therapies, as herbal medicine dates back 6,000 years. Aromatic plants have been used by those in many parts of the world, including India, China, North and South America, Greece, the Middle East, Australia, New Zealand, and Europe. According to the World Health Organization (WHO), today, more than 85% of the world population still relies on herbal medicine, and many of the herbs are aromatic (Sharma, Patel, & Chaturvedi, 2009).

The renaissance of modern aromatherapy appeared in France just before World War II. (This was around the time the first antibiotics were used.) A physician named Jean Valnet, a chemist named Maurice Gattefosse, and a surgical assistant by the name of Marguerite Maury were key figures in the rediscovery of this ancient art of healing. It is fascinating that they did not use aromatherapy for its nice smell, nor did they use it for stress reduction—two

> **KEY POINT**
>
> Aromatherapy is viewed as such an integral part of medicine in Germany that doctors and nurses there are tested in the use of essential oils in order to become licensed.

of the most popular ways aromatherapy is used today. Instead, they used it clinically, as they would use any natural medicine, to help wounds heal, fight infections, and to improve skin texture. This clinical approach to aromatherapy has survived in France, and many physicians still use essential oils as an alternative or enhancement to antibiotics today. In France, as in Germany, the use of plants medicinally (*phytomedicine*), including aromatherapy, is seen as an extension of orthodox medicine.

Aromatherapy does not just mean using aromas, but the therapeutic use of essential oils. Perhaps we should add the controlled therapeutic use as essential oils are not toys, but powerful tools that can help you stay healthy. Real essential oils are either steam distillates or expressed extracts from aromatic plants. Many of them have familiar smells, such as lavender, rose, and rosemary; however, things are not quite as simple as they seem at first glance.

Essential oils are highly volatile droplets created by a plant itself to help ward off infection (bacterial, fungal, or viral), to regulate growth and hormones, and to mend damaged tissues of the plant. These tiny reservoirs of plant medicine are stored in the plant's veins, glands, or sacs, and when they are broken by being crushed or rubbed, the essential oil is released along with the aroma. Lavender has a minimal scent until the flowering head is gently rubbed between two fingers. Some plants store large amounts of essential oils—some store very little. This, along with the difficulty of harvesting the essential oils, dictates the price. For example, more than 100 kilograms of fresh rose petals are needed to produce 60 grams of essential oils. (This means that rose is one of the most expensive essential oils on the market today and also one of the most frequently adulterated!)

There are a few important things to know about essential oils before you start using them for your health and well-being. These are the method of extraction, the botanical name (for clear identification, see **Exhibit 26-1**), methods of application, safety, storage, and contraindications.

Extraction

The method of extraction is crucial, as only steam-distilled or expressed extracts can legitimately be called essential oils. These two methods produce a pure product with no additional solvent or impurity. A bottle of essential

EXHIBIT 26-1   ESSENTIAL OILS MENTIONED IN THIS CHAPTER AND THEIR BOTANICAL NAMES

Aniseed	Pimpinella anisum
Basil	Ocimum basilicum
Chamomile German	Matricaria recutita
Chamomile Roman	Chamomelum nobile
Clary sage	Salvia sclarea
Coriander seed	Coriandrum sativum
Eucalyptus	Eucalyptus globulus
Fennel	Foeniculum vulgare
Geranium	Pelargonium graveolens
Ginger	Zingeber officinale
Hyssop	Hyssopus officinalis
Lavender True	Lavandula officinalis
Lemongrass	Cymbopogon citratus
Neroli	Citrus aurantium var amara
Palmarosa	Cymbopogon martini
Parlsey	Petroselinum sativum
Pennyroyal	Mentha pulegium
Peppermint	Mentha piperita
Rose	Rosa damascena
Rosewood	Aniba rosaeodora
Sage	Salvia officinalis
Sandalwood	Santalum album
Tarragon	Artemesia dracunculus
Wintergreen	Gaultheria procumbens
Ylang-ylang	Cananga odorata

oils should state that the contents are pure essential oils—steam distilled or expressed. (Only the peel from citrus plants, such as mandarin, lime, or lemon, produces an expressed oil.)

KEY POINT

Many of the essential oils on the market are solvents extracted using petrochemicals. These hexane-based residues may cause allergic or sensitive reactions.

<table>
<tr><td>

EXHIBIT 26-2   AROMATHERAPY JOURNALS

Aromatherapy Journal
www.naha.org

Aromatherapy Today
www.aromatherapytoday.com

International Journal of Clinical Aromatherapy
www.ijca.net

</td></tr>
</table>

Identification

There can be *many* different species of the same plant. For example, the genus (or surname) of thyme is *Thymus*, but there are more than 60 different species or varieties of thyme, each with different therapeutic effects. (There are 3 different species of lavender and 600 different species of eucalyptus.) It is very important to know the full botanical name of a plant so that you can use it correctly. (Exhibit 26-1 offers a list of safe essential oils to use. It gives both the botanical name and common name.) Do not buy anything that is labeled *lavender oil*, as you will have no way of knowing what you are buying. Which lavender is it? One lavender is a soothing, calming sedative oil that is exceptional for burns, but another lavender is a stimulant and expectorant (helps you cough up mucus and will not help you sleep or soothe your burns). You need to ask the following questions: Is it a true essential oil? How has it been extracted?

The simplest way to ensure that you are buying the real thing is to look in one of the professional journals (see **Exhibit 26-2**) and ask for an order form from one of the advertised suppliers. If they do not list the botanical name, method of extraction, country of origin, and part of the plant, then you cannot be certain of what they are selling.

Essential oils are common ingredients in the pharmaceutical, perfume, and food industries and, as such, are commonly used by most of the population on a daily basis. Pure essential oils rarely produce an allergic effect, unlike their synthetic cousins; however, many products on the market have been extended with synthetic fragrances and can cause a reaction.

How It Works

As mentioned, the term *aromatherapy* refers to the therapeutic use of essential oils that are the volatile organic constituents of plants. Essential oils are thought to work at psychological, physiological, and cellular levels; this means they can affect our body, our mind, and all of the delicate links in between. The effects of aroma can be rapid, and sometimes just thinking about a scent can be as powerful as the actual scent itself. Take a moment to think of your favorite flower. Then think about an odor that makes you feel nauseated. The effects of an aroma can be relaxing or stimulating depending on the previous experience of the individual (called the learned memory), as well as the actual chemical makeup of the essential oils used.

> **KEY POINT**
>
> Aromatherapy should not be used as a replacement for medical treatment. It is a complementary therapy and is most useful when integrated with conventional medicine.

How Scents Affect You

Olfaction

Essential oils are composed of many different chemical components or molecules. These different chemical components travel via the nose to the olfactory bulb. Nerve impulses travel to the limbic system of the brain, the oldest part of our brain. There the aroma is processed. The limbic part of our brain contains an organ called the *amygdala*. This is the organ that governs your emotional response. Valium is thought to have a calming, sedative effect on the amygdala; *Lavandula angustifolia* (true lavender) has a similar effect. The limbic system also contains another organ called the *hippocampus*. This organ is involved in the formation and retrieval of explicit memories. This is why an aroma can trigger memories that may have lain dormant for years.

REFLECTION

Does the scent of cinnamon buns baking, honeysuckle, or a particular cologne trigger specific memories for you? Can you recall unique scents associated with people or places you have known in the past? What are these?

The effect of scents on the brain has been mapped using computer-generated graphics. These brain electrical activity maps indicate how subjects, linked to an electroencephalogram, rated different odors presented to them.

> **KEY POINT**
>
> The loss of the sense of smell is called anosmia; however, even in a person who has lost the sense of smell, if the olfactory nerve is intact, the chemicals in the aroma will still be able to travel to the limbic part of the brain and have a therapeutic effect.

These maps have shown that scents can have a psychological effect even when the aroma is below the level of human awareness.

Absorption Through the Skin

Essential oils are absorbed through the skin through diffusion, in much the same way as medicines administered in patches. The two layers of the skin—the dermis and fat layers—act as a reservoir before the components within the essential oils reach the bloodstream. There is some evidence that massage or hot water enhances the absorption of at least some of the essential oil's components. Essential oils, because they are lipophilic (attracted to fat), can be stored in the fatty areas of the body and can pass through the blood–brain barrier.

Who Uses Aromatherapy?

Aromatherapy is commonly practiced in England, France, Germany, Switzerland, Sweden, Australia, New Zealand, and Japan. Its use has grown in the United States in the past 2 decades. Many nurses and other health professionals are receiving training in aromatherapy to enhance their care. In France, medical doctors and pharmacists use aromatherapy as part of conventional medicine, often for the control of infection.

Types of Aromatherapy

There are four basic types of aromatherapy:

1. *Esthetic.* Used purely for pleasure, such as in candles and soaps
2. *Holistic.* Used for general stress
3. *Environmental fragrance.* Used to manipulate mood or enhance sales
4. *Clinical.* Used for specific therapeutic outcomes that are measurable

Methods of Using Aromatherapy

Essential oils can be absorbed by the body in one of the following three ways:

1. *Inhalation*: 1–5 drops undiluted—without touch
2. *Through the skin*: 1%–12% diluted in a carrier oil, used with compresses or massage—via touch
3. *Orally*: 1–2 drops, which is considered aromatic medicine, requiring the training of a primary care provider who has prescribing privileges.

Inhalation

Direct inhalation means an essential oil is directly targeted to an individual by placing 1 to 5 drops on a tissue (or floated on hot water in a bowl) and being inhaled by that individual for 5–10 minutes.

Indirect inhalation includes the use of burners, nebulizers, and vaporizers that can use heat generated by battery or electricity and may or may not include the use of water. Larger portable aroma systems are available to control the release of essential oils on a commercial scale into rooms up to 1,500 square feet. This is similar to using environmental fragrance (using synthetics), which is common practice in hotels and department stores, and can be useful for mood enhancement and stress reduction.

Through the Skin

Baths

Essential oils can be used in baths by dissolving 4–6 drops of essential oil into a teaspoon of milk or salt and adding the oil–milk or oil–salt mixture to bath water. Essential oils do not dissolve in water and would float on the top if not dissolved in milk or salt, giving an uneven treatment. You then can relax in the bath for 10 minutes.

Compresses

To prepare a compress, add 4–6 drops of essential oil to warm water. Soak a soft cotton cloth in the mixture, wring it out, and apply to the affected area (contusion or abrasion). Cover the external surface with food plastic wrap to maintain moisture, cover with a towel, and keep in place for 4 hours.

Touch

Aromatherapy is often used in a gentle massage or the *M* technique. Use 1% to 12% essential oils diluted in a teaspoonful (5 ml) of cold-pressed vegetable oil, cream, or gel. Gentle friction and hot water enhance absorption of essential oils through the skin into the bloodstream. The amount of essential oils absorbed from an aromatherapy massage will normally be 0.025 to 0.1 ml; this is approximately 0.5 to 2 drops.

Aromatherapy and Women's Health

Aromatherapy, which involves the senses of smell and touch, is possibly the most feminine of complementary therapies and is ideally suited to women and their health concerns. (Maybe this is why so many trained aromatherapists are women!)

KEY POINT

As aromatherapy is one of the most nurturing therapies, it is hardly surprising that many nurses throughout the world are learning to use aromatherapy as an enhancement to their nursing care. Aromatherapy began its nursing debut in England, in geriatric care, when Helen Passant first brought it to the attention of the nursing community in Oxford. She used aromatherapy in the ward and reduced the ward's drug bill by one third. The hospital immediately responded by reducing her budget by one third!

Almost every aspect of a woman's life can be enhanced by aromatherapy. It should be noted that although most essential oils have no effect on orthodox medication, some essential oils can augment or diminish the effects of certain medications. Following are some women's health problems that can be successfully addressed with aromatherapy.

Problems with the Menstrual Cycle

The menstrual cycle is delicately balanced. Hormones, specifically estrogen, are easily thrown out of balance by stress, illness, poor diet, or overwork. *Pelargonium graveolens* (geranium) has been used for generations to balance the female hormonal system. This species of geranium acts on the adrenal cortex that regulates the endocrine system and therefore affects the menstrual cycle.

Premenstrual tension, irregular periods, and painful periods can be greatly helped by aromatherapy. Add three drops of geranium to a teaspoon of vegetable oil. The vegetable oil should be cold pressed and not one used for cooking. Gently massage into the lower abdomen and lumbar (lower back) areas. For optimum results, apply morning and evening. For very painful menstrual periods, add 1–2 drops of high-altitude *Lavandula angustifolia* (true lavender) and *Chamaemelum nobile* (Roman chamomile) to enhance the antispasmodic effect. Geranium may also encourage regular ovulation.

Menopausal Problems

Pelargonium graveolens (geranium) may be particularly useful during menopause as an adrenal regulator when estrogen supplies begin to dry up. A disruption in the supply of estrogen can lead to irritability, mood swings, and hot flashes. This essential oil is also excellent for mature, dry skin, and thread veins that can accompany menopause. Add 3–5 drops of geranium to a teaspoon of evening primrose oil. Rub anywhere on the body, or add 5 drops to a bath and soak for 10 minutes. *Foeniculum vulgare dulce* (sweet fennel), *Salvia sclarea* (sage), and *Pimpinella anisum* (aniseed) all contain a molecule similar to estradiol—the female hormone—and can be very beneficial at menopause also.

Hot flashes can be helped with a spritzer of geranium, clary sage, and peppermint. To make a spritzer, add 6 drops of each oil to 10 fluid ounces of water and shake vigorously before spraying the face and neck. Keep eyes closed while spraying.

Infertility

Where there is no physical reason why a woman cannot conceive, often the underlying reason can be one of tension. Trying to become pregnant can be emotionally draining, and the longer it takes the more stressful it can become. *Salvia sclarea* (clary sage), related to *S. officinalis* (common sage), can help a woman relax. Clary sage contains an alcohol (sclareol), which is similar in molecular form to estradiol. As well as having an estrogen-like effect, clary sage is very relaxing and antispasmodic (it contains 75% linalyl acetate, an ester).

Morning Sickness

Early morning nausea and vomiting often occur during the first 3 months of pregnancy, although they can sometimes last the whole 9 months. Nausea can be greatly alleviated with a brief inhalation of *Zingiber officinale* (ginger) or *Mentha piperita* (peppermint). Always remember that only 1 or 2 drops on a handkerchief is necessary, but you can repeat inhalation whenever the need arises. Do not use more than 2 drops of either essential oil at a time, as more may exacerbate the nausea, not alleviate it. The inhalation of either ginger or peppermint for nausea during pregnancy is a soothing and safe practice. Aromatherapy is not particularly advocated during the first 3 months of pregnancy, as essential oils do cross the placenta; however, many expectant mothers like to relax in an aromatic bath and the majority of essential oils will have no detrimental effect when used in this gentle way. Aromatherapy can be used during the remainder of a pregnancy, but use caution. Certain essential oils can also be safely used during labor. (Please refer to the books listed in the "Suggested Reading" section for guidance.)

Breastfeeding, Engorged Breasts, and Sore Nipples

Foeniculum vulgare (fennel), which mimics estrogen, has a good milk-producing action, which can help mothers who breastfeed. Fennel also has a soothing decongestant effect on engorged breasts, which will be a relief to those in the early stages of feeding. Apply 1–2 drops diluted in a carrier oil to the breasts, avoiding the nipple area, twice a day. Wash area immediately before feeding. As fennel is a gentle laxative, the baby may have looser stools.

KEY POINT

Because of their phytoestrogenic effect, sage, fennel, and aniseed are contraindicated in persons with cancer when the tumor is estrogen dependent.

Both *Matricaria recutita* (German chamomile) and *Chamaemelum nobile* (Roman chamomile) are used in some European commercial brands of ointment available for sore nipples. Both chamomiles are useful for swelling and congestion. Apply a diluted solution in a carrier oil (1 drop per 5 ml) after each feeding and be sure to wash off any residual solution before the next feeding. *Lavandula angustifolia* (true lavender) is another useful essential oil for sore nipples, as this lavender has recognized properties that enhance healing.

Postpartum Blues

Does one have postpartum blues or is she just down in the dumps? Aromatherapy can lift spirits and ease those moments when a mother feels that she cannot cope with yet another dirty diaper. *Citrus aurantium ssp bergamia* (bergamot) and *Melissa officinalis* (true Melissa) are ideal essential oils to make the day just a little brighter. *Melissa* is difficult to obtain unadulterated; however, there are some companies with integrity, and the real stuff can work miracles! It is used by putting 1–3 drops on a handkerchief and breathing deeply. Bergamot should not be used topically before sunbathing or using a sun bed, as this could result in skin photosensitivity and, in some cases, burns.

Vaginal Infections

Vaginal candidiasis, caused by the yeast *Candida albicans*, bacterial vaginosis caused by a bacteria, and *Trichomonas* vaginalis, caused by *Trichomonas*, are among the types of vaginitis that can create a nuisance in many women's lives. Sometimes a vaginal yeast infection may be a side effect of antibiotic treatment, and it can occur during pregnancy or after an illness. When it does occur, vaginal yeast infection is messy, uncomfortable, and embarrassing. Some forms are now resistant to many of the conventional preparations; however, one essential oil, called tea tree, can help eradicate this fungal infection in only a few days. Make sure you have the correct essential oil, as tea tree is the common name of many different types of plants. The right tea tree is called *Melaleuca alternifolia CT terpineol*. There is another, called *Melaleuca alternifolia*, that contains much higher percentages of terpineol, an oxide, which can produce discomfort when applied to the skin or mucous membrane. Put a teaspoon of carrier oil onto a saucer. Add 2–3 drops of essential oil and mix with a clean finger. Take a tampon, remove the applicator, and roll the tampon in the mixture until all the mixture has been absorbed. Insert the tampon into the vagina.

The tampon should be changed three times a day. Each time soak a new tampon in a fresh dilution of carrier oil with 2–3 drops of tea tree. The tampon needs to be kept in the vagina overnight. Commonly, tea tree will remove vaginal yeast infection in 3 days, regardless of how long the woman has had the infection.

Cystitis

Inflammation of the bladder can be due to infection, but this is not always the case. Cystitis can plague some unfortunate women for much of their lives, bringing pain and misery and taking a high toll on intimate relationships. Although many factors contribute to cystitis, such as tight clothing, insufficient fluid intake, and a diet of high sugar and refined food, stress does play an important role. Specific essential oils can help this condition. Choose an antispasmodic essential oil, which also has a strong antibiotic action, such as *Juniperus communis var. erecta* (juniper), *Cymbopogon citratus* (West Indian lemongrass), which contains myrcene and is very effective as a peripheral analgesic, or *Origanum majorana* (sweet marjoram). Apply 1–2 drops diluted in a teaspoon of cold-pressed vegetable oil to the lower abdomen and lower back (kidney area). Repeat up to 4 times a day while the attack lasts, and remember to drink at least 2 liters of water a day. Drinking cranberry juice can also help in the initial stages of cystitis.

REFLECTION

How do different scents affect your moods? How can you intentionally use this to enhance your health and well-being?

Aromatherapy for Common Complaints

Here are some essential oils that can be used for common complaints that typically are self-medicated. When using essential oils on the skin (topically), use 1–5 drops diluted in a teaspoon of carrier oil. When inhaling the essential oil, inhale 1–2 drops on a cotton ball, or add 1 to 2 drops to a basin of steaming hot water (method of application: I = inhalation, T = topical).

Psychological Issues

Insomnia: Lavender, ylang-ylang, clary sage, frankincense, neroli oil (I, T)

Depression: Bergamot, basil, lavender, geranium, neroli oil, angelica, rose, Melissa (I, T)

Stress and anxiety: Lavender, frankincense, Roman chamomile, mandarin, angelica, rose (I, T)

Anorexia: Rose, neroli oil, lemon, fennel (I, T)

Withdrawal from substance abuse: *Helichrysum*, angelica, rose (T)

Pain Relief

Migraine: Peppermint, lavender (T)

Osteoarthritis: Eucalyptus, black pepper, ginger, spike lavender, Roman chamomile, rosemary, myrrh (T)

Rheumatoid arthritis: German chamomile, lavender, peppermint, frankincense (T)

Lower back pain: Roman chamomile, black pepper, eucalyptus, lemongrass, rosemary, lavender, sweet marjoram (T)

Cramps: Roman chamomile, clary sage, lavender, sweet marjoram (T)

General aches and pains: Rosemary, lavender, lemongrass, clary sage, black pepper, lemon eucalyptus, spike lavender

Women's Problems

Menopausal symptoms: Clary sage, sage, fennel, aniseed, geranium, rose, cypress (I, T)

Menstrual cramping: Roman chamomile, lavender, clary sage (T)

Premenstrual syndrome; infertility with no physiological cause: Clary sage, sage, fennel, aniseed, geranium, rose (T)

Blood Pressure

Borderline high blood pressure (not on medication): Ylang-ylang, true lavender

Low blood pressure (can be caused by some antidepressants): Rosemary

Urinary Problems

Cystitis: Tea tree, palmarosa (T, especially sitz bath)

Water retention: Juniper, cypress, fennel (T)

Digestive Problems

Irritable bowel syndrome: Roman chamomile, clary sage, mandarin, cardamom, peppermint, mandarin, fennel, lavender

Constipation: Fennel, black pepper (T)

Indigestion: Peppermint, ginger (I)

Infections

Bacterial (MRSA, VRSA): Tea tree (I, T)

Other bacteria (depends on bacteria): Eucalyptus, niaouli, sweet marjoram, oregano, tarragon, savory, German chamomile, thyme, manuka (I, T)

Viral: ravensara, palmarosa, lemon, Melissa, rose, bergamot (I, T)

Fungal: Lemongrass, black pepper, holy basil, clove, cajuput, caraway (T), geranium, tea tree—particularly good for toenail fungus (apply twice daily undiluted for 3 months to nail bed)

Respiratory Problems

Bronchitis: Ravensara, *Eucalyptus globulus*, *Eucalyptus smithii*, tea tree, spike lavender (I)

Sinusitis: *Eucalyptus globulus*, lavender, spike lavender, rosemary (I)

Mild asthma: Lavender, clary sage, Roman chamomile (I; patch test on arm first)

Skin Problems

Mild acne: Tea tree, juniper, cypress, niaouli (T)

Mild psoriasis: Lavender, German chamomile (T)

Diabetic ulcers: Lavender, frankincense, myrrh (T)

Chemotherapy Side Effects

Nausea: Peppermint, ginger, mandarin (I)

Postradiation burns: Lavender, German chamomile with tamanu carrier oil (T)

Muscular Problems

Sports injuries: Spike lavender, rosemary, sweet marjoram, black pepper, lemongrass, frankincense (T)

Children's Care

Irritability: Mandarin, lavender, Roman chamomile, rose (I, T)

Colic: Roman chamomile, mandarin (T; gently massaged to the abdomen)

Diaper rash: Lavender, German chamomile (T)

Sleep problems: Lavender, rose, mandarin, ylang-ylang

Autism (to aid with social interaction): Rose, mandarin, lavender, sweet marjoram, clary sage, *Pinus sylvestris*

Geriatric Care

Memory loss: Rosemary, rose, eucalyptus, peppermint, bergamot (T, I)

Dry, flaky skin: Geranium, frankincense, oil of evening primrose (T)

Alzheimer's disease: Rosemary, lavender, pine, frankincense, rose (I)

End-of-Life Care: Pain Relief

Spiritual: Rose, angelica, frankincense

Physical: Lavender, peppermint, lavender, lemongrass, rosemary

Emotional: Geranium, pine, sandalwood

Relaxation: Lavender, clary sage, mandarin, frankincense, ylang-ylang

Bed sores: Lavender, tea tree, sweet marjoram, frankincense

Care of the Dying

Rites of passage: Choose selection of patient's favorite aromas or frankincense or rose

Bereavement: Rose, sandalwood, patchouli, angelica

Actions

The pharmacologically active components in essential oils work at psychological, physical, and cellular levels. Essential oils are absorbed rapidly through the skin—some essential oils are now being used to help the dermal penetration of orthodox medication. Essential oils are lipotropic and are excreted through respiration, kidneys, and skin.

> **KEY POINT**
>
> Because essential oils can produce physiological and psychological effects, consumers need to share their use with their healthcare providers and healthcare providers need to make inquiry into their use during every assessment.

Risks and Safety

Most essential oils have been tested by the food and beverage industry, as many essential oils are used as flavorings. Other research has been carried out by the perfume industry. Most of the commonly used essential oils in aromatherapy have generally been regarded as safe.

Aromatherapy is a very safe complementary therapy if it is used within recognized guidelines. People with atopic eczema should avoid use. Some essential oils have caused dermal sensitivity—mostly through impure extracts. Generally, essential oils that are high in esters and alcohols tend to be gentle in their action and are the most safe to use. Essential oils that are high in phenols tend to be more aggressive and should not be used over long periods of time. There is a list of banned or contraindicated essential oils to guide the novice (see **Exhibit 26-3**). Do not administer essential oils orally unless you are trained in this method.

Herbs can interact with medications. Some precautions are listed in **Exhibit 26-4**. Additional warnings to heed are as follows:

- Sage, clary sage, fennel, and aniseed should be avoided with estrogen-dependent tumors.
- Hyssop should be avoided in pregnancy.
- Rosemary should be avoided by those with high blood pressure.
- Hyssop should be avoided by those prone to seizures.
- Cinnamon may cause dermal irritation.
- Bergamot may cause dermal irritation or burns when used with sun beds or sunshine.
- Peppermint can increase the lungs' permeability to nicotine.
- Rosemary, eucalyptus, and bay laurel contain high levels of cineole, which can increase the metabolism of barbiturates and interfere with the metabolism of anesthesia (it is best to avoid inhaling these oils for at least a week prior to surgery).
- Essential oils high in methyl salicylate can interfere with anticoagulant therapy and trigger reactions in persons with asthma.

EXHIBIT 26-3   CONTRAINDICATED ESSENTIAL OILS

COMMON NAME	BOTANICAL NAME
Basil (exotic)	Ocimum basilicum
Birch	Betula lenta
Boldo	Peumus boldus
Buchu	Agothosma betulina
Cade	Juniperus oxycedrus
Calamus	Acorus clamus var angustatus
Camphor (brown)	Cinnamomum camphora
Camphor (yellow)	Cinnamomum camphora
Cassia	Cinnamomum cassia
Cinnamon bark	Cinnamomum zeylanicum
Costus	Saussurea costus
Elecampane	Inula helenium
Horseradish	Armoracia rusticana
Melaleuca	Melaleuca bracteata
Mustard	Brassica nigra
Pennyroyal	Mentha pulegium
Ravensara	Ravensara anisata
Sage (dalmation)	Salvia officinalis
Sassafras	Sassafras albidum
Tansy	Tanacetum vulgare
Tarragon*	Artemesia dracunculus
Thuja	Thuja occidentalis
Verbena*	Lippia citriadora (Aloysia triphylla)
Wintergreen	Gaultheria procumbens
Wormseed	Chenopodium ambrosiodes varanthelminticum
Wormwood	Artemesia absinthium

* Somewhat controversial. Many aromatherapists use tarragon and verbena.

EXHIBIT 26-4   DRUG INTERACTION WITH AROMATHERAPY

1. Avoid when using homeopathic remedies; strong aromas like peppermint and eucalyptus can negate homeopathic remedies.
2. Avoid chamomile if allergic to ragweed.
3. Eucalyptus globulus and cananga odorata may affect the absorption of 5-fluorouracil, a drug commonly used in chemotherapy.
4. Terpineol (a component of some essential oils) may decrease the narcotic effect of pentobarbital—mainly when used orally.
5. Cymbopogon citratus (West Indian lemongrass) can increase the effects of morphine (according to studies in rats and when given orally).
6. Peppermint may negate quinidine in atrial fibrillation.
7. Lavender may increase the effect of barbiturates.
8. The effect of tranquilizers, anticonvulsants, and antihistamines may be slightly enhanced by sedative essential oils.

Pregnancy and Lactation

Use caution during the first trimester. Although many women do use essential oils successfully and safely during their pregnancies, it is suggested that some essential oils should be avoided altogether, although the data are based on taking the essential oils orally, not inhaling or applying them topically. These include sage, pennyroyal, camphor, parsley, tarragon, wintergreen, juniper, hyssop, and basil. The following are thought to be safe in pregnancy: cardamom, chamomile (Roman and German), clary sage, coriander seed, geranium, ginger, lavender, neroli, palmarosa, patchouli, petitgrain, rose, rosewood, and sandalwood.

Warnings/Contraindications/Precautions When Using Essential Oils

- Do not take by mouth (unless guided by a person trained in aromatic medicine and in coordination with your healthcare provider).
- Do not touch your eyes with essential oils. If essential oils get into eyes, rinse out with milk or carrier oil (essential oils do not dissolve in water) and then water.

- Store away from fire or naked flame, as essential oils are highly volatile and highly flammable.
- Store in a cool place out of sunlight, in colored glass—amber or blue. Store expensive essential oils in a refrigerator.
- Many essential oils stain clothing—beware!
- Do not use essential oils undiluted on the skin.
- Keep away from children and pets.
- Only use essential oils from reputable suppliers who can supply the correct botanical name, place of origin, part of plant used, method of extraction, and batch number when possible. The term *lavender oil* by itself means absolutely nothing!
- Always close the container immediately.
- Use extra care during early pregnancy.
- Use extra care with people receiving chemotherapy.
- Be aware of which essential oils are photosensitive, for example, bergamot.
- Avoid use with individuals who have severe asthma or multiple allergies.

Adverse Reactions

There have been some rare cases of adverse skin reactions caused by sensitivity. The majority of cases were from extracts that contained residual petrochemicals. People with multiple allergies are more likely to be sensitive to aromas. Bergamot used in conjunction with sunshine or tanning beds can result in skin damage, ranging from redness to full thickness burns.

Administration

Essential oils can be used topically or inhaled. One to five drops of essential oils are diluted in 5 cc (a teaspoon) cold-pressed vegetable oil, such as sweet almond oil for topical application. For inhalation, inhale for 10 minutes as necessary. Use touch methods such as massage or the *M* technique when appropriate. Simple stress management can be incorporated into the everyday regime with the use of baths and foot soaks, vaporizers, and sprays.

Self-Help Versus Professional Help

Aromatherapy can be self-applied for stress management, but for more clinical uses, it is better to have some training and knowledge of the chemistry and

> ### KEY POINT
>
> Currently, there is no recognized national certification and no governing body for aromatherapy, but the Steering Committee for Educational Standards in Aromatherapy in the United States has established the Aromatherapy Registration Board, a nonprofit entity that is responsible for administering a national exam. This exam is not clinically based. Graduates have RA (registered aromatherapist) after their names. The largest professional body for aromatherapists is the National Association of Holistic Aromatherapy. Currently, there are no requirements to become certified or accredited, and training can range from one weekend to several years. For a clinical aromatherapist who is also a licensed health professional, look for CCAP (certified clinical aromatherapy practitioner) after his or her name.

extraction methods. Many essential oils are sold under their common names. *Origanum majorana* (sweet marjoram) is an excellent essential oil for insomnia; however, *Thymus mastichina* (Spanish marjoram) is frequently sold as marjoram, but it is not a marjoram and certainly will not help insomnia. *Lavandula angustifolia* and *L. latifolia* are both sold as lavender. *L. angustifolia* has sedative and antispasmodic properties; *L. latifolia* is a stimulant and expectorant.

The field of aromatherapy is vast. It can be fun to use essential oils in your home every day just for the pleasure they give. They can make your home smell more welcoming. They can help your stress level, calm you down, and many other things, as listed previously; however, if you wish to use essential oils for clinical conditions, such as a chronic health problem, it is best to visit a professional trained in clinical aromatherapy.

Aromatherapy is misunderstood because it is such a broad field; however, it encompasses many facets from pleasure through pain to infection, and it is not limited to scented candles and potpourri. Commercial use of the word *aromatherapy* to boost sales of products, such as shampoos, has confused the public. In fact, few if any cosmetic or pharmaceutical products currently include essential oils—they use synthetic fragrances as they are much cheaper; however, essential oils have been used safely for thousands of years. Aromatherapy used on a daily basis can be a useful component of healthy living.

Summary

Aromatherapy implies the therapeutic use of essential oils. Individuals can obtain the therapeutic effects of oils by inhaling them or absorbing them through the skin through baths, compresses, or touch; oils can be orally administered, but this must be done only by a trained professional. There are four basic types of aromatherapy: esthetic, holistic, environmental, and clinical. Special precautions must be considered when using aromatherapy. Essential oils are specific in their effects and must be selected based on their therapeutic benefit for the intended use and appropriateness for the specific individual.

References

Sharma, A., Patel, V. K, & Chaturvedi, A. N. (2009). Vibriocidal activity of certain medicinal plants used in Indian folklore medicine by tribals of Mahakoshal region of central India. India. *Journal of Pharmacology 41*(1), 129.

Suggested Reading

Buckle, J. (2000). The *M* technique. *Massage and Bodywork, 2*, 52–65.
Buckle, J. (2001). The role of aromatherapy in nursing care. *Nursing Clinics of North America, 36*(1), 57–71.
Buckle, J. (2003). *Clinical aromatherapy essential oils in practice*. London, England: Churchill Livingstone.
Cert, S. P., & Cert, L. P. (2012). *Aromatherapy for health professionals* (4th ed.). London, UK: Elsevier.
Clark, S. (2008). *Essential chemistry for aromatherapy*. New York, NY: Churchill Livingstone.
Cooksley, V. (2008). Aromatherapy research and practice. *AHNA Beginnings, 28*(3), 22–23.
Herz, R. S. (2009). Aromatherapy facts and fictions: A scientific analysis of olfactory effects on mood, physiology and behavior. *International Journal of Neuroscience, 119*(2), 263–290.
Keville, K., & Green, M. (2008). *Aromatherapy: A complete guide to the healing art*. New York, NY: Crossings Press.
Price, S., Price, L., & Daniel, P. (2006). *Aromatherapy for health professionals*. New York, NY: Churchill Livingstone.
Schnaubelt, K. (2011). *The healing intelligence of essential oils: The science of advanced aromatherapy*. Rochester, VT: Healing Arts Press.
Shutes, J., & Weaver, C. (2007). *Aromatherapy for bodyworkers*. New York, NY: Prentice Hall.
Smith, L. (2003). *Healing oils, healing hands*. Avada, CO: HTSM Press.
Suskind, P. (1987). *Perfume: The story of a murderer*. New York, NY: Penguin.
Tisserand, R. (2004). *The art of aromatherapy*. Rochester, VT: Healing Arts Press.
Valnet, J. (1990). *The practice of aromatherapy*. Rochester, VT: Healing Arts Press.
Worwood, S., & Worwood, V. A. (2003). *Essential aromatherapy*. Novato, CA: New World Press.
Worwood, V. A. (2003). *The complete book of essential oils and aromatherapy*. Novato, CA: New World Press.
Worwood, V. A. (2006). *Aromatherapy for the soul*. Novato, CA: New World Press.

Resources

Aromatherapy and Allied Practitioners Association
www.aapa.org

Aromatherapy Registration Council
www.aromatherapycouncil.org

National Association for Holistic Aromatherapy
www.naha.org

Mind–Body Therapies

OBJECTIVES

This chapter should enable you to

- Describe the mind–body therapies of meditation/relaxation, biofeedback, guided imagery, hypnotherapy, yoga, and T'ai Chi

Mind–body therapies engage the mind to promote health and affect physical changes in the body. They focus on interactions of the brain, mind, body, and behavior. Included in this category are meditation, relaxation, biofeedback, guided imagery, hypnotherapy, music, T'ai Chi, yoga, and acupuncture (which also falls within traditional Chinese medicine and is discussed in chapter 23). These practices have been used for thousands of years throughout the world.

> **KEY POINT**
>
> Brain and mind are terms often used interchangeably; however, there are differences. The brain is visible, with a definite size, shape, and components (e.g., nerve cells, blood vessels). The mind lacks those physical characteristics and is responsible for an individual's thought process and emotions.

Meditation

Meditation is an ancient art of focused attention of the mind and is a practice to help the body reach a relaxed state. It can take a variety of forms, including the relaxation response, mindfulness meditation, Transcendental Meditation, and Zen Buddhist meditation, among others. Yoga and T'ai Chi also include meditation to some degree. Although meditation is associated with some

> **KEY POINT**
>
> Some of the benefits of reduced stress include decreased heart rate, breathing rate, blood pressure, and brain waves; improved mood; increased awareness; and spiritual calm.

spiritual practices, it can be practiced outside of a religious context for health purposes.

How It Works

Two major forms of meditation are concentrative meditation and mindfulness meditation. The former involves focusing on a sound, an image, or one's own breathing. By doing this process, a person reaches a state of calm and deepens attention and awareness. An individual can do this independently or be led through the process by a guide.

Mindfulness meditation brings full awareness in the present moment. It teaches one to not allow outside distraction to interrupt focus on the present moment. If other thoughts or feelings enter the mind, they are observed and allowed to pass. This helps a person to slow down, become more relaxed, and have more insight into what is occurring in the immediate moment.

One of the positive aspects of meditation is that it requires no special clothing or equipment and can be done virtually anywhere.

REFLECTION

What would be your greatest obstacles to engaging in meditation? How could you overcome them?

What It Helps

Meditation has unlimited usage. Research has demonstrated that practicing mindfulness meditation is associated with measurable changes in the brain regions involved in memory, learning, and emotion (Hölzel et al., 2011). Previous research has demonstrated that mindfulness mediation may reduce symptoms of anxiety, depression, and chronic pain, but little is known about its effects on the brain (National Center for Complementary and Alternative Medicine, 2012).

Words of Wisdom/Cautions

Occasionally, negative emotions or thoughts can surface through meditation, indicating a need for referral for further professional consultation.

Biofeedback

Biofeedback is a process of learned control of physical responses of the body. Through this therapy, a person develops a deeper awareness and voluntary control over physical processes. Due to the ability to measure effects, there is evidence supporting the value of this therapy.

> **KEY POINT**
>
> With biofeedback, a person learns to use thought processes to control bodily processes.

How It Works

Through the use of instruments (**Exhibit 27-1**), a person is trained to relax and monitor changes through certain feedback devices. Meditation, relaxation, and visualization techniques are taught, and these techniques ultimately teach psychological control over physical processes.

What It Helps

Biofeedback is used to treat a variety of conditions and symptoms, including tension headaches, chronic pain, anxiety, asthma, high blood pressure, and fecal and urinary incontinence.

Words of Wisdom/Cautions

Biofeedback is a harmless technique. Biofeedback usually is not covered by health insurance and can take several sessions to have results achieved; in some circumstances, less expensive techniques may be equally effective and should be explored.

Some Christians may be concerned about leaving the mind open to demonic influences during meditation; this concern should be explored before this practice is recommended or offered so as not to cause spiritual distress. Also, not all biofeedback products or practitioners are reputable, so discernment of claims is important.

Guided Imagery

What It Is

Based on the belief that the mind can influence the body, guided imagery is used to relieve symptoms, heal disease, or promote relaxation. It is based on

---

EXHIBIT 27-1   EXAMPLES OF INSTRUMENTS USED IN BIOFEEDBACK

Electrocardiograph (ECG): A common screening test in the average adult physical exam that involves placing electrodes on the body, wrist, and legs to measure heart activity and can be used to provide feedback useful in treating noncardiac conditions

Electrodermograph: Uses electrodes placed over the fingers, hand, or wrist to measure electrical activity of the skin

Electroencephalograph (EEG): Measures electrical activity of the brain using electrodes placed on specific areas of the scalp

Electromyograph (EMG): Uses electrodes over specific muscles to measure muscle contraction

Feedback thermometer: Measures skin temperature with a device usually attached to a finger

Photoplethysmograph: Through the use of a hook and loop band attached to the fingers and to the temple, measures blood flow after an infrared light is transmitted to the area

Pneumograph: Measures expansion and contraction of the chest and abdomen through the use of a flexible sensor band placed around the area

Rheoencephalograph: Uses electrodes placed on certain points of the head to measure blood flow through the brain

---

the belief that the mind and body are interconnected and work together in the healing process.

How It Works

A person can use guided imagery alone or be led by a practitioner. Sessions can last 10–30 minutes. Different imaging is used as a person is guided through the process. A quiet place, free from distractions, is needed. Numerous studies have demonstrated physiologic and biochemical changes.

REFLECTION

Have you had a situation in which telling yourself that you could achieve something helped you to do so?

What It Helps

Almost any medical situation can benefit from guided imagery, especially if problem solving, relaxation, decision making, or symptom relief is useful. It is used frequently to prepare for surgery and to speed recovery after surgery. It can be used to enhance the immune system, reduce stress, lower blood pressure, control pain, and to induce a sense of well-being.

Words of Wisdom/Cautions

Guided imagery is a harmless healing technique. This technique may reduce the need for medications, but medications should not be adjusted or stopped without first checking with a primary healthcare professional.

Hypnotherapy

Hypnotherapy is a state of focused concentration or relaxation that is guided by a therapist. In this state, a person is open to suggestion.

Hypnosis dates back to ancient China and Egypt and was even included as part of surgical procedures. In the 18th century, Franz Mesmer, an Austrian physician, was known for a process of inducing trance states in people and is credited with introducing hypnotism into medicine. The term *mesmerize* was used to describe his process.

A surgeon, James Braid, developed the technique further in the mid-19th century and used it for pain control and as anesthesia in surgery; however,

---

KEY POINT

The U.S. Department of Labor, in its Dictionary of Occupational Titles (D.O.T. 079.157.010), describes a hypnotherapist as one who:

- Consults with a client to determine the nature of problem.
- Prepares a client to enter hypnotic states by explaining how hypnosis works and what will be experienced.
- Tests a client to determine the degrees of physical and emotional suggestibility.
- Induces a hypnotic state in client using individualized methods and techniques of hypnosis based on interpretation of test results and analysis of the client's problem.
- May train a client in self-hypnosis conditioning.

because over the years Vaudeville performers, magicians, and others exploited it, hypnosis became associated with superstition, quackery, and evil. It was not until 1958, when Milton Erickson, an American psychotherapist, demonstrated how psychosomatic symptoms could be resolved with hypnotherapy that the AMA finally accepted hypnotherapy.

How It Works

In a trance state, a state between sleep and waking, which is called an alpha state, a person is very relaxed. It is like awakening in the morning and not being fully conscious or fully connected with the surroundings. In this state, a person is very receptive to suggestions from a therapist. Everyone is unique in his or her receptivity to entering into a guided trance state. No one can enter this state by force, and in this state, there is full awareness of everything that is happening. In this hypnotic state, past events can be more easily remembered, and trauma and anxiety around such events can be resolved. When this happens, past events no longer affect present behavior negatively.

> ### KEY POINT
>
> Hypnosis can be used to implant a suggestion or to explore the root cause of a problem.

What It Helps

Hypnotherapy is a therapeutic tool that can be very helpful in managing many situations, including fears, anxiety, chronic pain, addictions, poor self-control, low self-esteem, and behavioral problems.

Words of Wisdom/Cautions

Situations that interfere with hypnotherapy are extreme fear, religious objections, skepticism, inability to trust the therapist, and inability to relax. Hypnotherapy is not suitable when there are serious psychiatric conditions.

Yoga

Yoga uses stretching, breathing, body postures, and relaxation/meditation to restore and promote good mental and physical health. It has been practiced in India for thousands of years, and although it was introduced in the United States in the 1890s, it did not become popular until the 1960s. The goal of yoga is to create balance between movement and stillness, which is said to be the state of a healthy body. Postures require little movement, but require

> **KEY POINT**
>
> Yoga consists of breathing exercises, various stretching postures, and meditation.

mental concentration. Originally yoga was developed as part of a spiritual belief system, but Western culture primarily uses it as a health practice for improved flexibility, strength, relaxation, and physical fitness.

How It Works

There are many styles of yoga, and most are performed in a class. Yoga consists of breathing exercises and various postures or poses (asanas) that promote stretching and toning. There are beginning, intermediate, and advanced stages of practices within the various styles of yoga. Each posture has specific benefits, and each session usually ends with some form of a relaxation exercise or meditation. People have been known to practice yoga well into their 80s.

What It Helps

Yoga helps with headaches, asthma, back pain, sciatica, insomnia, balance, coordination, circulation, concentration, flexibility, endurance, physical strength, range of motion, and immunity.

Words of Wisdom/Cautions

Certain positions can cause muscle injury if the body is forced into those positions. Knowing personal limitations and consistent practice are important keys to obtaining the most benefit from yoga.

T'ai Chi

What It Is

T'ai Chi is a discipline that has been practiced in China for centuries. It is a form of slow-moving exercise that assists in uniting the mind–body connection. It is a martial art form that has been described as meditation in motion. It combines physical movement, breathing, and meditation to bring about relaxation and a feeling of well-being. There are various styles of T'ai Chi, and some involve up to 108 different movements and postures. Much concentration and discipline are required, and it takes time to learn the proper motion and coordination. People of all ages and physical capabilities can practice the art form, and it develops endurance, and flexibility, decreases fatigue, and improves overall physical health.

How It Works

The Chinese believe T'ai Chi helps to increase the flow of qi (chi or universal life force) circulating throughout the body. The movements are learned in rhythmic coordinated patterns that slowly flow from one series of movements into another. Focus is placed on breath and the body's motion, which in turn rejuvenates, stretches, strengthens, releases tension, opens points, and calms and quiets the mind at the same time. The slow turning, twisting, and stretching allow every part of the body to be exercised without strain.

What It Helps

T'ai Chi helps with high blood pressure, nervous disorders, the immune system, balance, stress-related disorders, circulation, panic attacks, concentration, insomnia, muscle tone, dizziness, internal organs, fibromyalgia, and spine and back problems.

Words of Wisdom/Cautions

There should be no ill effects from doing T'ai Chi. To prevent falls, it is important to assure that people have sufficient balance to stand and change positions during the movements.

Summary

As anyone who has experienced fatigue, appetite changes, or headaches when feeling emotionally low understands, the mind can influence bodily functions and well-being. Mind–body therapies can capitalize on this by using the mind to affect positive results. Meditation, biofeedback, guided imagery, hypnotherapy, T'ai Chi, and yoga are popular mind–body therapies. Results can vary. The use of these therapies for health conditions should be done after a full examination to identify specific diagnoses so that appropriate treatment approaches are used.

References

Hölzel, B. K., Carmody, J., Vangel, M., Congleton, C., Yerramsetti, S. M., Gard, T., et al. (2011). Mindfulness practice leads to increases in regional brain gray matter density. *Psychiatry Research: Neuroimaging. 191*(1), 36–43.

National Center for Complementary and Alternative Medicine. (2012). Mindfulness meditation is associated with structural changes in the brain. *Research focus*. Retrieved from http://nccam.nih.gov/research/results/spotlight/012311.htm

Suggested Reading

Besant, A. (2011). *Introduction to yoga*. Seattle, WA: CreateSpace.

Carlson, L. K. (2002). Reimbursement of complementary and alternative medicine by managed care and insurance providers. *Alternative Therapies in Health and Medicine, 8*(1), 38–49.

Cerrato, P. L. (2001). Complementary therapies update. *RN, 61*(6), 549–552.

Chopra, D. (2001). *Perfect health: The complete mind/body guide*. New York, NY: Random House.

Decker, G. (1999). *An introduction to complementary and alternative therapies*. Pittsburgh, PA: Oncology Nursing Press, Inc.

Earthlink, Inc. (2000, June/July). Alternative healthcare: Is it the right alternative for you? *Blink*, p. 27.

Eisenberg, D. M. (1997). Advising patients who seek alternative medical therapies. *Annals of Internal Medicine, 127*(1), 61–69.

Hamilton, D. (2010). *How your mind can heal your body*. New York, NY: Hay House.

Hogan, K. (2011). *The new hypnotherapy handbook*. Eagan, MN: Network 3000 Publishers.

Huebscher, R., & Shuler, P. A. (2003). *Natural, alternative, and complementary health care practices*. St. Louis, MO: Mosby.

Khanna, P., & Carnes, D. (2011). Systematic review: Effectiveness of mind-body therapies in chronic musculoskeletal disorders. *Journal of Bone and Joint Surgery, 93-B*(Suppl. IV), 491.

Kirksey, K., Goodroad, B., Kemppainen, J., Holzemer, W., Bunch, E., Corless, I., et al. (2002). Complementary therapy use in persons with HIV/AIDS. *Journal of Holistic Nursing, 20*(3), 250–263.

Krohn, J., & Taylor, F. A. (2002). *Finding the right treatment. Modern and alternative medicine: A comprehensive reference guide that will help you get the best of both worlds*. Point Roberts, WA: Hartley & Marks Publishers.

McCaleb, R. S., Leigh, E., & Morien, K. (2000). *The encyclopedia of popular herbs: Your complete guide to the leading medicinal plants*. Roseville, CA: Prima Publishing.

Olshansky, E. (2000). *Integrated women's health: Holistic approaches for comprehensive care*. Gaithersburg, MD: Aspen Publications.

Skinner, S. E. (2001). *An introduction to homeopathic medicine in primary care*. Gaithersburg, MD: Aspen Publishers, Inc.

Smith, D. W., Arnstein, P., Rosa, K. C., & Wells-Felderman, C. (2002). Effects of integrating therapeutic touch into a cognitive behavioral pain treatment program. *Journal of Holistic Nursing, 20*(4), 367–387.

Stephenson, N. L. N., & Dalton, J. (2003). Using reflexology for pain management. *Journal of Holistic Nursing, 21*(2), 179–191.

Strovier, A. L., & Carpenter, J. E. (2006). *Introduction to alternative and complementary therapies*. Philadelphia, PA: Haworth Press.

Trivieri, L., & Anderson, J. W. (Eds.). (2002). *Alternative medicine: The definitive guide* (2nd ed.). Berkeley, CA: Celestial Arts.

Young, J. (2007). *Complementary medicine for dummies*. New York, NY: IDG Books Worldwide, Inc.

Resources

Biofeedback

Association of Applied Psychophysiology and Biofeedback and Biofeedback Certification Institute

10200 West 44th Avenue, Suite 304

Wheat Ridge, CO 80033-2840

800-477-8892

www.aapb.org

Guided Imagery

Academy for Guided Imagery
30765 Pacific Coast Highway, Suite 369
Malibu, CA 90265
800-726-2070
www.acadgi.com

Nurse Certificate Program in Imagery
Beyond Ordinary Nursing
P.O. Box 8177
Foster City, CA 94404
www.imageryinternational.org

Hypnotherapy

American Board of Hypnotherapy
P.O. Box 531605
Henderson, NV 89053
888-823-4823
www.abh-abnlp.com

American Society of Clinical Hypnosis
140 N. Bloomingdale Road
Bloomingdale, IL 60108
630-980-4740
www.asch.net

Meditation

American Meditation Institute
60 Garner Road
Averill Park, NY 12018
518-674-8714
www.americanmeditation.org

The Center for Mind-Body Medicine
5225 Connecticut Avenue, NW, Suite 414
Washington, DC 20015
202-966-7338
www.cmbm.org

Tai Chi

Yoga

American Yoga Association
P.O. Box 19986
Sarasota, FL 34236
800-226-5859
www.americanyogaassociation.org

Yoga Alliance
1701 Clarendon Boulevard, Suite 110
Arlington, VA 22209
888-921-YOGA (9642)
www.yogaalliance.org

Yogic Sciences Research Foundation
1228 Daisy Lane
East Lansing, MI 48823
517-351-3056
sites.google.com/site/ysrfyoga/

Manipulative and Body-Based Methods

OBJECTIVES

This chapter should enable you to

- Describe the manual healing methods of chiropractic, craniosacral therapy, energy medicine/healing, massage therapy, Trager approach, Feldenkrais method, and Alexander therapy

As the name implies, manipulative and body-based therapies focus on bodily structures, including the bones, joints, soft tissue, and circulatory and lymphatic systems. Although they may address similar conditions and share a holistic perspective, each of these methods has unique approaches. The effectiveness of these therapies in relieving musculoskeletal pain causes them to be popular.

Chiropractic

Approximately 8% of American adults have used the services of a chiropractor, usually for back pain (National Center for Complementary and Alternative Medicine, 2012). Chiropractors must meet specific training (typically a 4-year academic program that includes both classroom work and direct care experience) and licensing requirements.

Some chiropractors also complete a 2- to 3-year residency for training in specialized fields. Many health insurance plans cover the chiropractic treatments.

History tells us that manipulation as a healing technique was used as early as 2700 BC by the Chinese. The Greeks (in 1500 BC) and Hippocrates (460 BC) also used spinal manipulation to cure dysfunctions of the body.

Daniel Palmer founded chiropractic in the Midwest in 1895, and it is now the fourth largest health profession in the United States. Palmer believed that all body functions were regulated by the nervous system and that because nerves originate in the spine any displacement of vertebrae could disrupt nerve transmission (which he called subluxation). He hypothesized that almost all disease is caused by vertebral misalignment; therefore, spine manipulation could treat all disease. Today the theory has changed to what is being called intervertebral motion dysfunction. The key factor in this theory involves the loss of mobility of facet joints in the spine.

KEY POINT

Chiropractors believe that a strong, agile, and aligned spine is the key to good health.

How It Works

The spine is made up of 24 bones called vertebrae with discs of cartilage cushioning between each vertebra. The spinal cord runs through the middle of the vertebrae with many nerves branching off through channels in the vertebrae. Chiropractors believe that injury or poor posture can result in pressure on the spinal cord from misaligned vertebrae, and that this can lead to illness and painful movement.

The chiropractor identifies and corrects the misalignments through manipulation, which are called adjustments. Muscle work is also incorporated as muscles attach and support the spine. Manipulation and muscle work can be done by hand and/or be assisted by special treatment tables, application of heat or cold, or ultrasound. Some chiropractic physicians also advise about nutrition and exercise. The first visit includes a detailed medical history and examination of the spine. Sometimes X-rays of the spine are also obtained. The findings are reviewed by the chiropractor, and a plan is established with a suggested number of follow-up treatments.

What It Helps

Chiropractic is useful for lower back syndromes, muscle spasms, midback conditions, sports-related injuries, neck syndromes, whiplash and accident-related injuries, headaches, arthritic conditions, carpal tunnel syndrome, shoulder conditions, and sciatica.

> **KEY POINT**
>
> In the United States, chiropractic practitioners must meet the licensing and continuing education requirements of the state in which they practice. All states require practitioners to complete a doctor of chiropractic degree program at a properly accredited college.

Words of Wisdom/Cautions

With a conscientious, professionally trained chiropractor, there are few side effects; however, some soreness may be experienced for a few days after a spinal adjustment, and occasionally symptoms get worse. Manipulations are contraindicated in persons with osteoporosis and advanced degenerative joint disease as these might be worsened by spinal adjustment. Caution is needed when chiropractic is done in older adults due to risk of increased bone brittleness that can contribute to fractures.

Craniosacral Therapy

Craniosacral therapy was developed in the early 1900s and is an offshoot of osteopathy and chiropractic. At that time, it was called cranial osteopathy. The basic theory behind craniosacral therapy is that an unimpeded cerebrospinal fluid flow is the key to optimum health.

William Sutherland, an osteopathic physician, developed craniosacral therapy. He believed that the bones of the skull were movable and that they move rhythmically in response to production of cerebrospinal fluid in the ventricles of the brain. This belief contradicts the teachings of anatomy in Western medicine, which holds the bones of the skull fuse together at 2 years of age and are no longer movable after this point in the physical development of the body.

Craniosacral therapists also believe that by realigning the bones of the skull, free circulation of the cerebrospinal fluid is restored, and strains and stresses of the meninges (that surround the brain and spinal cord) are removed, which allows the entire body to return to good health. Sutherland researched his theory over 20 years and documented physical and emotional reactions to compression on the cranial bones.

Craniosacral therapy was further advanced by John Upledger, who performed scientific studies at Michigan State University from 1975 to 1983. His findings validated craniosacral therapy's capability to help evaluate and treat dysfunction and pain. The Upledger Institute in Palm Beach Gardens, Florida, trains practitioners in this discipline.

How It Works

Trained practitioners palpate the craniosacral rhythm by placing their hands on the cranium (skull) and sensing imbalances. This approach is painless as the practitioner uses gentle touch (less than the weight of a nickel) to sense the imbalances in the rhythm and stabilize it. Recipients report a release of tension and a state of deep relaxation and peace. Practitioners work in a quiet setting and use no needles, oils, or mechanical devices. They take a medical history, observe, and question about any symptoms. As with other alternative therapies, a person may experience a brief period of worsening symptoms after treatment (usually for only 24–48 hours) as the body adapts to the changes that occurred during the session.

What It Helps

Craniosacral therapy is promoted for the treatment of anxiety, headaches, central nervous system disorders, neck and back pain, chronic ear infections, chronic fatigue, motor coordination difficulties, facial pain, temporomandibular joint dysfunction, and sinusitis. Evidence supporting its effectiveness for these problems is weak at this point, although some research is being conducted as to the usefulness of craniosacral therapy for headaches.

Words of Wisdom/Cautions

As with any integrative therapy, use of craniosacral therapy with the exclusion of Western medical advice is not recommended. It is important to be discerning of claims of the benefit of this therapy for various health conditions due to the scarcity of scientific evidence supporting its effectiveness.

Massage Therapy

Massage is the third most common form of alternative treatment in the United States after relaxation techniques and chiropractic. It consists of the therapeutic practice of kneading or manipulating soft tissue and muscles with the intent of increasing health and well-being and assisting the body in healing.

There are many different types of massage, such as lymphatic massage, sports massage, Swedish massage, shiatsu massage, myofascial release, trigger point massage, Thai massage, and infant massage. A form of deep tissue massage that is known as structural integration is called Rolfing. This system works deeply into muscle tissue and fascia to stretch and release patterns of tension and rigidity and to return the body to a state of correct alignment.

How It Works

There have been few scientific studies that explain the mechanisms by which massage works, although it is known to bring relief and relaxation to recipients. It is understood that besides stretching and loosening muscle and connective tissue, the action of massage also

- Improves blood flow and the flow of lymph throughout the body.
- Speeds the metabolism of waste products.
- Promotes the circulation of oxygen and nutrients to cells and tissues.
- Stimulates the release of endorphins and serotonin in the brain and nervous system.

> **KEY POINT**
>
> Massage can be seen as a form of communication from the therapist that brings comfort, gentleness, connection, trust, and peace.

Touch
Emotional
Mental

What It Helps

Massage is good for health maintenance as well as an adjunct to healing. Research supports the benefits of massage for low back pain (Furlan et al., 2008), cancer pain (Kutner, Smith, Corbin, Hemphill, Benton, Mellis, et al., 2008), and chronic neck pain (Sherman, Cherkin, Hawkes, Miglioretti, & Deyo, 2009). Massage also can be useful for conditions that can benefit from relaxation.

Immune System

REFLECTION

What would it take for you to build regular massages into your life?

Words of Wisdom/Cautions

There are a few contraindications that will be screened by the massage therapist when taking a medical history at the first visit. This is a reason for choosing a well-trained and qualified massage therapist. Ask for credentials.

Note – caution areas

The Trager Approach

In the 1920s, a physician, Milton Trager, developed a method of passive, gentle movements with traction and rotation of extremities to help reeducate muscles and joints. Through this method, muscle tightness is relieved without pain, and the end result is a sense of freedom, flexibility, and lightness.

How It Works

In the Trager approach, through smooth joint movements and gentle rhythmic rocking of body parts, communication is made with the nerves that control muscle movement to release and reorganize old patterns of tension, pain, and muscle restriction. A session lasts 60–90 minutes, and after a session, instructions are given for a series of simple movements (called Mentastics) to help maintain the results of the treatment. Deep relaxation of mind and body is also promoted during these movements.

What It Helps

The Trager approach is promoted as a means to help chronic pain, muscle spasms, fibromyalgia, temporomandibular pain, headaches, plus many other neuromuscular disorders.

Words of Wisdom/Cautions

Scientific evidence supporting the benefit of the Trager approach to therapy is lacking, although with a trained practitioner, harmful side effects should not exist.

KEY POINT

Although scientific evidence supporting the effects and benefits of a therapy may be lacking, if it brings relaxation and comfort, has no side effects or risks, and does not substitute for known beneficial treatments, there should be little harm in its use.

Feldenkrais Method

Feldenkrais teaches a person how to alter the way the body is held and moved. It is a gentle method of bodywork that involves movement. Moshe Feldenkrais developed this method after suffering a knee injury. He studied and combined principles of anatomy, physiology, biomechanics, and psychology and integrated this knowledge with his own awareness of proper movement.

How It Works

In Feldenkrais, by developing awareness of body movement patterns and changing them through specific exercises, flexibility, coordination, and range of motion improve. Through instruction, a teacher guides a person through a series of movements, such as bending, walking, and reaching. These movements can help reduce stress and pain and improve self-image. It is believed these movements access the central nervous system. There are two types of sessions: (1) a set of movement lessons called awareness through movement learned with a group and (2) individual hands-on sessions called functional integration. The results benefit mind and emotion, as well as the physical body.

> ### KEY POINT
>
> Feldenkrais teaches a person to be aware of the way the body moves and to use proper movement.

What It Helps

Research is limited regarding the effects of the Feldenkrais method, although people have found it helpful in improving balance and in the treatment of musculoskeletal conditions and anxiety.

Words of Wisdom/Cautions

There are no known side effects or unsafe conditions when Feldenkrais is provided by a trained practitioner.

Alexander Therapy

Alexander therapy is an educational process that identifies poor posture habits and teaches conscious control of movements that underlie better body mechanics. Frederick Mathias Alexander, an Australian actor who lost his voice while performing, developed this therapy. Discouraged by only temporary relief from medical treatments, he began studying how posture affected his voice. After 9 years of study and perfecting his technique, he began to train others.

How It Works

Alexander therapy teaches simple exercises to improve balance, posture, and coordination. It is done with gentle hands-on guidance and verbal instruction. It results in release of excess tension in the body, lengthens the spine, and creates greater flexibility in movement. A session can last from 30 to 60 minutes; multiple sessions usually are necessary.

What It Helps

Many conditions that result from poor posture can be greatly helped with Alexander therapy. This technique is taught in many drama and music universities throughout the world. There is evidence that Alexander therapy can help with low back pain and chronic pain (Smith & Torrance, 2011).

Words of Wisdom/Cautions

Alexander therapy is safe therapy when taught and performed by a credentialed therapist.

Summary

With pain being a highly prevalent problem, therapies that can reduce this symptom without the use of medications have an important role. Although most of these therapies carry no serious risk of causing harm, those that manipulate body parts should be done only by trained and credentialed professionals.

References

Furlan, A. D., Imamura, M., Dryden, T., & Irvin, E. (2008). Massage for low-back pain. *Cochrane Database of Systematic Reviews,* (4), CD001929.

Kutner, J. S., Smith, M. C., Corbin, L., Hemphill, L., Benton, K., Mellis, B. K., et al. (2008). Massage therapy versus simple touch to improve pain and mood in patients with advanced cancer: A randomized trial. *Annals of Internal Medicine, 149*(6), 369–379.

National Center for Complementary and Alternative Medicine. (2012). *Chiropractic: An Introduction.* Retrieved from http://nccam.nih.gov/health/chiropractic/introduction.htm

Sherman, K. J., Cherkin, D. C., Hawkes, R. J., Miglioretti, D. L., & Deyo, R. A. (2009). Randomized trial of therapeutic massage for chronic neck pain. *Clinical Journal of Pain, 25*(3), 233–238.

Smith, B. H., & Torrance, N. (2011). Management of chronic pain in primary care. *Current Opinions Support Palliative Care, 5*(2), 137–142.

Suggested Reading

Bergman, T. F. (2010). *Chiropractic technique: Principles and procedures* (3rd ed.). St. Louis, MO: Mosby.

Gouveia, L. O., Castanho, P., & Ferreira, J. J. (2009). Safety of chiropractic interventions: A systematic review. *Spine, 34*(11), E405–E413.

Kanodia, A. K., Legedza, A. T., Davis, R. B., et al. (2010). Perceived benefit of complementary and alternative medicine (CAM) for back pain: A national survey. *Journal of the American Board of Family Medicine, 23*(3), 354–362.

Micozzi, M. (2010). *Fundamentals of complementary and alternative medicine* (4th ed.). St. Louis, MO: Saunders.

Stone, V. J. (2010). *The world's best massage techniques: The complete illustrated guide; innovative bodywork practices from around the globe for pleasure, relaxation, and pain relief.* Beverly, MA: Fair Winds Press.

Resources

Alexander Therapy

American Society for the Alexander Technique

P.O. Box 2307

Dayton, OH 45401-2307

800-473-0620

www.alexandertech.com

Chiropractic

American Chiropractic Association

1701 Clarendon Boulevard

Arlington, VA 22209

800-637-6244

www.acatoday.org

Federation of Chiropractic Licensing Boards

5401 W. 10th Street, Suite 101

Greeley, CO 80634

970-356-3500

www.fclb.org

World Chiropractic Alliance

2683 Via de La Valle, Suite G 629

Del Mar, CA 92014

866-789-8073

www.worldchiropracticalliance.org

Craniosacral Therapy

Upledger Institute

11211 Prosperity Farms Road

Palm Beach Garden, FL 33410

561-622-4706

www.upledger.com

The Feldenkrais Method

Feldenkrais Guild of North America
5436 N. Albina Avenue
Portland, OR 97217
800-775-2118
www.feldenkrais.com

Massage Therapy

American Massage Therapy Association
500 Davis Street
Evanston, IL 60201
877-905-0577
www.amtamassage.org

Associated Bodywork and Massage Professionals
800-458-2267
www.abmp.com

Massage Bodywork Resource Center
www.massageresource.com

The Trager Approach

Trager International
P.O. Box 3246
Courtenay, BC, Canada V9N 5N4
250-337-5556
www.trager.com

Energy Therapies

OBJECTIVES

This chapter should enable you to

- Describe the way touch has been used for healing throughout history
- Describe therapeutic touch, healing touch, and Reiki
- Discuss factors that promote healing benefits from touch

> *There is a mysterious healing power in touch*
> *that is beyond words and beyond our ideas about it.*
> —AILEEN CROW

Since the beginning of humanity, touch has had an important part in healing and survival. Almost instinctively, people physically reach out to those in pain, those who are suffering, those who are injured or sick, and those they love. Touch provides a soothing comfort and security.

Before the era of modern medicine as we know it, touch was a basic therapy used to compress a wound to stop it from bleeding, to caress the dying, and to welcome the newborn. Although other components of healing grew throughout time, touch remains constant.

Ancient Societies

Ancient carvings and pictures have been found throughout the world that show touch as a part of healing. In Egypt, rock carvings show hands-on healing for illness. All the ancient societies—Indian, Egyptian, Greek, Chinese, and Hebrew—used touch as part of the healing provided to their people (Graham, Litt, & Irwin, 1998). *Native Americans*

The Egyptians learned much of their healing techniques from the Yogis. The Greeks, in turn, learned from the Egyptians and the Indians. In ancient Athens, Aristophanes detailed the use of laying on of hands to heal blindness in a man and infertility in a woman (Sayre-Adams & Wright, 1995). In addition, the Greek priests used their hands to heal as did the physicians. As each society learned from the other, new healing modalities were added.

Native American Indians also used touch, with direct and nondirect body contact, as a method of healing. In nondirect body contact, healers hold their hands about 6 inches above the body or on opposite sides of the area needing healing. They incorporate ritual, such as cleansing the body, prayer, song, and dance, with the touch to promote healing. The Native American Indians brought to the healing process the sense of spirituality (Graham et al., 1998).

> **KEY POINT**
>
> Hands-on healing, or the laying on of hands, can be found in all major religions.

Spiritual Roots

Accounts of the laying on of hands for healing can be seen in both the old and new Judeo-Christian texts. For centuries, average people have been using this type of healing with their families, friends, and others in their community. Unfortunately, the works of Plato in the early 2nd century helped stir the thinking of separating the spirit from the body. As this common belief took hold, the church also followed in its thinking of the spirit and body as separate. Touch healing came under the domain of the religious leaders or the political leaders of that particular society. Because the political leaders had such a strong association with religion, it was natural for them to assume the same or better position of authority than the religious leaders (Graham et al., 1998).

There are many references to hands-on healing by religious and political figures. Ancient Druid priests were said to breathe on body parts and touch certain body areas in conjunction with prayer and ritual. St. Patrick helped to heal the blind, and St. Bernard healed the deaf and lame. Emperor Vespasian, Emperor Hadrian, and King Olaf were also said to have the ability to heal by touch. In fact, some of the early kings in England and France performed touch healings, which became known as the "King's touch" (Krieger, 1992, p. 15).

This power of hands-on healing began to lay solely with the priests and monarchs. Touch for healing soon began to disappear among the lay people.

> **KEY POINT**
>
> There have been times in history when women were suspected of practicing witchcraft when they demonstrated their healing powers. Sadly, it is estimated that millions of women and girls were executed in the Middle Ages after being accused of witchcraft (Ehrenreich & English, 1973).

Unfortunately, those lay persons who continued to do hands-on healing were often labeled as pretenders to the throne or witches. Women often were targeted as witches because they made up the majority of the healers.

Science and Medicine

Medicine began with the priests and philosophers of early Egyptian and Greek thinking. Over the centuries, medicine became the domain of men, and women were excluded from the profession. As medical science developed, touch as a healing act became viewed as within the domain of the church or monarchy.

Medicine considered touch to be within the realm of superstition rather than something with any scientific basis for healing. There was no hard proof that touch produced any therapeutic results; therefore, touching of patients was not needed other than that involved in performing a task. As the age of science developed, healing techniques and medicine became empirically based (proved by experiments). It was believed that if something could not be proved by science, it did not or could not work and had to be superstition or sorcery. Touch as a healing tool could not be proved. This same thinking continued into the 20th century and only recently has gained new awareness within the medical community.

We Have Come Full Circle

With the massive amount of information available in recent years, new knowledge has been gained about old methods of healing. With the world seemingly becoming smaller because of the mingling of cultures, computer networks, and advances in education, the old is being rediscovered. As more people turn to complementary and alternative therapies, mainstream medicine is forced to acknowledge the power of touch in healing.

The U.S. government, through the National Institutes of Health's National Center for Complementary and Alternative Medicine, is conducting research on many types of healing. One aspect of healing it is examining is the power of touch to heal. Among the touch therapies it recognizes are healing touch, therapeutic touch, and massage. In addition, a significant number of schools of medicine and nursing incorporate complementary therapies as part of their educational programs.

Therapeutic Touch

In the 1970s, Delores Krieger, a nurse scientist and professor at New York University, began studying touch as therapy with the help of Dora Kunz and Oskar Estebany. Kunz was not someone who would be considered a health-care provider based on the standards of having a license to heal someone. Rather, she was born with the ability to perceive energy around living things and describe them as accurately as a medical reference book.

Estebany was a retired colonel in the Hungarian cavalry who found he could heal animals. He was able to heal his own horse by spending the night in his stable, stroking and caressing the animal, talking to it and praying over it. Soon people brought other sick animals for him to heal, and then they began to bring children. He became known in his country as a hands-on healer. He decided to offer his services for research and moved to Canada where he met and joined Dora Kunz.

Dr. Krieger met the two of them and decided to do research after watching them in a healing session. She observed Estebany laying hands on people wherever he sensed they were needed. He would mostly sit in silence during the 20–25-minute sessions. Kunz would observe him and redirect him to another area if she perceived another area needing treatment. Dr. Krieger found that most people reported being relaxed and that most felt better. He treated those with emphysema, brain tumors, rheumatoid arthritis, and congestive heart disease (Krieger, 1992).

Estebany did not think hands-on healing could be taught to other people but that it was a gift that one has. Kunz disagreed and decided to begin workshops to teach people how to do hands-on healing. Dr. Krieger joined her in

---

KEY POINT

Therapeutic touch is based on the principle that people are energy fields who can transfer energy to one another to promote healing. The practitioner's hands do not directly touch the client, but are held several inches above the body surface within the energy field and positioned in purposeful ways. A core element of therapeutic touch is the mind-set or intent of the practitioner to help.

these workshops and learned how to do hands-on healing, too. Together they developed therapeutic touch and began an interesting trend in research and healing. As the research developed and the outcomes began to show significant responses, the power of touch was revealed. In the past 20 years, therapeutic touch has been offered in classes and workshops in more than 80 colleges and universities in the United States and over 70 countries worldwide.

Healing Touch

Another method of hands-on healing called healing touch was developed by nurse Janet Mentgen. This modality was inspired in part from therapeutic touch, the work of Brugh Joy, and the energy and philosophical concepts were from Rosalyn Bruyere and Barbara Brennan. It began as a pilot project at the University of Tennessee and in Gainesville, Florida, in 1989. By 1990, it became a certificate program, with certification beginning in 1993. This multilevel program combines healing techniques from various healers throughout the world.

Healing touch is done through a centered heart, bringing to it a spiritual aspect. There is no affiliation with any one religious belief, and it can be used by those of any religion. It complements traditional healing through modern medicine and psychotherapy, but it is not used as a substitute for them. Many hospitals are now incorporating healing touch as part of patient services through specific areas within the hospital, and the use of healing touch is being studied in people with cancer, pain, depression, HIV, cardiac problems, and diabetes.

> ### KEY POINT
>
> Healing touch is an energy-based therapy, which assists in healing the body–emotion–mind–spirit component of the self.

Reiki

Reiki, meaning universal life energy, is a Japanese-derived form of energy healing. It was developed by Dr. Mikao Usui, a Christian minister, in the middle of the 19th century. It consists of laying hands on the body or leaving them above the body and channeling energy to the recipient.

In a typical Reiki treatment, the Reiki practitioner instructs the recipient to lie down, usually on a massage table, and relax. The practitioner might

> **KEY POINT**
>
> Reiki is a form of healing in which the Reiki master channels energy to another individual.

take a few moments to enter a calm or meditative state of mind and mentally prepare for the treatment and then proceeds to place his or her hands in various positions. The hands are usually held in each position for several minutes.

Reiki is learned through a series of intensive sessions whereby a Reiki master passes on the knowledge to others by way of attunements. These attunements permit the new practitioner to be able to perform healings by touch. There is no formal certification process in the Usui system of Reiki. A student is deemed competent when a Reiki master decides so.

Although this therapy is popular in some circles, the scientific evidence supporting Reiki's effectiveness is lacking. Also, despite the fact that no physical harm can be done through a Reiki treatment alone, there is concern as to the spiritual implications. The Catholic Church has taken the position that Catholics should not engage in this therapy (Committee on Doctrine, 2012). There are additional kinds of hands-on healing, with more being developed all of the time. Examples of others include craniosacral therapy, polarity, chakra balancing, acupressure, and shiatsu. Some involve tissue or muscle manipulation, such as neuromuscular release, Rolfing, Trager, massage therapy, and reflexology (see **Exhibit 29-1**). Although they all have a slightly different belief or philosophy, they share the common denominator of touch.

Energy as a Component of Touch

There is considerable discussion today about energy. People state, "I don't have any energy today," "I feel energized," or "I have a lot of energy." You may have experienced these feelings and know that there is a difference between having high and low energy. After listening to an uplifting speech and hearing the thunderous applause, one may say the room was "energized" or "electrified." What is this energy that we are talking about?

REFLECTION

Do you have sensitivity to your body's energy? Are there particular experiences or people that seem to energize you and others that drain your energy?

EXHIBIT 29-1 TYPES OF TOUCH THERAPY AND PURPOSES

Acupressure. Used to stimulate the body's natural self-healing ability and to allow qi, the life energy of the body, to flow. The hands apply pressure to specific acupoints on the body similar to acupuncture.

Chakra balancing. An energy-based modality using the hands to balance the energy centers, or chakras, of the body.

Craniosacral therapy. Started in the early 1900s by Dr. William Sutherland, an osteopathic physician. He determined that the skull bones move under direct pressure. By working the skull, spine, and sacrum with gentle compression, the therapist aligns the bones and stretches underlying tissue to create balance and allow the spinal fluid to flow freely. This helps the body self-adjust.

Massage. Various types of light touch, percussion, and deep-tissue pressure by hands to assist in muscle relaxation, improved blood and lymph flow, and release of helpful chemicals naturally occurring in the body, such as endorphins.

Neuromuscular release. The goal is to assist a person in letting go by moving the limbs into and away from the body. It helps with circulation and emotional release.

Polarity. Created by Randolph Stone, polarity combines pressure-point therapy, diet, exercise, and self-awareness. It is based on the body having positive and negative charges. By varying hand pressure and rocking movements, the energy can be rebalanced.

Reflexology. By applying pressure to specific points on the body, energy movements to corresponding parts of the body are activated to clear and restore normal functioning. Reflexology can be done on the feet, hands, or ears.

Rolfing. By manipulation of muscle and connective tissue in a systematic way, the therapist helps the body readjust structurally to allow proper alignment of body segments.

Trager. Rhythmic rocking of one's limbs or whole body to aid in relaxing of muscles. This promotes optimal flow of blood, lymph fluid, nerve impulses, and energy.

EXHIBIT 29-2   CULTURAL NAMES FOR ENERGY

CULTURE	ENERGY NAME
Aborigine	Arunquiltha
Ancient Egypt	Ankh
Ancient Greece	Pneuma
China	Qi (chi)
General usage	Life force
India	Prana
Japan	Ki
Polynesia	Mana
United States	Bioenergy, biomagnetism, subtle energy

The terms for energy date back to the older traditions of ancient cultures. Every society had a term for the life force. For some cultures, caring for this energy influenced health and wellness; by blocking or disrupting energy, disease or death could result. Other societies merely referred to it in reference to religious or spiritual beliefs. For examples of these terms, see **Exhibit 29-2.**

The Role of Qi

The Chinese culture has used the idea of energy for thousands of years through the philosophy of qi, or life force, which is an energetic substance that flows from the environment into the body. The qi flows through the body by means of 12 pairs of meridians (energy pathways to provide life-nourishing and sustaining energy). The organs of the body are affected by the pairs of meridians. There needs to be a balance of qi flowing through each side of the paired meridian in order for balance to occur and health to be attained and/or maintained.

Science is beginning to get a better understanding of this energy, what it is, and how it functions. Some researchers have developed machinery to try to detect this energy. Although success has been limited for some, others have been somewhat successful.

The Energy Fields or Auras

In addition to the seven chakras discussed previously in this text, there are believed to exist seven layers of energy fields, or auric fields, around the body. These energy fields surround all living and nonliving matter. Science has begun to study and explain these fields over the last several years. The science of physics has done the most extensive work in explaining energy fields. Through Newtonian physics, field theory, and Einstein's theory of relativity, a better understanding of how energy works has been acquired, but it is quantum physics that has really helped explain the characteristics and behaviors of energy. The discussion of specifics is beyond this chapter; however, it is intended to show that science is now rethinking some of its early and persistent cause-and-effect notions.

One way energy fields have been viewed is through Kirlian photography, which was created in the 1940s by a Russian researcher, Semyon Kirlian. Kirlian photography (also called electrography and aura photography) uses special photographic techniques to create an image of the electrical charge being emitted from a body or object. By using this form of photography, he was able to measure changes in the energy fields of living systems. He found that various feelings and dysfunctions cause changes in the electromagnetic field around the body, producing different colors that are shown on the photograph. One of the best known experiments is that of the phantom leaf effect, whereby a portion of a leaf was cut away and Kirlian photography done on the amputated leaf. Amazingly, the leaf still appeared whole with an energy field present in the form and space of the cutaway portion.

In other important work, a Japanese researcher named Motoyama has developed a number of electrode devices to measure the human energy field at various distances from the body (Motoyama, 2003). Thus, it is now possible to measure the excess or lack of energy around the body and determine a person's health status. Future technology advances may allow a body to be scanned and the beginnings of disease detected before symptoms appear in the physical body. In fact, a new CT scan is being used for this very reason to detect early onset of heart disease and cancer.

Where exactly are these energy fields? As stated previously, there are seven energy fields that are generally accepted by most energy workers. Each field corresponds to a specific chakra; for example, the first energy field with the first chakra and so on; however, the fields lie one on top of another yet do not

> ### KEY POINT
>
> Energy fields of the body work similar to a television. When you change channels, you get a different picture. All of the pictures do not end up on the same channel, but they do come through one television set. Likewise, when listening to a radio, you change stations to hear different music, but you still have one radio. The same idea works for the energy fields of the body.

interfere with each other's functions. These fields interconnect yet maintain their own separateness.

When two people interact, their energy fields connect and they can pick up information about each other. For example, suppose you are riding on an elevator, and a stranger gets on. You do not say anything to the person, but you feel safe and secure. At another floor, a second person gets on. Immediately you feel uncomfortable with this person even though you have not spoken to him, either. What is occurring is the intermixing of energy fields, which is giving you subconscious information about these people. It has nothing to do with what they are wearing or how they look.

One of the leading causes of illness in our society is stress, yet stress is not internal. Stress originates outside the body and works its way inward through the energy fields and chakras. It is a well-known fact that long-term stress can weaken the immune system's response and cause heart attacks, high blood pressure, and depression. Of course, there are physical evidences for each of these, but the process starts long before any physical evidence is detected. How does this occur? It is thought that long-term stress causes changes in the energy fields and chakras that over time repattern themselves and begin to affect the cells of the body. Depending on the severity of the repatterning, illness and disease can occur. By addressing the stressors, one is able to fix or heal any changes made in the energy field or chakras.

Humans as Multidimensional Beings

This concept of energy fields and chakras intermixing contributes to humans being multidimensional. In other words, you are made up of more than just your parts. You not only contain energy, but you are energy that is constantly changing. The body uses energy to perform its functions—for nerves to stimulate muscles, for the beating of the heart, for lungs to exchange air, for cells to digest nutrients, and for the creation of an idea. There is nothing in the body that is not involved in some form of energy. When the normal energy flow is interrupted or destroyed, the body is unable to function to its fullest capacity, and illness, disease, and disability can develop. When all of the energy workings cease, death occurs.

Nevertheless, quantum physics has shown us that energy cannot be destroyed. Thus, when we die, it is just the physical part of ourselves that dies. The energy that aided us in our physical form transforms and is released into the environment; therefore, one does not die—the physical body is just transformed.

Touch Is Powerful

With this discussion of the energy systems of the body, you can see that a touch is not just a simple physical act. Something happens from human to human, human to animal, human to plant, and animal to animal when touch occurs. This exchange of energy affects not only the one receiving the touch, but also the one giving the touch.

> **KEY POINT**
>
> One of the most important aspects of touch is its intention.

There are neutral touches, such as tapping someone on the arm to get their attention. In addition, there are those touches out of anger that are meant to hurt someone, like a slap or kick. Then there are the loving touches of a hug or caress. The intention is what differentiates.

The way to obtain the power of touch for healing is through a centered heart. This is done in several ways. It occurs spontaneously for couples in love or in the love that a parent feels toward a child. One can feel the heart open and a strong connection to the other, even when the other person is not physically present. Another way to obtain this power is through the intention of helping someone. This can be spontaneous as in an accident or planned as in changing someone's wound dressing after surgery.

A third way is through inner focus and awareness in order to become calm and in a state of balance. You probably have experienced times when you are not in balance. Things around you are chaotic—you stumble and cannot seem to get anything done. Yet, in a state of balance you are able to accomplish many tasks regardless of what goes on around you.

> **KEY POINT**
>
> Belly breathing, sounds, and imagery can be used to help you center and focus.

You can learn to have a centered heart using several methods, one of which is belly breathing. Taking slow breaths in through the nose and out through the mouth allows you to become present in the moment. By having a rhythmic pattern of breathing, you become still and calm. In addition, belly breathing can help reduce your stress. To belly breathe, move your belly or abdomen outward on inspiration (the in breath) and retract inward on expiration (the out breath). Counting numbers during the in breath and out breath, along with belly breathing, helps quiet the mind and focus the awareness inward.

Another method of becoming more self-aware is through the use of sound. By focusing on a sound externally, like the sound of a bell chime, or one that is self-created, like *om*, you may be able to quickly focus and center. Some people find it easy to learn how to center by listening to quiet music, whether it be classical music, jazz, nature sounds set to music, religious music, or new age music. There is no one type of music that is perfect for all people. Generally, it is whatever you need to listen to in order to help you attain a sense of focus.

Imagery also can be used to center. The mind is a powerful tool and can be used to create a center of focus. The exercise in **Exhibit 29-3** is an example of a way to use imagery to focus.

With practice, you will be able to get into a centered state within a few seconds by using your breath, a sound, or imagery. You can combine any of these and see what works best for you. These techniques can be used in conjunction with touch healing or as a method to reduce stress at any time.

REFLECTION

When providing assistance or care to another person, do you give the person your undivided attention? Do you need to adopt a technique to help you get into a centered state?

The Practice of Touch

While centering is one of the most important aspects to any touch healing, intuition is another one. Intuition often is a heartfelt feeling of certainty. All people have intuition but may ignore it or not identify it as such. At one point or another, many people have said, "I knew that would happen." Some people can tell who is on the other end of the phone before or at the moment it rings. Others know when loved ones are in trouble or hurt. This is intuition. It is something often used in touch healing in order to allow for the best outcome for that person. You just seem to know where to place your hands.

When you get a burn on your finger, the first reaction is to pull away from the source of the burn and the second to place your hand over the spot that is

EXHIBIT 29-3   USING IMAGERY TO CENTER OR FOCUS

1. Find a quiet place to sit without distractions. Loosen clothes if necessary. Sit upright with back supported, feet flat on the ground, and hands in a comfortable position on your lap or the arms of a chair.

2. Close your eyes, and begin to breathe in through your nose, allowing your stomach to move outward. Then exhale slowly, allowing your stomach to retract naturally. Try to have the in breath equal the out breath and let this become a cycle. If you have difficulty, just keep practicing. There is no wrong way to do this.

3. Imagine your feet on the ground growing roots downward through the floor, like a tree. With each breath, allow the roots to move further and further into the earth until it reaches the center of the earth. Feel the warmth of the earth on your roots.

4. Using your breath, have the warmth move upward through the roots and into your feet. Feel the warmth. Continue moving this energy higher into your legs, hips, back, and stomach, into your chest and shoulders, your neck, your face, your head, and out the top of your head up to the sky.

5. Imagine a ball of white light way above your head shining downward. Using your breath, help this light come down and touch the top of your head. You may feel some pressure or heaviness on your head. This is normal.

6. Using the breath, allow this white light to move through the center of your body, out the base of your spine, and down your legs, through your feet and into your roots. Let the energy go down your roots to the center of the earth.

7. Leaving these images there, focus on your heart center in the middle of your chest. Imagine a pink, light, fluffy ball of light like cotton candy. With each breath, allow the ball to become bigger and bigger. Feel how warm and light this is.

8. Imagine a white or pink light coming from your heart center and running down your arms into your hands and fingers. Some of the sensations you may feel are warm, cold, tingly, prickly, vibrating, or any other sensation special to you. You may not even feel anything at all. It doesn't matter; just know it is there.

9. You are now centered. At this point, you can stay in this state and just enjoy it, or if there is someone who you are going to do touch healing on, open your eyes and begin.

[handwritten margin notes: Roots / White Light / Pink Fluffy Light]

> **KEY POINT**
>
> Intuition is the internal knowing of something, often without any external or visible means of proving its existence. One just knows.

burned. The covering hand helps reduce the pain. You know what to do. You do not even think about it. This is intuition.

The same thing happens in touch healing. Of course there are many distinct techniques one can learn step by step, but often this is not necessary. If you allow yourself to follow your intuition, you will know where to place your hands. It may be on the body in a specific area. It may be out in the field away from the body. You may use one hand, two hands together, or two hands in different places. Sometimes your hand may start to do a movement over an area for no particular reason. You may feel things while your hands are placed on the body or off the body, such as temperature, texture, vibration, density, or contour (shape). Just let go and allow the energy of touch to work. You do not need to do anything else, and when it is time to finish, you will just know to stop.

If the power of touch sounds easy, that is because it is. As long as you prevent yourself from thinking about what you are doing and allow the centered heart and energy to do its work, it happens naturally. The centered heart, with the intention of healing, is the greatest gift you can give anyone, including yourself. The energy used to help someone else benefits you, too. It is like you are a hose watering flowers. The flowers receive the water to drink and live and the hose stays wet inside; both benefit.

The way to master touch healing is to practice, practice, and practice. In addition, many schools and programs teach the various forms of touch healing for those interested in learning more specific techniques. For more information on various programs, see the "Resources" list at the end of the chapter.

The power of touch is a natural gift, ready to be developed. Some people are using it without knowing what they are really accomplishing. They just know it is a good thing. Others recognize this ability and use it help others heal and feel better, and some are afraid of the power their touch has on others. Just know that whenever touch is used with a centered heart and the intention to heal or help, it will not hurt anyone. It is a wonderful, loving act you are able to share with others or do for yourself. Touch has a powerful capacity to heal through a centered heart filled with love.

Cautions

As therapeutic and comforting as touch may be, it must be appreciated that not everyone will welcome touch or touch therapies. Cultural influences may cause people to be uncomfortable being touched by a stranger or to believe it is inappropriate for a person to be touched by a member of the opposite sex who is not a spouse. Some people have opposition to touch therapies because of their faith. For example, Christians who believe that the only spirit one should call on is Jesus Christ may be uncomfortable with a practitioner who channels energy from an unknown source. Sensitivity to different reactions to touch is warranted. It is beneficial for a practitioner to explain the procedure/process to an individual and obtain consent before initiating touch therapy. Consumers also should explore the spiritual foundation for any touch therapy used to assure it is compatible with personal belief systems.

Summary

All major religions have used some form of laying on of hands for healing, as have most ancient cultures. Therapeutic touch, healing touch, and Reiki are practices in which energy from one person transfers to another to stimulate healing. Other touch therapies include acupressure, craniosacral therapy, polarity, chakra balancing, neuromuscular release, Rolfing, Trager, massage, and reflexology. To obtain the most healing benefit from touch, one needs to have an intention to do good for the person being touched, have a centered heart, and be focused and aware. Self-awareness can be enhanced through belly breathing, focusing on a sound, and the use of imagery. Practice and experience enhance skill in using touch for healing. Caution is needed that these therapies are not used for healing conditions for which no evidence of effectiveness exists and that they do not conflict with a person's religion.

References

Committee on Doctrine, United States Conference of Catholic Bishops. (2012). *Guidelines for evaluating Reiki as an alternative therapy*. Retrieved from http://old.usccb.org/doctrine/Evaluation_Guidelines_finaltext_2009-03.pdf

Ehrenreich, B., & English, D. (1973). *Witches, midwives and nurses: A history of women healers*. New York, NY: The Feminist Press.

Graham, R., Litt, F., & Irwin, W. (1998). *Healing from the heart*. Winfield, BC, Canada: Wood Lake Books Publishing.

Krieger, D. (1992). *The therapeutic touch: How to use your hands to help or heal*. New York, NY: Simon & Schuster.

Motoyama, H. (2003). *Theories of the chakras.* Wheaton, IL: Theosophical Publishing House.

Sayre-Adams, J., & Wright, S. (1995). *The theory and practice of therapeutic touch.* New York, NY: Churchill Livingstone.

Suggested Reading

Alexander, S., & Schneider, A. (2010). *The pocket encyclopedia of healing touch therapies: 136 techniques that alleviate pain, calm the mind, and promote health.* Beverly, MA: Fair Winds Press.

Courcey, K. (2001). Investigating therapeutic touch. *Nurse Practitioner Forum, 26*(11), 12–15.

Engle, V., & Graney, M. (2000). Biobehavioral effects of therapeutic touch. *Journal of Nursing Scholarship, 32*(3), 287–289.

Halcon, L. (2002). Reiki. In M. Snyder & R. Lindquist (Eds.), *Complementary/alternative therapies in nursing* (pp. 197–203). New York, NY: Springer.

Hover-Kramer, D. (2011). *Healing touch: Essential energy medicine for yourself and others.* Boulder, CO: Sounds True.

Ives, J. A., & Jonas, W. B. (2011). Energy medicine. In M. Micozzi (Ed.), *Fundamentals of complementary and alternative medicine* (pp.130–142). Philadelphia, PA: Saunders.

Kunz, D., & Kreiger, D. (2004). *The spiritual dimension of therapeutic touch.* Rochester, VT: Bear & Co.

Mentgen, J. (2001). Healing touch. *Nursing Clinics of North America: Holistic Nursing Care, 36*(1), 142–145.

O'Mathuna, D. (2000). Evidence-based practice and review of therapeutic touch. *Journal of Nursing Scholarship, 32*(3), 277–279.

Paul, N. L. (2006). *Reiki for dummies.* Indianapolis, IN: Wiley & Sons.

Quest, P. (2010). *Reiki for life: The complete guide to Reiki practice for levels 1, 2 & 3.* New York, NY: Jeremy Tarcher.

Stein, D. (2007). *Essential Reiki teaching manual.* Berkeley, CA: Crossing Press.

Resources

American Holistic Nurses Association
100 SE 9th Street, Suite 3A
Topeka, KS 66612-1213
800-278-2462
www.ahna.org

Healing Touch International, Inc.
445 Union Blvd, Suite 105
Lakewood, CO 80228
303-989-7982
www.healingtouch.net

International Center for Reiki Training
21421 Hilltop St., Suite 28
Southfield, MI 48033
800-332-8112
www.reiki.org

International Society for the Study of Subtle Energies and
  Energy Medicine
11005 Ralston Road, Suite 100D
Arvada, CO 80004
303-278-2228
www.issseem.org

Nurse Healers and Professional Associates (Therapeutic Touch)
11250-8 Roger Bacon Drive, Suite 8
Reston, Virginia 20190
703-234-4149
www.therapeutic-touch.org

Appendix

Resources

A variety of resources exist to provide guidance and assistance in living a healthy, balanced life, and the savvy consumer needs to be equipped to use them. Some tips for developing a personal resource list—state, national, and international resources and important Internet sites, which address both conventional and complementary/alternative therapies—are offered in this appendix.

Tips for Developing a Personal Resource File

There is wisdom in having resources on hand before they are needed. This is similar to preparing for the proverbial rainy day or buying the hurricane shutters before the storm hits. Resources help people find information, and information offers more control of one's life. Here are some hints that can be beneficial to healthcare professionals and their clients.

- Purchase an accordion file that offers plenty of room for expansion. You can organize your resource file in any way that makes sense to you. One idea would be to divide the file into the following four sections: (1) local, (2) state, (3) national, and (4) international. This personal resource file can help to organize information you may need and enhance your self-care ability. There is value in having information when you face an actual health challenge or even before a challenge to your health occurs.
- Remember that information can come from a variety of sources. The trick is to be able to collect the information as soon as it is found and save it for when it is needed. Some places that you can find information are (1) friends who have had positive experiences with healthcare providers (keeper list), as well as less than adequate experiences (I will

have to think about it), (2) local newspapers, (3) special television programs that offer current information about a variety of topics (be sure to jot down the highlights of what was said), (4) your local library, (5) workshops and seminars in your community, (6) books, (7) journals, and (8) the Internet.

Special Guidelines for the Internet

- Do not assume that just because it is labeled "medical advice" that it is good advice. It is wise to check with a trusted healthcare provider before you take any remedies or discontinue prescribed medications or treatments.
- Always scroll to the end of the webpage to see whether the website is updated on a monthly basis. This will at least let you know that the information is current. Try more than one website pertaining to the topic.
- Recognize that commercial sites are just that. They are on the Web to capture your business. A more reliable source of information would come from a medical center; a university hospital; a college of nursing, medicine, physical therapy, social work, pharmacy, chiropractic, naturopathy, traditional Chinese medicine, or ayurveda; or a government health agency. A simple rule of thumb is this: certification and accreditation = highest standards.
- It is necessary to ask yourself these questions when visiting sites:
  - Is there any chance that the information that I am downloading is biased?
  - Is the information being shared for the greater good of all individuals?
  - Whose benefit does it serve?
  - Is the information based on opinion only and not referenced from current professional (nursing, medical, social work, pharmacological, complementary, and naturopathic) resources?
- Know that it is important for people to make sense of the information gathered so that they can discuss it with their healthcare providers for clarification. Clients can be guided to form questions, such as these:
  - "Doctor [or nurse], I found this on the Internet. It's about my condition and I want to know more about how this can help me."
  - "This condition is being helped by _____. What is being done in our area?"

- "I'd like to try _____. How does this affect the medicine I am taking now?"
- "I trust the treatment you and I agreed upon, but have an interest in trying something different as well. Can I use both of these approaches together?"

Local Resources

Tips for Finding Local Resources

- The American Holistic Nurses' Association has networks throughout the country. Call the national number, 800-278-AHNA, for information. This would be a valuable resource and well worth your effort to find out about what is happening at your local level. Meetings are open to members of the community who value holism in their lives.
- Always consider local healthcare information talent. Most communities have a community college or university that has a college or school of nursing, medicine, social work, physical therapy, dentistry, massage, acupuncture, or allied health, whose faculty may be excellent resources. These potentially valuable resources can be tapped by doing the following:
  1. Call the main number of the area you need (e.g., department of nursing).
  2. Identify yourself as an individual in the community.
  3. State your need (e.g., "I am interested in talking to someone on your faculty who knows about _____. I'd appreciate his or her ideas.")
  4. Leave a message so that faculty can get in touch with you
- Do not be intimidated if you have had little involvement with a college or university. The faculty often is eager to assist and enjoy the opportunity to put their knowledge to good use.
- Many individuals in the community experience similar issues/concerns and thus come together in support groups. Check local newspapers for health sections or community meeting announcements that list support group meetings. Even if you are not in need of a support group at present, it could be helpful to cut this information out and place it in your personal resource file for future reference. Do not forget the weekly columns on you and your health or pharmacology news. Become a resource finder.

- Libraries are rich resources not only for the books and journals they contain, but also for the workshops and printed material that are made available to the general public. Librarians can be a valuable guide to resources on the Internet, as well as reference material in their holdings and in other libraries.
- Check out DVDs on health-related content from the library.
- You probably have great resources that you may be overlooking at your fingertips: telephone directories. Directories typically have listings for community resources. Also, depending on your interest, simply look for a local agency (e.g., American Cancer Society or American Heart Association). Let the agency know specifically what your needs are (e.g., literature, schedule of classes).

Tips for Finding Local Resources on the Internet

- Go to your favorite search engine (e.g., Google.com, Search.com, Mama.com, Metacrawler.com, Dogpile.com, Healthfinder.gov), and type in the query that best describes your needs. Examples would be, "Palm Beach County, Arthritis Resources" or "Bucks County, Area Agency on Aging."
- Any time that you are looking for resources at the local level, try typing in the county where you live. This provides a more specific, local search. The reverse also may hold true. If you are not able to locate anything that fits your needs at the local level, you may need to start more globally and then work your way to the local level. An example might be looking for resources for the American Cancer Society and then linking until you find local resources.
- Please be patient! Consider yourself a detective who is looking for clues for resources to help you. It is worth the effort!

State Resources

Tip for Finding State Resources

- Check the state health department for your area of interest (e.g., AIDS Administration, Department of Mental Health, Nursing Home Licensing and Certification). State health departments also have divisions that license and monitor the practice of various professionals, such as nursing, medicine, nursing home administrators, dentists, acupuncturists, and massage therapists.

Tips for Finding State Resources on the Internet

- Go to your favorite search engine as you did in your local search (e.g., Dogpile.com, Google.com, Mama.com, Metacrawler.com, Health-finder.gov), and type in the query that best describes what it is you are looking for. An example would be New York State, Licensed Acupuncturists, or Oregon, Naturopaths. This should lead you to information pertinent to your state.

- Also consider the various state agencies that may lead you to the information that you need. An example would be Michigan, Department of Health, or Florida, AIDS resources.

- Think about the hospitals in your state that are major medical centers. An example of this would be Texas, Baylor University or Philadelphia, Jefferson University. This may open up information for you that you never suspected was available to the consumer. Another way of getting information about your major medical centers would be to type in the state and the term: Arizona, medical centers or Alabama, health resources.

- Keep trying, and with patience, you will find that the Internet can offer rich resources.

National Resources

National Institutes/Centers

National Cancer Institute
Building 31, Room 10A18
Bethesda, MD 20205
800-492-6600
www.nci.nih.gov

National Center for Complementary and Alternative Medicine of NIH
nccam.nih.gov
This is a valuable clearinghouse for information.

National Dissemination Center for Children With Disabilities
P.O. Box 1492
Washington, DC 20013
800-695-0285
www.nichcy.org

National Heart, Lung, and Blood Institute
P.O. Box 30105
Bethesda, MD 20824-0105
301-592-8573
www.nhlbi.nih.gov

National Institute on Aging
Building 31, Room 5C27
31 Center Drive, MSC 2292
Bethesda, MD 20892
301-496-1752
www.nia.nih.gov

National Institute on Alcohol Abuse and Alcoholism
301-443-3860
www.niaaa.nih.gov

National Institute of Arthritis and Musculoskeletal and Skin Diseases
301-495-4484
www.niams.nih.gov

National Institute of Child Health and Human Development
P.O. Box 3006
Rockville, MD 20847
800-370-2943
www.nichd.nih.gov

National Institute of Dental and Craniofacial Research
Building 45, Room 4AS19
45 Center Drive
Bethesda, MD 20892-6400
301-496-4261
www.nidcr.nih.gov

National Institute of Diabetes and Digestive and Kidney Diseases
Building 31, Room 9A04
31 Center Drive, MSC 2560
Bethesda, MD 20892-2560
301-654-3327
www.niddk.nih.gov

National Institutes of Health
www.nih.gov/health
This is a large resource for government agencies. This site has valuable
  publications that are free or at low cost.

U.S. Centers for Disease Control and Prevention
www.cdc.gov

U.S. Environmental Protection Agency
Ariel Rios Building
1200 Pennsylvania Avenue, NW
Washington, DC 20460
www.epa.gov

Warren Grant Magnuson Clinical Center
Clinical Center, NIH
Building 10, Room B1S234
Bethesda, MD 20892-1078
301-496-3311
www.cc.nih.gov
Requests can be made for Clinical Center nutrition education materials.

National and Professional Organizations

Academy for Guided Imagery
P.O. Box 2070
Malibu, CA 90265
800-726-2070
www.acadgi.com

Alzheimer's Association
225 N. Michigan Avenue, Suite 1700
Chicago, IL 60601-7633
800-272-3900
www.alz.org

American Academy of Pain Management
975 Morning Star Drive, Suite A
Sonora, CA 95370
209-533-9744
www.aapainmanage.org

American Association of Acupuncture and Oriental Medicine
9650 Rockville Pike
Bethesda, MD 20814
866-455-7999
www.aaaomonline.org

American Association of Kidney Patients
2701 N. Rocky Point Drive, Suite 150
Tampa, FL 33607
800-749-2257
www.aakp.org

American Association of Naturopathic Physicians
818 18th Street, NW, Suite 250
Washington, DC 20006
866-538-2267
www.naturopathic.org

American Botanical Council
6200 Manor Road
Austin, TX 78723
512-926-4900
www.herbalgram.org

American Chiropractic Association
1701 Clarendon Boulevard
Arlington, VA 22209
703-276-8800
www.acatoday.org

American College of Hyperbaric Medicine
6737 W. Washington Street, Suite 3265
West Allis, WI 53214
414-918-9300
www.hyperbaricmedicine.org

American Council on Exercise
4851 Paramount Drive
San Diego, CA 92123
858-279-8227
www.acefitness.org

American Diabetes Association, National Call Center
1701 N. Beauregard Street
Alexandria, VA 22311
800-342-2383
www.diabetes.org

American Heart Association
800-242-8721
www.heart.org

American Holistic Medical Association
27629 Chagrin Blvd., Suite 206
Woodmere, OH 44122
216-292-6644
www.holisticmedicine.org

American Holistic Nurses Association
100 SE 9th Street, Suite 3A
Topeka, KS 66612-1213
800-278-AHNA
www.ahna.org

American Liver Foundation
39 Broadway, Suite 2700
New York, NY 10006
800-465-4837
www.liverfoundation.org

American Lung Association
1301 Pennsylvania Ave. NW, Suite 800 Washington, DC 20004
202-785-3355
www.lungusa.org

American Massage Therapy Association
500 Davis Street, Suite 900
Evanston, IL 60201-4695
847-684-0123
www.amtamassage.org

Arthritis Foundation
1330 W. Peachtree Street, Suite 100
Atlanta, GA 30309
800-283-7800
www.arthritis.org

Asthma Hotline
800-222-LUNG

Ayurvedic Institute
11311 Menaul NE
Albuquerque, NM 87112
505-291-9698
www.ayurveda.com

Lighthouse International
111 E. 59th Street
New York, NY 10022
800-829-0500
www.lighthouse.org

Multiple Sclerosis Association of America
706 Haddonfield Road
Cherry Hill, NJ 08002
800-532-7667
www.msassociation.org

National Association for the Deaf
8630 Fenton Street, Suite 820
Silver Spring, MD 20910
301-587-1789
www.nad.org

National Center for Homeopathy
1760 Old Meadow Road, Suite 500
McLean, VA 22102
703-506-7667
www.homeopathic.org

National Hospice and Palliative Care Organization
1731 King Street
Alexandria, VA 22314
703-837-1500
www.nhpco.org

National Osteoporosis Foundation
1150 17th Street, NW, Suite 850
Washington, DC 20036
800-231-4222
www.nof.org

National Parkinson's Disease Foundation
1501 NW 9th Avenue
Bob Hope Road
Miami, FL 33136
800-327-4545
www.parkinson.org

National Stroke Association
9707 E. Easter Lane
Englewood, CO 80112
800-STROKES
www.stroke.org

North American Vegetarian Society
P.O. Box 72
Dolgeville, NY 13329
518-568-7970
www.navs-online.org

United Ostomy Association
P.O. Box 512
Northfield, MN 55057-0512
800-826-0826
www.uoa.org

Websites Related to Healthcare Information

ACTIS (Clinical Trials for HIV/AIDS)
800-TRIALS-A
www.actis.org

Adult Children of Alcoholics Worldwide Services Organization
www.adultchildren.org

Al-Anon Family Group Headquarters
www.al-anon-alateen.org

Anxiety and Depression Association of America
www.adaa.org

Association for Applied Psychophysiology and Biofeedback
www.aapb.org

CenterWatch
www.centerwatch.com
This is a resource for clinical trials on various diseases, a summary of
research, and numbers to contact.

Health on the Net
www.hon.ch/home.html
This site permits searches for available information/resources.

Healthfinder
www.healthfinder.gov
This is a gateway site that is sponsored by the government with links to
more than 1,400 health sites.

Healthy People
www.health.gov/healthypeople
This is a national health initiative coordinated by the Office of Disease
Prevention and Health Promotion and the U.S. Department of Health
and Human Services. It looks at health issues for the next decade.

Journal of the American Medical Association
jama.jamanetwork.com
Access to the patient page is a valuable resource for consumers.

Medhunt
www.hon.ch/HONsearch/Patients/medhunt.html
This is a medical search engine.

Medscape
www.medscape.com
This is a gateway to reviewed information from reputable medical journals.

National Council on Family Relations
www.ncfr.com

National Library of Medicine
www.nlm.nih.gov
This offers access to the world's largest biomedical library.

Self-Help Group Sourcebook Online
mentalhelp.net/selfhelp/
This is an online directory of thousands of medical and mental health support organizations around the world.

International Resources

Global Alliance for Women's Health
www.gawh.org

International Association of Yoga Therapists
PO Box 12890
Prescott, AZ 86304
928-541-0004
www.iayt.org

U.C. Berkeley, Public Health Library, International Health Resources
www.lib.berkeley.edu/PUBL/Inthealth.html

UN Aids Organization
www.unaids.org
This is a global source for HIV/AIDS information.

World Health Organization
www.who.int
This is a resource for issues around the globe.

Index

Exhibits, figures, and tables are indicated by exh, f, and t following page numbers.